ICM Millennium Lectures on Games

Springer
Berlin
Heidelberg
New York
Hong Kong
London
Milan
Paris
Tokyo

Leon A. Petrosyan
David W.K. Yeung
Editors

ICM Millennium Lectures on Games

With 48 Figures
and 27 Tables

Springer

Professor Leon A. Petrosyan
St. Petersburg State University
Faculty of Applied Mathematics & Controll Processes
Petrodyvorets
198904 St. Petersburg, Russian Federation

Professor David W. K. Yeung
Centre of Game Theory
Hong Kong Baptist University
Kowloon Tong
Hong Kong, PR China
and
Centre of Game Theory
St. Petersburg State University
Petrodyvorets
198904 St. Petersburg, Russian Federation

ISBN 3-540-00615-X Springer-Verlag Berlin Heidelberg New York

Bibliographic information published by Die Deutsche Bibliothek
Die Deutsche Bibliothek lists this publication in the Deutsche Nationalbibliografie;
detailed bibliographic data available in the internet at *http.//dnb.ddb.de*

Springer-Verlag Berlin Heidelberg New York
a member of BertelsmannSpringer Science + Business Media GmbH

http://www.springer.de

Printed in Germany

Cover design: Erich Kirchner, Heidelberg

SPIN 10916032 42/3130 – 5 4 3 2 1 0 – Printed on acid free paper

Preface

Since the first Congress in Zürich in 1897, the ICM has been an eagerly-awaited event every four years. Many of these occasions are celebrated for historic developments and seminal contributions to mathematics. 2002 marks the year of the 24th ICM, the first of the new millennium. Also historic is the first ICM Satellite Conference devoted to game theory and applications. It is one of those rare occasions, in which masters of the field are able to meet under congenial surroundings to talk and share their gathered wisdom.

As is usually the case in ICM meetings, participants of the ICM Satellite Conference on Game Theory and Applications (Qingdao, August 2002) hailed from the four corners of the world. In addition to presentations of high quality research, the program also included twelve invited plenary sessions with distinguished speakers. This volume, which gathers together selected papers read at the conference, is divided into four sections:

(I) Foundations, Concepts, and Structure.
(II) Equilibrium Properties.
(III) Applications to the Natural and Social Sciences.
(IV) Computational Aspects of Games.

The papers in Section I explore fundamental ideas, leading to new and analytically interesting analysis of current problems as well as new games and new modeling approaches in games. In Section II, seven papers discuss issues in the solution of games, and present a number of potentially very fruitful ideas regarding game equilibrium under different assumptions and conditions. Section III is devoted to applications. In particular, the articles on market structure and game-based computations would be of interest to researchers and practitioners in commerce, industry, banking and finance, and the public sector. Section IV focuses on the computational aspects of games: both computational algorithms and computability of equilibria.

Twelve invited plenary lectures were delivered in a feast of ideas. However, given that the distinguished lecturers were delivered before their audiences and that transcriptions from video recordings are likely to be insufficiently accurate to do justice to the subtle complexities of the reasoning, the editors have reluctantly decided to just list the lectures in this preface. Readers anxious to learn more about the great things discussed during the invited sessions can rest assured that they will be more formally published in good time.

Among the invited lectures, remarkable new ideas regarding solutions and equilibria in games were presented by Reinhard Selten, "On Two Behavioral Equilibrium Concepts", Lloyd Shapley, "Convexing the Pareto Set by Individual Order-Preserving Transformations", and Dov Samet, "Ordinal Solutions for Bargaining Problems". We can expect many analytically and practically interesting new games to follow from the insights offered by Robert Simon, "Games of Incomplete Information, Ergodic Theory, and the Measurability of Equilibria", Abraham Neyman, "Repeated Games with Bounded Descriptive Strategic Complexity", John Forbes Nash, "Projects Studying Cooperation in Games through Modelling in terms of Formally Non-cooperative action in a Repeated Game Context", and David Yeung, "Randomly Furcating Stochastic Differential Games: A Paradigm for Interactive Decision-Making under Sturcture Uncertainty". Finally, researchers of game modelling will discover a cornucopia of inspiration in the lectures of Robert John Aumann, "Bayes Rational Play", Michael Maschler, "Voting for Voters", Sergiu Hart, "Simple Adaptive Strategies", Leon Petrosjan, "Bargaining in Dynamic Games", and Sylvain Sorin, "The Operator Approach to Zero-Sum Repeated Games".

The editors would like to thank Qingdao University and the Organizing Committee of the Conference for hosting this historic event, and the Academic Committee for striving to maintain the great ICM tradition in the presentation of high quality papers and the dissemination of thought-provoking ideas.

We trust that the ICM Millenium Lectures on Games will prove to be a volume that researchers in game theory and applications will treasure and re-read in the years to come.

Leon A. Petrosyan and David W. K. Yeung
St. Petersburg, January 2003

Contents

Part II Equilibrium Properties

Part III Applications to the Natural and Social Sciences

Part I

Foundations, Concepts, and Structure

Stable Schedule Matching under Revealed Preference*

A. Alkan[1] and D. Gale[2]

[1] Sabanci University, 81474 Tuzla, Istanbul, Turkey
E-mail: alkan@sabanciuniv.edu
[2] U. C., Berkeley, Ca. 94720, USA
E-mail: gale@math.berkeley.edu

Summary. In a recent study Baiou and Balinski [3] generalized the notion of two-sided matching to that of schedule matching which determines not only what partnerships will form but also how much time the partners will spend together. In particular, it is assumed that each agent has a ranking of the agents on the other side of the market. In this paper we treat the scheduling problem using the more general preference structure introduced by Blair [5] and recently refined by Alkan [1, 2], which allows among other things for diversity to be a motivating factor in the choice of partners. The set of stable matchings for this model turns out to be a lattice with other interesting structural properties.

Key words: Stable Matching, Two-sided Market, Lattice, Diversity, Schedule
JEL Classification Numbers: C71, C78, D49

1 Introduction

The formulation of the Stable Matching Problem [7] was originally motivated by the real world problem of college admissions. It was an attempt to find a rational criterion for matching students with colleges which respected the preferences of both groups. The original approach was to first consider a special case, the so-called Stable Marriage Problem in which each college could accept only one student. The general case was then reduced to the marriage case by assuming that each college had a complete preference ordering on the set of students it was willing to admit as well as a *quota* giving an upper bound to the number of students that could be admitted. The model has applications

* Partial support by Turkish National Academy of Sciences is gratefully acknowledged as are useful comments from participants at Telaviv, Hebrew, Pennsylvaina, Northwestern, Columbia University seminars, SAET Ischia meeting, Istanbul NATO Advanved Research Workshop.

in other situations. A particularly natural application is the problem of hiring of workers by firms.[3] In general we refer to such a model as a *market* and the participants on the two sides as *agents*.

The present paper presents a broad generalization of the original model incorporating extensions in several directions.

(1) Agents on both sides of the market may form multiple partnerships.

(2) Preferences of agents over sets of possible partners are given by choice functions which are more general than those given by complete orderings of individuals. This is especially relevant for the college market where colleges are typically interested in the overall composition of an entering class, particularly these days as regards diversity.[4] A simple example will illustrate the point.

College A can admit two students. The applicants are two men m and m' and two women w and w'. A's first choice is the pair mw but if m (w) is not available the choice is $m'w$ (mw'). One sees at once that these choices are not possible from any strict ordering since A would have to order the pair m', w'. For example if the ordering was $m > w > m' > w'$ then it would mean that mm' was preferred to the diversified pair $m'w'$. Indeed, as regards diversity, in the algorithm which solves the original college admissions problem, there is nothing to prevent a college from ending up with a class which is either ninety percent male or female.

The remedy for this via choice functions simply formalizes what happens approximately in actual negotiations between colleges and students or firms and workers. Each agent is assumed to have a *choice function* C which, given a set P of agents on the other side of the market, picks out the most preferred subset $S = C(P)$ contained in P. S is then said to be *revealed preferred* to all other subsets of P. The case where colleges rank-order applicants is then a special case in which $C(P)$ consists of the q highest ranked applicants in P, but if, for example, the goal was gender balance one could choose, roughly, the highest ranked $q/2$ applicants of each sex or if, say there was an insufficient number of male applicants then choose all the men and fill the quota with the highest ranked women.

The choice function approach was first introduced by Blair [5] using somewhat different terminology, and a very recent variation on this idea has been given by Alkan [1, 2]. From now on we will refer to the case where agents are linearly ordered as the *classical* case.

It is worth pointing out that while our preference stucture is mathematically equivalent to that of Blair [5] we do not assume as he does that that

[3] See [8, 10].

[4] We quote Mr. Bollinger, the president of the University of Michigan: "Admissions is not and should not be a linear process of lining up applicants according to their grades and test scores and then drawing a line through the list. It shows the importance of seeing racial and ethnic diversity in a broader context of diversity, which is geographic and international and socio-economic and athletic and all the various forms of differences, complementary differences, that we draw on to compose classes year after year."

agents have a complete ordering of subsets of agents on the other side of the market. In our approach there is only a partial order on subsets. For example, if a college with quota 2 in the classical case ranks students $a > b > c$ then by revealed preference the pair ab is prefered to bc and ac, since given the triple abc the pair ab is chosen, but the pairs ac and bc are not comparable since neither of these pairs will be chosen from abc.

(3) Recently Baiou and Balinski [3] have generalized the notion of matching to that of a *schedule matching*. In the context of a set of workers W with members w and a set of firms F with members f, the idea is that a firm decides not only which workers it will hire but also how many man-hours of employment to give each of them. Similarly, the workers must decide how many of their available hours to allocate to each job. A schedule is then a $F \times W$ matrix X whose entries $x(fw)$ give the amount of time worker w works for firm f. The schedule matching is said to be *stable* (or pairwise stable) if there is no pair f and w who could make themselves better off by increasing the hours they work together while not increasing (possibly decreasing) the hours they work with their other partners.[5] This is the natural generalization of stability for ordinary (college admissions) matchings. In fact matchings correspond to the special case of schedule matchings where all entries of X are either 0 or 1. In [3] it is assumed that we are in the classical case where each agent has a strict ordering of the agents on the other side of the market and preferences on schedules are given by the condition that an agent, say a worker w, is made better off if he can increase the time he works for firm f by reducing the time he works for some less preferred firm f'.

The present paper treats schedules using the more general revealed preferences. Our main result shows that under appropriate conditions which include the classical case the set of stable matchings forms a distributive lattice with other interesting structural properties.

In the next section we develop the necessary material on the revealed preference ordering of an individual and show that if the preferences are *consistent* and *persistent* (to be defined) then the set of schedules becomes a (non-distributive) lattice with other important structural properties. These are then used in the following section to prove that stable schedules always exist. The lattice properties are then derived in the final sections.

2 The Revealed Preference Lattice

In the matching theory of later sections we will think of economic agents as firms and workers, or students and colleges, men and women, etc. However, the theory of revealed preference of the individual belongs to the general standard model of consumption or demand theory, and it will be presented in this context here.

[5] If coalitions other than pairs can form to block a matching, then stable matchings may not exist. See [11].

An *agent* (consumer) chooses (demands) amounts of n *items* (goods) from given availabilities of each item. This is formalized as follows.

Let R^n_+ be the nonnegative orthant, b an *upper bound* vector and $\mathrm{B} = \{x \in R^n_+ \mid x \leq b\}$. Let $\mathcal{B}$ be a subset of B which is closed under $\vee$ and $\wedge$ (the standard join and meet in R^n). A *choice function* is a map $C: \mathcal{B} \longrightarrow \mathcal{B}$ such that

$$C(x) \leq x$$

for all $x \in \mathcal{B}$. The elements $x = (x(1), \ldots, x(n))$ of the *domain* $\mathcal{B}$ will be called *choice vectors.* The range of C is denoted by $\mathcal{A}$ and its elements are called (acceptable) *schedules.*

The most relevant domains for our purposes are the *divisible* domain B itself and the *discrete* domain that consists of all the integer vectors in B. When all bounds are equal to 1, the discrete domain corresponds to the case of ordinary multipartner matching as in college admissions.

An important special case of our model is one in which the items can be measured in some common unit, for example, dollars worth for goods, or man-hours for services. In this case we denote the sum of the entries of a vector x by $|x|$ and call it the *size* of x. In such a model an agent may have a *quota* q which bounds the size of the schedule he can choose. For the college admissions case q is the maximum number of students a college can admit.

A choice function C is called *quota filling* if

$$|C(x)| = q \text{ if } |x| \geq q \text{ and } C(x) = x \text{ otherwise.}$$

Two interesting examples of quota filling choice functions are as follows.

Example 1. The items are ranked so that, say, item i is more desirable than $i+1$. Given a choice vector x with $|x| > q$, let j be the item such that $r = \sum^j x(i) \leq q$ and $r + x(j+1) > q$. Then

$$C(x) = (x(1), \ldots, x(j), q - r, 0, \ldots, 0) .$$

Thus, the agent fills as much of his quota as possible with the most desirable items. We will henceforth refer to this C as the *classical* choice function.

Example 2. The domain is B. Given a choice vector x with $|x| > q$, let r be the number such that $\sum_i r \wedge x(i) = q$. Then

$$C(x) = (r \wedge x(1), \ldots, r \wedge x(n)) .$$

In words, the agent tries to use all items as equally as possible. (On the discrete domain, as in college admissions, there may be more than one such schedule hence a tie-breaking criterion is necessary.) We will refer to C as the *diversifying* choice function.

As an illustration, suppose an agent with quota 5 is given the choice vector $(2,1,0,4,2)$. Then, the classical choice function chooses the schedule $(2,1,0,2,0)$ while the diversifying choice function chooses $(4/3,1,0,4/3,4/3)$.

Definition 1. *We say that $x \in \mathcal{A}$ is revealed preferred to $y \in \mathcal{A}$, and write $x \succeq y$, if $C(x \vee y) = x$. We write $x \succ y$ if $x \succeq y$ and $x \neq y$.*

We now impose some standard conditions on the choice function C.

Definition 2. *C is consistent if $C(x) \leq y \leq x$ implies $C(y) = C(x)$.*

This is a highly plausible assumption. Applied to college admissions, it says that if some set S of students is chosen for admission from a pool P then the same set will be chosen from any subset of P which contains S.

An immediate consequence of consistency is that $C(x) = x$ if and only if $x \in \mathcal{A}$. Without some further restrictions, revealed preference will not be transitive, hence not a partial ordering, as shown by the following example for the college admissions case.

Example 3. A college C can admit two students from two men m, m' and two women w, w'. The pair mw is $C's$ first choice, but if either w or m are not available then

$$\text{(i) } C(mm'w') = mw' ,$$
$$\text{(ii) } C(m'ww') = m'w' .$$

(In the case of college admissions, we will use the customary notation and represent a choice vector or schedule x by the set of all students s for whom $x(s) = 1$.) Transitivity fails because from (i) we have $mw' \succ m'w'$ and from (ii) $m'w' \succ m'w$ but mw' and $m'w$ are not comparable since $C(mw'm'w) = mw$.

To avoid this situation, we introduce the following condition of *persistence* (which extends the well known condition on the discrete domain that has been called *substitutability*).

Definition 3. *C is persistent if $x \geq y$ implies $C(y) \geq C(x) \wedge y$.*

For the college admissions problem, persistence means that if a college offers admission to a student from a given pool of applicants then it will also admit him if the pool of applicants is reduced. The condition is closely analogous to the *gross substitutes* condition in production theory where one assumes that an input which maximizes profits at given prices will remain profit maximizing if the price of some other good is increased [8]. It should be pointed out that there are natural choice functions in which persistence fails. Suppose as in Example 3 that mw is a college's first choice but it prefers not to separate m' and w' so that $C(mm'w') = C(m'ww') = m'w'$. This clearly violates persistence. In general, persistence rules out this sort of complementarity between items.

It is easy to verify that the classical and diversifying choice functions satisfy consistency and persistence.

An immediate consequence of persistence is that if $x \in \mathcal{A}$ and $x \geq y$ then $y \in \mathcal{A}$.

Definition 4. *C is subadditive if $C(x \vee y) \leq C(x) \vee y$ for all x, y.*

Lemma 1. *If C is persistent then it is subadditive.*

Proof. Since $C(x \vee y) \leq x \vee y$, we have

$$C(x \vee y) = C(x \vee y) \wedge (x \vee y) = (C(x \vee y) \wedge x) \vee (C(x \vee y) \wedge y) \tag{1}$$

by distributivity. Since $x \leq x \vee y$, we have $C(x \vee y) \wedge x \leq C(x)$ by persistence. Also $C(x \vee y) \wedge y \leq y$. Substituting these two inequalities in (1) gives subadditivity. □

Definition 5. *C is stationary if $C(x \vee y) = C(C(x) \vee y)$ for all x, y.*

Lemma 2. *If C is subadditive and consistent then it is stationary.*

Proof. By subadditivity $C(x \vee y) \leq C(x) \vee y$. Also $C(x) \vee y \leq x \vee y$. So $C(C(x) \vee y) = C(x \vee y)$ by consistency. □

It will be assumed from here on that all choice functions are consistent and persistent.

Notation. We write $x \curlyvee y$ for $C(x \vee y)$.

As immediate consequence of stationarity, we have

Corollary 1. *The relation $\succeq$ is transitive and $x \curlyvee y$ is the least upper bound of x and y.*

Proof. The operation $\curlyvee$ is associative: $(x \curlyvee y) \curlyvee z = C(C(x \vee y) \vee z) = C((x \vee y) \vee z) = C(x \vee (y \vee z)) = C(x \vee C(y \vee z)) = x \curlyvee (y \curlyvee z)$. Thus, if $x \succeq y$, $y \succeq z$ then $x \curlyvee z = (x \curlyvee y) \curlyvee z = x \curlyvee (y \curlyvee z) = x \curlyvee y = x$ so $x \succeq z$. Also, if $z \succeq x$, $z \succeq y$ then $z \curlyvee (x \curlyvee y) = (z \curlyvee x) \curlyvee y = z \curlyvee y = z$ so $z \succeq x \curlyvee y$. □

Thus, the set of schedules $\mathcal{A}$ is an upper-semilattice (with join $\curlyvee$) in the partial order given by $\succeq$. It is, in fact, a lattice and we will need an expression for its meet $\curlywedge$. First note, it follows at once from stationarity that if $C(x) = z$ and $C(y) = z$ then $C(x \vee y) = z$.

Definition 6. *The closure $\overline{x} \in \mathcal{B}$ of $x \in \mathcal{A}$ is $\sup\{y \in \mathcal{B} \mid C(y) = x\}$.*

In the classical college admissions case, $\overline{x}$ consists of x together with all students ranked below the least desired student in x. We henceforth assume that C is continuous. It then follows that $C(\overline{x}) = x$. Define $\overline{C}: \mathcal{A} \to \mathcal{B}$ by

$$\overline{C}(x) = \overline{x}\,.$$

Lemma 3 (Isomorphism). *The mapping $\overline{C}$ is a lattice isomorphism from $(\mathcal{A}, \succeq)$ to $(\mathcal{B}, \geq)$.*

Proof. It suffices to prove that $\overline{C}$ is order preserving: suppose $x \succeq y$. Then $x = C(x \vee y) = C(\overline{x} \vee \overline{y})$ by stationarity. So by definition of closure $\overline{x} \vee \overline{y} \leq \overline{x}$ thus $\overline{y} \leq \overline{x}$. Conversely, if $\overline{x} \geq \overline{y}$ then $\overline{x} \vee \overline{y} = \overline{x}$, so $x = C(\overline{x}) = C(\overline{x} \vee \overline{y}) = C(x \vee y)$ by stationarity, that is, $x \succeq y$. □

Lemma 4. *The revealed preference meet is given by* $x \curlywedge y = C(\overline{x} \wedge \overline{y})$.

Proof. We must show (i) $C(\overline{x} \wedge \overline{y}) \preceq x$ (and $C(\overline{x} \wedge \overline{y}) \preceq y$) and (ii) $z \preceq x$ and $z \preceq y$ implies $z \preceq C(\overline{x} \wedge \overline{y})$.

By definition (i) is true if and only if $C(C(\overline{x} \wedge \overline{y}) \vee x) = x$. By stationarity this is equivalent to $C(x') = x$ where $x' = (\overline{x} \wedge \overline{y}) \vee x$ and, since $x \leq x' \leq \overline{x}$ and $C(\overline{x}) = x$, the result follows by consistency.

To prove (ii) we must show that $C(C(\overline{x} \wedge \overline{y}) \vee z) = C((\overline{x} \wedge \overline{y}) \vee z)$ (by stationarity) $= C(\overline{x} \wedge \overline{y})$, so note that $z \preceq x$ means $C(x \vee z) = x$, hence by definition of closure $x \vee z \leq \overline{x}$, so $z \leq \overline{x}$ and similarly $z \leq \overline{y}$ so $z \leq \overline{x} \wedge \overline{y}$ so $(\overline{x} \wedge \overline{y}) \vee z = \overline{x} \wedge \overline{y}$ and the result follows. □

Note that in college admissions, $x \curlywedge y$ may include students who are neither in x nor y: suppose there are four students $1, 2, 3, 4$ ranked in that order, and $x = \{1, 3\}, y = \{2, 3\}$. Then $x \curlywedge y = \{3, 4\}$.

We will need some further properties of the revealed preference lattice.

Lemma 5. $x \curlywedge y \geq x \wedge \overline{y}$.

Proof. Since $\overline{x} \geq \overline{x} \wedge \overline{y}$, we have from persistence $x \curlywedge y = C(\overline{x} \wedge \overline{y}) \geq C(\overline{x}) \wedge \overline{x} \wedge \overline{y} = x \wedge \overline{x} \wedge \overline{y} = x \wedge \overline{y}$. □

Lemma 6. $(x \curlywedge y) \wedge (x \curlyvee y) \leq x \wedge y$.

Proof. Since $\overline{x} \vee \overline{y} \geq \overline{x}$, we have from persistence $C(\overline{x}) = x \geq C(\overline{x} \vee \overline{y}) \wedge \overline{x} = (x \curlyvee y) \wedge \overline{x}$ from stationarity, and similarly $y \geq (x \curlyvee y) \wedge \overline{y}$, so $x \wedge y \geq (x \curlyvee y) \wedge (\overline{x} \wedge \overline{y}) \geq (x \curlyvee y) \wedge C(\overline{x} \wedge \overline{y}) = (x \curlyvee y) \wedge (x \curlywedge y)$ from Corollary 4. □

2.1 Satiation

The following definition is basic.

Definition 7. *A schedule x is i-satiated if $C_i(y) \leq x(i)$ for all $y \geq x$.*

In words, x is i-satiated if the agent would not choose more of item i if it were offered. To illustrate, in the classical case, x is i-satiated if i is the highest ranked item with $x(j) = 0$ for $j > i$. For the diversifying choice function, x is i-satiated if $x(i) = \max_j \{x(j)\}$. The following properties will be needed in the next section.

Lemma 7. *x is i-satiated if there exists $y \geq x$, $y(i) > x(i)$ such that $C(y) \leq x$.*

Proof. Suppose $z \geq x$ and $z(i) > x(i)$ (otherwise there is nothing to prove). Let $y' = z \wedge y$ and note that $y'(i) > x(i)$. Now $y \geq y' \geq x$ so by consistency $C(y') = C(y) \leq x$. Also $z \geq y'$ so by persistence $x \geq C(y') \geq C(z) \wedge y'$ so $x(i) \geq C_i(z) \wedge y'(i)$ but since $y'(i) > x(i)$ we have $C_i(z) \leq x(i)$. □

Lemma 8. *x is i-satiated if and only if $\overline{x}(i) = b(i)$.*

Proof. If $x(i) = b(i)$ there is nothing to prove so suppose $x(i) < b(i)$. If $\overline{x}(i) = b(i)$ then x is i-satiated by the previous Lemma. If x is i-satiated then let $y = x \vee b^i$ where b^i is the vector with ith entry $b(i)$ and others 0. Then from satiation $C_i(y) \leq x(i)$ and since $C_j(y) \leq x(j)$ for $j \neq i$ we have $C(y) \leq x \leq y$ so by consistency $C(y) = C(x) = x$ so $y \leq \overline{x}$ so $\overline{x}(i) = b(i)$. □

Lemma 9. *Suppose $x \succeq y$.*
(i) If y is i-satiated then x is i-satiated.
(ii) If $x(i) > y(i)$ then y is not i-satiated.

Proof. (i) By the isomorphism lemma, $\overline{x} \geq \overline{y}$ so $\overline{x}(i) \geq \overline{y}(i) = b(i)$ so x is i-satiated by the previous Lemma. (ii) Since $x \succeq y$ we have $x \vee y \geq y$ and $C(x \vee y) = x$ so $C_i(x \vee y) = x(i) > y(i)$ so y is not i-satiated. □

Lemma 10. *(i) If x or y is i-satiated then $x \curlyvee y$ is i-satiated.*
(ii) If x and y are i-satiated then $x \curlywedge y$ is i-satiated.

Proof. (i) Say x is i-satiated. Then since $x \curlyvee y \succeq x$ the conclusion follows from Lemma 9(i). (ii) We have $C((x \curlywedge y) \vee b^i) = C(C(\overline{x} \wedge \overline{y}) \vee b^i) = C((\overline{x} \wedge \overline{y}) \vee b^i)$ (by stationarity) $= C((\overline{x} \vee b^i) \wedge (\overline{y} \vee b^i)) = C(\overline{x} \wedge \overline{y})$ (using Lemma 8, since x and y are i-satiated) $= x \curlywedge y$. □

3 Stable Matchings

We now consider two finite sets of agents which we interpret as *firms*, F, with members f, and *workers*, W, with members w, having respectively the choice functions C_f, C_w, with ranges $\mathcal{A}_f, \mathcal{A}_w$. We write $\curlyvee_f, \curlywedge_f, \succeq_f$ for the join, meet, preference ordering for f, and similarly for w.

A *matching* X is a nonnegative $F \times W$ matrix whose entries, written $x(fw)$, represent the amount of time w works for f. We write $x(f)$ for the f-row and $x(w)$ for the w-column of X. We assume all matchings X are bounded above by some positive matrix B. The choice functions C_F, C_W are defined from C_f, C_w in the natural way.

The revealed preference ordering for agents translates in an obvious way to an ordering on matchings.

Definition 8 (Group Preference). *The matching X is preferred to Y by F, written $X \succeq_F Y$, if $x(f) \succeq_f y(f)$ for all f in F.*

Definition 9 (Acceptability). *A matching X is* F-acceptable *if $x(f) \in \mathcal{A}_f$ for all f, and it is* W-acceptable *if $x(w) \in \mathcal{A}_w$ for all w. It is acceptable if it is both F and W-acceptable.*

The fundamental stability notion is now formalized as follows.

Definition 10 (Stability). *An acceptable matching X is stable if, for every pair fw, either $x(f)$ is w-satiated or $x(w)$ is f-satiated (or both).*

3.1 Existence, Polarity, Optimality, Comparative Statics

We will show that stable matchings always exist by constructing a sequence of alternately F and W-acceptable matchings which converge to a stable matching.[6] The method is the standard one in which the firms make offers of employment to the workers who then choose (via their choice functions) their most preferred schedule. Firms whose offers have been declined then make alternative offers. Of course the proof must make use of persistence of all firms' and workers' choice functions since counterexamples exist if this condition is not satisfied (see Section 4).

Theorem 1 (Existence). *There exists a stable matching.*

Proof. Define the sequences $(B^k), (X^k), (Y^k)$ by the following recursion rule:

$$B^0 = B\,, \quad X^k = C_F(B^k)\,, \quad Y^k = C_W(X^k)\,,$$

and B^{k+1} is obtained from B^k as follows:

$$b^{k+1}(fw) = b^k(fw) \;\text{ if }\; y^k(fw) = x^k(fw)\,,$$
$$b^{k+1}(fw) = y^k(fw) \;\text{ if }\; y^k(fw) < x^k(fw)\,.$$

The matrices B^k will be called the choice matrix for the firms. Note that (B^k) is a nonincreasing nonnegative sequence and hence converges, so by continuity of C_F it follows that (X^k) converges, and hence by continuity of C_W it follows that (Y^k) converges. Call the limits $\widehat{B}, \widehat{X}, \widehat{Y}$. Note that if X^n is W-acceptable for some n then the sequence becomes stationary so this will be included as a special case. We will show,

(i) $\widehat{X} = \widehat{Y}$ and hence it is acceptable,
(ii) $\widehat{X} (= \widehat{Y})$ is stable.

To prove (i), note that $Y^k \le X^k \le B^k$. If, for some fw, $\widehat{x}(fw) - \widehat{y}(fw) > \epsilon$, then $x^k(fw) - y^k(fw) > \epsilon$ for infinitely many k and therefore from the recursion rule $b^k(fw) - b^{k+1}(fw) > \epsilon$ which is impossible since B^k converges.

[6] In the discreet case the sequence actually terminates after a finite number of iterations. It is not known whether this is also true for the general continuous case treated here.

To prove (ii), we first show that $Y^{k+1} \succeq_W Y^k$, thus workers are "better off" after each step of the recursion. From the recursion rule $Y^k \leq B^{k+1} \leq B^k$, so from persistence we have

$$C_F(B^{k+1}) = X^{k+1} \geq C_F(B^k) \wedge B^{k+1} \geq X^k \wedge Y^k = Y^k,$$

so $Y^{k+1} = C_W(X^{k+1})$ is revealed preferred to Y^k. It follows by continuity that

$$\widehat{Y} \succeq_W Y^k. \tag{2}$$

Now suppose $\widehat{y}(f)$ is not w-satiated. Then from Lemma 8 $\widehat{y}(fw) < b(fw)$ so from the recursion rule, for some k, $y^k(fw) < x^k(fw)$ so, since $y^k(w) = C_w(x^k(w)) \leq x^k(w)$, from Lemma 7 we have $y^k(w)$ is f-satiated and from (2) $\widehat{y}(w) \succeq_w y^k(w)$, so from Lemma 9($i$) $\widehat{y}(w)$ is f-satiated. This proves stability of $\widehat{Y}$. □

The following are extensions of familiar properties of the marriage market (see [10, 6]).

Lemma 11. *Let X be a stable matching and let Y be an F-acceptable matching such that $Y \succeq_F X$. Then $C_W(X \vee Y) = X$.*

Proof. If the conclusion is false, then there is some w such that $C_w(x(w) \vee y(w)) = z(w) \neq x(w)$. Hence, $z(w) \succ_w x(w)$, so $z(fw) > x(fw)$ for some f, hence from Lemma 9(ii) $x(w)$ is not f-satiated, but $z(fw) \leq y(fw)$ so $x(fw) < y(fw)$ and by hypothesis $y(f) \succeq_f x(f)$ so again by Lemma 9(ii) $x(f)$ is not w-satiated, contradicting stability of X. □

Corollary 2 (Polarity). *If X, Y are stable matchings then $X \succeq_F Y \iff Y \succeq_W X$.*

Theorem 2 (Optimality). *If $\widehat{X}$ is the matching given by the Existence Theorem and X is any other stable matching then $\widehat{X} \succeq_F X$.*

Proof. Let X be a stable matching. We will show that $X \leq \widehat{B}$ where $\widehat{B}$ is the matrix given in the Existence Theorem. Since $\widehat{X} = C_F(\widehat{B})$, the conclusion follows. So suppose not. Then there is an index k such that $B^k \geq X^k$ but $b^{k+1}(fw) < x(fw)$ for some fw. From the recursion rules, this means that

$$y^k(fw) < x(fw) \text{ and } y^k(fw) < x^k(fw). \tag{3}$$

Now $X^k = C_F(B^k)$ so $X^k \succeq_F X$ so from Lemma 11 $C_W(X \vee X^k) = X$. But $X \vee X^k \geq X^k$ so from persistence $Y^k = C_W(X^k) \geq X^k \wedge C(X \vee X^k) = X^k \vee X$ so $y^k(fw) \geq x^k(fw) \wedge x(fw)$ for all fw which contradicts (3). □

Suppose a new firm or a new worker enters the market. The following theorem shows that, in the firm-optimal matching, in the first case no firm is better off and no worker worse off, while in the second case no worker is better off and no firm worse off. Formally, let $\widehat{X}$ be the firm-optimal matching in the original $F \times W$ market, and let $\widehat{X^\phi}$ denote the $F \times W$ component of the firm-optimal matching with the additional firm ϕ and let $\widehat{X^\omega}$ be the same for the additional worker ω.

Theorem 3 (Comparative Statics).
(i) $\widehat{X} \succeq_F \widehat{X^\phi}$ *and* $\widehat{X^\phi} \succeq_W \widehat{X}$*; (ii)* $\widehat{X^\omega} \succeq_F \widehat{X}$ *and* $\widehat{X} \succeq_W \widehat{X^\omega}$.

Proof. To prove (i), we continue the algorithm of the Existence Theorem. The new firm ϕ offers an employment schedule x_ϕ which gives a new offer schedule X' to W where $X' \geq X$ and since workers get no worse off with each step of the recursion they are at least well off under $\widehat{X}^\phi$ as under $\widehat{X}$. The firms are no better off since their choice matix can never exceed $\widehat{B}$.

To prove (ii), we suppose the original market includes ω but $b_\omega = 0$. We denote by $(B^k), (X^k), (Y^k)$ and $(B'^k), (X'^k), (Y'^k)$, respectively, the sequences in the Existence Theorem recursion for the original and new market. Note $B' \geq B$. It suffices to show that $B'^k \geq B^k$ and $x'^k(w) \leq x^k(w)$ for all k and $w \neq w'$. Assume this is true up to k. Since $B'^k \geq B^k$, we have by persistence $X^k = C_F(B^k) \geq C_F(B'^k) \wedge B^k = X'^k \wedge B^k$ so $x^k(fw) \geq x'^k(fw) \wedge b^k(fw)$ but for $w \neq w'$ we have $b^k(fw) = b'^k(fw)$, hence $x^k(w) \geq x'^k(w)$. This shows that no W-worker is better off in the new market.

To show that no firm is worse off, we show that $B'^k \geq B^k$ for all k. Since $x^k(w) \geq x'^k(w)$ for $w \neq w'$, we have by persistence $y'^k(w) \geq y^k(w) \wedge x'^k(w)$. There are two cases. If $x'^k(fw) \leq y^k(fw)$ then $y'^k(fw) = x'^k(fw)$ so from the recursion rule $b'^{k+1} = b'^k \geq b^k \geq b^{k+1}$. If on the other hand $x^k(fw) \geq x'^k(fw) > y^k(fw)$ then by consistency $y'^k(fw) = y^k(fw)$ so from the recursion rule $b^{k+1}(fw) - y^k(fw) = y'^k(fw) \leq b'^{k+1}(fw)$, completing the proof. □

3.2 The Stable Matching Lattice

Let X, X' be an arbitrary pair of matchings fixed throughout this section. We write $X^F = X \curlyvee_F X'$ for the matching whose f-row, $x^F(f)$, is $x(f) \curlyvee_f x'(f)$ and write $X_F = X \curlywedge_F X'$ for the matching whose f-row, $x_F(f)$, is $x(f) \curlywedge_f x'(f)$. We define X^W, X_W via w-columns similarly.

Note that if X, X' are acceptable then X^F is of course F-acceptable but not in general W-acceptable.

The following is a key result.

Lemma 12. *If X and X' are stable matchings then $X^F \leq X_W$.*

Proof. We must show that $x^F(fw) \leq x_W(fw)$ for all fw.

Case (i) $x^F(fw) \leq x(fw) \wedge x'(fw)$. Then, since by Lemma 4 $x(w) \wedge x'(w) \leq x(w) \curlywedge_w x'(w) = x_W(w)$, the conclusion follows.

Case (ii) $x(fw) < x^F(fw) \leq x'(fw)$. Then, since $x^F(f) \succeq_f x(f)$, we have by Lemma 8(*ii*), $x(f)$ is not w-satiated, so by stability $x(w)$ is f-satiated, so from Lemma 7 $\overline{x(w)}(f) = b(fw)$, so $x^F(fw) \le \overline{x(w)}(f) \wedge x'(fw) = (\overline{x(w)} \wedge x'(w))(f) \le (x(w) \curlywedge_w x'(w))(f)$ (again from Lemma 4) $= x_W(fw)$. □

In order to make the above inequality to an equation, it is necessary to make some further assumption. We will assume that the entries of a schedule are measured in some common unit so that it makes sense to add them up. The following condition extends the condition of "cardinal monotonicity" for the discrete case introduced in [2].

Definition 11. *The choice function C is size monotone if $x \le y$ implies $|C(x)| \le |C(y)|$ for all x, y in $\mathcal{A}$.*

Remark Note that size monotonicity implies that if $x \succeq y$ then $|x| \ge |y|$ since $x \vee y \ge y$.

The condition means, for example, that if a worker is forced to cut down on the hours allocated to some firm, then he may choose to work longer for other firms, but he will not increase his total working hours. In the discrete model the condition says that if a firm loses the services of one worker it will replace him by at most one worker. Note that if C is quota filling then it is automatically size monotone. From size monotonicity, we get

Theorem 4 (Lattice Polarity). *If all choice functions are size monotone then $X^F = X_W$.*

Proof. First, since for all w, $x^W(w) \succeq x_W(w)$, it follows from the remark above that $|x_W(w)| \le |x^W|$ so $|X_W| = \sum_w |x_W(w)| \le \sum_w |x^W(w)| = |X^W|$, and similarly $|X_F| \le |X^F|$. From the previous Lemma $|X^F| \le |X_W|$, so now $|X_F| \le |X^F| \le |X_W| \le |X^W| \le |X_F|$, so $|X^F| = |X_W|$, so the conclusion follows, and also for any agent, say w,

$$|x^W(w)| = |x_W(w)| \,. \tag{4}$$

□

Theorem 5. *The set of stable matchings is a lattice under the orderings $\succeq_F$ and $\succeq_W$.*

Proof. It suffices to show that X^F is a stable matching. By definition X^F is F-acceptable and, since $X^F = X_W$, it follows that X^F is also W-acceptable. It remains to show stability, so suppose $x^F(f)$ is not w-satiated. Then by Lemma 9(i) $x(f), x'(f)$ are not w-satiated. So by stability $x(w), x'(w)$ are f-satiated, so by Lemma 9(ii) $x_W(w)$ is f-satiated, but by Theorem 4 $x_W(w)$ is the w-column of X^F, so X^F is stable. □

3.3 Properties of the Stable Matching Lattice

The following property, which says $|x(w)|=|x'(w)|$ for all w, generalizes a result for the classical model.

Theorem 6 (Unisize). *The schedules that an agent may have in any stable matching all have the same size.*

Proof. Note $x^W(w) \curlyvee_w x(w) = x^W(w)$ and $x^W(w) \curlywedge_w x(w) = x(w)$ so from (4) $|x^W(w)| = |x(w)|$ and similarly $|x^W(w)| = |x'(w)|$. □

An immediate consequence is the following result which was first shown by Roth and Sotomayor [9] for the classical college admissions model.

Corollary 3. *If the choice function of an agent is quota filling and he does not fill his quota in a stable matching then he has the same schedule in all stable matchings.*

Proof. Suppose $x(f) \neq x'(f)$ and $|x(f)| = |x'(f)| = c < q$. Then $|x(f) \vee x'(f)| > c$, so by quota filling $|x^F(f)| = |x(f) \vee x'(f)| > c$, contradicting Theorem 6. □

A striking structural property of stable matchings is that, for all pairs fw, $\{x^F(fw), x_F(fw)\} = \{x(fw), x'(fw)\}$, stated equivalently in the following form.

Theorem 7 (Complementarity). *If X and X' are stable matchings $X^F \vee X_F = X \vee X'$ and $X^F \wedge X_F = X \wedge X'$.*

Proof. Let f be any firm. First, from Lemma 6 we have

$$x^F(f) \wedge x_F(f) \leq x(f) \wedge x'(f) . \tag{5}$$

Secondly, for all w, by lattice polarity (Theorem 4) $x_F(fw) = x^W(fw) = (x(w) \curlyvee_w x'(w))(f) \leq (x(w) \vee x'(w))(f) = x(fw) \vee x'(fw)$, thus $x_F(f) \leq x(f) \vee x'(f)$ so, since $x^F(f) = x(f) \curlyvee_f x'(f) \leq x(f) \vee x'(f)$, we have

$$x^F(f) \vee x_F(f) \leq x(f) \vee x'(f) , \tag{6}$$

so $|x^F(f)| + |x_F(f)| - |x^F(f) \wedge x_F(f)| = |x^F(f) \vee x_F(f)| \leq |x(f) \vee x'(f)| = |x(f)| + |x'(f)| - |x(f) \wedge x'(f)|$, but from the unisize property (Theorem 6) $|x^F(f)| = |x_F(f)| = |x(f)| = |x'(f)|$ so

$$|x^F(f) \wedge x_F(f)| \geq |x(f) \wedge x'(f)| , \tag{7}$$

therefore (5) and (7) are equations, hence (6) also is an equation. □

Complementarity implies that the lattice of stable matchings is distributive.

Definition 12. *A lattice $\mathcal{L}$, with join $\curlyvee$ and meet $\curlywedge$, is distributive if $z \curlyvee (z' \curlywedge z'') = (z \curlyvee z') \curlywedge (z \curlyvee z'')$ and $z \curlywedge (z' \curlyvee z'') = (z \curlywedge z') \curlyvee (z \curlywedge z'')$ for all z, z', z'' in $\mathcal{L}$.*

Remark A standard fact in lattice theory (Corollary to Theorem II.13 in [4]) is that a lattice $(\mathcal{L}, \curlyvee, \curlywedge)$ is distributive if and only if the following *cancellation law* holds:

$$\text{if } z \curlyvee z' = z \curlyvee z'' \text{ and } z \curlywedge z' = z \curlywedge z'' \text{ then } z' = z'' \text{ for all } z, z', z'' \text{ in } \mathcal{L}. \quad (8)$$

Theorem 8 (Distributivity). *The* $(\curlyvee_F, \curlywedge_F)$ *and* $(\curlyvee_W, \curlywedge_W)$ *lattices of stable matchings are distributive.*

Proof. Let X, X', X'' be any three stable matchings. If $X \curlyvee_F X' = X \curlyvee_F X''$ and $X \curlywedge_F X' = X \curlywedge_F X''$ then $(X \curlyvee_F X') \vee (X \curlywedge_F X') = (X \curlyvee_F X'') \vee (X \curlywedge_F X'')$ and $(X \curlyvee_F X') \wedge (X \curlywedge_F X') = (X \curlyvee_F X'') \wedge (X \curlywedge_F X'')$, hence by complementarity (Theorem 7) $X \vee X' = X \vee X''$ and $X \wedge X' = X \wedge X''$, so by distributivity of $\vee, \wedge$ using cancellation $X' = X''$. Thus the cancellation law holds for $\curlyvee_F, \curlywedge_F$, similarly for $\curlyvee_W, \curlywedge_W$, and the theorem follows from the remark above. □

An important theorem in the classical case asserts that for stable matchings the schedules $x(f)$ and $x'(f)$ are comparable, that is either they are identical or f prefers one to the other. This was proved for college admissions in [10] and for schedules in [3]. This result does not hold in the general case as we show in the next section. However, we will here show that, for classical agents, it is a direct consequence of complementarity and the unisize property.

Corollary 4. *In the classical case let x and y be schedules where $x \succ y$. Then $x(i) > 0$ implies $x(j) \geq y(j)$ for $j < i$.*

Proof. If $y(j) > x(j)$ then for some $\epsilon > 0$ define the schedule $x_\epsilon \leq x \vee y$ by $x_\epsilon(i) = x(i) - \epsilon$, $x_\epsilon(j) = x(j) + \epsilon$, $x_e(k) = x(k)$ otherwise. Then $x_\epsilon \succ x$ contradicting $C(x \vee y) = x$. □

Theorem 9. *In the classical case if X and X' are stable matchings then either $x(f) \succ_f x'(f)$, $x(f) = x'(f)$, or $x(f) \prec_f x'(f)$.*

Proof. Let $y(f) = x_F(f) = x(f) \curlywedge_f x'(f)$. By the unisize property we cannot have $y(f) < x(f)$ or $y(f) < x'(f)$. Therefore, if $y(f)$ is distinct from $x(f)$ and $x'(f)$ then by complementarity there is a w such that $y(fw) = x(fw) > x'(fw)$ and there is a w' such that $y(fw') = x'(fw') > x(fw')$. But if, say, w' is preferred by f to w then since $x(f) \succ_f y(f)$ and $x(fw) > 0$ it follows from Corollary 4 that $x(fw') \geq y(fw')$, contradiction. □

4 Examples

In this section we will show by examples the need for our various assumptions. All examples are in the context of the special case of college admissions.

Example 4. If choice functions are consistent and size monotone but not persistent then stable matchings may not exist.

There are two colleges A and B and four students m, w, m', w'. College A has quota 2 and the choice function as in Section 1 so that

$$mw \succ_A mw' \succ_A m'w' \succ_A m'w\,. \tag{9}$$

College B has quota 1 and prefers m to w and will not admit m' or w'. Student m prefers A to B while student w prefers B to A. In the table below, the preferences of B, m, w are indicated by the arrows.

$$\begin{array}{ccccc} & m & w & m' & w' \\ A & Am & Aw & Am' & Aw' \\ & \uparrow & \downarrow & & \\ B & Bm & \longleftarrow Bw & & \end{array}$$

For every assignment of students to A there is a blocking as seen from (9) and the table above:

(A, mw)	is blocked by	B and m,
(A, mw')	"	A and w,
$(A, m'w)$	"	A and w',
$(A, m'w')$ and (B, m)	"	B and w,
$(A, m'w')$ and (B, w)	"	A and m.

Example 5. If preferences are consistent and persistent but not size monotone then stable matchings may not form a lattice. More precisely, the sup of stable matchings may not be stable.

There are colleges $A, \ldots, E$ and students $a, \ldots, e$. Preferences are given by the table below. Note that the preferences of A and B violate size monotonicity.

A	B	C	D	E	a	b	c	d	e
a^*	$b^\#$	c^*	$d^\#$	e	$C^\#$	D^*	$A^\#$	B^*	$A^\#$
$ce^\#$	de^*	$a^\#$	b^*	-	A^*	$B^\#$	C^*	$D^\#$	B^*
-	-	-	-	-	-	-	-	-	E

One easily verifies that the entries marked $*$ and those marked $\#$ correspond to stable matchings. Namely, in each matching, where a college is matched with its second choice, the preferred student is matched with her first choice. But in the matching which is the college supremum of $*$ and $\#$, both E *and* e are unmatched, hence they block and the college supremum is therefore unstable. Note that the unisize condition is also violated for A and B in the matchings $*$ and $\#$.

The following two examples show that certain results for the classical model do not generalize to the model with revealed preference (with consistent, persistent and size monotone choice functions).

Example 6. The college optimal stable matching may not be Pareto optimal for colleges.

There are colleges A, B, Z with quotas $1, 1, 2$, male students m, m' and female students w, w'.

Z chooses mw if all four students are available and otherwise chooses the sexually diverse pair.

The other preferences are given by the table below where the left entry in each pair is the college's ranking of the student and the right entry is the student's ranking of the college.

	m	w	m'	w'
Z	$(-,2)$	$(-,2)^{\#}$	$(-,1)^{*\#}$	$(-,1)^{*}$
A	$(1,3)^{\#}$	$(2,1)^{*}$	$(4,3)$	$(3,2)$
B	$(2,1)^{*}$	$(3,3)$	$(4,2)$	$(1,3)^{\#}$

The only stable matching is the student optimal matching $*$. One sees this by checking from the algorithm that it is also the college optimal matching. But the matching # makes all colleges (strictly) better off. Of course # is unstable, being blocked by Z and m.

Example 7. As shown in Corollary 5, in the classical model all stable matchings are comparable for each agent. This need not be so in the non-classical model.

Colleges A and B have quota 2. Students are m, w, m', w', and the choice function of A and B are as in Example 3 except A most prefers mw and B most prefers $m'w'$. For the students, m and w prefer B to A, m' and w' prefer A to B.

One easily verifies that all four ways of allocating diverse pairs to A and B are stable and also that mw' and $m'w$ are noncomparable in the preferences of both A and B.

References

1. Alkan, A. (2001): On Preferences over Subsets and the Lattice Structure of Stable Matchings. Review of Economic Design **6**, 99–111
2. Alkan, A. (2002): A Class of Multipartner Matching Models with a Strong Lattice Structure. Economic Theory (to appear)
3. Baiou, M., Balinski, M. (2000): The Stable Scheduling (or Ordinal Transportation) Problem. Laboratoire d'Econometrie, Ecole Polytechnique, Paris (mimeo)
4. Birkhoff, G. (1973): Lattice Theory. In: American Mathematical Society Colloquium Publications, Vol. XXV
5. Blair, C. (1988): The Lattice Structure of the Set of Stable Matchings with Multiple Partners. Mathematics of Operations Research **13**, 619–628
6. Crawford, V. P. (1991): Comparative Statics in Matching Markets. Journal of Economic Theory **54**, 389–400

7. Gale, D., Shapley, L. (1962) College Admissions and the Stability of Marriage. American Mathematical Monthly **69**, 9–15
8. Kelso, A. S., Crawford, V. P. (1982): Job Matching, Coalition Formation and Gross Substitutes. Econometrica **50**, 1483–1504
9. Roth, A. E., Sotomayor, M. (1989): The College Admissions Problem Revisited. Econometrica **57**, 559–570
10. Roth, A. E., Sotomayor, M. (1990): Two-Sided Matching: A Study in Game-Theoretic Modeling and Analysis. Cambridge University Press, Cambridge
11. Sotomayor, M. (1999): Three Remarks on the Many-to-Many Stable Matching Problem. Mathematical Social Sciences **38**, 55–70

Banzhaf Permission Values for Games with a Permission Structure

René van den Brink

Department of Econometrics, Free University,
De Boelelaan 1105, 1081 HV Amsterdam, The Netherlands
E-mail: jrbrink@feweb.vu.nl

Summary. A *game with a permission structure* describes a situation in a cooperative game with transferable utility in which cooperation possibilities are limited because some players need permission from other players before they are allowed to cooperate. In the *conjunctive approach*, it is assumed that every player needs permission from *all* direct superiors. In the *disjunctive approach*, it is assumed that each player only needs permission from *at least one* direct superiors. Both approaches yield modified games which take account of limited cooperation possibilities. Applying the *Banzhaf value* to these modified games yields allocation rules that distribute the payoffs over the individual players. This paper provides axiomatic characterizations of these *Banzhaf permission values*.

Key words: cooperative TU-game, Banzhaf value, directed graph, permission structure

1 Introduction

A situation in which a finite set of players $N \subset \mathbb{N}$ can generate certain payoffs by cooperation can be described by a *cooperative game with transferable utility* (or simply a TU-game), being a pair (N,v) where $v\colon 2^N \to IR$ is a *characteristic function* on N satisfying $v(\emptyset) = 0$. The collection of all characteristic functions on a particular player set N is denoted by $\mathcal{G}^N$.

In a TU-game there are no restrictions on the cooperation possibilities of the players, i. e., every coalition $E \subset N$ is formable and can generate a payoff. Examples of models in which there are restrictions on the possibilities of cooperation are the *games in coalition structure* in which it is assumed that the set of players is partitioned into disjoint sets which represent social groups such that for a particular player it is more easy to cooperate with players in its own group than to cooperate with players in other groups (see, e. g., [1], [20], [15] and [23]), and the games with *limited communication structure* in which the edges of an undirected graph on the set of players represent binary communication links between the players such that players can cooperate only if they are connected (see, e. g. [18], [16], [21] and [3]).

This paper considers *games with a permission structure* in which it is assumed that players in a TU-game are part of a hierarchical organization in which there are players that need permission from other players before they are allowed to cooperate within a coalition. Thus the possibilities of coalition formation are determined by the positions of the players in this so-called *permission structure*. Various assumptions can be made about how a permission structure affects the cooperation possibilities. In the *conjunctive approach*, as developed in [12] and [6], it is assumed that every player needs permission from *all* its direct superiors before it is allowed to cooperate with other players. Alternatively, in the *disjunctive approach*, as considered in [11] and [4] it is assumed that every player needs permission from *at least one* of its direct superiors before it is allowed to cooperate with other players.

Given a game and a permission structure a modified game is defined which takes account of the limited cooperation possibilities. The conjunctive and disjunctive approach yield different modified games. An *allocation rule* for games with a permission structure is a function that assigns to every game with a permission structure a distribution of the payoffs that can be obtained by cooperation over the individual players. Applying solutions for TU-games (being functions that assign a payoff distribution to every TU-game) to these modified games yields allocation rules for games with a permission structure. Applying, for example, the *Shapley value* [22] to these modified games yields the *conjunctive* and *disjunctive (Shapley) permission values* which have been characterized in [6] and [4], respectively. In [5] it has been shown that these two permission values only differ with respect to the *fairness* axiom that is used. The disjunctive (Shapley) permission value satisfies *disjunctive fairness* implying that deleting the relation between a player and one of its direct superiors changes the payoffs of these two players by the same amount (under the condition that the player has at least one other direct superior). Instead, the conjunctive (Shapley) permission value satisfies *conjunctive fairness* implying that deleting the relation between a player and one of its direct superiors changes the payoffs of this player and each of its *other* direct superiors by the same amount.

In this paper we study another well known solution for TU-games, namely the *Banzhaf value* which is based on the Banzhaf index for voting games [2], and is generalized to arbitrary games by, e. g., [19] and [10]. A potential approach to the Banzhaf value has been given by Dragan [9]. Applying the Banzhaf value to the modified games in the conjunctive and disjunctive approach yields the *conjunctive*, respectively, *disjunctive Banzhaf permission value*. In this paper we give axiomatic characterizations of these Banzhaf permission values.

New properties that are satisfied by both Banzhaf permission values are predecessor fairness and two neutrality properties. *Predecessor fairness* states that the payoffs of two direct superiors of a player change in opposite direction if we delete its relation with one of the two superiors, and moreover, the absolute values of the changes in payoffs of these two direct superiors are the same. The *neutrality* properties state that, if a new player enters the game with permission structure and "merges" with a player that is already present, then these two players share the payoff that was earned by the player that was already present in the original game. We distinguish between two such neutrality properties. *Vertical neutrality* considers the situation in which the new player enters the game as a null player, and enters the permission structure as a direct superior of the old player. *Horizontal neutrality* considers the situation in which the new player enters the game as a veto player for the old player, and enters the permission structure at the same level as the old player.

The paper is organized as follows. In Section 2 we state some preliminaries on games with a permission structure. In Section 3 we discuss predecessor fairness. In Section 4 we introduce the neutrality axioms and characterize the two Banzhaf permission values. Finally, in Section 5 we show how predecessor fairness and the neutrality axioms can be adapted to give new characterizations of the (Shapley) permission values.

2 Preliminaries: Games with a Permission Structure

We assume that players who participate in a TU-game are part of a hierarchical organization in which there are players that need permission from certain other players before they are allowed to cooperate. For a finite set of players $N \subset \mathbb{N}$ such a hierarchical organization is represented by a pair (N, S), where the mapping $S\colon N \to 2^N$ is called a *permission structure* on N. (So the set $\{(i,j) \in N \times N \mid i \in N,\ j \in S(i)\}$ describes a directed graph on N.) The players in $S(i)$ are called the *successors* of player $i \in N$ in S. The players in $S^{-1}(i) := \{j \in N \mid i \in S(j)\}$ are called the *predecessors* of i in S. By $\widehat{S}$ we denote the *transitive closure* of the permission structure S, i.e., $j \in \widehat{S}(i)$ if and only if there exists a sequence of players $(h_1, \ldots, h_t)$ such that $h_1 = i$, $h_{k+1} \in S(h_k)$ for all $1 \le k \le t-1$, and $h_t = j$. The players in $\widehat{S}(i)$ are called the *subordinates* of i in S, and the players in $\widehat{S}^{-1}(i) := \{j \in N \mid i \in \widehat{S}(j)\}$ are called the *superiors* of i in S. In this paper we restrict our attention to *hierarchical* permission structures being permission structures $S\colon N \to 2^N$ that are

1) *acyclic*, i.e., $i \notin \widehat{S}(i)$ for all $i \in N$, and
2) *quasi-strongly connected*, i.e., there exists an $i \in N$ such that $\widehat{S}(i) = N \setminus \{i\}$.

We denote the collection of all hierarchical permission structures on a particular player set N by $\mathcal{S}_H^N$. A triple (N, v, S) with $N \subset \mathbb{N}$, $v \in \mathcal{G}^N$ and $S \in \mathcal{S}_H^N$ is called a *game with a (hierarchical) permission structure*. In a hierarchical permission structure there exists a *unique* player i_0 such that $\widehat{S}(i_0) = N\backslash\{i_0\}$. Moreover, $S^{-1}(i_0) = \emptyset$ for this player. We call this player the *top-player* in the permission structure. Since we only consider hierarchical permission structures we will often refer to these simply as permission structures. (The results in this paper could be stated for acyclic permission structures that not necessarily are quasi-strongly connected. For notational convenience we restrict attention to permission structures satisfying both conditions (i) and (ii).)

2.1 Disjunctive and Conjunctive Restrictions

In the *disjunctive approach*, as developed in [11] and [4], it is assumed that each player needs permission from *at least one* of its predecessors before it is allowed to cooperate with other players. Consequently, a coalition is formable if and only if every player in the coalition, except the top-player i_0, has a predecessor who also belongs to the coalition. Thus, the formable coalitions are the ones in the set

$$\Phi^d_{N,S} := \left\{E \subset N \mid S^{-1}(i) \cap E \neq \emptyset \text{ or } S^{-1}(i) = \emptyset \text{ for all } i \in E\right\} . \tag{1}$$

Note that $E \in \Phi^d_{N,S}$ implies that $i_0 \in E$. The coalitions in $\Phi^d_{N,S}$ are called the *disjunctive autonomous* coalitions in S. The largest disjunctive autonomous subset of $E \subset N$ in $S \in \mathcal{S}^N_H$ is denoted by $\sigma^d_{N,S}(E) = \cup\{F \in \Phi^d_{N,S} \mid F \subset E\}$, and is called the *disjunctive sovereign part* of E in S. It consists of those players in E that can be reached by a directed "permission path" starting from the top-player such that all players on this path belong to coalition E. Using this concept we can transform the characteristic function v into a modified characteristic function which takes account of the limited cooperation possibilities as determined by the permission structure S as follows. Given a game with a permission structure (N, v, S), the *disjunctive restriction* of v on S is the characteristic function $r^d_{N,v,S}\colon 2^N \rightarrow I\!R$ given by $r^d_{N,v,S}(E) = v(\sigma^d_{N,S}(E))$ for all $E \subset N$.

Alternatively, in the *conjunctive approach*, as developed in [12] and [6], it is assumed that each player needs permission from *all* its predecessors before it is allowed to cooperate. This implies that a coalition E is formable if and only if for every player in the coalition it holds that all its predecessors belong to the coalition. The set of formable coalitions in this approach thus is given by

$$\Phi^c_{N,S} := \left\{E \subset N \mid S^{-1}(i) \subset E \text{ for all } i \in E\right\} . \tag{2}$$

The coalitions in the set $\Phi^c_{N,S}$ are called the *conjunctive autonomous* coalitions in S. The largest conjunctive autonomous subset of E is denoted by $\sigma^c_{N,S}(E) = \cup\{F \in \Phi^c_{N,S} \mid F \subset E\}$, and is refered to as the *conjunctive sovereign part* of E in S. It consists of all players in E whose superiors all belong to E. Given game with permission structure (N, v, S), the *conjunctive restriction* of v on S is the characteristic function $r^c_{N,v,S}\colon 2^N \rightarrow I\!R$ given by $r^c_{N,v,S}(E) = v(\sigma^c_{N,S}(E))$ for all $E \subset N$.

Example 1. Consider the game with permission structure (N, v, S) with $N = \{1, 2, 3, 4\}$, $v \in \mathcal{G}^N$ given by $v(E) = 1$ if $4 \in E$, $v(E) = 0$ otherwise, and $S \in \mathcal{S}_H^N$ given by $S(1) = \{2, 3\}$, $S(2) = S(3) = \{4\}$, $S(4) = \emptyset$.

The disjunctive and conjunctive restrictions of v on S, respectively, are given by

$$r^d_{N,v,S}(E) = \begin{cases} 1 & \text{if } E \in \{\{1,2,4\},\ \{1,3,4\},\ \{1,2,3,4\}\}\,, \\ 0 & \text{otherwise}\,, \end{cases}$$

and

$$r^c_{N,v,S}(E) = \begin{cases} 1 & \text{if } E = \{1,2,3,4\}\,, \\ 0 & \text{otherwise}\,. \end{cases}$$

2.2 The (Shapley) Permission Values

An *allocation rule* for games with a permission structure is a function that assigns to every game with a permission structure (N, v, S) a distribution of the payoffs that can be obtained by cooperation according to v taking into account the limited cooperation possibilities determined by S. The *disjunctive (Shapley) permission value* φ^d is obtained by appyling the *Shapley value* [22] to the disjunctive restricted games, while the *conjunctive (Shapley) permission value* φ^c is obtained by applying the Shapley value to the conjunctive restricted games, i. e.,

$$\varphi^d(N, v, S) = Sh(N, r^d_{N,v,S}) \text{ and } \varphi^c(N, v, S) = Sh(N, r^c_{N,v,S})\,,$$

where

$$Sh_i(N, v) = \sum_{\substack{E \subset N \\ i \in E}} \frac{(|N| - |E|)!(|E| - 1)!}{|N|!}(v(E) - v(E \setminus \{i\}))\,, \quad \text{for all } i \in N\,.$$

Example 2. The disjunctive and conjunctive (Shapley) permission values of the game with permission structure given in Example 1 are $\varphi^d(N, v, S) = \frac{1}{12}(5, 1, 1, 5)$ and $\varphi^c(N, v, S) = \frac{1}{4}(1, 1, 1, 1)$.

The disjunctive (Shapley) permission value is axiomatized in [4], while axiomatizations of the conjunctive (Shapley) permission value can be found in [6] and [5]. We refer to these papers for a discussion on the axioms presented below. The first two axioms are straightforward generalizations of the efficiency and additivity axioms of solutions for TU-games.

Axiom 1 (Efficiency). *For every $N \subset \mathbb{N}$, $v \in \mathcal{G}^N$ and $S \in \mathcal{S}_H^N$, it holds that $\sum_{i \in N} f_i(N, v, S) = v(N)$.*

Axiom 2 (Additivity). *For every $N \subset \mathbb{N}$, $v, w \in \mathcal{G}^N$ and $S \in \mathcal{S}_H^N$, it holds that $f(N, v + w, S) = f(N, v, S) + f(N, w, S)$, where $(v + w) \in \mathcal{G}^N$ is defined by $(v + w)(E) = v(E) + w(E)$ for all $E \subset N$.*

Player $i \in N$ is *inessential* in game with permission structure (N, v, S) if i and all its subordinates are *null* players in (N, v), i.e., if $v(E) = v(E \setminus \{j\})$ for all $E \subset N$ and $j \in \{i\} \cup \widehat{S}(i)$.

Axiom 3 (Inessential player property). *For every $N \subset \mathbb{N}$, $v \in \mathcal{G}^N$ and $S \in \mathcal{S}_H^N$, if $i \in N$ is an inessential player in (N, v, S) then $f_i(N, v, S) = 0$.*

The next two axioms are stated for monotone characteristic functions. A characteristic function $v \in \mathcal{G}^N$ is *monotone* if $v(E) \leqq v(F)$ for all $E \subset F \subset N$. The class of all monotone characteristic functions on N is denoted by $\mathcal{G}_M^N$. Player $i \in N$ is called *necessary* in game (N, v) if $v(E) = 0$ for all $E \subset N \setminus \{i\}$.

Axiom 4 (Necessary player property). *For every $N \subset \mathbb{N}$, $v \in \mathcal{G}_M^N$ and $S \in \mathcal{S}_H^N$, if $i \in N$ is a necessary player in v then $f_i(N, v, S) \geqq f_j(N, v, S)$ for all $j \in N$.*

We say that player $i \in N$ dominates player $j \in N$ "completely" if all directed "permission paths" from the top-player i_0 to player j contain player i. We denote the set of players that player i dominates "completely" by $\overline{S}(i)$, i.e.,

$$\overline{S}(i) = \left\{ j \in \widehat{S}(i) \;\middle|\; \begin{array}{c} i \in \{i_0, i_1, \ldots, i_{t-1}\} \text{ for every sequence of nodes} \\ i_0, i_1, \ldots, i_t \text{ such that} \\ i_t = j \text{ and } i_k \in S(i_{k-1}),\ k \in \{1, \ldots, t\}, \end{array} \right\}.$$

We also define $\overline{S}^{-1}(i) = \{j \in \widehat{S}^{-1}(i) \mid i \in \overline{S}(j)\}$.

Axiom 5 (Weak structural monotonicity). *For every $N \subset \mathbb{N}$, $v \in \mathcal{G}_M^N$ and $S \in \mathcal{S}_H^N$, if $i \in N$ and $j \in \overline{S}(i)$ then $f_i(N, v, S) \geq f_j(N, v, S)$.*

Weak structural monotonicity is a weaker version of *structural monotonicity* as introduced in [6] which states that for a monotone game with a permission structure, if $i \in N$ and $j \in S(i)$ then $f_i(N, v, S) \geq f_j(N, v, S)$. (This stronger structural monotonicity is satisfied by the conjunctive (Shapley) permission value but is not satisfied by the disjunctive (Shapley) permission value.)

The five axioms given above are satisfied by both the disjunctive and conjunctive (Shapley) permission values. These two permission values differ with respect to the fairness axiom that is used. *Disjunctive fairness* states that deleting the relation between two players h and $j \in S(h)$ (with $|S^{-1}(j)| \geq 2$) changes the payoffs of players h and j by the same amount. Moreover, also the payoffs of all players i that "completely" dominate player h, in the sense that $i \in \overline{S}^{-1}(h)$, change by this same amount.

For $S \in \mathcal{S}_H^N$, $h \in N$ and $j \in S(h)$ we denote $S_{-(h,j)}(h) = S(h) \setminus \{j\}$ and $S_{-(h,j)}(i) = S(i)$ for $i \in N \setminus \{h\}$.

Axiom 6 (Disjunctive fairness). *For every $N \subset \mathbb{N}$, $v \in \mathcal{G}^N$ and $S \in \mathcal{S}_H^N$, if $h \in N$ and $j \in S(h)$ with $|S^{-1}(j)| \geq 2$, then*

$$f_i(N, v, S) - f_i(N, v, S_{-(h,j)}) = f_j(N, v, S) - f_j(N, v, S_{-(h,j)})$$

$$\text{for all } i \in \{h\} \cup \overline{S}^{-1}(h).$$

Disjunctive fairness is related to *fairness* as introduced in [18] for games with a limited communication structure. In that model fairness means that deleting a communication relation between two players in an undirected communication graph has the same effect on both their payoffs. (Note that in our fairness property we require that the successor on the relation to be deleted has at least two predecessors.)

Theorem 7 (van den Brink [4]). *An allocation rule f is equal to the disjunctive (Shapley) permission value φ^d if and only if it satisfies efficiency, additivity, the inessential player property, the necessary player property, weak structural monotonicity and disjunctive fairness.*

The conjunctive (Shapley) permission value does not satisfy disjunctive fairness. However, it satisfies the alternative *conjunctive fairness* which states that deleting the relation between two players h and $j \in S(h)$ (with $|S^{-1}(j)| \geq 2$) changes the payoffs of player j and any *other* predecessor $g \in S^{-1}(j) \setminus \{h\}$ by the same amount. Moreover, also the payoffs of all players that 'completely' dominate the other predecessor g change by this same amount.

Axiom 8 (Conjunctive fairness). *For every $N \subset \mathbb{N}$, $v \in \mathcal{G}^N$ and $S \in \mathcal{S}_H^N$, if $h, j, g \in N$ are such that $h \neq g$ and $j \in S(h) \cap S(g)$, then*

$$f_i(N, v, S) - f_i(N, v, S_{-(h,j)}) = f_j(N, v, S) - f_j(N, v, S_{-(h,j)})$$

$$\text{for all } i \in \{g\} \cup \overline{S}^{-1}(g).$$

Theorem 9 (van den Brink [5]). *An allocation rule f is equal to the conjunctive (Shapley) permission value φ^c if and only if it satisfies efficiency, additivity, the inessential player property, the necessary player property, weak structural monotonicity and conjunctive fairness.*

3 Predecessor Fairness and Banzhaf Permission Values

The two fairness axioms that are mentioned in the previous section compare the effects of deleting the relation between players h and $j \in S(h)$ on the payoffs of players h and j, and on the payoffs of players $g \in S^{-1}(j) \setminus \{h\}$ and j, respectively. These properties do not compare the change in payoffs of players h and $g \in S^{-1}(j) \setminus \{h\}$ after deleting the relation between h and j. For monotone games the changes in the disjunctive and conjunctive (Shapley) permission values of these players are opposite.

Proposition 1. *For every* $N \subset \mathbb{N}$, $v \in \mathcal{G}^N$, $S \in \mathcal{S}_H^N$ *and* $h, g, j \in N$ *with* $h \neq g$ *and* $j \in S(h) \cap S(g)$:

1) $\varphi_j^d(N,v,S) \geq \varphi_j^d(N,v,S_{-(h,j)})$, $\varphi_h^d(N,v,S) \geq \varphi_h^d(N,v,S_{-(h,j)})$ *and* $\varphi_g^d(N,v,S) \leq \varphi_g^d(N,v,S_{-(h,j)})$;

2) $\varphi_j^c(N,v,S) \leq \varphi_j^c(N,v,S_{-(h,j)})$, $\varphi_h^c(N,v,S) \geq \varphi_h^c(N,v,S_{-(h,j)})$ *and* $\varphi_g^c(N,v,S) \leq \varphi_g^c(N,v,S_{-(h,j)})$.

The proof of this proposition is straightforward and is therefore omitted. As a consequence we have that deleting the relation between player h and $j \in S(h)$ (with $|S^{-1}(j)| \geq 2$) in a monotone game with a hierarchical permission structure changes the disjunctive and conjunctive (Shapley) permission values of players h and $g \in S^{-1}(j) \setminus \{h\}$ in opposite direction. The fairness properties discussed in the previous section hold for all games with a hierarchical permission structure. This is not the case for Proposition 1 and the "opposite change" consequence described here.

Example 3. Consider the game with permission structure (N, v, S) given by $N = \{1,2,3,4,5\}$, $v = u_{\{4,5\}} - \frac{7}{10}u_{\{4\}}$, where u_T denotes the unanimity game of $T \subset N$ (i.e., $u_T(E) = 1$ if $E \supset T$, and $u_T(E) = 0$ otherwise) and $S(1) = \{2,3,5\}$, $S(2) = S(3) = \{4\}$ and $S(4) = S(5) = \emptyset$. In this example, $\varphi_2^d(N,v,S) - \varphi_2^d(N,v,S_{-(2,4)}) = -\frac{1}{120} - 0 < 0$ and $\varphi_3^d(N,v,S) - \varphi_3^d(N,v,S_{-(2,4)}) = -\frac{1}{120} - \frac{1}{60} = -\frac{1}{40} < 0$. Thus, deleting the relation between players 2 and 4 changes the disjunctive (Shapley) permission values of player 4's predecessors 2 and 3 in the same direction. (By disjunctive fairness the change in the disjunctive (Shapley) permission value of player 4 is the same as for player 2: $\varphi_4^d(N,v,S) - \varphi_4^d(N,v,S_{-(2,4)}) = \frac{1}{120} - \frac{1}{60} = -\frac{1}{120}$.)

For the conjunctive (Shapley) permission values $\varphi_2^c(N,v,S) - \varphi_2^c(N,v,S_{-(2,4)}) = \frac{1}{40} - 0 > 0$ and $\varphi_3^c(N,v,S) - \varphi_3^c(N,v,S_{-(2,4)}) = \frac{1}{40} - \frac{1}{60} = \frac{1}{120} > 0$. Thus, deleting the relation between players 2 and 4 also changes the conjunctive (Shapley) permission values of player 4's predecessors 2 and 3 in the same direction. (By conjunctive fairness the change in the conjunctive (Shapley) permission value of player 4 is the same as for player 3: $\varphi_4^c(N,v,S) - \varphi_4^c(N,v,S_{-(2,4)}) = \frac{1}{40} - \frac{1}{60} = \frac{1}{120}$.)

It turns out that applying the *Banzhaf value* to the conjunctive and disjunctive restricted games yields allocation rules for games with a permission structure that satisfy the "opposite change" property for every game with a hierarchical permission structure. The Banzhaf value for TU-games is given by

$$B_i(N,v) = \frac{1}{2^{|N|-1}} \sum_{\substack{E \subseteq N \\ i \in E}} (v(E) - v(E \setminus \{i\})), \quad \text{for all } i \in N\,. \tag{3}$$

The *disjunctive Banzhaf permission value* β^d and the *conjunctive Banzhaf permission value* β^c are obtained by applying the Banzhaf value to the disjunctive and conjunctive restricted games, respectively, yielding

$$\beta^d(N,v,S) = B(N,r^d_{N,v,S}) \text{ and } \beta^c(N,v,S) = B(N,r^c_{N,v,S}) \,.$$

Example 4. The disjunctive and conjunctive Banzhaf permission values of the game with permission structure given in Example 1 are $\beta^d(N,v,S) = \frac{1}{8}(3,1,1,3)$ and $\beta^c(N,v,S) = \frac{1}{8}(1,1,1,1)$.

The disjunctive and conjunctive Banzhaf permission values of players h and g, $h \neq g$, always change in opposite direction after deleting the relation between players h and $j \in S(h) \cap S(g)$. Moreover, the absolute values of the changes in the Banzhaf permission values of these two players are the same.

Axiom 10 (Predecessor fairness). *For every $N \subset \mathbb{N}$, $v \in \mathcal{G}^N$ and $S \in \mathcal{S}^N_H$, if $h,g,j \in N$ are such that $h \neq g$ and $j \in S(h) \cap S(g)$, then*

$$f_h(N,v,S) - f_h(N,v,S_{-(h,j)}) = f_g(N,v,S_{-(h,j)}) - f_g(N,v,S) \,.$$

Theorem 11. *The disjunctive and conjunctive Banzhaf permission values satisfy predecessor fairness.*

Proof. Let $N \subset \mathbb{N}$, $v \in \mathcal{G}^N$, $S \in \mathcal{S}^N_H$ and $h,g,j \in N$ be such that $j \in S(h) \cap S(g)$ and $h \neq g$. Since $\sigma^d_{N,S}(E) = \sigma^d_{N,S_{-(h,j)}}(E)$ if $h \notin E$ or $g \in E$, it follows that

$$\begin{aligned}
&\beta^d_h(N,v,S) - \beta^d_h(N,v,S_{-(h,j)}) = B_h(N,r^d_{N,v,S}) - B_h(N,r^d_{N,v,S_{-(h,j)}}) \\
&= \frac{1}{2^{|N|-1}} \sum_{\substack{E \subset N \\ h \in E}} \Big(v(\sigma^d_{N,S}(E)) - v(\sigma^d_{N,S}(E \setminus \{h\})) - v(\sigma^d_{N,S_{-(h,j)}}(E)) \\
&\quad + v(\sigma^d_{N,S_{-(h,j)}}(E \setminus \{h\}))\Big) \\
&= \frac{1}{2^{|N|-1}} \sum_{\substack{E \subset N \\ h \in E}} \Big(v(\sigma^d_{N,S}(E)) - v(\sigma^d_{N,S_{-(h,j)}}(E))\Big) \\
&= \frac{1}{2^{|N|-1}} \sum_{\substack{E \subset N \\ h \in E,\ g \notin E}} \Big(v(\sigma^d_{N,S}(E)) - v(\sigma^d_{N,S_{-(h,j)}}(E))\Big) \\
&= \frac{1}{2^{|N|-1}} \sum_{\substack{E \subset N \\ g \notin E}} \Big(v(\sigma^d_{N,S}(E)) - v(\sigma^d_{N,S_{-(h,j)}}(E))\Big) \\
&= \frac{1}{2^{|N|-1}} \sum_{\substack{E \subset N \\ g \in E}} \Big(v(\sigma^d_{N,S}(E \setminus \{g\})) - v(\sigma^d_{N,S_{-(h,j)}}(E \setminus \{g\}))\Big) \\
&= \frac{1}{2^{|N|-1}} \sum_{\substack{E \subset N \\ g \in E}} \Big(v(\sigma^d_{N,S_{-(h,j)}}(E)) - v(\sigma^d_{N,S_{-(h,j)}}(E \setminus \{g\})) - v(\sigma^d_{N,S}(E)) \\
&\quad + v(\sigma^d_{N,S}(E \setminus \{g\}))\Big) \\
&= B_g(N,r^d_{N,v,S_{-(h,j)}}) - B_g(N,r^d_{N,v,S}) = \beta^d_g(N,v,S_{-(h,j)}) - \beta^d_g(N,v,S) \,,
\end{aligned}$$

showing that β^d satisfies predecessor fairness.

To show that β^c satisfies predecessor fairness note that $\sigma^c_{N,S}(E) = \sigma^c_{N,S_{-(h,j)}}(E)$ if $h \in E$ or $g \notin E$. But then, in a similar way as above for β^d, it follows that

$$\begin{aligned}
&\beta^c_h(N,v,S) - \beta^c_h(N,v,S_{-(h,j)}) = B_h(N,r^c_{N,v,S}) - B_h(N,r^c_{N,v,S_{-(h,j)}})\\
&= \frac{1}{2^{|N|-1}} \sum_{\substack{E\subset N\\ h\in E}} \Big(v(\sigma^c_{N,S}(E)) - v(\sigma^c_{N,S}(E\setminus\{h\})) - v(\sigma^c_{N,S_{-(h,j)}}(E))\\
&\quad + v(\sigma^c_{N,S_{-(h,j)}}(E\setminus\{h\}))\Big)\\
&= \frac{1}{2^{|N|-1}} \sum_{\substack{E\subset N\\ h\in E}} \Big(v(\sigma^c_{N,S_{-(h,j)}}(E\setminus\{h\})) - v(\sigma^c_{N,S}(E\setminus\{h\}))\Big)\\
&= \frac{1}{2^{|N|-1}} \sum_{\substack{E\subset N\\ \{h,g\}\subset E}} \Big(v(\sigma^c_{N,S_{-(h,j)}}(E\setminus\{h\})) - v(\sigma^c_{N,S}(E\setminus\{h\}))\Big)\\
&= \frac{1}{2^{|N|-1}} \sum_{\substack{E\subset N\\ g\in E, h\notin E}} \Big(v(\sigma^c_{N,S_{-(h,j)}}(E)) - v(\sigma^c_{N,S}(E))\Big)\\
&= \frac{1}{2^{|N|-1}} \sum_{\substack{E\subset N\\ g\in E}} \Big(v(\sigma^c_{N,S_{-(h,j)}}(E)) - v(\sigma^c_{N,S}(E))\Big)\\
&= \frac{1}{2^{|N|-1}} \sum_{\substack{E\subset N\\ g\in E}} \Big(v(\sigma^c_{N,S_{-(h,j)}}(E)) - v(\sigma^c_{N,S_{-(h,j)}}(E\setminus\{g\}))\\
&\quad - v(\sigma^c_{N,S}(E)) + v(\sigma^c_{N,S}(E\setminus\{g\}))\Big)\\
&= B_g(N,r^c_{N,v,S_{-(h,j)}}) - B_g(N,r^c_{N,v,S}) = \beta^c_g(N,v,S_{-(h,j)}) - \beta^c_g(N,v,S)\ ,
\end{aligned}$$

showing that β^c satisfies predecessor fairness. □

4 Axiomatizations of Banzhaf Permission Values

Main purpose of this paper is to axiomatize the Banzhaf permission values.

4.1 The Disjunctive Banzhaf Permission Value

Compared to the disjunctive (Shapley) permission value the disjunctive Banzhaf permission value satisfies the axioms stated in Theorem 7 except efficiency. However, it satisfies efficiency for one player games with a permission structure. Note that $S \in \mathcal{S}^N_H$ with $|N| = 1$ implies that $S(i) = \emptyset$ for $i \in N$, and thus $r^d_{N,v,S} = v$ in that case.

Axiom 12 (One player efficiency). *For every $N \subset \mathbb{N}$ with $|N| = 1$, $v \in \mathcal{G}^N$ and $S \in \mathcal{S}_H^N$, it holds that $f_i(N,v,S) = v(\{i\})$ for $i \in N$.*

Replacing efficiency in Theorem 7 by one player efficiency and predecessor fairness is not sufficient to characterize the disjunctive Banzhaf permission value as can be seen from Example 5.2 and 5.3 at the end of this subsection. To characterize the disjunctive Banzhaf permission value we introduce two more axioms that, in particular, will be applied to permission *trees*, i. e., hierarchical permission structures in which each player has at most one predecessor. These two axioms are related to *amalgamation neutrality* and *collusion neutrality* as used, respectively, in [17] and [13] to characterize the Banzhaf value for TU-games. A similar neutrality property is used here and is defined in the appendix of this paper.

First, suppose that all subordinates of player $j \in N$ are null players (note that this does not imply that j is an inessential player because j need not be a null player itself). Now, let $h \in \mathbb{N} \setminus N$ be a new player whose only task is to supervise player j, i. e., h will be a null player in the new game and in the permission structure he becomes the only predecessor of j and gets all previous predecessors of j as its predecessors. Then the sum of the payoffs of players h and j in the new game with permission structure is equal to the payoff of player j in the original game with permission structure.

Axiom 13 (Vertical neutrality). *For every $N \subset \mathbb{N}$, $v \in \mathcal{G}^N$, $S \in \mathcal{S}_H^N$ and $h \in \mathbb{N} \setminus N$, if $j \in N$ and all $i \in \widehat{S}(j)$ are null players in (N,v), then*

$$f_j(N \cup \{h\}, v', S') + f_h(N \cup \{h\}, v', S') = f_j(N, v, S) ,$$

where $v' \in \mathcal{G}^{N \cup \{h\}}$ is given by $v'(E) = v(E \cap N)$ for all $E \subset N \cup \{h\}$, and $S' \subset \mathcal{S}_H^{N \cup \{h\}}$ is given by

$$S'(i) = \begin{cases} \{j\} , & \text{if } i = h , \\ (S(i) \setminus \{j\}) \cup \{h\} , & \text{if } i \in S^{-1}(j) , \\ S(i) , & \text{if } i \in N \setminus S^{-1}(j) . \end{cases}$$

Fig. 1. An illustration of permission structures S and S' as described in Axiom 13

For the second neutrality property, suppose that player $j \in N$, with $S(j) = \emptyset$ and $S^{-1}(j) \neq \emptyset$, is split in two in the sense that a new player h enters who gets the same successors (none) and predecessors as j, and in the new game h and j veto each other. Then, again, the sum of the payoffs of players h and j in the new game with permission structure is equal to the payoff of player j in the original game with permission structure.

Axiom 14 (Horizontal neutrality). *For every $N \subset \mathbb{N}$, $v \in \mathcal{G}^N$, $S \in \mathcal{S}_H^N$ and $h \in \mathbb{N} \setminus N$, if $S(j) = \emptyset$ and $S^{-1}(j) \neq \emptyset$ for $j \in N$, then*

$$f_j(N \cup \{h\}, v'', S'') + f_h(N \cup \{h\}, v'', S'') = f_j(N, v, S) ,$$

where $v'' \in \mathcal{G}^{N \cup \{h\}}$ is given by

$$v''(E) = \begin{cases} v(E \backslash \{j\}), & \text{if } E \subset N , \\ v(E \cap N), & \text{if } E \subset N \cup \{h\} \text{ with } h \in E , \end{cases}$$

and $S'' \in \mathcal{S}_H^{N \cup \{h\}}$ is given by $S''(i) = \begin{cases} \emptyset, & \text{if } i = h , \\ S(i) \cup \{h\}, & \text{if } i \in S^{-1}(j) , \\ S(i), & \text{if } i \in N \setminus S^{-1}(j) . \end{cases}$

Fig. 2. An illustration of permission structures S and S'' as described in Axiom 14

Replacing efficiency in Theorem 7 by one player efficiency, predecessor fairness, vertical neutrality and horizontal neutrality characterizes the disjunctive Banzhaf permission value β^d. First, we show that β^d satisfies the axioms.

Theorem 15. *The disjunctive Banzhaf permission value β^d satisfies one player efficiency, vertical neutrality, horizontal neutrality, additivity, the inessential player property, the necessary player property, weak structural monotonicity, disjunctive fairness and predecessor fairness.*

Proof. Proving that the disjunctive Banzhaf permission value satisfies additivity, the inessential player property, the necessary player property and weak structural monotonicity is along the same lines as this is shown for the disjunctive (Shapley) permission value in [4]. One player efficiency is evident. Predecessor fairness follows from Theorem 11.

Vertical neutrality follows from Proposition 2 (see Appendix) and the fact that $r^d_{N,v',S'} = (r^d_{N,v,S})_{hj}$ where v' and S' are as given in Axiom 13, and v_{hj} is as given in equation (16) of the appendix. Similarly, horizontal neutrality follows from Proposition 2 and the fact that $r^d_{N,v'',S''} = (r^d_{N,v,S})_{hj}$ where v'' and S'' are as given in Axiom 14.

So, we are left to show that β^d satisfies disjunctive fairness. Let $S \in \mathcal{S}^N_H$ and $h, j \in N$ be such that $j \in S(h)$ and $|S^{-1}(j)| \geq 2$. Further, let $i \in \{h\} \cup \overline{S}^{-1}(h)$.

Since $\{i, j\} \not\subset E$ implies that $\sigma^d_{N,S}(E) = \sigma^d_{N,S_{-(h,j)}}(E)$, it follows that

$$\beta^d_i(N, v, S) - \beta^d_i(N, v, S_{-(h,j)}) = B_i(N, r^d_{N,v,S}) - B_i(N, r^d_{N,v,S_{-(h,j)}})$$

$$= \frac{1}{2^{|N|-1}} \sum_{\substack{E \subset N \\ i \in E}} \Big(v(\sigma^d_{N,v,S}(E)) - v(\sigma^d_{N,v,S}(E \setminus \{i\})) - v(\sigma^d_{N,v,S_{-(h,j)}}(E))$$

$$+ v(\sigma^d_{N,v,S_{-(h,j)}}(E \setminus \{i\}))\Big)$$

$$= \frac{1}{2^{|N|-1}} \sum_{\substack{E \subset N \\ i \in E}} \Big(v(\sigma^d_{N,v,S}(E)) - v(\sigma^d_{N,v,S_{-(h,j)}}(E))\Big)$$

$$= \frac{1}{2^{|N|-1}} \sum_{\substack{E \subset N \\ \{i,j\} \subset E}} \Big(v(\sigma^d_{N,v,S}(E)) - v(\sigma^d_{N,v,S_{-(h,j)}}(E))\Big)$$

$$= \frac{1}{2^{|N|-1}} \sum_{\substack{E \subset N \\ j \in E}} \Big(v(\sigma^d_{N,v,S}(E)) - v(\sigma^d_{N,v,S}(E \setminus \{j\})) - v(\sigma^d_{N,v,S_{-(h,j)}}(E))$$

$$+ v(\sigma^d_{N,v,S_{-(h,j)}}(E \setminus \{j\}))\Big) = \beta^d_j(N, v, S) - \beta^d_j(N, v, S_{-(h,j)}),$$

showing that β^d satisfies disjunctive fairness. □

Next we show that the axioms mentioned in Theorem 15 characterize the disjunctive Banzhaf permission value. We do this in three steps. First, we prove a lemma for positively scaled unanimity games with a permission tree that has no inessential players. We denote by $\mathcal{S}^N_{tree} = \{S \in \mathcal{S}^N_H \mid \text{ for all } i \in N, |S^{-1}(i)| \leq 1\}$ the class of all permission trees on N. For $T \subset N$, $T \neq \emptyset$, and $c_T > 0$, the *positively scaled unanimity game* $w_T = c_T u_T$ is given by

$$w_T(E) = \begin{cases} c_T\,, & \text{if } E \supset T\,, \\ 0, & \text{otherwise}\,. \end{cases} \tag{4}$$

For $S \in \mathcal{S}^N_H$ and $h \in N$ the permission structure $S_{-h} \in S^{N \setminus \{h\}}$ is given by

$$S_{-h}(i) = \begin{cases} (S(i) \setminus \{h\}) \cup S(h)\,, & \text{if } i \in S^{-1}(h)\,, \\ S(i)\,, & \text{if } i \in N \setminus (\{h\} \cup S^{-1}(h))\,. \end{cases} \tag{5}$$

Lemma 1. *If the allocation rule f satisfies one player efficiency, vertical neutrality, horizontal neutrality, the necessary player property, weak structural monotonicity and predecessor fairness, then $f(N, w_T, S)$ is uniquely determined for all $S \in \mathcal{S}^N_{tree}$ and $w_T = c_T u_T$ for some $c_T > 0$ and $T \subset N$ satisfying $T \cup \widehat{S}^{-1}(T) = N$.*

Fig. 3. An illustration of permission structures S_{-h}

Proof. Suppose that the allocation rule f satisfies the six axioms. Consider a hierarchical permission (tree) structure $S \in \mathcal{S}^N_{tree}$ and the monotone characteristic function $w_T = c_T u_T$, $T \subset N$, $c_T > 0$. Suppose that $T \cup \widehat{S}^{-1}(T) = N$. (So, there are no inessential players.) Clearly, f satisfying the necessary player property implies that there exists a constant $c^* \in IR$ such that $f_i(N, w_T, S) = c^*$ for all $i \in T$, and $f_i(N, w_T, S) \leq c^*$ for all $i \in N \setminus T$. But then weak structural monotonicity and the fact that $\widehat{S}(i) = \overline{S}(i)$ for all $i \in N$ and $S \in \mathcal{S}^N_{tree}$ imply that $f_i(N, w_T, S) = c^*$ for all $i \in T \cup \overline{S}^{-1}(T) = T \cup \widehat{S}^{-1}(T) = N$. So, we have determined $f(N, w_T, S)$ if we determine c^*. We do this by induction on $|N|$.

If $|N| = 1$ then one player efficiency implies that $f_i(N, w_T, S) = c_T$ for $i \in N$.

Proceeding by induction assume that $f_i(\widetilde{N}, w_{\widetilde{T}}, \widetilde{S}) = c_{\widetilde{T}}/(2^{|\widetilde{N}|-1})$ for all $(\widetilde{N}, w_{\widetilde{T}}, \widetilde{S})$ with $\widetilde{S} \in \mathcal{S}^{\widetilde{N}}_{tree}$, $c_{\widetilde{T}} > 0$ and $|\widetilde{N}| < |N|$.

We distinguish the following three cases (of which at least one must occur).

1. Suppose there exist $h, g, j \in N$ such that $S(h) = S(j) = \emptyset$ and $\{h, j\} \subset S(g)$. (Note that this case can only occur if $|N| \geq 3$.) By the assumption that $T \cup \widehat{S}^{-1}(T) = N$, we have $\{h, j\} \subset T$. Horizontal neutrality implies that
$$f_j(N, w_T, S) + f_h(N, w_T, S) = f_j(N \setminus \{h\}, c_T u_{T \setminus \{h\}}, S_{-h}) \,.$$
With the induction hypothesis and the fact that $f_j(N, w_T, S) = f_h(N, w_T, S) = c^*$ this yields that $2c^* = c_T/(2^{|N|-2})$, and thus $f_i(N, w_T, S) = c^* = c_T/(2^{|N|-1})$ for all $i \in N$.
2. Suppose that there exist $j \in T$ and $h \in N \setminus T$ with $S(j) = \emptyset$ and $S(h) = \{j\}$. Vertical neutrality implies that
$$f_j(N, w_T, S) + f_h(N, w_T, S) = f_j(N \setminus \{h\}, w_T, S_{-h}) \,.$$
With the induction hypothesis and the fact that $f_j(N, w_T, S) = f_h(N, w_T, S) = c^*$ this yields that $2c^* = c_T/(2^{|N|-2})$, and thus $f_i(N, w_T, S) = c^* = c_T/(2^{|N|-1})$ for all $i \in N$.
3. Suppose that there exist $h, j \in T$ with $S(j) = \emptyset$ and $S(h) = \{j\}$. Then take a $g \in \mathbb{N} \setminus N$, and define $S', S'' \in \mathcal{S}^{N \cup \{g\}}$ by

$$S'(i) = \begin{cases} \{h\}, & \text{if } i = g, \\ (S(i) \setminus \{h\}) \cup \{g\}, & \text{if } i \in S^{-1}(h), \\ S(i), & \text{if } i \in N \setminus S^{-1}(h) \end{cases}$$

and

$$S''(i) = \begin{cases} \{h, j\}, & \text{if } i = g, \\ (S(i) \setminus \{h\}) \cup \{g\}, & \text{if } i \in S^{-1}(h), \\ S(i), & \text{if } i \in N \setminus S^{-1}(h). \end{cases}$$

(Note that $S'' \notin \mathcal{S}_{tree}^{N\cup\{g\}}$.) The necessary player property and weak structural monotonicity imply that

$$f_h(N \cup \{g\}, w_T, S'') = f_j(N \cup \{g\}, w_T, S'') = f_g(N \cup \{g\}, w_T, S'') . \quad (6)$$

Since $((N \cup \{g\}) \setminus \{h\}, c_T u_{T\setminus\{h\}}, S''_{-h}) \in \mathcal{S}_{tree}^{(N\cup\{g\})\setminus\{h\}}$ (with $|(N \cup \{g\} \setminus \{h\})| = |N|$) is as considered in case (ii) it follows from that case that

$$\begin{aligned} &f_j((N \cup \{g\}) \setminus \{h\}, c_T u_{T\setminus\{h\}}, S''_{-h}) \\ &\quad = f_g((N \cup \{g\}) \setminus \{h\}, c_T u_{T\setminus\{h\}}, S''_{-h}) = \frac{c_T}{2^{|N|-1}} . \end{aligned}$$

But then horizontal neutrality implies that

$$\begin{aligned} &f_h(N \cup \{g\}, w_T, S''_{-(h,j)}) + f_j(N \cup \{g\}, w_T, S''_{-(h,j)}) \\ &\quad = f_j((N \cup \{g\}) \setminus \{h\}, c_T u_{T\setminus\{h\}}, S''_{-h}) = \frac{c_T}{2^{|N|-1}} . \end{aligned}$$

With the necessary player property and weak structural monotonicity this yields

$$\begin{aligned} &f_h(N \cup \{g\}, w_T, S''_{-(h,j)}) = f_j(N \cup \{g\}, w_T, S''_{-(h,j)}) \\ &\quad = f_g(N \cup \{g\}, w_T, S''_{-(h,j)}) = \frac{c_T}{2^{|N|}} . \end{aligned} \quad (7)$$

Now, predecessor fairness implies that

$$\begin{aligned} &f_h(N \cup \{g\}, w_T, S'') - f_h(N \cup \{g\}, w_T, S''_{-(h,j)}) \\ &\quad = f_g(N \cup \{g\}, w_T, S''_{-(h,j)}) - f_g(N \cup \{g\}, w_T, S'') , \end{aligned}$$

which with (6) and (4) gives

$$\begin{aligned} &f_h(N \cup \{g\}, w_T, S'') = f_g(N \cup \{g\}, w_T, S'') \\ &\quad = f_j(N \cup \{g\}, w_T, S'') = \frac{c_T}{2^{|N|}} . \end{aligned} \quad (8)$$

Again applying predecessor fairness yields

$$\begin{aligned} &f_g(N \cup \{g\}, w_T, S'') - f_g(N \cup \{g\}, w_T, S') \\ &\quad = f_h(N \cup \{g\}, w_T, S') - f_h(N \cup \{g\}, w_T, S'') , \end{aligned}$$

which with the necessary player property, weak structural monotonicity and (3) yields

$$f_g(N \cup \{g\}, w_T, S') = f_h(N \cup \{g\}, w_T, S') = f_j(N \cup \{g\}, w_T, S') = \frac{c_T}{2^{|N|}}.$$

Finally, vertical neutrality yields

$$c^* = f_h(N, w_T, S) = f_h(N \cup \{g\}, w_T, S') + f_g(N \cup \{g\}, w_T, S') = \frac{c_T}{2^{|N|-1}},$$

and thus $f_i(N, w_T, S) = c^* = c_T/(2^{|N|-1})$ for all $i \in N$.

So, for $S \in S^N_{tree}$ and $w_T = c_T u_T$ with $c_T > 0$ and $T \cup \widehat{S}^{-1}(T) = N$, we conclude that $f_i(N, w_T, S) = c^* = c_T/(2^{|N|-1}) = \beta^d_i(N, w_T, S)$ for all $i \in T \cup \widehat{S}^{-1}(T) = N$. □

The next step is to show that adding the inessential player property and disjunctive fairness to the axioms implies that f is uniquely determined for all positively scaled unanimity games with a permission tree.

Lemma 2. *If the allocation rule f satisfies one player efficiency, vertical neutrality, horizontal neutrality, the inessential player property, the necessary player property, weak structural monotonicity, disjunctive fairness and predecessor fairness, then $f(N, w_T, S)$ is uniquely determined whenever $S \in S^N_{tree}$ and $w_T = c_T u_T$ for some $T \subset N$, $T \neq \emptyset$, and $c_T > 0$.*

Proof. Suppose that the allocation rule f satisfies the eight axioms. Consider a permission (tree) structure $S \in S^N_{tree}$ and monotone characteristic function $w_T = c_T u_T$ as given in (4) for some $c_T > 0$. Denote $\alpha_S(T) = T \cup \widehat{S}^{-1}(T)$.

Since all players in $N \setminus \alpha_S(T)$ are inessential players in (N, w_T, S), the inessential player property implies that $f_i(N, w_T, S) = 0$ for all $i \in N \setminus \alpha_S(T)$. Further, f satisfying the necessary player property and weak structural monotonicity and the fact that $\widehat{S}(i) = \overline{S}(i)$ for all $i \in N$ and $S \in S^N_{tree}$ imply that there exists a constant $c^* \in IR$ such that $f_i(N, w_T, S) = c^*$ for all $i \in \alpha_S(T)$. So,

$$f_i(N, w_T, S) = \begin{cases} c^*, & \text{if } i \in \alpha_S(T) = T \cup \widehat{S}^{-1}(T), \\ 0, & \text{if } i \in N \setminus \alpha_S(T). \end{cases} \tag{9}$$

We prove that c^* (and thus $f(N, w_T, S)$) is uniquely determined by induction on $|N \setminus \alpha_S(T)|$.

If $|N \setminus \alpha_S(T)| = 0$ then c^* (and thus $f(N, w_T, S)$) is uniquely determined by Lemma 1.

Proceeding by induction assume that $c^* = f_{i_0}(\widetilde{N}, w_{\widetilde{T}}, \widetilde{S})$ is uniquely determined for all $(\widetilde{N}, w_{\widetilde{T}}, \widetilde{S})$ with $\widetilde{S} \in S^{\widetilde{N}}_{tree}$, $c_{\widetilde{T}} > 0$ and $|\widetilde{N} \setminus \alpha_{\widetilde{S}}(\widetilde{T})| < |N \setminus \alpha_S(T)|$.

Since $N \setminus \alpha_S(T) \neq \emptyset$ there exists a $j \in N \setminus \alpha_S(T)$ with $S(j) = \emptyset$. We distinguish the following three cases (of which at least one must occur).

1. Suppose there exists an $h \in \alpha_S(T) \cap S(i_0)$ such that $j \notin \widehat{S}(h)$. (Remember that i_0 denotes the unique top-player in the permission structure.) Define $S' \in S^N_H$ by

$$S'(i) = \begin{cases} \{h\}, & \text{if } i = j, \\ S(i), & \text{otherwise.} \end{cases}$$

(Note that $S' \notin \mathcal{S}^N_{tree}$.) Disjunctive fairness implies that

$$f_{i_0}(N, w_T, S') - f_{i_0}(N, w_T, S) = f_j(N, w_T, S') - f_j(N, w_T, S). \quad (10)$$

Predecessor fairness implies that

$$f_{i_0}(N, w_T, S) - f_{i_0}(N, w_T, S') = f_j(N, w_T, S') - f_j(N, w_T, S). \quad (11)$$

So, from (10) and (11) it follows that

$$f_{i_0}(N, w_T, S) = f_{i_0}(N, w_T, S') \text{ and } f_j(N, w_T, S) = f_j(N, w_T, S'). \quad (12)$$

With the inessential player property it then follows that

$$f_j(N, w_T, S') = f_j(N, w_T, S) = 0. \quad (13)$$

Predecessor fairness also yields that

$$\begin{aligned} f_{i_0}(N, w_T, S') &- f_{i_0}(N, w_T, S'_{-(i_0,h)}) \\ &= f_j(N, w_T, S'_{-(i_0,h)}) - f_j(N, w_T, S'). \end{aligned} \quad (14)$$

Since $|N \setminus \alpha_{S'_{-(i_0,h)}}(T)| < |N \setminus \alpha_S(T)|$ and $S'_{-(i_0,h)} \in \mathcal{S}^N_{tree}$, the induction hypothesis, (13) and (14) yield that $f_{i_0}(N, w_T, S') = f_{i_0}(N, w_T, S'_{-(i_0,h)}) + f_j(N, w_T, S'_{-(i_0,h)})$ is uniquely determined. So, with (12) also $c^* = f_{i_0}(N, w_T, S)$ is uniquely determined.

2. Suppose that there exists an $h \in \alpha_S(T) \cap S(i_0)$ such that $j \in \widehat{S}(h)$. Take $g \in \widehat{S}(h) \cap (S^{-1}(j) \cup \{j\})$, and define $S' \in \mathcal{S}^N_H$ by

$$S'(i) = \begin{cases} S(i) \cup \{g\}, & \text{if } i = i_0, \\ S(i), & \text{otherwise.} \end{cases}$$

(Note that again $S' \notin \mathcal{S}^N_{tree}$.) Disjunctive fairness implies that

$$f_{i_0}(N, w_T, S') - f_{i_0}(N, w_T, S'_{-(h,g)}) = f_h(N, w_T, S') - f_h(N, w_T, S'_{-(h,g)}),$$

while predecessor fairness implies that

$$f_{i_0}(N, w_T, S'_{-(h,g)}) - f_{i_0}(N, w_T, S') = f_h(N, w_T, S') - f_h(N, w_T, S'_{-(h,g)}).$$

So, $f_{i_0}(N, w_T, S') = f_{i_0}(N, w_T, S'_{-(h,g)})$ and $f_h(N, w_T, S') = f_h(N, w_T, S'_{-(h,g)})$. Since $S'_{-(h,g)}$ is as in case (i), we determined $f_{i_0}(N, w_T, S')$ and $f_h(N, w_T, S')$.
With predecessor fairness we then also determined

$$c^* = f_{i_0}(N, w_T, S) = f_h(N, w_T, S') - f_h(N, w_T, S) + f_{i_0}(N, w_T, S').$$

3. Finally, we must consider the case $\alpha_S(T) = \{i_0\}$. Define the game with permission structure $(N \cup \{h\}, v', S')$ where, for some $h \in \mathbb{N} \setminus N$, $v' \in \mathcal{G}^{N\cup\{h\}}$ is given by $v'(E) = v(E \setminus \{i_0\}) = 0$ for all $E \subset N$, and $v'(E) = v(E \cap N)$ for all $E \subset N \cup \{h\}$ with $h \in E$, and $S' \in \mathcal{S}_{tree}^{N\cup\{h\}}$ is given by $S'(h) = \{i_0\}$ and $S'(i) = S(i)$ for all $i \in N$. Vertical neutrality, the necessary player property and weak structural monotonicity imply that $f_{i_0}(N, v, S) = f_{i_0}(N \cup \{h\}, v', S') + f_h(N \cup \{h\}, v', S') = 2f_{i_0}(N \cup \{h\}, v', S')$.
   Since $(N \cup \{h\}, v', S')$ is as considered in case (ii), we determined $c^* = f_{i_0}(N, v, S)$.

So, in all three cases we uniquely determined c^* and thus, with (9) we uniquely determined $f(N, w_T, S)$. □

Finally, adding additivity to the axioms of Lemma 2, we state the main result.

Theorem 16. *An allocation rule f is equal to the disjunctive Banzhaf permission value β^d if and only if it satisfies one player efficiency, vertical neutrality, horizontal neutrality, additivity, the inessential player property, the necessary player property, weak structural monotonicity, disjunctive fairness and predecessor fairness.*

Proof. With Theorem 15 we only have to show that there can be at most one allocation rule that satisfies the nine axioms stated in the theorem. Therefore, suppose that allocation rule f satisfies the nine axioms.

Consider the hierarchical permission structure $S \in \mathcal{S}_H^N$ and the monotone characteristic function $w_T = c_T u_T$, $c_T \geq 0$, as given in (4). If $c_T = 0$ then the inessential player property implies that $f_i(N, w_T, S) = 0$ for all $i \in N$.

Now suppose that $c_T > 0$. Again, we denote by $\alpha_S(T) = T \cup \widehat{S}^{-1}(T)$ the set consisting of all players in T and all their superiors. For $S \in \mathcal{S}_H^N \setminus \mathcal{S}_{tree}^N$ it need not hold that $\widehat{S}_i^{-1}(T)$ equals $\overline{S}_i^{-1}(T)$ for all $i \in N$. Therefore, by $\gamma_S(T) := \{i \in \alpha_S(T) \mid T \cap (\{i\} \cup \overline{S}(i)) \neq \emptyset\}$ we denote the set of those players in $\alpha_S(T)$ who belong to T or have subordinates in T that they dominate "completely".

Again, the inessential player property implies that $f_i(N, w_T, S) = 0$ for all $i \in N \setminus \alpha_S(T)$. Further, f satisfying the necessary player property and weak structural monotonicity implies that there exists a constant $c^* \in IR$ such that $f_i(N, w_T, S) = c^*$ for all $i \in \gamma_S(T)$. We prove that c^* and all $f_i(N, w_T, S)$, $i \in \alpha_S(T)$, are uniquely determined by induction on the number $\sum_{i\in N} |S(i)|$.

If $\sum_{i\in N} |S(i)| = |N| - 1$ then $S \in \mathcal{S}_{tree}^N$ and c^* and all $f_i(N, w_T, S)$, $i \in \alpha_S(T)$, are uniquely determined by Lemma 2. (Note that $\sum_{i\in N} |S(i)| \geq |N| - 1$ for all $S \in \mathcal{S}_H^N$.)

Proceeding by induction assume that $f(N, w_T, \widetilde{S})$ is uniquely determined for all $\widetilde{S} \in \mathcal{S}_H^N$ with $\sum_{i\in N} |\widetilde{S}(i)| < \sum_{i\in N} |S(i)|$.

Next we recursively define the sets L_k, $k \in \{0\} \cup \mathbb{N}$, by

$$L_0 := \emptyset \text{ and } L_k := \left\{ i \in N \setminus \bigcup_{t=1}^{k-1} L_t \;\middle|\; S(i) \subset \bigcup_{t=1}^{k-1} L_t \right\}, \quad \text{for all } k \in \mathbb{N}.$$

In [7] it is shown that for hierarchical permission structures there exists an $M < \infty$ such that the sets $L_1, \ldots, L_M$ form a partition of N consisting of non-empty sets only.

Next we describe a procedure which determines the values $f_i(N, w_T, S)$ as linear functions of the constant c^* for all $i \in N$.

STEP 1 For every $i \in L_1$ one of the following two conditions is satisfied.

1. If $i \in N \setminus \alpha_S(T)$ then $f_i(N, w_T, S) = 0$ as mentioned before.
2. If $i \in \alpha_S(T)$ then $i \in T$ since $S(i) = \emptyset$. Thus $f_i(N, w_T, S) = c^*$.

Let $k = 2$.

STEP 2 If $L_k = \emptyset$ then STOP.
Else, for every $i \in L_k$ one of the following three conditions is satisfied.

1. If $i \in N \setminus \alpha_S(T)$ then $f_i(N, w_T, S) = 0$.
2. If $i \in \gamma_S(T)$ then $f_i(N, w_T, S) = c^*$.
3. If $i \in \alpha_S(T) \setminus \gamma_S(T)$ then by definition of $\alpha_S(T)$ and $\gamma_S(T)$ there exists an $h \in \{i\} \cup \overline{S}(i)$ and a $j \in S(h)$ such that $|S^{-1}(j)| \geq 2$. Disjunctive fairness then implies that

$$f_i(N, w_T, S) - f_i(N, w_T, S_{-(h,j)}) = f_j(N, w_T, S) - f_j(N, w_T, S_{-(h,j)}).$$

Using the induction hypothesis and the fact that $j \in \widehat{S}(i)$ implies that we already determined $f_j(N, w_T, S)$ as a linear function of c^* (since $j \in L_l$ with $l < k$), it follows that we have determined

$$\begin{aligned} f_i(N, w_T, S) &= f_j(N, w_T, S) - f_j(N, w_T, S_{-(h,j)}) + f_i(N, w_T, S_{-(h,j)}) \\ &= c^* + c_i, \end{aligned}$$

where $c_i = c_j + f_i(N, w_T, S_{-(h,j)}) - f_j(N, w_T, S_{-(h,j)})$ is known (for the particular choice of h and j).

STEP 3 Let $k = k + 1$. GOTO STEP 2.

Since there exists an $M < \infty$ such that the sets $L_1, \ldots, L_M$ form a partition of N consisting of non-empty sets only the procedure described above determines the values $f_i(N, w_T, S)$ as linear functions of c^* for all $i \in \alpha_S(T)$, and $f_i(N, w_T, S) = 0$ for all $i \in N \setminus \alpha_S(T)$.

Next, we distinguish the following two cases.

1. Suppose there exists $j \in \alpha_S(T)$ with $|S^{-1}(j)| \geq 2$. Then take $h, g \in N$, $h \neq g$, such that $j \in S(h) \cap S(g)$. In the procedure above, we used disjunctive fairness to show that

$$f_h(N, w_T, S) = c^* + c_j - f_j(N, w_T, S_{-(h,j)}) + f_h(N, w_T, S_{-(h,j)}), \quad (15)$$

where the induction hypothesis determines c_j, $f_j(N, w_T, S_{-(h,j)})$ and $f_h(N, w_T, S_{-(h,j)})$.
Predecessor fairness implies that

$$f_h(N, w_T, S) - f_h(N, w_T, S_{-(g,j)}) = f_g(N, w_T, S_{-(g,j)}) - f_g(N, w_T, S) ,$$

which with disjunctive fairness yields

$$f_h(N, w_T, S) - f_h(N, w_T, S_{-(g,j)}) = f_j(N, w_T, S_{-(g,j)}) - f_j(N, w_T, S) ,$$

and thus

$$f_h(N, w_T, S) = f_j(N, w_T, S_{-(g,j)}) - c^* - c_j + f_h(N, w_T, S_{-(g,j)}) .$$

With (15) this yields

$$c^* = \frac{1}{2}(-2c_j + f_h(N, w_T, S_{-(g,j)}) - f_h(N, w_T, S_{-(h,j)})$$
$$+ f_j(N, w_T, S_{-(g,j)}) + f_j(N, w_T, S_{-(h,j)})) ,$$

and thus with the induction hypothesis c^* is uniquely determined.

2. Suppose that $|S^{-1}(j)| = 1$ for all $j \in \alpha_S(T) \setminus \{i_0\}$. Then, by assumption, there exists $j \in N \setminus \alpha_S(T)$ with $|S^{-1}(j)| \geq 2$. Take $h \in S^{-1}(j)$. Then disjunctive fairness and the inessential player property imply

$$f_{i_0}(N, w_T, S) - f_{i_0}(N, w_T, S_{-(h,j)}) = f_j(N, w_T, S) - f_j(N, w_T, S_{-(h,j)}) = 0 .$$

So, $f_{i_0}(N, w_T, S) = f_{i_0}(N, w_T, S_{-(h,j)})$ is uniquely determined by the induction hypothesis. Since we already determined $f_{i_0}(N, w_T, S)$ as a linear function of c^*, we have determined c^*.

In both cases we determined c^*. Since we already determined all values $f_i(N, w_T, S)$, $i \in N$, as linear functions of c^*, we uniquely determined $f(N, w_T, S)$.

Now, let $S \in \mathcal{S}_H^N$ and consider the characteristic function $w_T = c_T u_T$ with $c_T < 0$. (Note that we cannot apply the null player property and weak structural monotonicity to this game with permission structure, because $c_T u_T$ is not monotone if $c_T < 0$.) Let $v_0 \in \mathcal{G}^N$ denote the *null game*, i. e., $v_0(E) = 0$ for all $E \subset N$. From the inessential player property it follows that $f_i(N, v_0, S) = 0$ for all $i \in N$. Since $-w_T = -c_T u_T$ with $-c_T \geq 0$ is monotone (and thus $f(N, -w_T, S)$ is uniquely determined as shown above) and $r^d_{N,w_T,S} + r^d_{N,-w_T,S} = r^d_{N,v_0,S}$ for all $T \subset N$, additivity of f implies that $f(N, w_T, S) = f(N, v_0, S) - f(N, -w_T, S) = -f^d(N, -w_T, S)$. Since $f^d(N, -w_T, S)$ is uniquely determined, $f^d(N, w_T, S)$ is uniquely determined.

Finally, since every characteristic function $v \in \mathcal{G}^N$ can be expressed as a linear combination of unanimity games it follows with additivity of f that $f(N, v, S)$ is uniquely determined for every $v \in \mathcal{G}^N$ and $S \in \mathcal{S}_H^N$. □

The axioms stated in Theorem 16 are logically independent.

Example 5. We show logical independence of the axioms of Theorem 16 by the following nine allocation rules for games with a hierarchical permission structure.

1. The allocation rule f given by $f_i(N,v,S) = 0$ for all $i \in N$ satisfies all axioms except one player efficiency.
2. The allocation rule f given by $f_{i_0}(N,v,S) = v(\{i_0\})$, and $f_i(N,v,S) = 0$ for all $i \in N \setminus \{i_0\}$ satisfies all axioms except vertical neutrality. (Remember that i_0 denotes the top-player in (N,S).)
3. The allocation rule f given by $f(N,v,S) = \beta^d(N, \sum_{i \in N} v(\{i\})u_{\{i\}}, S)$ satisfies all axioms except horizontal neutrality.
4. For game (N,v) define $d(N,v) = \max\{\Delta_{(N,v)}(T) \mid T \subset N\}$, where $\Delta_{(N,v)}(T) = \sum_{F \subset T}(-1)^{|T|-|F|}v(F)$ are the Harsanyi dividends (see [14]), and $D(N,v) = \{T \subset N \mid \Delta_{(N,v)}(T) = d(N,v)\}$. The allocation rule f given by $f(N,v,S) = \beta^d(N, \sum_{T \in D(N,v)} \Delta_{(N,v)}(T)u_T, S)$ satisfies all axioms except additivity.
5. The allocation rule f given by $f_i(N,v,S) = v(N)/(2^{|N|-1})$ for all $i \in N$, satisfies all axioms except the inessential player property.
6. The allocation rule f given by $f_{i_0}(N,v,S) = v(N)$ and $f_i(N,v,S) = 0$ for all $i \in N \setminus \{i_0\}$ satisfies all axioms except the necessary player property. (Remember that i_0 denotes the top-player in (N,S).)
7. The allocation rule f given by $f(N,v,S) = B(N,v)$ satisfies all axioms except weak structural monotonicity.
8. The conjunctive Banzhaf permission value β^c satisfies all axioms except disjunctive fairness.
9. For $S \in \mathcal{S}_H^N$ and $T \subset N$ let $\mathcal{Z}_T^S = \{Z \in \mathcal{S}_H^N \mid Z^{-1}(i) = S^{-1}(i)$ for all $i \in N \setminus \alpha_S(T)$, $Z^{-1}(i) \subset S^{-1}(i)$ for all $i \in \alpha_S(T)$, and $|Z^{-1}(i)| = 1$ for all $i \in \alpha_S(T) \setminus \{i_0\}\}$. The allocation rule f given by $f(N,v,S) = \sum_{T \subset N} \sum_{Z \in \mathcal{Z}_T^S} \beta_i^d(N, \Delta_{(N,v)}(T)u_T, Z)$ satisfies all axioms except predecessor fairness.

4.2 The Conjunctive Banzhaf Permission Value

As mentioned in Example 5.8, the conjunctive Banzhaf permission value satisfies all axioms stated in Theorem 16 except disjunctive fairness. Similarly as with the (Shapley) permission values, we can characterize the conjunctive Banzhaf permission value by replacing disjunctive fairness by conjunctive fairness.

Theorem 17. *An allocation rule f is equal to the conjunctive Banzhaf permission value β^c if and only if it satisfies one player efficiency, vertical neutrality, horizontal neutrality, additivity, the inessential player property, the necessary player property, weak structural monotonicity, conjunctive fairness and predecessor fairness.*

Proof. Proving that the conjunctive Banzhaf permission value satisfies additivity, the inessential player property, the necessary player property and weak structural monotonicity is along the same lines as this is shown for the conjunctive (Shapley) permission value in [6]. One player efficiency is evident. Vertical and horizontal neutrality of the conjunctive Banzhaf permission value follow from similar arguments given for the disjunctive Banzhaf permission value in the proof of Theorem 15. Predecessor fairness follows from Theorem 11.

So, we are left to show that β^c satisfies conjunctive fairness. Let $S \in \mathcal{S}_H^N$ and $h, g, j \in N$ be such that $j \in S(h) \cap S(g)$, $h \neq g$. Further, let $i \in \{g\} \cup \overline{S}^{-1}(g)$. The proof is similar to the corresponding proof for disjunctive fairness of the disjunctive Banzhaf permission value, but with the roles of h and g exchanged. Since $\{i, j\} \not\subset E$ implies that $\sigma^c_{N,S}(E) = \sigma^c_{N,S_{-(h,j)}}(E)$, it follows that

$$\beta^c_i(N, v, S) - \beta^c_i(N, v, S_{-(h,j)}) = B_i(N, r^c_{N,v,S}) - B_i(N, r^c_{N,v,S_{-(h,j)}})$$

$$= \frac{1}{2^{|N|-1}} \sum_{\substack{E \subset N \\ i \in E}} \Big(v(\sigma^c_{N,S}(E)) - v(\sigma^c_{N,S}(E \setminus \{i\})) - v(\sigma^c_{N,S_{-(h,j)}}(E))$$

$$+ v(\sigma^c_{N,S_{-(h,j)}}(E \setminus \{i\})) \Big) = \frac{1}{2^{|N|-1}} \sum_{\substack{E \subset N \\ i \in E}} \Big(v(\sigma^c_{N,S}(E)) - v(\sigma^c_{N,S_{-(h,j)}}(E)) \Big)$$

$$= \frac{1}{2^{|N|-1}} \sum_{\substack{E \subset N \\ \{i,j\} \subset E}} \Big(v(\sigma^c_{N,S}(E)) - v(\sigma^c_{N,S_{-(h,j)}}(E)) \Big)$$

$$= \frac{1}{2^{|N|-1}} \sum_{\substack{E \subset N \\ j \in E}} \Big(v(\sigma^c_{N,S}(E)) - v(\sigma^c_{N,S}(E \setminus \{j\})) - v(\sigma^c_{N,S_{-(h,j)}}(E))$$

$$+ v(\sigma^c_{N,S_{-(h,j)}}(E \setminus \{j\})) \Big) = \beta^c_j(N, v, S) - \beta^c_j(N, v, S_{-(h,j)}) \,,$$

showing that β^c satisfies conjunctive fairness.

Uniqueness can be shown along the same lines as the uniqueness part in the proof of Theorem 16 and the corresponding lemma's. □

We end this section by remarking that we can restate the axiomatization for games with an acyclic permission structure (which is not necessarily quasi-strongly connected). Then we can also adapt the axioms so that the disjunctive and conjunctive Banzhaf permission values are axiomatized for games with a fixed player set. For notational convenience we discussed games with a variable player set and a hierarchical permission structure.

5 Predecessor Fairness and Neutrality Properties for the (Shapley) Permission Values

In Section 2 we mentioned some known characterizations of the (Shapley) permission values. In Section 4 we gave characterizations of the Banzhaf permission values. From the new axioms introduced in Sections 3 and 4 both (Shapley) permission values satisfy one player efficiency. They do not satisfy predecessor fairness, vertical neutrality nor horizontal neutrality. However, these three axioms can easily be adapted so that they are satisfied by the (Shapley) permission values. Instead of stating that the change in payoffs of players h and g, $h \neq g$, are the same in absolute value but have opposite sign after deleting the relation between players h and $j \in S(h) \cap S(g)$, the (Shapley) permission values satisfy the property that after deleting the relation between players h and $j \in S(h)$, the payoffs of player h and the sum of the payoffs of *all* other players are the same in absolute value but have opposite sign.

Axiom 18 (Shapley predecessor fairness). *For every $N \subset \mathbb{N}$, $v \in \mathcal{G}^N$ and $S \in \mathcal{S}_H^N$, if $h \in N$ and $j \in S(h)$ with $|S^{-1}(j)| \geq 2$, then*

$$f_h(N,v,S) - f_h(N,v,S_{-(h,j)}) = \sum_{i \in N \setminus \{h\}} \left(f_i(N,v,S_{-(h,j)}) - f_i(N,v,S)\right).$$

In a similar way we adapt vertical and horizontal neutrality. Let player $h \in \mathbb{N} \setminus N$ be a new player in the game with permission structure $(N \cup \{h\}, v', S')$ given in Axiom 13, respectively, $(N \cup \{h\}, v'', S'')$ given in Axiom 14. Then the sum of the payoffs of all players (including player h) in the new game with permission structure is equal to the sum of the payoffs of all players (excluding player h) in the original game with permission structure.

Axiom 19 (Shapley vertical neutrality). *For every $N \subset \mathbb{N}$, $v \in \mathcal{G}^N$, $S \in \mathcal{S}_H^N$ and $h \in \mathbb{N} \setminus N$, if $j \in N$ and all $i \in \widehat{S}(j)$ are null players in (N,v), then*

$$\sum_{i \in N \cup \{h\}} f_i(N \cup \{h\}, v', S') = \sum_{i \in N} f_i(N,v,S).$$

Axiom 20 (Shapley horizonal neutrality). *For every $N \subset \mathbb{N}$, $v \in \mathcal{G}^N$, $S \in \mathcal{S}_H^N$ and $h \in \mathbb{N} \setminus N$, if $S(j) = \emptyset$ and $S^{-1}(j) \neq \emptyset$ for $j \in N$, then*

$$\sum_{i \in N \cup \{h\}} f_i(N \cup \{h\}, v'', S'') = \sum_{i \in N} f_i(N,v,S).$$

Theorem 21. *An allocation rule f is equal to*

1) the disjunctive (Shapley) permission value φ^d if and only if it satisfies one player efficiency, Shapley vertical neutrality, Shapley horizontal neutrality, additivity, the inessential player property, the necessary player property, weak structural monotonicity, disjunctive fairness and Shapley predecessor fairness;

2) the conjunctive (Shapley) permission value φ^c if and only if it satisfies one player efficiency, Shapley vertical neutrality, Shapley horizontal neutrality, additivity, the inessential player property, the necessary player property, weak structural monotonicity, conjunctive fairness and Shapley predecessor fairness.

Proof. Since $v'(N) = v''(N) = v(N)$ for every $N \subset \mathbb{N}$, $v \in \mathcal{G}^N$, $S \in \mathcal{S}_H^N$, $j \in N$ and $h \in \mathbb{N} \setminus N$ such that $j \in S(h)$ and $|S^{-1}(j)| \geq 2$, φ^d and φ^c satisfying efficiency implies that they satisfy one player efficiency, Shapley vertical neutrality and Shapley horizontal neutrality. Efficiency of φ^d and φ^c also implies that they satisfy Shapley predecessor fairness. The Shapley permission values satisfying the other axioms mentioned in the Theorem follows directly from the results of Section 2.

Showing that there can be at most one allocation rule satisfying the axioms can be done in a similar way as in the proof of Lemma's 1 and 2 and Theorem 16. □

Appendix

We introduce a similar neutrality property as amalgamation and collusion neutrality as defined in [17] and [13], respectively. Suppose that player $h \in \mathbb{N} \setminus N$ is entering the game as a veto player for player $j \in N$. Then the sum of the Banzhaf values of players h and j in the new game is equal to the Banzhaf value of player j in the original game. To be more precize, for game (N, v) and players $j \in N$ and $h \in \mathbb{N} \setminus N$, let the characteristic function $v_{hj} \in \mathcal{G}^{N \cup \{h\}}$ be given by

$$v_{hj}(E) = \begin{cases} v(E \setminus \{j\}), & \text{if } E \subset N, \\ v(E \cap N), & \text{if } E \subset N \cup \{h\},\ h \in E. \end{cases} \tag{16}$$

Proposition 2. *If $N \subset \mathbb{N}$, $j \in N$, $h \in \mathbb{N} \setminus N$ and $v \in \mathcal{G}^N$, then*

$$B_h(N \cup \{h\}, v_{hj}) + B_j(N \cup \{h\}, v_{hj}) = B_j(N, v),$$

where $v_{hj} \in \mathcal{G}^{N \cup \{h\}}$ is given by (16).

Proof. If $N \subset \mathbb{N}$, $j \in N$, $h \in \mathbb{N} \setminus N$ and $v \in \mathcal{G}^N$, then $v_{hj}(E) - v_{hj}(E \setminus \{h\}) = 0$ if $j \notin E$, and $v_{hj}(E) - v_{hj}(E \setminus \{j\}) = 0$ if $h \notin E$. Thus

$$B_h(N \cup \{h\}, v_{hj}) + B_j(N \cup \{h\}, v_{hj}) =$$

$$= \frac{1}{2^{|N|}} \Bigg(\sum_{\substack{E \subset N \cup \{h\} \\ h \in E}} (v_{hj}(E) - v_{hj}(E \setminus \{h\})) + \sum_{\substack{E \subset N \cup \{h\} \\ j \in E}} (v_{hj}(E) - v_{hj}(E \setminus \{j\})) \Bigg)$$

$$= \frac{1}{2^{|N|}} \Bigg(\sum_{\substack{E \subset N \cup \{h\} \\ \{h,j\} \subset E}} (v_{hj}(E) - v_{hj}(E \setminus \{h\})) + \sum_{\substack{E \subset N \cup \{h\} \\ \{h,j\} \subset E}} (v_{hj}(E) - v_{hj}(E \setminus \{j\})) \Bigg)$$

$$= \frac{1}{2^{|N|}} \sum_{\substack{E \subset N \cup \{h\} \\ \{h,j\} \subset E}} (v(E \cap N) - v((E \cap N) \setminus \{j\}) + v(E \cap N) - v((E \cap N) \setminus \{j\}))$$

$$= \frac{1}{2^{|N|-1}} \sum_{\substack{E \subset N \\ j \in E}} (v(E) - v(E \setminus \{j\})) = B_j(N, v) . \qquad \square$$

References

1. Aumann, R. J., Drèze, J. H. (1974): Cooperative Games with Coalition Structure. International Journal of Game Theory **3**, 217–237
2. Banzhaf, J. F. (1965): Weighted Voting Doesn't Work: A Mathematical Analysis. Rutgers Law Review **19**, 317-343
3. Borm, P., Owen, G., Tijs, S. (1992): On the Position Value for Communication Situations. SIAM Journal on Discrete Mathematics **5**, 305–320
4. Brink, R. van den (1997): An Axiomatization of the Disjunctive Permission Value for Games with a Permission Structure. International Journal of Game Theory **26**, 27–43
5. Brink, R. van den (1999): An Axiomatization of the Conjunctive Permission Value for Games with a Hierarchical Permission Structure. In: Swart, H. de (Ed.): Logic, Game Theory and Social Choice. 125–139
6. Brink, R. van den, Gilles, R. P. (1996): Axiomatizations of the Conjunctive Permission Value for Games with Permission Structures. Games and Economic Behavior
7. Brink, R. van den, Gilles, R. P. (1994): A Social Power Index for Hierarchically Structured Populations of Economic Agents. In: Gilles, R. P., Ruys, P. H. M. (Eds.): Imperfections and Behaviour in Economic Organizations. Kluwer, Dordrecht, 279–318
8. Derks, J., Peters, H. (1993): A Shapley Value for Games with Restricted Coalitions. International Journal of Game Theory **21**, 351–366
9. Dragan, I. (1996): New Mathematical Properties of the Banzhaf Value. European Journal of Operational Research **95**, 451–463
10. Dubey, P., Shapley, L. S. (1979): Mathematical Properties of the Banzhaf Power Index. Mathematics of Operations Research **4**, 99–131
11. Gilles, R. P., Owen, G. (1994): Games with Permission Structures: The Disjunctive Approach. Department of Economics, Virginia Polytechnic Institute and State University, Blacksburg, Virginia (mimeo)
12. Gilles, R. P., Owen, G., Brink, R. van den (1992): Games with Permission Structures: The Conjunctive Approach. International Journal of Game Theory **20**, 277–293

13. Haller, H. (1994): Collusion Properties of Values. International Journal of Game Theory **23**, 261–281
14. Harsanyi, J. C. (1959): A Bargaining Model for Cooperative n-Person Games. In: Tucker, A. W., Luce, R. D. (Eds.): Contributions to the Theory of Games. Princeton UP, Princeton, Vol. IV, 325–355
15. Hart, S., Kurz, M. (1983): Endogeneous Formation of Coalitions. Econometrica **51**, 1047–1064
16. Kalai, E., Postlewaite, A., Roberts, J. (1978): Barriers to Trade and Disadvantageous Middlemen: Nonmonotonicity of the Core. Journal of Economic Theory **19**, 200–209
17. Lehrer, E. (1988): An Axiomatization of the Banzhaf Value. International Journal of Game Theory **17**, 89–99
18. Myerson, R.B. (1977): Graphs and Cooperation in Games. Mathematics of Operations Research **2**, 225–229
19. Owen, G. (1975): Multilinear Extensions and the Banzhaf Value. Naval Research Logistics Quarterly **22**, 741–750
20. Owen, G. (1977): Values of Games with a Priori Unions. In: Henn, R., Moeschlin, O. (Eds.): Essays in Mathematical Economics and Game Theory. Springer–Verlag, Berlin, 76–88
21. Owen, G. (1986): Values of Graph-Restricted Games. SIAM Journal of Algebraic Discrete Methods **7**, 210–220
22. Shapley, L. S. (1953): A Value for n-Person Games. In: Kuhn, H. W., Tucker, A. W. (Eds.): Contributions to the Theory of Games. Annals of Mathematics Studies. Princeton UP, Princeton, No. 28, Vol. 2, 307–317
23. Winter, E. (1989): A Value for Cooperative Games with Levels Structure of Cooperation. International Journal of Game Theory **18**, 227–240

Moral Hazard in Teams Revisited

Baomin Dong*

Department of Economics, Concordia University
Montréal, Québec H3G 1M8,
E-mail: don@alcor.concordia.ca

Summary. This paper addresses the classic free riding question in a two-person managerial team. Unlike pure moral hazard models, we assume that individual entrepreneurial abilitie also affects team output. Using a two-period model, we show that in an Alchian–Demsetz firm, even in a finite period game setting, effort levels of both team members higher than commonly perceived can be achieved. We argue that this is due to partial mutual observability between the team members. We then show that the existence of a self-enforcing mechanism in managerial teams alleviates free riding, and this is one reason why team structures persist. Comparison with classic capitalistic firms where group performance evaluation is abandoned yields the result that the optimal incentive piece rate should be lower in a team. This may explain the Jensen–Murphy puzzle.

Key words: Moral Hazard, Adverse Selection, Team
JEL Classification Numbers: D73, L22, D82, D83

> "The team concept is a nice idea, but when you put the teams under pressure, it becomes a damn effective way to divide workers."
> — a dissident union leader in the NUMMI[2] plant.

1 Introduction

The study of teams can be traced back to [4], where in order to eliminate or minimize free-riding, monitoring is essential and this role is performed by a

* I thank Greg LeBlanc, Karine Gobert for their comments and discussions on this research however, all remaining errors are mine.

[2] New United Motor Manufacturing, Inc. (NUMMI), the Toyota-GM joint venture in California is a big success in automobile assembly industry where the success is attributed to team production. Indeed, there is some evidence that the entire plant functions as one team where monitoring is very much done by peers or coworkers.

monitor who becomes a residual claimant. Alchian and Demsetz [4] show how team production results in an organizational structure commonly refered to as a "classic firm". They emphasize that in team production, the jobs done by workers are not perfectly separable. It is hence not feasible to compensate workers based on their marginal contributions. The workers therefore have an incentive to shirk, to free-ride on other team members' efforts. The well known reason is that the cost of shirking is born by the entire team. The result is that since market contracting with individuals is not possible, the free-riding problem is overcome by the "classic firm" structure in which workers are paid a wage and a principal supervises or monitors the workers. It is then optimal to let the principal be the residual claimant. By this arrangement, the incentive compatibility of the residual claimant is satisfied and team members' free-riding is attenuated as the monitor is given the authority to dismiss any slackers. This is, however, only one possible solution to the incentive problem, there may be others.

Holmstrom [18] offers alternative reasons for the existence of a residual claimant to prevent the team from falling apart due to free-riding: a credible group penalty can induce an efficient effort level even if there is no monitoring *per se*. Holmstrom shows that collective punishments and rewards may provide the necessary incentives for workers to exert the desired level of effort without the need of a monitor. The feature is that the penalties or rewards required to ensure efficient production are not budget-balancing. Thus the role of the principal of a firm or the owner of it is to break the budget-balancing constraint rather than to act as a monitor. The conclusion is that the capitalistic firms in which the owners do not provide labor services have advantages over proprietorship because of the owners' ability to finance budget-breaking schemes. It offers a deeper understanding on the difference between a team and a commune. Moreover, it explains why the residual claimant generally is not involved in supervision while the supervisor who fulfills the monitoring role is commonly salaried.

Rasmusen [47], McAfee and McMillan [37] follow this approach. However, the organizational design suggestions and pay structure recommendations in this research is uncommon. For instance, the scapegoat penalty in [47] or the large lump-sum upfront commitment fee payment in [37] are almost never seen in practice[3].

In neither of these approaches is mutual monitoring among the agents addressed and discussed. In the performance comparison literature[4], mutual monitoring is considered but the mechanism proposed to correct incentives in teams is to make each team member's pay not only depend on his own output but also on his peer's output if they are correlated. This is potentially

[3] The main reason here that we understood is that the access to credit market is highly restricted so that despite an individual's risk attitude, persons who do not pose entrepreneurship can not acquire capital to finance himself upfront.

[4] See [38, 39] for details.

problematic for two reasons. First, the individual contribution in a managerial team cannot usually be distinguished by an outsider if she is not involved in the production process; second, teams appear to emphasize cooperation and synchronization instead of harsh competition. Indeed, Proposition 1 of [39] shows a sufficient and necessary condition for relative performance evaluation to be efficient. Meyer and Vickers also noticed that a ratchet effect, which may arise, would cause further welfare loss, stated in their Proposition 3. Moreover, because of the lack of strategic interactions between the agents, relative performance evaluation could cause a loss of efficiency. In pro-tournament models, tournaments cause collusion and hence are not coalition proof.

Recently, some authors have considered reputational and career concerns in teams. Glazer and Segendorff [14] consider a reputational effect in teams and they conclude that if production must be carried out by a two person team, it is better to have a low ability partner if one cares about one's reputation. Their model is simple and illustrative however the self-formed partnerships studied in their paper account only a small proportion of firms. In most cases, managerial teams are not self-selected, instead, they are appointed by the Board of Directors and the agents are obligated to work together as a corporate norm.

The technical setup of Breton, St-Amour and Vencatachellum [6] is closely related to ours. In [6] the wages of the agents are directly reputation-based in a dynamic stochastic framework. Three organizational forms, compulsory individual performance evaluation, compulsory teams, and elective teams are studied. Also the authors differentiate between common shocks and idiosyncratic shocks in production. Welfare analysis is also presented. The analysis of [6] is fairly comprehensive, especially when teams are formed endogenously and inter-generational grouping is available given *de facto* reputation for each candidate is common knowledge. For instance, in academic research in a university, no scheme is levied either to foster or impair team formation. However, in [6], all agents are risk neutral and the compensation is a random binary variable. These assumptions make the model different from the facts documented or commonly seen. Also, the lack of risk aversion in team members leaves no room for an insurance effect against correlated shocks, which, in our understanding, is an important factor in organizational structure determination.

In fact, relationships between managers or workers, in the real world, appear much more cooperative than portrayed by the existing theory on incentives, which emphasizes competition among the agents or candidates. Our model shows that this cooperative behaviour is consistent with the noncooperative, competing nature of the parallel team members instead of induced or instructed by the principal.

Secondly, much of the existing literature on incentives models the principal-agent relationship in a static setting, neglecting the repeated nature of most employment relationships, procurement and partnership. The relationship either between the owner of the business and its managers, or, the colleagues themselves, are long term and repeated rather than one-shot.

Thirdly, adverse selection and moral hazard are rarely considered together, in fact, often a firm's output is a function of management's abilities and effort exerted, not one of those alone. Therefore, both hidden information and hidden actions taken by the managers as well as the interaction between the two will affect the aggregate output.

Finally, in an optimal incentive scheme, not only are there explicit incentives, but also numerous types of implicit incentives as well. Therefore a scheme with a low pay-performance sensitivity may not be as inefficient as one might perceive. A low powered pay scheme and a simple multiagent firm may have virtues to approximate a first best outcome with implicit incentives. Careless policy suggestions that neglect this fact could very possibly undermine the optimal or suboptimal incentives already provided in the scheme.

This essay addresses the above four issues together by focusing on mutual observability between peers and the repeated nature of working relationships in firms. These four elements:

A. the cooperative outlook of a team powered by the implicit incentives;

B. repeated environment;

C. combination of hidden action and hidden information;

D. the very passive role that a principal plays in a team,

combined together, enable us to explain how cooperative behaviors among agents can be sustained by a self-enforcing mechanism. Our model is consistent with many stylized features of the team-oriented profit organizations observed. We consider that this is largely due to the combined assumptions we imposed. We will find that reputational or career concerns play a major role in the implementation of optimal compensation scheme.

To this end, we consider a two period model of managerial teams where only the total output is observable to the principal but individual output for each period is known by the team members by the end of the period. This information structure essentially captures the nature of mutual monitoring and mutual observability in a small size team. Output is additive and is aggregate on effort, ability and random shocks. The team members have the option to quit and work on their own at the beginning of the second period if the information provided through output level indicates that her colleague shirked and/or is a low ability type.

Unlike the usual two period model in which the subgame perfect Nash equilibrium outcome yields an inefficient output level in the second period and free riding in the first period, the latter will not occur in our model. The reason is that learning of the agents' ability and quitting as a credible threat makes first period suboptimal output level possible. It then, in some degree, prevents teams from collapsing because of free riding, and hence also makes team formation more attractive than individual production because teaming provides more insurance relative to individual production[5].

We invoke the career concern model framework[6], but our model differs from others in the sense that career concerns here are for peers reciprocally, not for the principal who does not participate in the production but is just a residual claimant. Early research on single agent career concerns provides a benchmark case for our model. However our model is not a trivial extension of early career concern models because of our focus on the interaction between the agents. Therefore the paper attempt to provide an answer to fundamental questions in the theory of firm rather than employees' tenure dynamics.

Our findings are summarized as follows.

First, we find that a group output based compensation scheme could approximate first best production in the initial period. We eliminate technological synergy between the teammates in teams to isolate the incentive due to career concerns and make the problem nontrivial. By treating teams just as aggregations of individual production, we are hoping to provide other robust reasons for teams to survive, such as an insurance effect. Many researchers realized that in a repeated setting, a competitive scheme, such as a tournament, will not be optimal as coalitions (even without side-payment) will arise. In our case, since individual output can not be observed, tournaments are naturally ruled out, but similar schemes, such as a penalty introduced when output falls below some critical level, will not be optimal either[7]. Indeed, a moderate, passive scheme in which strategic collusion between agents will not be beneficial and hence muted, will perform well. More importantly, joint performance evaluation in a repeated environment creates not only an incentive for agents to monitor each other but, with abilities attributed to the output, career concerns will force agents to work harder in their earlier careers than that of a one-shot relationship.

Second, as in the single agent career concern models, we find that the piece rate of the optimal contract increases over time to balance out the implicit incentives restored by reputational concerns between team members. This is a natural corollary from the first finding. However, further comparisons between team setting and a single agent setting give us some other results. Our findings are: 1, second period incentive power in team performance evaluation is lower than that of individual performance evaluation if output fluctuations are positively correlated; 2, the team setting provides an insurance effect compared with the single agent case, and this is a Pareto improvement as the trade-off of incentive and insurance between the principal and the agent is improved by adding another agent.

---

[5] The insurance effect by teaming is not so novel as aggregation also provides insurance effect. Think for an extreme case, when an insurance company insures all the individuals in a sufficiently large economy, it will face no risk.

[6] [19], [9], [12], etc.

[7] We say it is similar because agents can communicate and form a coalition. Therefore tournament and group penalty/reward are similar in this sense.

Third, our findings imply that decentralizing authority to a certain degree could be beneficial to the firm. In [1], decentralization is good if it facilitates the agent's investment in acquiring information about decision alternatives if the principal and agent's interests are sufficiently congruent. We agree with this conjecture. Here the problem is two-fold, first, how congruent are these interests and do they diverge with decentralization, second, will moral hazard arise and become severe with the decentralization. The optimal level of decentralization can be then characterized for future research with these considerations.

Our innovation is powered by the combination of the four major considerations mentioned above. We believe that only a careful study with a more realistic environment can lead to sensible policy recommendations on optimal compensation scheme design, grouping, multitask design, and optimal allocation of the authority. We try to provide an alternative explanation for why an Alchian–Demsetz[8] firm would persist in the real world and why the pay-to-performance ratio is low in explicit contracts. Nevertheless, our approach can be extended into a variety of directions including grouping and authority allocation and policy suggestions can be carried out in further research.

2 The Model

2.1 Preliminaries

Our model's formulation is similar to that of Gibbons and Murphy [12]. There are many firms that are engaged in perfect competition in the labor market. A representative firm consists of an owner (principal) and two managers (agents). The firm (or a project) is operated by this two person managerial team, each of whom contributes with an effort and intrinsic ability. The intrinsic ability follows a certain probabilistic distribution, and the distribution is known to the principal and the agents *a priori*. The value of own intrinsic ability is known to the agents themselves after the first period output is realized and the posterior belief on their counterparts' ability is improved. The principal, usually the owner of the firm, does not participate in production or decision making, except for setting up a linear incentive compensation scheme for the agents, sometimes called team members.

The project lasts for two periods in which random shocks enter production additively. Only the total output for each period is revealed and verifiable to the principal and the market by the end of each period. Individual output is not observable or verifiable to the principal in either period but is observable to the team members themselves immediately after each period ends. We assume that agents are free to accept offers from other firms and therefore no

[8] However the firm in our model differs slightly from [4] in the sense that the role of monitoring the effort is replaced by peers instead of the principal.

long term binding contract can be imposed on individual agents. It is natural to have only short term contracts considered and offered in our model.

Furthermore, given that the principal cannot acquire private information about agents' abilities even after the first period ends, and the competitiveness of the labour market, the principal has no reason to offer a contract that pays more for one agent and underpays the other. In fact, the principal treats every agent identically, in our terms, given compensation contingent on aggregate output, the intra-period incentive sensitivities of the contracts ought to be the same across individual team members.

Therefore the principal offers two short term contracts (one for each period) that can differ in piece rate across periods. After the first period has elapsed, the total output is revealed to all players; agents have the option to continue in this team or quit and work alone for another employer. The market has the same information as the current principal. The output of the team is simply the aggregation of individual outputs. Finally, because of the competitiveness of the labour market, the principal is forced to make zero expected profits in each period.

By assumption, the managers learn their own ability by the end of first period[9] when it is still unknown by the principal. The agents also know how much effort they exert in equilibrium, therefore they have improved beliefs about their colleague's ability.

Each of the agents, after the first-period outputs are realized, will have an updated belief on his partner's ability and makes a decision on whether or not to stay in this team. If both decide to stay, a second-period contract is offered. If one of the agents decides to quit in the second period, we say that a separation has occured. For simplicity, we assume that when separation occurs, the agents decide to work alone in the second period, they also leave the firm and do not team up with other agents. If agent i decides to stay, his second period expected payoff is $EU_{i2}|_{stay}$, and $EU_{i2}|_{quit}$ otherwise, where $EU_{it}|_k$ is agent i's expected utility at period t given event (or decision) k.

Team member i's contribution to output[10] at period t is,

$$y_{it} = \eta_i + a_{it} + \xi_{it}\,, \qquad i = 1,2\,;\ t = 1,2\,, \tag{1}$$

where η_i is the agent's inborn entrepreneurial ability, a_{it} is the i's effort exerted in period t, ξ_{it} is the individual shock at t, and $a_{it} \in \mathbb{R}^1_+$, ξ_{it}, $\eta_i \in \mathbb{R}^1$. The principal's prior belief on team members' entrepreneurial abilities follow

$$\eta_i \sim N(m_0, \sigma_0^2)\,, \quad \text{with correlation} \quad corr(\eta_i, \eta_j) = 0\,, \quad \text{for } i \neq j\,,$$

and random shocks in each period follow

$$\xi_{it} \sim N(0, \sigma_\xi^2)\,, \quad \text{with correlation} \quad corr(\xi_{it}, \xi_{jt}) = \rho\,, \quad \text{for } i \neq j\,.$$

[9] This assumption is not so strong as it seems natural for one to realize his entreprenurial ability only after a period of business practice. It is also due to this, agents would over exert effort in the first period to "impress" others.

[10] This production technology follows [9].

Team members' utility is negative exponential, i. e.,

$$u_i(\cdot) = \sum_{t=1}^{2} \delta^{t-1}\{-\exp(-r\{[w_{it} - g(a_{it})]\})\} , \tag{2}$$

where w_{it} is the earnings of agent i at t and $g(a_{it})$ is the disutility of effort for agent i at t. We assume as usual that $g'(\cdot) > 0$, $g''(\cdot) > 0$ and $g'(0) = 0$, $g'(\infty) = \infty$.

Agents can choose to leave the team (firm) in the second period. The intuition is that if one of the agents finds his partner's ability is too low relative to his, he will have an incentive to quit and work alone in the second period. The firms, as well as other agents in the economy, on the other hand, have the same information as the current firm, therefore they cannot distinguish between any two agents in terms of abilities and it will be too risky for others to team up with one of the separated agents. Therefore, we assume that the two agents will work under individual performance evaluation scheme, given separation occurs.

The market, as well as the principal, can not distinguish the agents' types as the agents can costlessly mimic the high ability by announcing quitting at the same time. We show below the invariance of market perception on agents' abilities.

2.2 Market Perception on the Agents' Abilities

In a two period model, the team members can decide whether to continue work in a team or separate and work alone. The separation decision depends on the two team members' ability difference. When first period output is realized, team members can update their beliefs on their counterparts' abilities and because the sharing rule is equal sharing between the two agents, the agent with relatively higher ability will have an incentive to work alone if the benefit from pooling the risk is can not compensate the free riding. However since individual outputs are private information, the market, as well as the principal, can not distinguish the 2 agents in terms of their abilities. The question arises is that whether there is an asymmetry between separation and no separation in terms of the market's belief on the sum of the two agents' abilities.

Proposition 1. *Market's beliefs on the aggregate ability of the team are the same under separation and no separation.*

Proof. We prove by contradiction. Suppose that the proposition is false.

1. We assume that market belief of the aggregate ability, given separation has occurred, denoted by $E^M[\eta_1 + \eta_2 \mid \mathit{Separation}, Y_1]$, is lower than market belief of the aggregate ability, given no separation has occurred, denoted by $E^M[\eta_1 + \eta_2 \mid \mathit{No\ Separation}, Y_1]$. Suppose there is a positive number ε, such that

$$E^M[\eta_1 + \eta_2 \mid \mathit{Separation}, Y_1] + \varepsilon = E^M[\eta_1 + \eta_2 \mid \mathit{No\ Separation}, Y_1] , \tag{3}$$

then there is a distortion in incentives for any $\varepsilon > 0$. Suppose agent 1 is the high ability agent, relative to agent 2, therefore instead of deciding to separate for any ability difference greater than the critical value $D^c = (\eta_1 + \eta_2)^c$, agent 1 will separate only when the difference is greater than $D^c + \frac{\varepsilon}{2} = (\eta_1 + \eta_2)^c + \frac{\varepsilon}{2}$. Obviously, this overestimation of the aggregate ability will distort the agents' decision to separate downwards. Agent 1, now, instead of only considering the tradeoff between risk pooling and free riding, will have to also consider a negative impact on his ability perception given his decision to separate. This will cause over pooling in the second period (too few separation in the second period).

2. Next we assume that market belief of the aggregate ability, given separation has occurred is higher than market belief of the aggregate ability, given no separation has occurred. Once again, suppose there is a negative number ε, such that

$$E^M[\eta_1 + \eta_2 \mid \textit{Separation}, Y_1] + \varepsilon = E^M[\eta_1 + \eta_2 \mid \textit{No Separation}, Y_1], \quad (4)$$

then there is a distortion in incentives for any $\varepsilon < 0$. Suppose agent 1 is the high ability agent, relative to agent 2, therefore instead of deciding to separate for any ability difference greater than the critical value $D^c = (\eta_1 + \eta_2)^c$, agent 1 will separate only when the difference is greater than $D^c + \frac{\varepsilon}{2} = (\eta_1 + \eta_2)^c + \frac{\varepsilon}{2}$. This is contrary to the first case and obviously, this overestimation of the aggregate ability will distort the agents' decision to separate upwards. Agent 1, now, instead of only considering the tradeoff between risk pooling and free riding, will have to also consider a positive impact on his ability perception given his decision to separate. This will cause over separating in the second period.

Now we prove that in equilibrium, $\varepsilon = 0$.

Suppose for case 1, because the impact of underestimation is negative to the higher ability agent, another employer, using $\varepsilon \leq 0$, can beat the incumbent employer (principal) be offering a better contract. For case 2, a negative expected profit will incur to those employers who use $\varepsilon < 0$ if at least one employer in the market is using $\varepsilon \geqslant 0$. Given the market is perfectly competitive, any $\varepsilon < 0$ is not credible, therefore case 2 is not subgame perfect. Hence, $\varepsilon = 0$.

2.3 Timing and Structure

Figures 1, 2 depict the timing of all events.

In this model, the principal moves first, at date 0, two managers are hired and a team is formed and a linear short term compensation scheme is proposed by the principal. Agents with uncorrelated distributed abilities accept the first period contract and exert efforts simultaneously. Nature chooses two correlated random shocks that additively enter output. After the agents observe the first period individual outputs at date 1, they will have updated beliefs about their colleague's abilities and will consequently decide whether

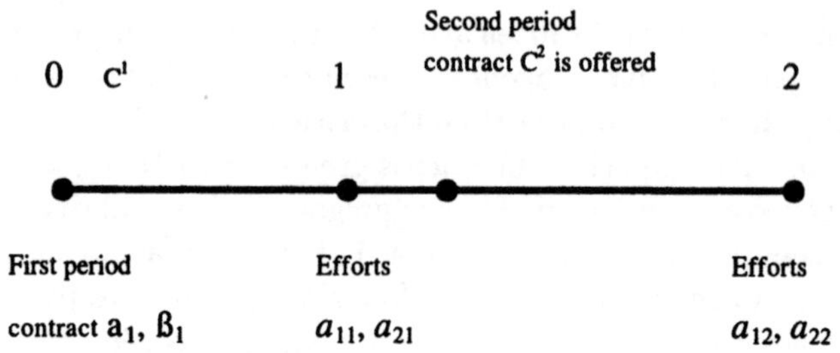

Fig. 1. Continuation decision node for each manager

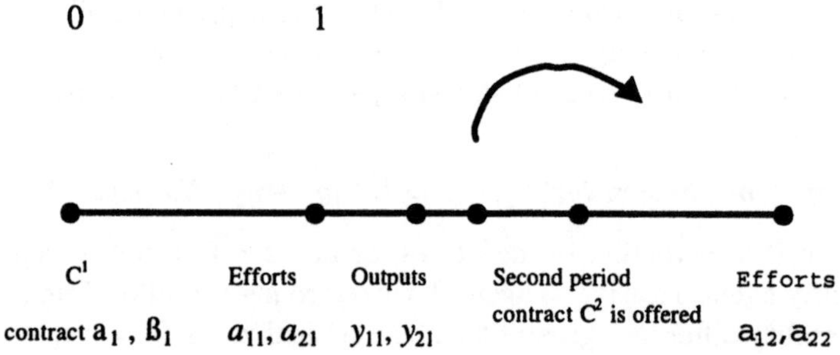

Fig. 2. Individual outputs observed by managers while only aggregate output observed by the market and the principal

to continue working in this team or to exit. If they decide to continue, a second period contract is proposed by the principal and both agents exert second-period efforts. Note that at any time, information on individual outputs can not be made available to the principal, i. e., the principal and the market can only observe aggregate outputs by the end of each period.

The principal makes zero expected profits because of the perfectness of labour market competition: this, in term, implies that the expected output will be assigned to the agents by the compensation scheme[11]. Indirectly, the principal's objective is to maximize agents' aggregate utility given market beliefs.

In our model, for reasons of tractabilty, we assume a linear incentive scheme.

[11] Note the zero-profit condition is not very straighforward to see given the overlapping environment. As we will find, separation can occur under certain conditions. This implies that individual performance evaluation and team performance evaluation can co-exist in the same period: principals of team structured firms may make strictly positive profits to take advantage of the insurance effect on the agents while competition of the market results zero profit for the firms that adopt individual performance evaluation. Nonetheless, if we allow other firms to freely acquiring the whole team through the market by offering more attractive compensation, zero profit still holds.

Linear schemes are also robust in the real world, by the argument of Holmstrom and Milgrom [20]. A simple linear incentive scheme may perform well across a wide range of environments and as well has low writing costs. Nonlinear schemes fine-tuning complex incentive schemes are not very realistic for at least two reasons: firstly, they would not maintain their optimality for even a slightest change in the information structure or technological change; secondly, intricate schemes are hard to implement: agents would take advantage of the complexity of the contracts by arbitrage-taking like behaviours, and this situation can be exaggerated by multiple agents because collusion and deviating from optimal reciprocal help[12] level that the principal desires will be their options[13].

Linear incentive compensation schemes do not require precise knowledge of the agents' preferences and the production technologies they control, and they automatically adjust to environmental changes without renegotiation[14]. In a step function scheme, a slight change in the industry's benchmark will immediately make the current scheme inefficient. One can approximate other nonlinear schemes by polynomials of higher order terms. Positive coefficients on these terms can have two effects: an incentive-improving effect and an insurance-dampening effect. Furthermore, they substantially reduce the transaction costs associated with customized contingent schemes.

Next, consider the total output for the team is

$$Y_1 = y_{11} + y_{21} = \eta_1 + \eta_2 + a_{11} + a_{21} + \xi_{11} + \xi_{21} \tag{5}$$

for period one and

$$Y_2 = y_{12} + y_{22} = \eta_1 + \eta_2 + a_{12} + a_{22} + \xi_{12} + \xi_{22} \tag{6}$$

for period two.

The total expected utility for agent i is then

$$\begin{aligned} E[u^i(\cdot)] = & -E[\exp\{-r[\alpha_1 + \beta_1 Y_1 - g(a_{i1})]\}] \\ & - \delta E[\exp\{-r[\alpha_2 + \beta_2 Y_2 - g(a_{i2})]\}] , \end{aligned} \tag{7}$$

[12] In multiagent incentive contract design, especially when multitask is introduced, it is usually better to have certain level of helps between the agents when they are heterogeneous.

[13] If free disposal of the output is feasible, a practical scheme then has to be monotone despite any other argument on either information structure or stochastic production technology it may have.

[14] Mirrlees (1974) shows that instead of a simple linear scheme, the first-best outcome can be approached arbitratily close by a suitable chosen sequence of step functions if output is lognormally distributed. However since a solution does not exist, it is called the "Mirrlees Problem". The optimality of the step functions proposed by Mirrlees depends on several strong assumptions: 1, output is lognormally distributed; 2, it is assumed that the first-order-approach is valid; 3, an economy with large number of agents. Moreover, transaction costs must be sufficiently small. See Mirrlees (1974): *Notes on Welfare Economics, Information and Uncertainty*, in: Essays on Economic Behaviour under Uncertainty.

if no separation occurs in the second period.

If one of the agents quits, both agents will work alone and because quitting implies individual performance evaluation in the second period, we have

$$EU_{i2}|_{quit} = \max_{a_{i2}} - \mathrm{E}[\exp\{-r(\widetilde{\alpha}_2 + \widetilde{\beta}_2(\eta_i + a_{i2} + \xi_{i2}) - g(a_{i2})) \mid Y_1, \mathrm{Sep}, y_{i1}\}] ,$$

where Sep represents the fact that separation has occurred and ~ refers to the case of individual performence evaluation linear scheme in the second period[15]. We characterize the case of individual performance evaluation in section 3.

The linear contract in each period is $\alpha_t + \beta_t Y_t$[16]. The competitiveness of the managerial market induces the principal (could be shareholders) to make zero expected profits in each period, i.e., $\alpha_t = (\frac{1}{2} - \beta_t)E[Y_t \mid \Omega]$, $t = 1, 2$ and $\Omega = \{Y_1, \mathrm{NS/Sep}, \rho\}$ is the principal's information set by the end of the first period which includes the first period aggregate output, the fact that whether separation occured or not, and correlation of the transient production noise terms. The expected profit is zero within period (by the argument of perfect competition) instead of lifetime because there is no binding constraint to restrict the agent from quitting the firm, nor punitive penalties. Again this is due to the competitiveness of the market, suppose if a contract overpays the agent in the first period, he will quit in the second period and join a new firm[17], or if he is underpaid in the first period, he will not accept the offer anyhow, but seek a fair deal in the market[18]. In other words, the principal can neither pay the agents below nor above the market expected output level in each period.

The intercept α and incentive term β are set by the principal but not arbitrarily, in fact, they are endogenous as we will see below. We use backwards induction to find a subgame perfect equilibrium. The principal observes aggregate output by the end of period one and she uses it in learning about aggregate ability; team members will exert effort noncooperatively and this fact is common knowledge. Assuming peers will observe individual output by the end of the first period but not the principal, quitting and working alone in the subsequent period would be a credible punishment to their partner if the partner ever shirks in the first period relative to equilibrium level or his ability is too low.

15 It can be written also as $\max_{a_{i2}} - \exp[-r\{m_0 + \widetilde{a}_{i2} - g(\widetilde{a}_{i2}) - \frac{1}{2}r\widetilde{\beta}_2^2\sigma_\xi^2\}]$ because the agent is rewarded with the expected output and quitting implies individual evaluation.

16 If the agents continue to work in the team, their income is $w_{it} = \alpha_t + \beta_t Y_t$ for agent i at t. However if either one of the agents decides to quit from team work, both will work alone and the second period scheme will be $w_{i2} = \alpha_2 + \beta_2 y_{i2}$.

17 Here is a simple moral hazard problem between the firm and the agents.

18 This is a simplest adverse selection problem.

3 Equilibrium

3.1 Continuation Game with No Separation

Using backward induction, we solve for second period non-separation case. The problem turns to a principal's problem in choosing α_2, β_2 in the second period. To see it explicitly, see the following decision tree.

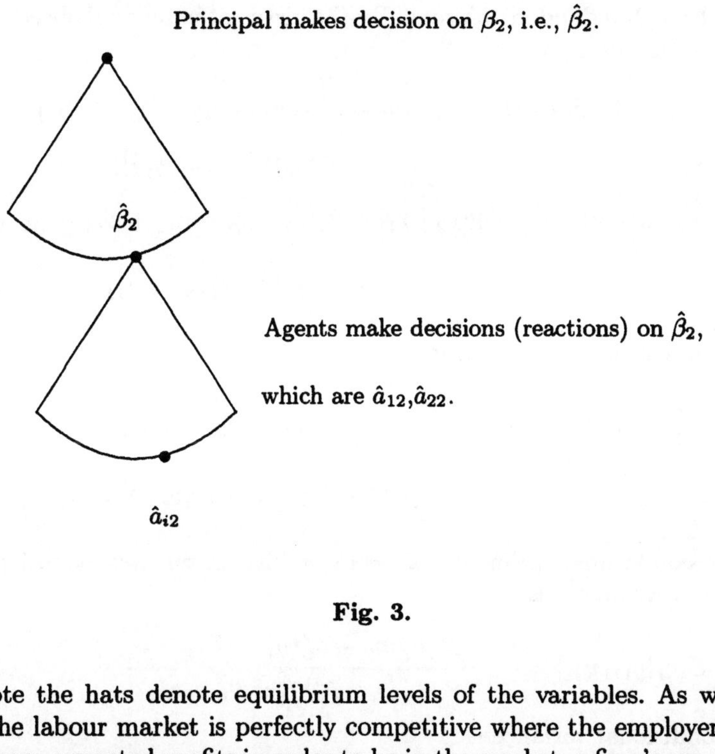

Fig. 3.

Note the hats denote equilibrium levels of the variables. As we assume that the labour market is perfectly competitive where the employers have to make zero-expected profits in order to be in the market, a firm's maximization problem is equivalent to maximize the agents' expected utility given market information, otherwise the managers will be bid away.

Now we formally characterize the equilibrium in the second period in the case of no separation.

We use backward induction to derive the second period effort level and the parameters of the optimal incentive contractual form. In the second period, which is also the last period, the team members will exert effort noncooperatively and they maximize the payoff for that period. Because the decision is symmetric, we study the case for agent 2. Agent 2 maximizes the following when the second period commences and assuming he decides to stay in this team[19]:

[19] Note the first period decision has no impact on the second period□s. In our model it is assumed that the agents capture all the expected output, therefore the first

$$\max_{a_{22}} - \mathrm{E}[\exp\{-r(\alpha_2 + \beta_2(\eta_1 + \eta_2 + a_{12} + a_{22} + \xi_{12} + \xi_{22}) - g(a_{12})) \mid Y_1, y_i, y_j\}] . \tag{8}$$

By the end of the first period, the team member's individual contribution to the output is revealed to himself and because production technology is additive, his peer's output is therefore observed as well. As team members learn their own ability perfectly by the end of first period, the efficient contract for the team in the last period is deduced from a subgame perfect equilibrium effort level through individual learning. By the virtue of individual observability, we have the following for agent 2,

$$\begin{aligned} &\max_{a_{22}} - \mathrm{E}[\exp\{-r(\alpha_2 + \beta_2(\eta_1 + \eta_2 + a_{12} + a_{22} + \xi_{12} + \xi_{22}) \\ &\qquad - g(a_{22})) \mid Y_1, y_i, y_j\}] \\ &= \max_{a_{22}} - \mathrm{E}[\exp\{-r\Big(\frac{1-2\beta_2}{2}\mathrm{E}[Y_2 \mid Y_1] + \beta_2(\eta_1 + \eta_2 + a_{12} + a_{22} + \xi_{12} + \xi_{22}) \\ &\qquad - g(a_{22})\Big) \mid Y_1, y_i, y_j)\}] , \end{aligned} \tag{9}$$

which is equivalent to the following,

$$\begin{aligned} \max_{a_{22}} - r\Big\{&\frac{1}{2}(\eta_2 + m_3) + \beta_2(\eta_1 - m_3) + \frac{1}{2}(\widehat{a}_{12} + \widehat{a}_{22}) - g(\widehat{a}_{22}) \\ &- \frac{1}{2} r \beta_2^2 \mathrm{var}(\eta_1 + \xi_{12} + \xi_{22} \mid Y_1, y_i, y_j)\Big\} . \end{aligned} \tag{10}$$

where the conditional mean of his peer's ability given first period private information revealed[20], is

$$m_3 \equiv E[\eta_1 | y_{11}, y_{21}] = \frac{(1-\rho^2)\sigma_\xi^2 m_0 + \sigma_0^2(y_{11} - \widehat{a}_{11} - \rho\widehat{\xi}_{21})}{(1-\rho^2)\sigma_\xi^2 + \sigma_0^2} , \tag{11}$$

and the conditional variance is

$$\sigma_T^2 \equiv \mathrm{var}(\eta_1 + \xi_{12} + \xi_{22} | \Omega) = \frac{(1-\rho^2)\sigma_\xi^2\sigma_0^2}{(1-\rho^2)\sigma_\xi^2 + \sigma_0^2} + (2+\rho)\sigma_\xi^2 . \tag{12}$$

From the first order condition of (9), we immediately have the following identity,

$$g'(a_{22}) = \beta_2 . \tag{13}$$

---

period choice of effort affects second period only in utility level through the intercept term of a linear compensation scheme. This is why even a non-time-separable preference does not yield correlated efforts across periods.

20 The agents perfectly learns his own ability by the end of first period, and hence perfectly learns the random transient shock $\widehat{\xi}_{i1}$. As the correlation coefficient ρ is common knowledge, he then improves his belief on his colleague's ability *ex post* by the improved posterior distribution on $\widehat{\xi}_{j1}$.

Similarly, the maximization with respect to agent 1 yields $g'(a_{12}) = \beta_2$. Indeed, (13) is agent 2's reaction function (implicit) with respect to contract piece rate. It is worth mentioning that apart from aggregate output of the first period, Y_1, the principal knows that the agents know each other's individual output.

The conditional mean and variance in (11) and (12) are derived from [8] and the following lemma.

Lemma 1 (Mao et al. [36]). *Let a vector of random variables*

$$\mathbf{X} = \begin{pmatrix} \mathbf{X}_1 \\ \mathbf{X}_2 \end{pmatrix} \sim N\left(\begin{pmatrix} \mu_1 \\ \mu_2 \end{pmatrix}, \begin{pmatrix} \Sigma_{11} & \Sigma_{12} \\ \Sigma_{21} & \Sigma_{22} \end{pmatrix} \right).$$

Then given that $\mathbf{X}_2 = \mathbf{x}_2$, *the conditional distribution of* $\mathbf{X}_1$ *is*

$$N\left(\mu_1 + \Sigma_{12}\Sigma_{22}^{-1}(\mathbf{x}_2 - \mu_2), \Sigma_{11} - \Sigma_{12}\Sigma_{22}^{-1}\Sigma_{12}'\right).$$

Given agents' reaction functions are common knowledge, the principal solves maximization problem in a subgame perfect manner. The principal's problem in second period can be then written as the following maximand,

$$\begin{aligned}\max_{\alpha_2,\beta_2} - \mathrm{E}[\exp\{-r(\alpha_2 + \beta_2(\eta_1 + \eta_2 + a_{12}(\beta_2) + a_{22}(\beta_2) + \xi_{12} + \xi_{22})\\ -g(a_{12})) + \exp\{-r(\alpha_2 + \beta_2(\eta_1 + \eta_2 + a_{12}(\beta_2)\\ + a_{22}(\beta_2) + \xi_{12} + \xi_{22}) - g(a_{22})) \mid \Omega\}] \quad (14)\end{aligned}$$

subject to zero expected profit condition $\alpha_t = (\frac{1}{2} - \beta_t)E[Y_t]$.

By the virtue of individual learning and linearity of the contract, we can rewrite the above maximand as

$$\begin{aligned}\max_{\beta_2} - 2\mathrm{E}[-r\{\frac{1}{2}(\eta_2 + m_3) + \beta_2(\eta_1 - m_3) + \frac{1}{2}(\widehat{a}_{12}(\beta_2) + \widehat{a}_{22}(\beta_2))\\ - g(\widehat{a}_{22}) - \frac{1}{2}r\beta_2^2\mathrm{var}(\eta_1 + \xi_{12} + \xi_{22} \mid \Omega)\}] , \quad (15)\end{aligned}$$

assuming that agent 2 is the representative agent.

The first order condition yields

$$\widehat{\beta}_2 = \frac{1}{2 + 2r\sigma_T^2 g''(a_{i2})} , \quad (16)$$

where

$$\sigma_T^2 \equiv \mathrm{var}(\eta_1 + \xi_{12} + \xi_{22}|\Omega) = \frac{(1-\rho^2)\sigma_\xi^2\sigma_0^2}{(1-\rho^2)\sigma_\xi^2 + \sigma_0^2} + (2+\rho)\sigma_\xi^2 . \quad (17)$$

Note that the solution to β_2 is not

$$\frac{\eta_1 - m_3 + 1}{2 + 2r\sigma_T^2 g''(a_{i2})} \quad (18)$$

because η_1 and m_3 are agents' private information and can not be retrieved by the principal and it is the principal who makes the compensation package proposals, not the agents or one of the agents. Therefore given no improved information on individual ability η_i, she can only offer an unbiased scheme with $\widehat{\beta}_2$.

The principal then solves the following,

$$\max_{\beta_2}\left(\begin{array}{l}\max_{a_{22}} - \mathrm{E}[\exp\{-r\{\frac{1}{2}(\eta_2+m_3)+\beta_2(\eta_1-m_3)+\frac{1}{2}(\widehat{a}_{12}+\widehat{a}_{22}) \\ \qquad -g(\widehat{a}_{22})-\frac{1}{2}r\beta_2^2\mathrm{var}(\eta_1+\xi_{12}+\xi_{22}\mid Y_1,y_i,y_j)\}\}]\end{array}\middle|\Omega\right) \tag{19}$$

Notice that in (19), there is an information iteration. Since we have that the principal's information set Ω is a subset of any of the agents' information set, (19) can be rewritten as

$$\max_{\beta_2}\left(\begin{array}{l}\max_{a_{22}} - \mathrm{E}[\exp\{-r\{\frac{1}{2}(\eta_2+m_3)+\beta_2(\eta_1-m_3)+\frac{1}{2}(\widehat{a}_{12}+\widehat{a}_{22}) \\ \qquad -g(\widehat{a}_{22})-\frac{1}{2}r\beta_2^2\mathrm{var}(\eta_1+\xi_{12}+\xi_{22}|\Omega\}\}]\end{array}\right). \tag{20}$$

The first order condition to the utility maximization yields

$$\widehat{\beta}_2=\frac{1}{2+2r\sigma_T^2g''(a_{22})}\,. \tag{21}$$

This first order condition makes the problem interesting in another way, say, with an improved incentive in team work. To better understand this, note that $\sigma_T^2\equiv\mathrm{var}(\eta_1+\xi_{12}+\xi_{22}|\Omega)$ instead of $\sigma_2^2\equiv\mathrm{var}(\eta_1+\eta_2+\xi_{12}+\xi_{22}\mid Y_1,NS)$ where in the latter the principal does not make use of her knowledge on the correlation between the two random shocks on individual outputs. We show in next subsection that any "mechanically passive" principal is irrational under the environment in our model.

3.2 Equilibrium Scheme versus "Passive" Scheme

If instead, the principal is a "passive" principal and only propose a second period contract based on the aggregate output, the slope of the linear contract is then

$$\beta_2'=\frac{1}{2+2r\sigma_2^2g''(a_{i2})}\,, \tag{22}$$

where $\sigma_2^2\equiv\mathrm{var}(\eta_1+\eta_2+\xi_{12}+\xi_{22}\mid Y_1)>\sigma_T^2$.

$\widehat{\beta}_2$ provides more incentive than β_2' does as one can easily establish $\widehat{\beta}_2>\beta_2'$. The reason is that $\widehat{\beta}_2$ fully exploits information available to the principal including the correlation between the two random shocks, ρ, which is a common knowledge.

As for the first period contract, we show below in next section that $\widehat{\beta}_2>\widehat{\beta}_1$. This implies that optimal explicit incentives should be lower in their earlier careers even in a team setting. This is consistent with [12] for the single agent case but contradicts the common wisdom that incentive in earlier careers in a team should not be systematically different from the one in their later careers because reputation can hardly be established as only aggregate output is available to the principal and the market.[21]

Next we compare how individual learning differs from a "mechanically passive" principal's learning and the impact on the piece rate in a compensation scheme. We show that a "mechanically passive" principal is not rational in our environment and hence does not exist in a perfectly competitive market.

Suppose the principal now only uses aggregate indice and ignores the agents' mutual observability because she thinks that individual outputs are not verifiable to her whatsoever.

By the assumption that individual output can be observed by the agents at the end of each period, the principal's learning can be only on the aggregate terms, again from [8], we have the conditional mean of aggregate ability is,

$$\mathrm{E}\{\eta_1+\eta_2 \mid Y_1, \mathrm{NS}\} \equiv 2m_1 = \frac{(2+\rho)\sigma_\xi^2 m_0 + 2\sigma_0^2(Y_1 - \widehat{a}_{11} - \widehat{a}_{21})}{(2+\rho)\sigma_\xi^2 + 2\sigma_0^2}\,. \tag{23}$$

The conditional variance of principal's perception on $\eta_1+\eta_2$ is then

$$Var[\eta_1+\eta_2 \mid Y_1, \mathrm{NS}] \equiv \frac{2(2+\rho)\sigma_\xi^2\sigma_0^2}{(2+\rho)\sigma_\xi^2 + 2\sigma_0^2}\,. \tag{24}$$

Denote $\sigma_1^2 = \frac{2(2+\rho)\sigma_\xi^2\sigma_0^2}{(2+\rho)\sigma_\xi^2+2\sigma_0^2}$, the variance of $(\eta_1+\eta_2+\xi_{12}+\xi_{22} \mid Y_1, \mathrm{NS})$ is then

$$\sigma_2^2 \equiv \sigma_1^2 + (2+\rho)\sigma_\xi^2\,, \tag{25}$$

therefore the above maximand[22] can be rewritten as

$$\begin{aligned}
&\max_{\beta_2} - \mathrm{E}[\exp\{-r(\alpha_2(\beta_2) + \beta_2(\eta_1+\eta_2+a_{12}+a_{22}+\xi_{12}+\xi_{22}) \\
&\qquad\qquad - g(a_{12}(\beta_2)) \mid Y_1, \mathrm{NS}\}] \\
&= -\exp\{-r(m_1(Y_1, \widehat{a}_{11}, \widehat{a}_{21}) + \frac{1}{2}\widehat{a}_{12}(\beta_2) + \frac{1}{2}\widehat{a}_{22}(\beta_2) \\
&\qquad\qquad - g(\widehat{a}_{12}(\beta_2)) - \frac{1}{2}r\beta_2^2\sigma_2^2)\}\,,
\end{aligned} \tag{26}$$

where $\widehat{a}_{12}(\beta_2)$ and $\widehat{a}_{22}(\beta_2)$ are equilibrium levels and $\widehat{a}_{12}(\beta_2) = \widehat{a}_{22}(\beta_2)$.

First order condition of the utility maximization problem yields

$$\beta_2' = \frac{1}{2 + 2r\sigma_2^2 g''(a_{12})}\,. \tag{27}$$

[21] In our model, the contract is only based on market information, not individual information, namely, y_i, y_j. It is true that the market cannot distinguish the agents' types but the market will fully exploit the agents' mutual observability.

[22] Readers only interested in our main results can skip this part and go to (21) directly. Steps here leading to the intermediate result (27) are for the purpose of illustrating the impact of individual observability on optimal incentive provision.

It is easy to show that $\sigma_2^2 > \sigma_T^2$ when $\rho > 0$, therefore, obviously,

$$\beta_2' < \widehat{\beta}_2 \quad \text{if } \rho > 0 . \tag{28}$$

The reason that why β_2' is not a correct contract piece rate is that the principal did not use ρ, which is public information.

Proposition 2. $\beta_2' < \widehat{\beta}_2$. *Team incentive is powered partially by mutual observability.*

Remark For an illustrative example of Proposition 2, consider a central government that cares about the units of a good medicine per dollar. It may be desirable for the public hospitals and research units to be pooled together and make the funding contingent on the patients' duration or survival rate on new medicines. For example, Royal Victoria of McGill University and Jewish General Hospital team on AIDS R & D, if Roal Victoria finds that Jewish General is completely ignorant about AIDS, they might very possible to quit from this team even on the cost of accepting a lower powered incentive compensation scheme alone. Also, from the moral hazard point of view, even if Jewish General is good at the pathology of AIDS but they are constantly reluctant in R & D investment or participation de facto which cannot be observed by the government, Royal Victoria will force them to improve just by the fact they can observe Jewish General's input on this matter using quitting as a credible threat. Therefore, although the government does not see the effort they exert in this joint research team, it still can use a high powered incentive scheme based on aggregate results because the government knows that the individual team members can essentially monitor each other.

The same logic even applies to a parent who works in the day hiring her kids to mow her lawn regularly in the summer, the remuneration could be based on the final visual effect in the backyard, not having kids' individual contribution in mind. Also, for the case of two-man policeman teams, promotion can be based on the pair's aggregate output, not contingent only on who cuffs the suspect in a crime scene. A detective who is unwilling to risk for danger, or always miss the target in gun shooting apparently not ideal to his partner and the partner may file a complaint. Eventually he shall be regrouped with someone comparable to him. Therefore the sheriff does not have to worry distinguishing who indeed (in a two-man team) was essential in capturing the suspect. On the contrary, incentive is low in one shot projects. For example, in organizing celebration parties or community event services, or one-shot fund raising, although team work is also essential but the compensation is often flat rate or even voluntary. On the other hand, for a mutual fund broker, a type of fund raiser, the salary is contingent on his clients' contribution and his team mates, often subordinates. The reason that we consider is that this business is routine and repeated, relative to one shot fund raising.

So far we have completed the derivation of equilibrium strategy set and gained some intuition about mutual observability and informativeness in a dynamic managerial team. However these presentations are still incomplete. Now we consider the case of separation.

3.3 Continuation Game under Separation

The key point for teams to survive and prevail, in our understanding, is the right to quit. Reflected in our model, it is the continuation decision. As we will find out below, the continuation decision heavily depends on $m_3 \equiv \mathrm{E}_i[\eta_j|y_{i1}, y_{j1}]$.

Agent i will decide to separate from j in the second period if

$$EU_{i2}|_{stay} < EU_{i2}|_{quit} ,$$

therefore there is an incentive for agent j to not shirk in the initial period because agent i would leave if agent j shirks in some degree in the first period. The rationale is that because each agent takes half of the output which is a monotonic positive function of aggregate abilities, and shirking is off-equilibrium, agent i will deduce that j is a low ability agent and quiting of i is the method to eliminate j's free riding on i's ability. We give the conditions for separation in the second period in section 5. Intuitively, given i's ability is lower than some certain level relative to j's, i will quit in the second period.

Consider for a certain player with some particular entrepreneurial ability, he faces 3 strategies in the first period: to exert effort according to what β_1 implicitly instructs; deviating by shirking in the first period; deviating by overworking in the first period. In either of the latter two cases, he is hoping the gain from deviation in the first period could cover the loss in the second period, in expected terms.

Suppose agent 1 is the one who deviates. We consider the case where agent 1 shirks in the first period in an attempt for instant benefit. Because adverse selection only occurs when the second period begins, agent 2 does not change his learning rule and therefore his estimate on his peer's ability after the first period is lower than if he would not have shirked. Agent 2 then might find working alone beneficial rather than staying in this team with agent 1. Therefore there is an incentive for agent 2 to leave if agent 1 shirks in the first period.

Separation implies either someone has shirked or someone is a low ability type however cannot be distinguished between the two members. Indeed, Proposition 1 shows that the principal and the market still apply learning on gross output given separation occurs, therefore in the second period no other individual agent in the economy will risk himself to team up with any one of the member. Hence, there is another contract for individual evaluation in the second period in the case of separation. It is obtained by the following maximization,

$$\max_{a_{22}} - \mathrm{E}[\exp\{-r(\widetilde{\alpha}_2 + \widetilde{\beta}_2(\eta_2 + a_{22} + \xi_{22}) - g(a_{22})|\Omega, y_{i1}\}]$$
$$= \max_{a_{22}} - \exp\left[-r\left\{m_1 + \widetilde{a}_{22} - g(\widetilde{a}_{22}) - \frac{1}{2}r\widetilde{\beta}_2^2\sigma_\xi^2\right\}\right],$$

subject to

$$\widetilde{\alpha}_2 = (1-\widetilde{\beta}_2)\left[\frac{1}{2}\frac{(2+\rho)\sigma_\xi^2 m_0 + 2\sigma_0^2(Y_1 - \widehat{a}_{11} - \widehat{a}_{21})}{(2+\rho)\sigma_\xi^2 + 2\sigma_0^2} + \widetilde{a}_{22}\right].$$

Because

$$\sigma_\xi^2 < \frac{(1-\rho^2)\sigma_\xi^2\sigma_0^2}{(1-\rho^2)\sigma_\xi^2 + \sigma_0^2} + 2(1+\rho)\sigma_\xi^2,$$

we easily establish

$$\widetilde{a}_{22} > \widehat{a}_{22},$$

where $\widetilde{a}_{22}$ is the effort level that agent 2 exerts if he decides to work on his own in the second period.

The following formally characterizes the second period separation contract:

$$\widetilde{\beta}_2 = \frac{1}{1 + r\sigma_\xi^2 g''(a_{12})} \tag{29}$$

and

$$\widetilde{\alpha}_2 = (1-\widetilde{\beta}_2)\left[\frac{1}{2}\frac{(2+\rho)\sigma_\xi^2 m_0 + 2\sigma_0^2(Y_1 - \widehat{a}_{11} - \widehat{a}_{21})}{(2+\rho)\sigma_\xi^2 + 2\sigma_0^2} + \widetilde{a}_{22}\right]. \tag{30}$$

We have

$$EU_{22}|_{stay} = \frac{\eta_2}{2} + \frac{m_3}{2} + \widehat{a}_{22} - g(\widehat{a}_{22}) - \frac{1}{2}r\widehat{\beta}_2^2\sigma_T^2, \tag{31}$$

and

$$EU_{22}|_{quit} = \eta_2 - \widetilde{\alpha}_{22} - g(\widetilde{a}_{22}) - \frac{1}{2}r\widetilde{\beta}_2^2\sigma_\xi^2. \tag{32}$$

Team member 2's decision on whether to leave or stay with his peer depends on if $EU_{22}|_{stay} \gtreqless EU_{22}|_{quit}$ and it is equivalent to the following by subtracting (32) from (31),

$$\frac{\eta_2}{2} + \frac{m_3}{2} + \widehat{a}_{22} - g(\widehat{a}_{22}) - \frac{1}{2}r\widehat{\beta}_2^2\sigma_T^2 - \eta_2 - \widetilde{\alpha}_{22} - g(\widetilde{a}_{22}) + \frac{1}{2}r\widetilde{\beta}_2^2\sigma_\xi^2 \gtreqless 0, \tag{33}$$

where m_3, agent 2's posterior belief on his colleague's ability conditional on the first period statistics being revealed, is

$$m_3 \equiv E[\eta_1|y_{11}, y_{21}] = \frac{(1-\rho^2)\sigma_\xi^2 m_0 + \sigma_0^2(y_{11} - \widehat{a}_{11} - \rho\widehat{\xi}_{21})}{(1-\rho^2)\sigma_\xi^2 + \sigma_0^2}.$$

Condition (33) is obtained by is indeed an implicit function of agent 2's ability, η_2. The condition implies that depending on agent 1's first period performance, agent 2 learns if his peer's ability is too low to work with in the sense that his partner takes too much of his expected contribution of the output.

The short run expected gain from myopic behaviour (shirking in the first period) for agent 1 is $EU_{11}(\underline{a}_{11}) - EU_{11}(\widehat{a}_{11})$ which is equivalent to

$$\frac{\underline{a}_{11}}{2} - \frac{\widehat{a}_{11}}{2} - g(\underline{a}_{11}) + g(\widehat{a}_{11}) ,$$

where $\underline{a}_{11}$ is the short run optimal private effort level for agent 1 and it is obtained from the following utility maximization,

$$\begin{aligned}\max_{a_{11}} &- \mathrm{E}[\exp\{-r(\alpha_1 + \beta_1(\eta_1 + \eta_2 + a_{11} + \widehat{a}_{21} + \xi_{11} + \xi_{21}) - g(a_{11})\}] \\ &-(1-\pi)\delta\mathrm{E}[\exp\{-r(\widehat{\alpha}_2(\widehat{\beta}_2) + \widehat{\beta}_2(\eta_1 + \eta_2 + \widehat{a}_{12} + \widehat{a}_{22} + \xi_{12} + \xi_{22}) - g(\widehat{a}_{12}))\}] \\ &- \pi\delta\mathrm{E}[\exp\{-r(\widetilde{\alpha}_2(\widetilde{\beta}_2) + \widetilde{\beta}_2(\eta_1 + \widetilde{a}_{12} + \xi_{12}) - g(\widetilde{a}_{12}))\}] . \end{aligned} \tag{34}$$

$\underline{a}_{11}$, the solution to (34), solves the following,

$$g'(\underline{a}_{11}) = \widehat{\beta}_1 + \delta\frac{\partial\pi}{\partial a_{11}}\left[\frac{E\widehat{U}_{12}}{EU_{11}} - \frac{E\widetilde{U}_{12}}{EU_{11}}\right] , \tag{35}$$

The expected loss to agent 1 at the first period's perspective is

$$\begin{aligned}- \Pr\big(EU_{22}|_{stay} < EU_{22}|_{quit}\big)\delta\Big[&\exp\Big(-r\big(\frac{\eta_1}{2} + \frac{m_0}{2} + \widehat{a}_{12} - g(\widehat{a}_{12}) \\ &- \frac{1}{2}r\widehat{\beta}_2^2\sigma_T^2\big)\Big) - \exp(-r(\eta_1 + \widetilde{\alpha}_{12} - g(\widetilde{a}_{12}) - \frac{1}{2}r\widetilde{\beta}_2^2\sigma_\xi^2))\Big] ,\end{aligned}$$

where $\Pr\left(EU_{22}|_{stay} < EU_{22}|_{quit}\right)$ is the probability of $EU_{22}|_{stay} < EU_{22}|_{quit}$, and it is obtained by converting η_1 into a standard normal random variable and checking cumulative Z table for (33). Denote this probability by π such that $\pi \equiv \Pr\left(EU_{22}|_{stay} < EU_{22}|_{quit}\right) = \Pr(\mathit{Separation})$.

As we discussed in this chapter, cooperative behaviours are endogenous in a dynamic team. This may explain why the team as an organizational choice, with apparent drawbacks in the first glance, like exposure to free-riding, still robust in real corporate life. Indeed, we show that a team is a device to eliminate free-riding through mutual monitoring by its members and the illusion of potential free-riding in teams is because previous research neglected the dynamic nature of business partnerships. In fact, team could be more efficient than ordinary principal supervision in the existence of mutual observability as people in the play have a more precise vision of the business and his partner's entrepreneurship than outsiders or even some pseudo insiders like the members of the board of directors or major shareholders. In other words, because production technology is additive and team members are risk averse, working alone will expose too much risk relative to working in a team. Therefore the team members' first period choice on hidden actions is a trade-off between free-riding and loss of control plus team insurance effect. It is then possible that team members exert efforts at the level that the principal dictates through the incentive contract.

3.4 Equilibrium in the Initial Period

Now consider the expected utility for the representative agent, agent 1, at the first period perspective, which is

$$\begin{aligned}
&-\mathrm{E}[\exp\{-r(\alpha_1+\beta_1(\eta_1+\eta_2+a_{11}+a_{21}+\xi_{11}+\xi_{21})-g(a_{11}))\}]\\
&-(1-\pi)\delta\mathrm{E}[\exp\{-r(\widehat{\alpha}_2(\widehat{\beta}_2)+\widehat{\beta}_2(\eta_1+\eta_2+\widehat{a}_{12}+\widehat{a}_{22}+\xi_{12}+\xi_{22})-g(\widehat{a}_{12}))\}]\\
&\quad-\pi\delta\mathrm{E}[\exp\{-r(\widetilde{\alpha}_2(\widetilde{\beta}_2)+\widetilde{\beta}_2(\eta_1+\widetilde{a}_{12}+\xi_{12})-g(\widetilde{a}_{12}))\}], \qquad (36)
\end{aligned}$$

where

$$\widehat{\alpha}_2=\frac{1-2\widehat{\beta}_2}{2}\left[\frac{(2+\rho)\sigma_\xi^2 m_0+2\sigma_0^2(Y_1-\widehat{a}_{11}-\widehat{a}_{21})}{(2+\rho)\sigma_\xi^2+2\sigma_0^2}+\widehat{a}_{12}+\widehat{a}_{22}\right]$$

and

$$\widetilde{\alpha}_2=(1-\widetilde{\beta}_2)\left[\frac{1}{2}\frac{(2+\rho)\sigma_\xi^2 m_0+2\sigma_0^2(Y_1-\widehat{a}_{11}-\widehat{a}_{21})}{(2+\rho)\sigma_\xi^2+2\sigma_0^2}+\widetilde{a}_{12}\right].$$

Note that second period contract is history dependent, however like in the one agent case, this dependence is only captured in α_2, not in β_2. It is also natural to assume that $\pi\equiv\pi(Y_1)$, $\partial\pi/\partial a_{i1}>0$ and $\partial\pi/\partial Y_1<0$.

The first order condition with respect to a_{11} is then obtained accordingly to the following,

$$\begin{aligned}
g'(\widehat{a}_{11})=\widehat{\beta}_1+\delta\frac{\sigma_0^2}{(2+\rho)\sigma_\xi^2+2\sigma_0^2}&\left[(1-\pi)(1-2\widehat{\beta}_2)\frac{E\widehat{U}_{12}}{EU_{11}}\right.\\
&\left.+\pi(1-\widetilde{\beta}_2)\frac{E\widetilde{U}_{12}}{EU_{11}}\right]+\delta\frac{\partial\pi}{\partial a_{11}}\left[\frac{E\widehat{U}_{12}}{EU_{11}}-\frac{E\widetilde{U}_{12}}{EU_{11}}\right], \qquad (37)
\end{aligned}$$

where the second term of RHS depicts the career concern effect in a team. The common factor, $\delta(\sigma_0^2/[(2+\rho)\sigma_\xi^2+2\sigma_0^2])$, depicts the regular career concerns, which is positive and increases with the uncertainty about ability, σ_0^2, and decreases with the uncertainty about production, σ_ξ^2. The factor $[(1-\pi)(1-2\widehat{\beta}_2)(E\widehat{U}_{12}/EU_{11})+\pi(1-\widetilde{\beta}_2)(E\widetilde{U}_{12}/EU_{11})]$ constitutes the team career concern effect. Note it is a tradeoff between loss of utility due to exerting effort in the first period and taking more risk for the second period in the sense of higher probability of separation continuation game.

Maximizing the representative agent's utility, i.e.,

$$\begin{aligned}
&\max_{\beta_1}\mathrm{E}[-\exp\{-r\{m_0+a_{11}(\beta_1)-g(a_{11}(\beta_1))+\tfrac{1}{2}(\xi_{11}+\xi_{21})\}\}]\\
&\quad+(1-\pi)\delta\mathrm{E}[-\exp\{-r\{m_0+\widehat{a}_{12}(\widehat{\beta}_2)-g(\widehat{a}_{12}(\widehat{\beta}_2))+\tfrac{1}{2}(\xi_{12}+\xi_{22}\}\}]\\
&\qquad+\pi\delta\mathrm{E}[-\exp\{-r\{m_0+\widetilde{a}_{12}(\widetilde{\beta}_2)-g(\widetilde{a}_{12}(\widetilde{\beta}_2))+\xi_{12}\}\}]. \qquad (38)
\end{aligned}$$

Expand (38), we have,

$$-\exp\{-r[m_0+a_{11}(\beta_1)-g(a_{11}(\beta_1))]\}\cdot\exp\Big\{\frac{1}{2}r^2\sigma_f^2\Big\}$$
$$-(1-\pi)\delta\exp\{-r[m_0+\widehat{a}_{12}(\widehat{\beta}_2)-g(\widehat{a}_{12}(\widehat{\beta}_2))]\}\cdot\exp\Big\{\frac{1}{2}r^2\sigma_s^2\Big\}$$
$$-\pi\delta\exp\{-r[m_0+\widetilde{a}_{12}(\widetilde{\beta}_2)-g(\widetilde{a}_{12}(\widetilde{\beta}_2))]\}\cdot\exp\Big\{\frac{1}{2}r^2\sigma_\xi^2\Big\}. \quad (39)$$

where $\sigma_f^2 \equiv var[\beta_1(\eta_1+\eta_2)+\beta_1(\xi_{11}+\xi_{21})]+\beta_1^2 Cov[\xi_{11},\xi_{21}]$ and

$$\sigma_s^2 \equiv \delta^2 var\Bigg[\frac{1-2\beta_2}{2}\Bigg(\frac{(1+\rho)\sigma_\xi^2 m_0+\sigma_0^2(Y_1-\widehat{a}_{11}-\widehat{a}_{21})}{(2+\rho)\sigma_\xi^2+2\sigma_0^2}+\widehat{a}_{12}\Bigg)$$
$$+\frac{1}{2}\beta_2(\xi_{12}+\xi_{22})\Bigg]+\frac{1}{4}\delta^2(1-2\beta_2)^2\Bigg(\frac{\sigma_0^2}{(2+\rho)\sigma_\xi^2+2\sigma_0^2}\Bigg)^2 Cov[\xi_{11},\xi_{21}]$$
$$+\frac{1}{4}\delta^2\beta_2^2\,Cov[\xi_{12},\xi_{22}]$$

and σ_f^2, σ_s^2, σ_ξ^2denote the variances of the the first, second and last term of (38), respectively.

We assume that the random disturbances are not serially correlated, i. e., $Cov[\xi_{i1},\xi_{i2}]=0$ and $Cov[\xi_{i1},\xi_{j2}]=0$, where $i,j=1,2$ and $i\neq j$. Notice also $Y_1-\widehat{a}_{11}-\widehat{a}_{21}=\eta_1+\eta_2+\xi_{11}+\xi_{21}$, we have

$$\sigma_f^2 \equiv 2\beta_1^2(\sigma_\xi^2+\sigma_0^2)+\beta_1^2\rho\sigma_\xi^2$$
$$+\delta^2\Bigg((1-2\beta_2)^2\Bigg(\frac{\sigma_0^2}{(2+\rho)\sigma_\xi^2+2\sigma_0^2}\Bigg)^2+\beta_2^2\Bigg)(\sigma_\xi^2+\sigma_0^2)\,.$$

Maximizing (39) with respect to β_1 gives the following first order condition

$$a_{11}'(\beta_1)-g'(a_{11})a_{11}'(\beta_1)-\frac{1}{2}r\frac{\partial\sigma_f^2}{\partial\beta_1}=0\,. \quad (40)$$

Note that π, the ratios $E\widehat{U}_{12}/EU_{11}$ and $E\widetilde{U}_{12}/EU_{11}$ do not enter (40) directly as β_1 appears in the first term of the maximization only.

Note also in the first period, the principal and team members have the same knowledge over the team members abilities, therefore the only conflict between the principal and the team members is that the principal maximizes aggregate objectives while team members tend to maximize private objectives causing possible discrepancies in effort levels.

Evaluating the derivative of variance σ_f^2 with respect to β_1, we have

$$\frac{\partial\sigma_f^2}{\partial\beta_1}=4\beta_1\sigma_0^2+2(2+\rho)\beta_1\sigma_\xi^2\,. \quad (41)$$

Recall that the first order condition to (36) with respect to a_{i1} is

$$g'(\widehat{a}_{11}) = g'(\widehat{a}_{21}) = \widehat{\beta}_1 + \delta \frac{\sigma_0^2}{(2+\rho)\sigma_\xi^2 + 2\sigma_0^2}\left[(1-\pi)(1-2\widehat{\beta}_2)\frac{E\widehat{U}_{12}}{EU_{11}} + \pi(1-\widetilde{\beta}_2)\frac{E\widetilde{U}_{12}}{EU_{11}}\right] + \delta\frac{\partial \pi}{\partial a_{11}}\left[\frac{E\widehat{U}_{12}}{EU_{11}} - \frac{E\widetilde{U}_{12}}{EU_{11}}\right]. \quad (42)$$

Substituting (41) and (42) into (40) yields

$$\begin{aligned}\widehat{\beta}_1 = & \frac{1}{2+2r[(2+\rho)\sigma_\xi^2 + 2\sigma_0^2]g''(a_{11})} \\ & -\frac{1}{2}\delta\frac{E\widehat{U}_{12}}{EU_{11}}\frac{(1-\pi)(1-2\widehat{\beta}_2)\sigma_0^2 + \frac{\partial\pi}{\partial a_{11}}((2+\rho)\sigma_\xi^2 + 2\sigma_0^2)}{(1+r[(2+\rho)\sigma_\xi^2 + 2\sigma_0^2]g''(a_{11}))((2+\rho)\sigma_\xi^2 + 2\sigma_0^2)} \\ & -\frac{1}{2}\delta\frac{E\widetilde{U}_{12}}{EU_{11}}\frac{\pi(1-\widetilde{\beta}_2)\sigma_0^2 - \frac{\partial\pi}{\partial a_{11}}((2+\rho)\sigma_\xi^2 + 2\sigma_0^2)}{(1+r[(2+\rho)\sigma_\xi^2 + 2\sigma_0^2]g''(a_{11}))((2+\rho)\sigma_\xi^2 + 2\sigma_0^2)}. \quad (43)\end{aligned}$$

Lemma 2. *The agents' myopic (Cournot) effort level is lower than the cooperative level, i. e., $\widehat{a}_{11} > \underline{a}_{11}$.*

Proof. Comparing (35) and (37), we have $g'(\widehat{a}_{11}) > g'(\underline{a}_{11})$. By the convexity of the disutility function, we then have $\widehat{a}_{11} > \underline{a}_{11}$. □

Proposition 3. $\widehat{\beta}_1 < \widehat{\beta}_2$.

Proof. $\beta_2 < \widehat{\beta}_2$ is proven earlier. To prove $\widehat{\beta}_1 < \widehat{\beta}_2$, it suffices to prove $\widehat{\beta}_1 < \beta_2$, and it is sufficient to show

$$(2+\rho)\sigma_\xi^2 + 2\sigma_0^2 \geq \sigma_2^2$$

given the assumption that $g'''(\cdot) \geq 0$. The relationship that $(2+\rho)\sigma_\xi^2 + 2\sigma_0^2 \geq \frac{(1-\rho^2)\sigma_\xi^2\sigma_0^2}{(1-\rho^2)\sigma_\xi^2 + \sigma_0^2} + (2+\rho)\sigma_\xi^2$ is obvious. □

4 Comparison with Classical Capitalistic Firms

We refer the term classical capitalism as individual performance evaluation where team based evaluation and any teamwork are abandoned.

Proposition 4. *Second period incentive power in team performance evaluation is lower than that of individual performance evaluation if output fluctuations are positively correlated, i. e., $2\widehat{\beta}_2 < b_2$ if $\rho > 0$ where b_2 is the second period piece rate in individual performance evaluation case.*

Proof. Notice that

$$b_2 = \frac{1}{1 + r\left(\sigma_\xi^2 + \frac{\sigma_\xi^2\sigma_0^2}{\sigma_\xi^2 + \sigma_0^2}\right)g''(a_2)},$$

and it is straightforward to show

$$\sigma_\xi^2 + \frac{\sigma_\xi^2\sigma_0^2}{\sigma_\xi^2+\sigma_0^2} < \frac{(1-\rho^2)\sigma_\xi^2\sigma_0^2}{(1-\rho^2)\sigma_\xi^2+\sigma_0^2} + (2+\rho)\sigma_\xi^2 .$$

□

Lemma 3. *First period incentive power in team performance evaluation is lower than that of individual performance evaluation if correlation ρ is small enough and market compensation levels to the agents of both capitalistic firms and Alchian–Demsetz firms are comparable, that is,*

$$2\widehat{\beta}_1 < b_1 \quad \textit{if} \quad \rho < -\frac{1}{2}\left(1+\frac{\sigma_0^2}{\sigma_\xi^2}\right) \quad \textit{and} \quad \frac{(1-\pi)E\widehat{U}_{i2}+\pi E\widetilde{U}_{i2}}{EU_{i1}} = \frac{EU_2}{EU_1},$$

where b_t, EU_t are the piece rate and expected utility level at period t in single agent case, respectively.

Proof. Note that in the single agent case,

$$b_1 = \frac{1}{1+r[\sigma_\xi^2+\sigma_0^2]g''(a_1)} - \frac{EU_2}{EU_1}\frac{\delta(1-b_2)\frac{\sigma_0^2}{\sigma_\xi^2+\sigma_0^2}}{1+r[\sigma_\xi^2+\sigma_0^2]g''(a_1)} . \tag{44}$$

If $\dfrac{(1-\pi)E\widehat{U}_{i2}+\pi E\widetilde{U}_{i2}}{EU_{i1}} = \dfrac{EU_2}{EU_1}$, $\rho < -\dfrac{1}{2}(1+\dfrac{\sigma_0^2}{\sigma_\xi^2})$, then,

$$\frac{EU_2}{EU_1}\delta(1-b_2)\frac{\sigma_0^2}{\sigma_\xi^2+\sigma_0^2} > \frac{1}{2}\frac{E\widehat{U}_{12}}{EU_{11}}\frac{\delta(1-\pi)(1-2\widehat{\beta}_2)\sigma_0^2}{(2+\rho)\sigma_\xi^2+2\sigma_0^2} - \frac{1}{2}\frac{E\widetilde{U}_{12}}{EU_{11}}\frac{\pi(1-\widetilde{\beta}_2)}{(2+\rho)\sigma_\xi^2+2\sigma_0^2} .$$

Therefore, because

$$2r[(2+\rho)\sigma_\xi^2+2\sigma_0^2] > r[\sigma_\xi^2+\sigma_0^2]$$

and

$$1+r[(2+\rho)\sigma_\xi^2+2\sigma_0^2]g''(a_{11}) > 1$$

given the two sufficient conditions above, we have $2\widehat{\beta}_1 < b_1$. □

Remark Note conditions $EU_{12}/EU_{11} = EU_2/EU_1$ and $\rho < -(1/2)(1+(\sigma_0^2/\sigma_\xi^2))$ are just one set of sufficient conditions. It does not mean that it has to be such restrictive to have the relationship $2\widehat{\beta}_1 < b_1$ to hold. Unfortunately, when $\rho > 0$, conditions for $2\widehat{\beta}_1 < b_1$ to hold are hard to establish, therefore we are unable to provide a comparison on the tenure dynamics between team oriented firms and individual performance evaluation oriented firms.

5 Concluding Remarks

Apart from career concerns that we discussed throughout, we emphasize the informativeness and authority allocation within a firm. Our findings are consistent with Kim's [31] claim that complete retrieval of information is not always necessary. Individual output, even though available to the principal, may not be desired as if the correlation of the random shocks are not known to the principal, it will levy more risk to the agents which worsen the trade-off between incentive and insurance effect. Therefore, although detailed accounting information may be acquired under a low cost, it may not be efficient as making the pay contingent on coarser information might induce the team members to work harder. Our model also suggests that if the principal's and agents' objectives or interests are sufficiently congruent, it may be beneficial to let the agents have more control rights, in particular, mutual monitoring might be Pareto superior than supervision in a traditional capitalistic firm. Our findings are robust because our model does not rely on synergy or synchronization, nor any other strong conditions imposed. We find that the repeated nature of partnership is the reason why teams sustain.

References

1. Aghion, P., Tirole, J. (1997): Formal and Real Authority in Organizations. Journal of Political Economy, 1–29
2. Aggarwal, R., Samwick, A. (1996): Executive Compensation, Strategic Competition, and Relative Performance Evaluation: Theory and Evidence. Working Paper No. 5648, NBER
3. Aggarwal, R., Samwick, A. (1999): Performance Incentives Within Firms: the Effect of Managerial Responsibility. Working Paper No. 7334, NBER
4. Alchian, A., Demsetz, H. (1972): Production, Information Costs, and Economic Organization. A. E. R. **62**, 777–795
5. Baker, G., Gibbons, R., Murphy, K. (1997): Implicit Contracts and the Theory of the Firm. Working Paper No. W6177, NBER
6. Breton, M., St-Amour, P., Vencatachellum, D. (2001): Dynamic Production Teams with Strategic Behavior. Working Paper, Hautes Etudes Commerciales, Canada
7. Che, Y.-K., Yoo, S. (2001): Optimal Incentives for Teams. American Economic Review
8. DeGroot, M. (1982): Optimal Statistical Decisions. McGraw-Hill, NewYork
9. Dewatripont, M., Jewitt, I., Tirole, J. (1999): The Economics of Career Concerns, Part I: Comparing Information Structures. Review of Economic Studies **66**, 183–198
10. Effinger, M., Polborn, M. (2001): Herding and Anti-Herding: A Model of Reputational Differentiation. European Economic Review **45(3)**, 385–403
11. Gibbons, R. (1998): Incentives in Organizations. Journal of Economic Perspectives **12(4)**, 115–132

12. Gibbons, R., Murphy, K. (1992): Optimal Incentive Contracts in the Presence of Career Concerns: Theory and Evidence. Journal of Political Economy **100(3)**, 468–505
13. Glazer, A. (2001): Allies as Rivals: Internal and External Rent Seeking. Journal of Economic Behavior and Organization (to appear)
14. Glazer, A., Segendorff, B. (2001): Reputation in Team Production. Working Paper, University of California, Irvine
15. Harris, M., Holmstrom, B. (1982): A Theory of Wage Dynamics. Rev. Econ. Studies **49**, 315–333
16. Hart, O., Holmstrom, B. (1987): The Theory of Contract. In: Bewley, T. (Ed.): Advances in Economic Theory: Fifth World Congress. Cambridge University Press
17. Holmstrom, B. (1979): Moral Hazard and Observability. Bell J. of Economics **13**, 324–340
18. Holmstrom, B. (1982): Moral Hazard in Teams. Bell J. Econ. **7**, 324–340
19. Holmstrom, B. (1999): Managerial Incentive Problems: A Dynamic Perspective. Review of Economic Studies **66**, 169–182
20. Holmstrom, B., Milgrom, P. (1987): Aggregation and Linearity in the Provision of Intertemporal Incentives. Econometrica **55**, 303–328
21. Holmstrom, B., Milgrom, P. (1990): Regulating Trade Among Agents. Journal of Institutional and Theoretical Economics **146**, 85–105
22. Holmstrom, B., Milgrom, P. (1991): Multiproject Principal-Agent Analyses: Incentive Contracts, Asset Ownership, and Job Design. Journal of Law, Economics and Organization **7**, 524–552
23. Holmstrom, B., Milgrom, P. (1994): The Firm As an Incentive System. AER **84**, 972–991
24. Itoh, H. (1991): Incentives to Help in Multi-Agent Situation. Econometrica **59**, 611–636
25. Itoh, H. (1992): Cooperation in Hierarchical Organizations: an Incentive Perspective. Journal of Law, Economics, and Organization **8**, 321–345
26. Itoh, H. (1993): Coalitions, Incentives, and Risk Sharing. Journal of Economic Theory **60**, 410–427
27. Jensen, M., Murphy, K. (1990): Performance Pay and Top-Management Incentives. J. P. E. **98**, 225–264
28. Jeon, S. (1996): Moral Hazard and Reputational Concerns in Teams: Implications for Organizational Choice. In'l J. Industrial Org. **14**, 297–315
29. Jeon, S. (1998): Reputational Concerns and Managerial Incentives in Investment Decisions. European Economic Review **42(7)**, 1203–1229
30. Jewitt, I. (1999): Information and Principal-Agent Problems. Unpublished manuscript, University of Bristol
31. Kim, S. (1995): Efficiency of an Information System in an Agency Model. Econometrica **10**, 74–91
32. Laband, D., Piette, M. (1995): Team Production in Economics: Division of Labor or Mentoring. Labour Economics **2**, 33–40
33. Laffont, J.-J., Martimort, D. (1997a): The Firm As a Multicontract Organization. Journal of Economics and Management Strategy **6(2)**, 201–234
34. Laffont, J.-J., Martimort, D. (1997b): Collusion Under Asymmetric Information. Econometrica **65(4)**, 875–911
35. Laffont, J.-J., Martimort, D., Tirole, J. (1990): Adverse Selection and Renegotiation in Procurement. Review of Economic Studies **57**, 597–625

36. Mao, S., Wang, J., Pu, X. (1998): Advanced Mathematical Statistics. China Higher Education Press, Beijing, and Springer–Verlag, Berlin, Heidelberg (in Chinese)
37. McAfee, P., McMillan, J. (1991): Optimal Contracts for Teams. Int'l Econ. Rev. **32(3)**, 561–577
38. Meyer, M. (1994): The Dynamic of Learning with Team Production: Implications for Task Assignment. Q. J. E., 1157–1184
39. Meyer, M., Vickers, J. (1997): Performance Comparisons and Dynamic Incentive. J. P. E. **105**, 547–581
40. Mixon, F. (1997): Team Production in Economics: A Comment and Extension. Labour Economics **4**, 185–191
41. Murphy, K. (1986): Incentives, Learning, and Compensation: A Theoretical and Empirical Investigation of Managerial Labor Contracts. Rand J. Econ. **17**, 59–76
42. Nalbantian, H. R., Schotter, A. (1997): Productivity Under Group Incentives: An Experimental Study. AER **87(3)**, 314–341
43. Prat, A. (2001): Should a Team be Homogeneous? European Economic Review (to appear)
44. Predergast, C. (1999): The Provision of Incentives in Firms. Journal of Economic Literature **37**, March, 7–63
45. Prendergast, C., Topel, R. (1996): Favoritism in Organizations. J. P. E. **104**, 958–978
46. Ramakrishnan, R. T. S., Thakor, A. V. (1991): Cooperation Versus Competition in Agency. Journal of Law, Economics, and Organization **7**, 248–283
47. Rasmusen, E. (1987): Moral Hazard in Risk-averse Teams. Rand J. Econ. **18**, 428–435
48. Schmidt, K., Schnitzer, M. (1995): The Interaction of Implicit and Explicit Contracts. Economics Letters **48**, 193–199
49. Sherstyuk, K. (1998): Efficiency in Partnership Structure. Journal of Economic Behavior and Organization **36**, 331–346
50. Sjostrom, T. (1996): Implementation and Information in Teams. Economic Design **1**, 327–341
51. Swank, O. (2000): Policy Advice, Secrecy, and Reputational Concerns. European Journal of Political Economy **16(2)**, 257–271
52. Tirole, J. (1992): Collusion and the Theory of Organizations. In: Advances in Economic Theory, World Congress of the Economic Society
53. Trueman, B. (1994): Analyst for Forecasts and Herding Behavior. Review of Financial Studies **7(1)**, 97–124
54. Valsecchi, I. (1996): Policing Team Production Through Job Design. J. of Law, Eco., and Org. **12(2)**, 361–375
55. Veen, T. (1995): Optimal Contracts for Teams: A Note on the Results of McAfee and McMillan. Int'l Econ. Rev. **36**, 1051–1056

Endogenous Determination of Utility Functions: An Evolutionary Approach

Alexander A. Vasin

Moscow State University,
Faculty of Computational Mathematics and Cybernetics,
Vorobievi Gori, Moscow, 119899, Russia
E-mail: vasin@cs.msu.su

Summary. The assumptions on utility functions in economic models are questionable in many cases. The present paper studies whether it is possible to endogenize determination of payoff functions, and the mechanisms governing changes in the latter. The main conclusion is that "natural" payoff functions concern reproduction of individuals. We confirm this conclusion by known results on the Replicator Dynamics (RD). Another model shows that altruistic or cooperative behavior among relatives is evolutionarily stable if it maximizes the total fitness of the family. Finally, we discusses why the mentioned models in question do not apply to social populations, the mechanisms influencing individual behavior, and who or what changes individual utility functions.

Key words: population dynamics, evolutionary games, optimality

1 Introduction

The basic models of the game theory and economics include individual utility or payoff functions. These models describe an individual's behavior as aimed to maximize his or her payoff function. For instance, in a normal form game and other general models of the game theory each player, or participant of the conflict is characterized by his set of strategies and arbitrary exogenously given payoff function. Each player independently sets his strategy which influences not only his payoff, but also payoffs of other participants. The theory studies methods of the rational choice of strategies. A standard assumption is that each player knows payoff functions of all participants.

The case with incomplete information on the payoff functions of other players is also studied in the literature. Another direction considers methods of the utility function specification on the partial ordering of possible outcomes of the game (see [6]). Note that in any case, the payoff functions are exogenously given and do not change.

Economic literature typically considers more concrete payoff or utility functions. In the market models producers usually aim to maximize their profits, and consumer's utility depends on the vector of purchased goods [10]. If a model includes the labor market, then the utility function also includes the time of labor and decreases in this variable.

Let us note that actual behavior of some part of population in the market does not correspond to the mentioned assumptions. These people consider some kinds of job as interesting or important for the society and are ready to work for a smaller wage than they could earn at another place. For other individuals the most important incentive to work is not the income value but its range in the distribution of incomes in the society. Of course, applicability of traditional models depends on the field. Besides the mentioned examples, standard assumptions on utility functions are questionable in models of electoral behavior. They are in contradiction with the phenomenon of altruistic behavior and poorly agree with the ethics of several world religions.

The social practice of the 20th century brought a new challenge to the standard view on payoff and utility functions. It showed that these functions are variable and the most efficient way to reach the purpose of the participant of the conflict may be to change payoff functions of other participants. The typical examples of the impact on utility functions are aggressive advertising on TV and drug distribution. In particular, the change in the utility function of an individual after several doses of the drug are well studied.

Thus, the standard assumptions on utility functions in economic models are questionable in many cases. Moreover, payoff or utility functions may change during the interaction between the agents.

The present paper studies whether it is possible to endogenize determination of payoff functions and what are the mechanisms of their changing. Section 2 considers evolutionary games that describe behavior dynamics in self-reproducing populations. The main conclusion of this investigation is that "natural" payoff functions concern reproduction of individuals irrespective of the kind of a population and interaction. More precisely, an evolutionary stable distribution over strategies that includes only those strategies which maximize the fitness of individuals. We confirm this conclusion by the known results on the Replicator Dynamics (RD). However, the difficulty is that the endogenous utility function in general depends on the particular evolutionary mechanism. In order to solve this problem we construct and study the model of natural selection of evolutionary mechanisms. The model shows that if replication is among competing evolutionary mechanisms then the evolutionary stable behavior maximizes the individual fitness. Only those mechanisms that are compatible with fitness maximization do survive under the competition.

Another result relates to evolution of altruistic and cooperative behavior. We consider a model where an individual distinguishes his relatives (sibs, cousins, etc.) and can alternate his behavior depending on this character. The model shows that altruistic or cooperative behavior among relatives is evolutionary stable if it maximizes the total fitness of the family. We also discuss factors that may limit the spread of such behavior.

Demographic data [5] shows that behavior in modern social populations maximizes neither individual nor family or population fitness. Section 3 discusses why the mentioned models do not work for social populations, what mechanisms provide a possibility to influence individual behavior and who or what changes individual utility function.

2 Evolutionary Models of Behavior in Self-Reproducing Populations

The evolutionary game theory (EGT) was primarily developed with respect to biological populations [9]. However, its models and approaches are also of interest for social modeling. This section considers several models of behavior dynamics in a population of interacting individuals. The result of interaction in every time period determines birth and death rates for individuals depending on their strategies. Let S denote the set of possible strategies, $\pi = (\pi_s, s \in S)$ is a distribution of individuals over strategy, N_s and π_s are respectively the number and the share of individuals with strategy s, $N = \sum_s N_s$ is the total size of the population. For any strategy s fertility $fer_s(\pi, N)$ and viability $v_s(\pi, N)$ of individuals playing this strategy depend on the distribution π and the size N in a given period. The basic model of the EGT is the replicator dynamics. It assumes that every individual does not change his strategy during his life and new individuals inherit strategies from their parents. The evolutionary dynamics of such population is described by equations

$$N_s(t+1) = N_s(t) f_s(\pi(t), N(t)) \,, \tag{1}$$

where $f_s(\pi, N) = fer_s(\pi) + v_s(\pi)$ is the fitness function of a strategy s.

Note that if the function has a form $f_s(\pi, N) = a(\pi, N)\overline{f}(\pi)$ then system (1) implies the autonomous model of behavior dynamics

$$\pi_s(t+1) = \pi_s(t) f_s(\pi(t)) \Big/ \sum_r \pi_r(t) f_r(\pi(t)) \,, \quad s \in S \,. \tag{2}$$

The main content of the following results is that the fitness turns out to be an endogenous payoff function of individuals in the given model. In particular, if a strategy provides the greater fitness than another one under any distribution π then the share of the worse strategy in the population tends to 0 as the time tends to infinity. Moreover, if the strategy distribution $\pi(t)$ tends to some steady point π^* then only those strategies s which maximize the fitness under π^* are included in this distribution with positive shares $\pi_s^* > 0$.

In order to present formal results, we need to recall or introduce several concepts of evolutionary game theory.

2.1 Population Game. Nash Equilibrium. Dominance Set of Strategies

The model of single interaction of individuals in a population at an instant or in a period of time is a population game (an analog of a game in the normal form in the classical game theory). Formally, a population game G is given by a collection

$$G = \langle S,\, f_s(\pi, w),\, s \in S,\, \pi \in \Pi,\, w \in \Omega \rangle \,, \tag{3}$$

where S is a set of strategies of the participants of this game; $\pi = (\pi_s)_{s \in S}$ is a distribution of the players in strategies; $\Pi = \{\pi \mid \pi_s \geq 0, \sum_{s \in S} \pi_s = 1\}$ is a standard simplex; and $f_s(\pi, \omega)$ is a payoff function for players that use strategy s under the distribution π over strategies and other parameters of the model ω (e. g., a total population size and environmental conditions). For biological populations, the gain is usually the fitness or a function that is consistent with the fitness. For social populations,the payoff function usually corresponds to consumer utility, the income, or the profit. Consider main static optimality principles employed for the investigation of the population interactions.

Definition 1. *A Nash equilibrium for the population game of form (3) is a distribution π such that*

$$\forall \omega \in \Omega,\ \forall s \in S:\ \pi^*_s > 0 \Longrightarrow s \in \arg\max_{s \in S} f_i(\pi^*, w)\,. \tag{4}$$

Assume that the payoff function for the game G has the following additive form:

$$f_s(\pi, \omega) = a(\pi, \omega)\overline{f}_s(\pi) + b(\pi, \omega)\,, \quad \text{where } a(\pi, \omega) > 0\,. \tag{5}$$

In other words, the term that depends on the strategy chosen by a player is independent of the parameter of the model ω. Then, condition (4) is equivalent to condition (6), which is free of the parameter ω:

$$\forall s \in S:\ \pi^*_s > 0 \Rightarrow s \in \arg\max_{s \in S} \overline{f}_i(\pi^*)\,. \tag{6}$$

If a game describes interaction in a biological population and the payoff function characterizes the fitness of the employed strategies, then the concept of Nash equilibrium completely corresponds to the Darwin principle of natural selection: only the strategies that fit best a given distribution are present in π with positive probabilities (i. e., survive).

The concept of Nash equilibrium was introduced in [11] and is the most known optimality criterion used in the behavior modeling. However, among Nash equilibria, there exist unstable states that actually are never realized. For this reason, we also describe other optimality criteria.

Definition 2. *Distribution π^* is a strict equilibrium for a population game G if*

$$\begin{gathered} \exists \varepsilon > 0,\ \exists s \in S \ \textit{ such that} \\ \pi^*_s = 1 \ \textit{and} \ \forall i \neq s,\ \omega \in \Omega:\ f_s(\pi^*, \omega) \geq f_i(\pi^*, \omega) + \varepsilon\,. \end{gathered} \tag{7}$$

For the payoff functions $f(\pi, \omega)$ of the general form, Nash equilibria do not necessary exist. Another optimality principle also associated with the Darwin's concept of natural selection is the *dominance.*

Definition 3. *Strategy s dominates strategy i* ($s \succeq i$) *on the set of distributions* $\Pi' \subseteq \Pi$ *if*

$$\exists \varepsilon \geq 0 : \forall \omega \in \Omega, \; \forall \pi \in \Pi' \;\; f_s(\pi, \omega) \geq f_i(\pi, \omega) + \varepsilon \,, \tag{8}$$

i. e., for any distribution from the set Π', *the strategy s provides a greater gain than the strategy i does.*

Definition 4. *The set* $S' \subseteq S$ *is called a dominating set if it can be obtained by iterative elimination of dominated strategies; i. e.,*

$$\begin{gathered} S' = S_T \subset S_{T-1} \subset \ldots \subset S_1 = S \,, \\ \text{where } \forall k \in \{1, \ldots, T-1\}, \; \forall i \in S_k \backslash S_{k+1} \; \exists s \in S_{k+1} : \; j \succeq i \text{ on } \Pi_k \,, \end{gathered} \tag{9}$$

where $\Pi_k = \{\pi \in \Pi \mid \pi_s = 0, \forall s \notin S_k\}$.

The described procedure for iterative elimination of dominated strategies can be considered as a quasi-dynamic model of the behavior microevolution in a population. Indeed, this procedure describes a sequential reduction of the set of strategies used by players: at each stage, more efficient (better fitted) strategies are substituted for less efficient ones.

If, in the definitions, we require the condition $\varepsilon > 0$, then *s strictly dominates i* ($s \succ i$) and S is called a *strictly dominating set.*

The concepts of dominating by mixed strategies and a set dominating in mixed strategies are introduced in a similar way. The gain of the mixed strategy π is determined as the expectation $f_\pi(\pi', \omega) = \sum_{s \in S} \pi_s f_s(\pi', \omega)$. (In more detail, the domination is described in [3].)

2.2 Stability of Solutions

Special concepts of stability and asymptotic stability of a distribution over strategies consistent with the concept of Lyapunov stability are necessary for dynamic models of population behavior. Consider these definitions in terms of [21] for a dynamic model of the general form given by the system

$$x(t+1) = \Phi(x(t)) \,, \quad \pi(t) = h(x(t)) \,. \tag{10}$$

Let $\pi(t, x^0)$ denote the distribution in the population over strategies at time t if the initial state of the system is x^0, and $\pi^0 = \pi(0, x^0)$ denote the corresponding distribution at the initial period.

Distribution π^0 is called *stationary* for dynamic model (10) if

$$\exists x^0 : \; \pi(t, \pi^0) \equiv \pi^0 \,, \;\; \forall t \geq 0 \,, \tag{11}$$

i. e., there exists an initial state of the system such that the distribution over strategies within the population remains invariant.

Distribution π^0 is called *stable* for dynamic model (10) if it is stationary and, in addition,

$$\forall \varepsilon > 0 \, \exists \delta > 0 : \ \forall x \in O_\delta(x^0) \ \pi(t, x) \in O_\varepsilon(\pi^0) , \ \ \forall t \geq 0 , \tag{12}$$

i. e., for small deviations from the initial state, the distribution does not "go far" from the initial one.

Distribution π^0 is called *asymptotically stable* for dynamic model (10) if it is stable and, in addition,

$$\exists \delta^0 > 0 : \ \lim_{t \to \infty} \pi(t, x) = \pi^0 , \ \ \forall x \in O_{\delta^0}(x^0) , \tag{13}$$

i. e., for sufficiently small deviations from the initial state, the distribution over strategies always tends to π^0.

For continuous-time models, the corresponding concepts are introduced similarly.

Note that, for *autonomous models*, in which $x = \pi$, these concepts of stability and asymptotic stability are equivalent to the ordinary definitions of the Lyapunov stability and asymptotic stability (see [14]).

Theorems on relation of static optimality principles with solution stability of RD are obtained in different variants in [12, 17, 18, 20]. We give the formulations of these theorems following to [20, 21].

Theorem 1 (On relation of Nash distributions with stable points of RD). *Assume that the fitness function f_s is representable in additive form (5). Then, the following propositions are valid:*

(1) any stable distribution π of system (1) is a Nash equilibrium in the population game $G = \langle J, f_s(\pi, N), s \in S \rangle$;

(2) if for a certain trajectory $\overline{N}(t)$, the initial distribution $\overline{N}(0) > 0$ and $\exists \lim\limits_{t \to \infty} \pi(\overline{N}(0), x) = \pi^*$, then π^* is a Nash equilibrium of the specified population game.

Theorem 2 (On asymptotic stability of a strict equilibrium). *Assume that π is a strict equilibrium of the population game $G = \langle S, f_s(\pi, N), s \in S \rangle$. Then, π is an asymptotically stable distribution of system (1).*

Theorem 3 (On relation of dominating sets of strategies with the behavior dynamics). *Assume that $\overline{S}$ is a strictly dominating set of strategies in the game $G' = \langle S, \ln f_s(\pi, N), s \in S \rangle$. Then, for any $s \notin \overline{S}$ and any $\overline{N}(0) > 0$, $\lim_{t \to \infty} \pi_j(\overline{N}(0), t) = 0$ on the corresponding trajectory of system (1).*

Note that, in the case of the dominance by pure strategies, Theorem 3 is also valid for the standard population game $G = \langle S, f_s(\pi, N), s \in S \rangle$. However, in the case of dominance by mixed strategies (i. e., by distributions), it is necessary to consider the logarithm of the fitness as the payoff function, because there exist examples, where a strategy strictly dominated by a distribution in the game $G = \langle S, f_s(\pi, N), s \in S \rangle$ is not eliminated on the trajectories of the discrete-time RD.

In many studies, a continuous variant of RD is considered (see [18, 7], etc.). The population size in this model varies continuously. Function $f_s(\pi, N)$ determines the specific variation-rate of the size of a group of individuals employing the strategy $s \in S$, whereas other assumptions are similar to the discrete variant. As a result, the dynamic equations take the form $\dot{N}_s = N_s f_s(\pi, N)$ and, for functions in the additive form (5), by changing the time $d\tau = a(\pi, N)\, dt$, one obtains the autonomous system

$$\dot{\pi}_s = \pi_s(\overline{f}_s(\pi) - \sum_{i \in S} \pi_i \overline{f}_i(\pi)) \,. \tag{14}$$

Analogs of the above-presented theorems are also proved for these models (see [4, 24, 18, 15]).

Of course, these results strongly depend on the evolutionary mechanism of replicators. As an alternative, consider a model of random imitation where a new individual follows a strategy of a randomly chosen adult. Then the population dynamics meets equations

$$N_s(t+1) = N_s(t)v_s(t) + \sum_r N_r(t) fer_r(t) \frac{N_s(t)v_s(t)}{\sum_r N_r(t)v_r(t)}, \quad s \in S \,. \tag{15}$$

Note that the right-hand side is of the form $N_s v_s(t) V(t)$ where $V(t)$ does not depend on strategy s. Under general assumptions, such dynamics is coordinated with the function $v_s(t)$ in the sense of theorems 1–3. Thus, the viability turns out to be an endogenous payoff function of individuals in the corresponding dynamical process instead of fitness.

2.3 A Model of Evolutionary Mechanisms' Selection

Proceeding from the previous example it seems that we changed arbitrariness in the choice of payoff functions for arbitrariness in the choice of evolutionary mechanisms. However, the actual evolutionary mechanisms are under the natural selection. Only the most efficient mechanisms survive in the process of competition.

Consider the corresponding model of a society including several populations that differ only in their evolutionary mechanisms. Individuals of all populations interact and do not distinguish population characters in this process. Thus, the evolutionary mechanism of an individual is an unobservable character. Fertility and viability functions $fer_s(\pi, N), v_s(\pi, N)$ describe the outcome of the interaction and depend on the total distribution over strategies and the size of the society. The set of strategies S and the functions are the same for all populations. Let L denote the set of populations, N_l denote the size of population l, $\pi^l = \{\pi^l_s, s \in S\}$ be a distribution over strategies in population l. Then the total distribution over strategies is $\pi = \sum_l \frac{N^l}{N} \pi^l$.

Assume that the dynamics of distribution π^l is determined by the operator Φ^l corresponding to the evolutionary mechanism of population l. Then dynamics of the society meets equations

$$N^l(t+1) = N^l(t)\sum_s \pi_s^l(t) f_s(\pi(t), N(t)), \tag{16}$$

$$\pi^l(t+1) = \Phi^l(\pi^k(t), N^k(t), k \in L)\ , \tag{17}$$

where $f_s(\pi, N)$ is a fitness function of strategy s.

Theorem 4. *Let there exists a population of replicators in the society. Then the total distribution over strategies meets the following analogs of theorems 1 and 2:*

1) any stable distribution π of system (15) is a Nash equilibrium of the population game $G = \langle S, f_s(\pi, N), s \in S\rangle$;

2) if initial distribution $\pi_0 > 0$ and $\exists \pi = \lim_{t\to\infty} \pi(\overline{N}_0, t)$ then π is a Nash equilibrium of the population game G;

3) if π is a strict equilibrium of the game G then π is asymptotically stable distribution for system (15).

Proof. Let us show that tendency of $\pi(t)$ to $\overline{\pi}$ implies that π is a Nash equilibrium. Suppose from the contrary that $\overline{\pi}_s > 0$ for non-optimal strategy $s \notin \arg\max_{i\in J} \overline{f}_i(\overline{\pi})$. Consider a neighborhood of $\overline{\pi}$. Due to continuity of f_i, the gain of strategy s in a small enough vicinity $O_\delta(\overline{\pi})$ is still strictly less than for any optimal strategy $r \in \arg\max_i \overline{f}_i(\overline{\pi})$. Moreover, $\overline{f}_r(\pi) > \overline{f}_d(\pi)$ under $\pi \in O_\delta(\overline{\pi})$, if $d_s > \overline{\pi}_s - \delta$, in particular, for any $d \in O_\delta(\overline{\pi})$. In a population of replicators consider those individuals who apply strategy r which is the optimal response to $\overline{\pi}$. The growth rate of this group is higher than the average growth rate for the total society and a share of this group in the total society will increase until distribution π leaves vicinity $O_\delta(\overline{\pi})$. We obtain a contradiction with assumption that $\overline{\pi}$ is a limit point for the positive initial distribution over populations and strategies. The rest propositions are proved in a similar way. □

Generalization of theorem 3 on elimination of dominated strategies is possible under more strict assumptions on the variety of evolutionary mechanisms. For any evolutionary mechanism Φ and a pair of strategies s, r, let us call an s, r-substitute of mechanism Φ a mechanism $\Phi_{s,r}$ such that for strategies other than s, r the shares of individuals who apply these strategies change as under mechanism Φ and instead of strategy s they always play r. According to [21], if for any s, r, l the set of mechanisms includes all possible substitutes $\Phi_{s,r}$, then any strictly dominated strategies $\pi_s(t) \to 0$ as $t \to \infty$.

Thus, the evolutionary mechanisms selection model confirms fitness to be an endogenous utility function for self-reproducing populations.

Does this result correspond to the actual behavior? First, consider behavior in biological populations. A general opinion of biologists is that, on the whole, the principle of individual fitness maximization does not contradict to the actual behavior (see [8]). However, there exist the known examples of behavior that do not correspond to this principle, that are altruistic and cooperative behavior. The concepts of altruistic and cooperative behavior may be illustrated by different variants of the known "Prisoner's dilemma" (see [12]). In this symmetric two-player game each player has two possible strategies: to cooperate (C) or to be selfish (S). As an example we propose the following payoff matrix:

$$\begin{array}{ccc} & C & S \\ C & (5,5) & (1,6) \\ S & (6,1) & (2,2) \end{array}$$

.

In general case we assume that under any behavior of the other player the selfish strategy is more profitable ($u_{SS} > u_{CS}, u_{SC} > u_{CC}$), at the same time the total gain is maximal when both players cooperate ($u_{CC} > (u_{CS} + u_{SC})/2, u_{SS} > u_{CC}$). In this game there exists a unique Nash equilibrium, which corresponds to selfish behavior and is also a dominance solution. However, it is known that actually in such situations players often cooperate.

Altruistic behavior deviates even more from individual fitness maximization. Consider the following payoff matrix (A denotes altruistic behavior, S denotes selfish behavior):

$$\begin{array}{ccc} & A & S \\ A & (5,5) & (1,10) \\ S & (10,1) & (2,2) \end{array}$$

.

The altruistic behavior of one player together with selfish behavior of the other corresponds to the total fitness maximization. Meanwhile, the altruist obtains less than his guaranteed payoff which he could get under Nash equilibrium.

As an example of cooperative behavior in biological populations, let us note behavior of animals which in turn stand on guard or predators which participate in joint hunting. As to altruism, it may be observed in interaction between relatives. The altruistic behavior of parents towards children is rather widespread and does not contradict to the concept of fitness maximization, since fitness is equal to the sum of fertility and viability. The interesting examples of altruism are those which do not relate to individual fitness maximization. For instance, the behavior of social insects. The explanation is that the individuals in families of social insects are close relatives. The shortage of the model of direct inheritance is that it takes into account only relation type "parent-child" and does not consider "brother-sister" relations, etc. Taking this relations in account, it is possible to explain the spread of the altruistic behavior in the sense of fitness maximization of the group of relatives.

Let us describe the corresponding model. As above, assume that interaction in a population is characterized by a set of strategies S and fitness functions $f_s(\pi)$, $s \in S$. But in contrast to the previous models, individuals can distinguish sibs, that is brothers and sisters, from other members of population and choose a strategy depending on this characteristic. Thus, the strategy (s, s') includes a component s which is applied to sibs and s' for other individuals. At a time period an individual interacts with sibs with some frequency $\lambda_r \in (0,1)$, and with other members of populations with frequency $1-\lambda_r$. The total fitness additively depends on the results of interactions with relatives and the rest part of the population:

$$\overline{f}_{(s,s')}(\pi') = \lambda_r f_s^r(s) + (1-\lambda_r) f_{s'}(\pi'),$$

where $f_s^r(s)$ determines the result of the interaction with relatives, π' is a distribution over component s'. It is assumed that all sibs apply the same strategies. Thus, the interaction is characterized by a population game

$$\overline{G} = \langle \overline{S} = \{(s,s') \in S \times S\}, \overline{f}_{(s,s')}(\overline{\pi}) = \lambda_r f_s^r(s) + (1-\lambda_r) f_{s'}(\pi')\rangle,$$

where $\overline{\pi}$ is a distribution over full strategies.

Theorem 5. *Any strategy (s, s') where $s \notin \arg\max_i f_i^r(i)$ is strictly dominated by strategy (s^*, s') where $s^* \in \arg\max_i f_i^r(i)$. The distribution $\overline{\pi}$ is a Nash equilibrium iff for all specified non-optimal strategies $\pi_{ss'} = 0$, and the correspondent distribution π' is a Nash equilibrium of the game $\langle S, f_{s'}(\pi')\rangle$.*

Thus, in relations similar to "Prisoner's dilemma", sibs play the cooperative strategy with respect to each other. In order to explain altruistic behavior, consider the following modification of the model. Assume that in the interaction sibs may be in two different roles (for instance, a dominating individual α and a suppressed one β), let a strategy and a fitness function depend on the role. Then the full strategy with respect to relatives is determined by pair $s = (s_\alpha, s_\beta) \in S \times S$. Only strategies $s^* = (s_\alpha^*, s_\beta^*) \to \max_{(s_\alpha, s_\beta)} (f^\alpha(s) + f^\beta(s))$ which provide the maximum of the total fitness survive elimination of strictly dominated strategies. Proceeding from theorems 4, 5, we may conclude that evolution in self-reproducing populations leads to behavior which maximizes the total fitness of sibs. Though the last model contains an implicit restriction on the evolutionary mechanism that determines the distribution over strategies with respect to relatives, this restriction is not important: any other mechanism would lose in competition with the given which realizes the optimal strategy s^* specified in theorem 5.

Note that these results may be generalized for relations between cousins, etc. Let a strategy include variants of behavior $s_1, \ldots, s_k$ with respect to relatives of different degrees $1, \ldots, k$ and s' with respect to other individuals. The fitness function is of the form

$$f_s(\pi) = \sum_i \lambda_i f_{s_i}^i(s_i) + (1 - \sum_i \lambda_i) f_{s'}(\pi'), \tag{18}$$

where $\lambda_i \in (0,1)$ characterizes the frequency of interaction with relatives of degree i. As above, it is assumed that all relatives apply the same strategy with respect to each other, that corresponds to the model of direct inheritance and does not restrict generality in context of the proposed model of evolutionary mechanisms selection. Then elimination of strictly dominated strategies leads to optimization of relations between relatives: there remain such strategies s^* that $s_i^* \in \arg\max_{s_i} f_{s_i}^i(s_i)$, $i = 1, \ldots, k$. In biological populations as well as in social populations, individuals usually have common ancestors in the 4-5-6 generation. According to the mentioned results, we should expect a widespread of cooperative and altruistic behavior that maximizes the total fitness. However, actual behavior often does not correspond to these expectations. The examples of severe competition between close relatives (up to eating the offsprings of each other) are well known. One of the reasons of the limited prevalence of cooperative behavior is that it is not invasion-proof against selfish mutants.

Imagine a population with cooperative behavior where all individuals play strategy s^* that maximizes the total fitness. Consider the following mutation: there appears an individual that maximizes his individual fitness by playing $s' \to \max_s f_s(s^*)$. In the first generation such individual will get the reproductive advantage since all other members cooperate wish him and he is selfish. In the second generation the mutant's offsprings would be in less favorable position as they interact with each other. In the general case, the outcome depends on the relation between the coefficients λ_i. The share of the mutants will increase in the population until their relatives up to the sufficiently large degree turn out to be the mutants. Thus, the optimal strategy is not invasion-proof.

Note that the problem of cooperative and altruistic behavior prevalence is studied in many papers (see [2] and the bibliography of this book). For instance, the theory of repeated games explains the prevalence of cooperation in repeated conflict situations proceeding from individual fitness optimization (see [19, 22]). The repetition provides a possibility to punish those individuals that evade cooperative behavior. Taking into account the punishment, cooperative behavior turns out to be individually profitable. It is not true in the considered models. Nevertheless, cooperative behavior spreads due to the selection of evolutionary mechanisms.

3 Peculiarities of Behavior Evolution in Social Populations

Actual behavior in modern social populations maximizes neither individual, nor group fitness. Russia, where the reproduction conditions are unfavorable during the last decade, is a special case. Let us take social-welfare states such as Sweden or Germany. Favorable conditions of survival and normal biological development are guaranteed there to any new-born citizen independent on his/her social origin. However, the statistics witnesses that most of population ignore the advantages in reproduction opportunities. For instance, the questionnaire of Stockholm inhabitants showed that 70% of adult population have no children and do not plan to have them. The fertility rate in Sweden and Germany is almost as low as in Russia: respectively 10 and 9 per 1000. According to the demographic forecast by the UN, the population size in Germany will decrease from 82 millions in 1999 to 80 millions in 2025 (see [5]).

By what reasons the proposed models do not work for social populations? Let us note three factors that explain this discrepancy.

1) It is impossible to realize the inheritance of parent strategies by their children in social populations. The behavior strategies are so complicated that individual would spend most of his life trying to teach his descendants. The differentiation of labor, in particular, the institute of teaching, appeared at the early stage of the mankind development. This institute plays a crucial role in formation of behavior of new generations [11]. Later the government, the church, etc. realized the importance of this institute and effectively used it to form a desirable behavior.

2) The problem of evaluation of different behavior strategies is rather complicated even for theoretical analysis. In practice, decision-making persons are usually unable to carry out such an evaluation in the reasonable time. Biological evolution formed different mechanisms that facilitate fast decisions close to the optimal in the sense of fitness.

One of such mechanisms is the feeling of pleasure or satisfaction. In nature actions that are pleasant or aim to reach pleasure are usually rational in the sense of individual reproduction. In particular, provision and consumption of food and some other resources are necessary for reproduction. However, the absolutization of this objective function may impede reproduction. The known example of such sort for ecological system is the interaction between a beetle lamehuza and ants who feed on the secret produced by the beetle. Sometimes the secret has a drug effect on ants: they through the queen out of the ant-hill, put lamehuza at its place and take care of it in order to get more secret. In some time this ant family perishes ...

3) Another auxiliary widespread mechanism of strategy determination is imitation. It promotes learning and permits to coordinate actions of individuals in groups and thus increase their fitness. At the same time it creates an additional possibility for behavior manipulation by choosing the appropriate leader.

The mentioned methods of behavior manipulation (control over learning process, usage of pleasure incentives and the imitation mechanism) were practiced since ancient times. But the situation has dramatically changed with the development of mass media in the 20th century. Currently, television provides a possibility to influence billions of people playing the role of teacher and creating models for imitation.

Now consider who or what changes the objective functions in order to form certain behavior in social populations.

In the above mentioned examples for ecosystems, we found two different variants. In the case of lamehuza, the manipulator is an individual of another population. The interaction between the populations is of "predator-prey" type (see [23]).

In the case of social insects individual behavior is formed by a self-reproducing superindividual – the family of ants, bees or other social insects. While suppressing individual reproduction of some part of some part of the population the evolutionary mechanism provides efficient reproduction of superindividuals and the population as a whole.

In social populations we meet similar and more complicated variants of behavior regulation. Besides superindividuals of biological nature, there exists self-reproducing superindividuals of social-economic nature, for instance, corporations, scientific schools, public and religious organizations. Instead of replication, superindividuals can grow involving new human and material resources or collapse.

Thus, superindividuals use the basic population as a resource for their own reproduction and growth, and influence population behavior for this purpose. The relevant theory of social and economic behavior should reflect the processes of interaction and selection of self-reproducing individuals at different levels.

References

1. Axelrod, R. (1984): The Evolution of Cooperation. Basic Books, New York
2. Bomze, I. M. (1986): Non-Cooperative Two Person Games in Biology: a Classification. International Journal of Game Theory **15**, 31–57
3. Fudenberg, D., Tirole, J. (1991): Game Theory. MIT Press, Cambridge
4. Germeyer, U. B. (1976): Games with Non-Antagonistic Interests. Nauka, Moscow (in Russian)
5. Hofbauer, J., Sigmund, K. (1988): Dynamical Systems and the Theory of Evolution. Cambridge University Press, Cambridge
6. Maynard Smith, J. (1982): Evolution and the Theory of Games. Cambridge University Press, Cambridge
7. Myles, G. (1995): Public Economics. Cambridge University Press, Cambridge
8. Moiseev, I. (1999): To Be or Not to Be for the Mankind. Nauka, Moscow (in Russian)
9. Nachbar, J. H. (1990): Evolutionary Selection Dynamics in Games: Convergence and Limit Properties. International Journal of Game Theory **19**, 59–89
10. Naselenie i obschestvo (1998): **20** (in Russian)
11. Nash, J. (1951): Non-Cooperative Games. Annals of Mathematics **54**
12. Owen, G. (1974): Game Theory. Mir, Moscow (in Russian)
13. Pontryagin, L. S. (1980): Differential Equations. Nauka, Moscow (in Russian)
14. Samuelson, L., Zhang, J. (1992): Evolutionary Stability in Asymmetric Games. Journal of Economic Theory **57**, 363–391

15. Schuster, P., Sigmund, K. (1983): Replicator Dynamics. Journal of Theoretical Biology **100**
16. Schuster, P., Sigmund, K., Hofbauer, J., Wolf, R. (1981): Self-Regulation of Behavior in Animal Societies. Games between Two Populations without Self-Interaction. Biology and Cybernetics **40**, 9–15
17. Taylor, P., Jonker, L. (1978): Evolutionary Stable Strategies and Game Dynamics. Mathematical Biosciences **40**
18. Van Damme, E. (1987): Stability and Perfection of Nash Equilibrium. Springer, Berlin
19. Vasin, A. (1989): Models of Collective Behavior Dynamics. Moscow University Press, Moscow
20. Vasin, A. (1995): On Some Problems of the Theory of Collective Behavior. Obozrenie prikladnoy i promishlennoy matematiki **2** (in Russian)
21. Vasin, A. (1997): The Folk Theorem in Contex of Evolution of Cooperation. In: IIASA Workshop "Growth and Adaptation in Evolutionary Systems". Laxemburg, Austria
22. Volterra, V. (1931): Lecons sur la Theorie Mathematique de la Lutte pour la Vie, Gauthier-Villars Editeur, Paris
23. Weibull, J. (1996): Evolutionary Game Theory. MIT Press, Cambridge
24. III Congress of the ESEB (1991): "Abstracts". Debrecen, Hungary

N-person Prisoner's Dilemma with Mutual Choice by Agent-based Modeling

Tomohisa Yamashita[1] and Azuma Ohuchi[2]

[1] Graduate School of Engineering, Hokkaido University,
Kita 13 Nishi 8, Kita-ku, Sapporo, Hokkaido, 060-8628, Japan
E-mail: tomohisa@complex.eng.hokudai.ac.jp

[2] Graduate School of Engineering, Hokkaido University,
Kita 13 Nishi 8, Kita-ku, Sapporo, Hokkaido, 060-8628, Japan
E-mail: ohuchi@complex.eng.hokudai.ac.jp

Summary. Recent research in economics, sociology, psychology and computer science has attampted to model partner selection in dilemma situation. Also, social interactions are often characterized by game partner choice. How do partner choice mechanisms influence the behaviors of players in N-IPD? To answer these questions, we proposed a N-IPD game with unilateral choice, mutual choice and extended mutual choice. Agent-based modeling in which an agent considered as a player with bounded rationality is applied to describe the N-IPD with partner choice mechanisms. As a result of agent-based simulations with an evolutionary approach, we confirmed that extended mutual choice or mutual choice with the rule on split of a group accelerates and maintains the emergence of collective cooperation in the N-IPD.

1 Introduction

The question of how collective cooperation among altruistic individuals is emerged has been taken up by researchers in many fields, e. g., sociology, economics, social psychology and computer science and so on [13]. Recently, many researchers have used the Iterated Prisoner's Dilemma (IPD) to explore the potential emergence of mutually cooperative behavior among non-altruistic individuals. The emergence of cooperation in dilemma situation has been considered as the resolution of this problem [3, 7]. Although Tit-for-Tat strategy often emerges cooperation in a 2-person Iterated Prisoner's Dilemma (2-IPD), it does not effectively emerge cooperation in an N-person Iterated Prisoner's Dilemma (N-IPD) which represents the general dilemma situation [2, 20].

Most studies of the 2-PD and N-PD assume that individual players have no control over which opponents they play. Players are matched as game partners either randomly or by means of a deterministic mechanism such as round-robin or neighborhood. However, in the dilemma situations of real world, individuals are not always prisoners that have no alternative but to play their assigned games. Social interactions are also often characterized by choosing the game partner, that is, the rupture of interaction, e. g., the rupture of diplomatic relations in the international society, the suspension of business transactions in the corporate community and the breach of friendships in the human relation. If one individual keeps to choose a defect, the others may cease to play the game with him rather than they may keep to choose defect against his defect. It is important for each player to choose not only to cooperate or defect but also together with whom he plays the game.

In general dilemma models, each player has no choice about with whom he plays a game. In order to deal with such dilemma situations including the partner choice, it is necessary to consider that players have a choice about which opponent they play. New models of games are needed to effectively deal with the choosing its opponents under social dilemma situations. Therefore, we propose a game model which include not only the alternative of to cooperate or defect but also that of with whom he plays a game.

In next section, we refer multiple 2-IPD with the partner choice mechanism in previous research.

2 Prisoner's Dilemma with Partner Choice Mechanism

2.1 Classification of Partner Choice Mechanism

The partner matching mechanisms can be classified into two types. One is an exogenous matching mechanism such as round robin, random matching and grid neighborhood. The other is an endogenous matching mechanism, i. e., a partner choice mechanism. Furthermore, the existing researches on a partner choice mechanism can be classified into four types:

Exit Option and Opt Out.

Exit option and opt out of games are the choice between playing and not playing the PD game after matched [5, 9, 10, 16, 19].

Schluessler [16] examines global interaction in which finite automata are randomly matched to play the prisoner's dilemma game but can exit the match after every round. The whole model is assumed to represent the conditions for egoistical cooperation in a large and anonymous society, which is based mainly on exchange relationships and voluntary cooperation between individuals. The strategies have no memory, effects of reputation do not exist, and defectors cannot be identified, traced and held responsible for their actions. If a player exits, both return to the pool of unmatched players and are randomly matched again. This scenario abandons the preconditions usually regarded as necessary for establishing cooperation. Simulations show the evolutionary success of a cooperative machine that exits immediately once it is defected against.

Ostracism.

Ostracism is the practice of excluding disapproved individuals from interaction with a social group [6, 12].

Hirshleifer and Rasmussen [12] find that group ostracism can permit cooperative agents to protect themselves form defectors. The most important point of the argument is in the fact that players can credibly threaten to exclude a defector in the last round since such a move does not entail any cost for them. The possibility to inflict a real punishment at no cost to the punisher has the effect of breaking the logic of the backward-induction argument underlying the non-cooperative outcome of the finitely repeated PD game. If it is moreover assumed, in violation of assumption that the voting procedure used by the group to exclude a violator does not entail any direct or indirect benefit, that players gain a little bit of pleasure from excluding a violator, that is if they have a little morality, the above strategy is the unique subgame-perfect equilibrium of the game. As pointed out by the authors, the strategy "Always Defect, Never Ostracize" is no longer an equilibrium because players would raise their payoffs by ostracizing defectors in the next-to-last round. In this case, Morality achieves cooperation because some players want to reduce the welfare of others, if those others behave wrongfully.

Mutual choice.

In mutual choice applied as partner choice, the PD game is played between two players only if both players mutually agree to player the game [1, 17, 18].

Stanley *et al.* [17] develop a choice and refusal mechanism for as small-scale society in which players leave the realm of anonymity. This article extends the traditional iterated prisoner's dilemma with round-robin partner matching by permitting players to choose and refuse partners in each iteration on the basis of continually updated expected payoffs. Players can make game offers to specific players and accept or reject offers they received. The decision is reached by comparing expected payoff if playing with an exogenously given fixed tolerance level. IPD strategies are represented by finite automata and evolve by genetic algorithms. Comparative computer experiments are reported that indicate the introduction of partner choice and refusal accelerates the emergence of mutual cooperation in the IPD relative to round-robin partner matching. As a result of the introduction of mutual choice, the emergence of cooperation is accelerated. Although full cooperation is the most frequently observed population structure, several other population structure, including societies were everyone is inactive, arise, one of which is completely immune to invasion by mutants.

Unilateral Choice.

In unilateral choice, the PD game has to be played if at least one partner chooses to do play the PD game [11].

Hauk and Nagel [11] find that unilateral choice is better sanction than mutual choice by not using finite automaton in computer simulation but test subjects in psychological experiment. Their unilateral choice setup allowed them to sturdy players' intentions in the PD game when they choose the outside option by examining their behavior in undesired matches. Unilateral choice leads to low defection levels and less singles than with mutual choice. If mutual choice agreement is necessary, cooperation might not arise because a potential cooperator has a lower chance to signal his cooperative intentions. If the potential partner does not want to play, the possibility of future cooperation is foregone. Furthermore, with the unilateral choice setup they are able to show that intending defectors are more likely to try to avoid a match than intending cooperators. They compare the results of multiple games with single game PD-experiments and find no difference in aggregate behavior. Studying the unilateral choice setup would correct for this shortcoming and allow a better understanding whether or not the false consensus idea is valid. If it were valid, moderate cooperators, i. e., players who cooperate sometimes but not always would have an evolutionary advantage even the one-shot game with exit.

2.2 Application of Mutual Choice for N-IPD

There are two types of game models to represent N-person group under dilemma situation. One is the game model based on the multiple 2-person Prisoner's Dilemma game [2, 3, 7], which represents the dilemma situation with plural pair-wise interactions. In this model, each individual plays the PD game one by one. Most of above-mentioned game models with partner matching mechanisms correspond with this type. The other is the game model based on the N-person Prisoner's Dilemma game [15]. In most of previous researches, all individuals participate in the same social dilemma game simultaneously. There are few researches treating N-PD with an endogenous partner matching mechanism [14]. There are notable differences between 2-IPD and N-IPD in some context. For example, in evolutionary context, although Tit-for-Tat strategy often emerges cooperation in a 2-IPD, it does not effectively emerge cooperation in an N-IPD which represents the general dilemma situation[20, 2]. Therefore, we have a question, "What behavior of players does an endogenous partner matching mechanism emerge in the N-IPD?"

In this paper we use mutual choice as the partner choice mechanism because mutual choice is the most appropriate mechanism from the point of view of representing the rupture of interaction among individuals. The exit option and the ostracism are more like the relations between the individual and the group. Does the partner choice mechanism accelerate the emergence of cooperation similarly to the case of 2-IPD? What rules of mutual choice are necessary to emerge cooperation? In order to answer these questions, we propose N-person Iterated Prisoner's Dilemma with mutual choice, where players choose a group as the opponents of social dilemma game and then play the

N-IPD game with members of the group. To analyze the N-IPD with mutual choice, we use agent-based modeling because this game situation is too complicated to solve by game theory. Through an agent-based simulation with an evolutionary approach, we examine how mutual choice influence the behaviors of the players.

3 N-IPD with Mutual Choice

In this section, the mutual choice mechanism for the N-person game is proposed because previous mutual choice schemes were designed as matching mechanisms when two players play the 2-IPD game if both agree to play. At first, we introduce the concept of "group formation" based on mutual choice. A group is defined as a subset of the overall player set. Each player can join only one group at the same time. Only members of its group as game partners play the N-IPD. The result of the N-IPD in a group is independent of the decision making of other group. The group formation is the process that player set is divided into subset (group), where players choose one group and join the group.

In group formation, if player A chooses one player or one group including some players as game partners, and if the player or the group chosen by player A also chooses player A, that is, if the condition of mutual choice for playing with N-IPD is satisfied, they form a new group. After group formation, players interact (play N-IPD) only with members of the group. According to group formation based on mutual choice, we propose the N-IPD with mutual choice.

At first, we propose the procedure of group formation based on unilateral choice. The reason is that unilateral choice is simple mechanism included in mutual choice, that is, unilateral choice is equal to mutual choice without the opportunity that the player or the group chosen by a player chooses it oppositely.

3.1 Group Formation Based on Unilateral Choice

In the partner choice using unilateral choice, the player applied for the 2-IPD has to play if at least one partner chooses to do play the 2-IPD game. Therefore, the alternatives in group formation based on unilateral choice are forming a new group or joining an existing group. Suppose the order of decision-making on unilateral choice is set as random, the procedure of group formation based on unilateral choice is as follows.

1. Let $N = \{1, .., i, .., n\}$ be the player set. The first player in the decision-making of group formation cannot join an existing group and has to form a new group. The players make decisions after the first player chooses one group k out of the group set $G = \{G_1, .., G_k, .., G_m\}$, where G_k is the set of the players that have already made a decision on group formation. Player

i chooses one group then player i makes a game offer to group k which is most desirable for player i. Player i can join group k surely because group formation is based on unilateral choice. Player i is added to the group k as $G_k \cup \{i\}$, which is the new group k.
If there is no desirable group for player i, player i does not make a game offer and player i forms a new group, $G_{m+1} = i$. New group $m+1$ including only player i is added to group set G, and then group set G is modified as $G = \{G_1, .., G_k, .., G_m, G_{m+1}\}$.
2. After the decision-making for group formation, the players in groups of more than two players play N-IPD with the players in the same group. Because the player whom group including only player itself after group formation, i. e., the player not joining any group, can not play N-IPD, such a player acquire the payoff for alone players, P_{alone}, instead of the payoff of N-PD.

3.2 Group Formation Based on Mutual Choice

In the partner choice using mutual choice, the 2-IPD game is played between both players making a game offer and the player receiving, only if both players mutually agree to player the game. Therefore, refusing or accepting a game offer are added as new alternative in group formation based on mutual choice to the alternatives with unilateral choice, forming a new group, or joining an existing group. Suppose the order of decision-making on unilateral choice is set as random, the procedure of group formation based on mutual is as follows.

1. The first player in the decision-making of group formation cannot join an existing group and has to form a new group. In the same as group formation based on unilateral choice, player i chooses one group then player i makes a game offer to group k which is most desirable for player i. If there is no desirable group for player i, player i does not make a game offer.
2. After the group choice of player i, the group k chosen by player i is given an opportunity to refuse or accept the game offer of player i. The players in group k decide whether to refuse or accept the game offer of player i. The consensus of the group about whether to refuse or accept is built by accumulating the decision-makings of individual players. If the consensus of the group agree to accept player i, group k accepts the game offer of player i and then player i joins group k. Player i is added to the group k as $G_k \cup \{i\}$, which is the new group k.
3. If group k refuses player i, player i makes a game offer to the second most desirable group l. Player i continues making game offers until a group accepts his/her game offer or until all desirable groups refuse. If player i is refused by all desirable groups, player i forms a new group, G_{m+1}. New group $m+1$ including only player i is added to group set G, and then group set G is modified as $G = \{G_1, .., G_k, .., G_m, G_{m+1}\}$.

4. After the decision-making for group formation, the players in groups of more than two players play N-IPD with the players in the same group.

The outline of the decision-making of player i on group formation based on mutual choice is shown in Fig. 1.

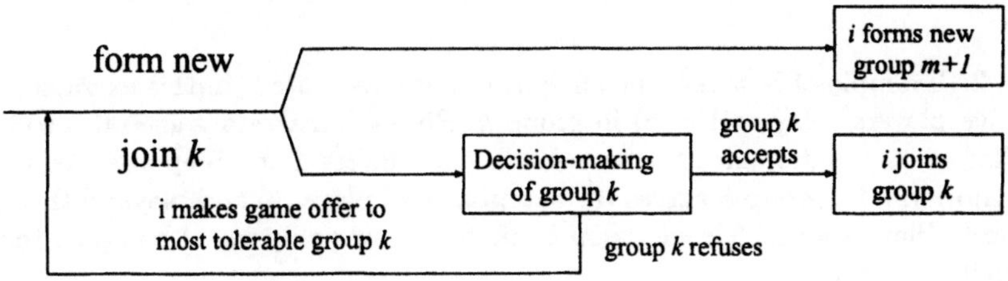

Fig. 1. Outline of decision-making of player i in group formation based on mutual choice

3.3 Group Formation Based on Extended Mutual Choice

Finally, we propose the extended mutual choice in order to reflect the decision of the players opposing the acceptance of player i. The extended mutual choice is the combination of original mutual choice with the rule on split of a group. which is the rule that the players opposing the acceptance of player i leave from the group that they belong. The rule on split of a group is applied after group k accepting the game offer of player i.

When player i makes a game offer to group k, the players in group k decide whether to accept or refuse player i. If the majority of players agree the acceptance of player i, the players opposing the acceptance of player i consider the players agreeing the acceptance of player i as players accepting a defector into the group. In group k, the players agreeing the acceptance of player i is defined as G_k^{agree} and the players opposing the acceptance of player i is defined as G_k^{oppose}, where $G_k = G_k^{agree} \cup G_k^{oppose}$ and $G_k^{agree} \cap G_k^{oppose} = \phi$. The players opposing the acceptance of player i leave group k and form the new group G_k^{oppose} the rule on split of a group. G_k^{oppose} is assigned to G_{m+1}, and G is modified as $G = \{G_1, .., G_k, .., G_m, G_{m+1}\}$. Then, player i joins group k ($G_k = G_k^{agree} \cup \{i\}$). If G_{m+1} includes only player j, player j makes a game offer to his/her tolerable groups based on the above-described procedure of unilateral choice. An example of how the rule on split of a group is used is shown in Fig. 2.

3.4 N-Person Prisoner's Dilemma

The N-PD is one of game representing the social dilemmas [8]. The social dilemmas is the situation, where players achieve Pareto-inferior outcome al-

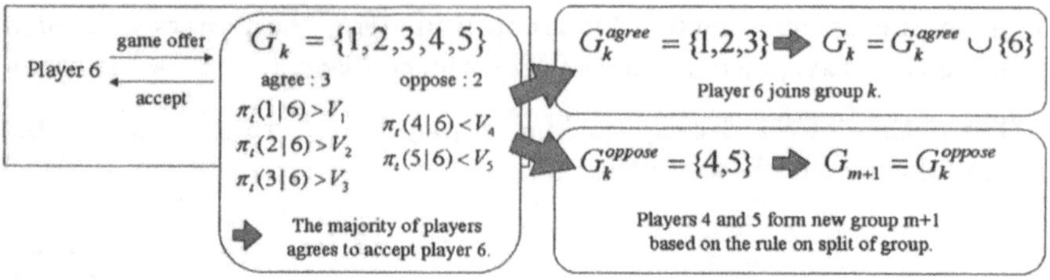

Fig. 2. Example of how the rule on split of a group is used. In this example, there are five players (players 1 $\sim$ 5) in group k. Player 6 makes a game offer to group k. Players 1, 2 and 3 agree the acceptance and players 4 and 5 oppose. As a result of majority vote, group k agrees the acceptance of player 6, and player 6 then joins group k. Players 4 and 5 leave group k and form new group $m+1$ based on the rule on split of a group

though players make decisions to maximize their own payoffs and a Pareto-superior alternative exists. In order to move closer to Pareto-superior outcome, it is important to find a way to avoid Pareto-inferior outcome i. e., to prevent individuals from becoming free-riders.

In the N-PD, each player has two possible moves, "cooperate (C)" or "defect (D)" like the PD. Suppose $D(m)$ be the payoff for a defector when m players choose C in group k, and suppose $C(m)$ be the payoff for a cooperator, the payoff function of the N-PD is used as follows.

$$\begin{aligned} C(m) &= m/|G_k| \,, \\ D(m) &= B + m/|G_k| \,, \end{aligned} \tag{1}$$

where $|G_k|$ is the number of players in group k and B is a positive constant. This payoff function is constructed based on the condition of social dilemma [8]. Furthermore, the payoff function of the N-PD satisfies the condition that the payoffs increase linearly as the ratio of cooperators, i. e., a linear relationship between and the ratio (or the number) of cooperators and the payoff of $C(m)$ and $D(m)$. Nash equilibrium of the N-PD is that all players choose D and that no players choose C because the move D dominates the move C based on the payoff function of the N-PD. If collective outcome is evaluated by the amount of payoffs of all players, collective outcome is the best when all players choose C Therefore, individual rationality that all players choose D leads to a deficient equilibrium less beneficial to all players. If all players choose C, although these choices are individual irrational, these choices results rational collective outcome. Therefore, Nash equilibrium of the N-PD is non-Pareto, sub-optimal, and irrational collective.

It would be interesting to adopt the "bandwagon effect [4]" based on group size to the payoff function of the N-PD game. For example of bandwagon effect, the payoff of a large group is more than that of a small group although the ratios of cooperating players is equal in a large and a small group . Because our purpose is to examine the influence of the partner choice mechanisms, the bandwagon effect is not adopted to simplify our model. Therefore, the payoff function of the N-PD game in each group depends on only the ratio of cooperating and defecting players.

4 Agent-Based Modeling

Game theory is suitable for treating a comparatively simple game situation. However, it is difficult to analyze a complex social behavior of heterogeneous individuals by game theory. In recent years, agent-based modeling has attracted attention. Agent-based modeling is one of methods for treating complicated game situations. The rapid growth of computing power and the increase of simulation research have created agent-based modeling paradigm and associated tools.

In the N-IPD with mutual choice, because group formation based on mutual choice and extended mutual choice is complicated for the analysis with using game theoretical rationality. Therefore, agent-based modeling is applied for describing the N-IPD with mutual choice. Each agent considered as a player makes a decision for N-IPD and group formation based on mutual choice.

4.1 Architecture of Agent

In this paper, player's strategy has two dimensions, boldness and vengefulness.The strategies of player i are described as Boldness B_i and Vengefulness V_i, where both B_i and V_i are real numbers from 0.0 to 1.0. B_i represents player i's degree of boldness to defect, i. e., the probability of defecting. Vengefulness V_i represents player i's degree of vengefulness to defection associated with the other players. The higher vengefulness of player i is, the more player i does not forgive the defection of the others.

In the previous researches of 2-IPD with evolutionary approach, a finite state machine and an automaton have been often used to describe the strategy of 2-IPD, that is, mechanism to decide whether to cooperate or defect in each round of iterated game. If collective cooperation is emerged with evolutionary automata, we can distinguish the emergence of cooperation by mutual choice from that by evolution of automata. For example, if the strategies of all players are Tit-for-Tat, cooperation is emerged whether with mutual choice or not. Therefore, the strategies are described by a real number.

4.2 Behavior of Agent

Player i makes a decision for group formation based on Vengefulness V_i and the *expected cooperation*, and for N-IPD based on Boldness B_i. The procedures of the decision-making of agent are about making, and refusing or accept a game offer, updating expected cooperation and using the rule on split of a group, choosing to cooperate or defect.

Making a Game Offer.

At the beginning of group formation, each player has two alternatives about group formation, forming a new group, or joining an existing group. Each player makes a game offer based on the history of N-IPD, i. e., the *expected cooperation* as follows.

At the t-th iteration of group formation, player i chooses one group based on the *expected cooperation* with each other player $j(\in N)$, and this is denoted by $\pi_t(i|j)$ [1, 18]. This expected cooperation is used to determine which group is most tolerable. Given any player i, group k is tolerable for player i in iteration t, only if

$$\frac{\sum^{|G_k|} \pi_t(i|j)}{|G_k|} \geq V_i \,. \tag{2}$$

The groups satisfying condition (2) are define as "tolerable groups" for player i. If any groups are tolerable for player i, then player i makes a game offer to group k, whose average expected cooperation for player i is the highest.

Refusing and Accepting a Game Offer.

After player i's choosing group k, the group k chosen by player i is given an opportunity to refuse or accept the game offer of player i. The group makes a decision based on accumulation of the decision-makings of the group members as follows.

After receiving the game offer of player i, a majority vote is used to build the consensus of group k about whether to refuse or accept the game offer of player i. The player j in group k agrees to accept player i only if $\pi_t(j|i) \geq V_j$. Otherwise, player j opposes. If the majority of players agree to accept player i, group k accepts the game offer of player i and then player i joins group k. Player i is added to the group k as $G_k \cup \{i\}$, which is the new group k.

Updating Expected Cooperation.

In the initial iteration of group formation, prior to any interaction, all players have the same initial expected cooperation value π_0 for each player. Expected cooperations are updated whenever each N-PD is played. Considering any player in group k, if player j is not in the group k that includes player i in the current iteration t, the expected cooperation value of $\pi_t(i|j)$ is not changed. On the other hand, if player j is in the group k that includes player i, they both play N-IPD in group k. In the N-IPD of group k, player i can observe the other player j's decision and denote it as $S(i|j)$. If player j cooperates at rate s in all iterations of N-IPD, player i denotes the decision making history of player j as cooperation rate $S(i|j) = s$ $(0 \leq s \leq 1)$. Player i's expected cooperation value for player j is updated by taking the weighted average over player i's decision-making history with player j,

$$\pi_{t+1}(i|j) = w\pi_t(i|j) + (1-w)S(i|j) , \tag{3}$$

where the memory weight w controls the relative weighting of the distance to the recent decision-making. Players can observe the decisions of the other players in N-PD and update the expected cooperation values only if the players are in the same group.

Using the Rule on Split of a Group.

After the majority vote in group k, if the number of players agreeing the acceptance of player i is more than that opposing, player i joins group k. Based on the rule on split of a group, the players opposing the acceptance of player i, i.e., each player j ($\in G_k^{oppose}$) of these satisfying the condition $\pi_t(j|i) \le V_j$, leave group k and form a new group. If the member of G_k^{oppose} is only player j, player j makes a game offer to most tolerable group of player j until a group accepts the game offer of player j or until all desirable groups refuse.

Choosing C or D in N-PD.

In each round of N-IPD, player i chooses C or D based on B_i, which is player i's the probability of choosing D.

4.3 Evolutionary Approach

In our simulations, genetic algorithm is applied to evolve the player's strategies. The genetic algorithm is a model of machine learning that derives its behavior from a metaphor of the processes of evolution in nature. The dynamic behavior of agent-based modeling with genetic algorithm is regarded as a powerful method for modeling social learning.

In regard to implementations of the encoding of the chromosomes, a strategy of a individual player is represented by a binary bit string. The two dimensions of a strategy, boldness B_i and vengefulness V_i, are each divided into $2^x - 1$ equal levels, from 0.0 to 1.0. Because $2^x - 1$ levels are represented by x binary bits, a player's strategy needs a total of $2x$ bits: x bits for boldness B_i and x bits for vengefulness V_i.

Each simulation is initialized with a population of all players. A simulation consists of a sequence of generations inter-spaced with genetic phases. Each generation consists of an iteration of the N-IPD with unilateral choice, mutual choice and extended mutual choice in which players make, refuse, and accept game offers for the group formation, and then play N-IPD. At the beginning of the genetic phase, each player's strategy in a population is assigned a fitness equal to its average payoff given per payoff received. A partner for crossover is selected by means of roulette wheel selection. Uniform crossover is accomplished between the strategies of a player and a partner to obtain a new strategy for one offspring if the fitness of the partner is better than that of the player in tournament selection. After that, the strategy of this offspring is subjected to mutation, where each bit is flipped one bit with a certain probability.

5 Simulation Setup

In this paper, because our purpose is to examine how the N-IPD with unilateral choice, mutual choice and extended mutual choice influence the behaviors of players, we concentrate on the emergence of cooperation. The simulations for the N-IPD with unilateral choice, mutual choice and extended mutual choice under the initial condition are conducted where each player forgive the defection of others and is completely bold, i. e., $V_i = 0.0$ and $B_i = 1.0$ ($\forall i \in N$).

If $V_i = 1.0$ and $B_i = 0.0$ ($\forall i \in N$), all players never make and accept a game offer to the players defecting at least only one time in group formation and always cooperate in N-PD game. However, it is impossible to realize such a state constant due to high mutation rate 0.05. Therefore, in this paper, we define the term "collective cooperation" as the condition that both average boldness of all player is smaller than 0.1 and average vengefulness is bigger than 0.9.

Under this condition, we examine whether the partner choice mechanism accelerate the emergence of cooperation similarly to the case of 2-PD and what rules of mutual choice are necessary for the emerge of cooperation. In this simulation, One trial is defined as the one process of the N-IPD with unilateral choice, mutual choice or extended mutual choice from 0 to 5000 generations. The important parameters and the payoff function of the N-PD game are shown in Table 1.

Table 1. Common parameters in the simulations of the N-IPD with unilateral choice, mutual choice and extended mutual choice

Number of players	50
Number of generations	10000
Number of partner choices per generation	200
Number of N-PDs per partner choice	20
Coefficient of payoff function B	0.6
Payoff for alone player P_{alone}	0.1
Initial expected cooperation value π_0	1.0
Memory weight w	0.8
Mutation rate	0.05
Binary bits for B_i and V_i	5

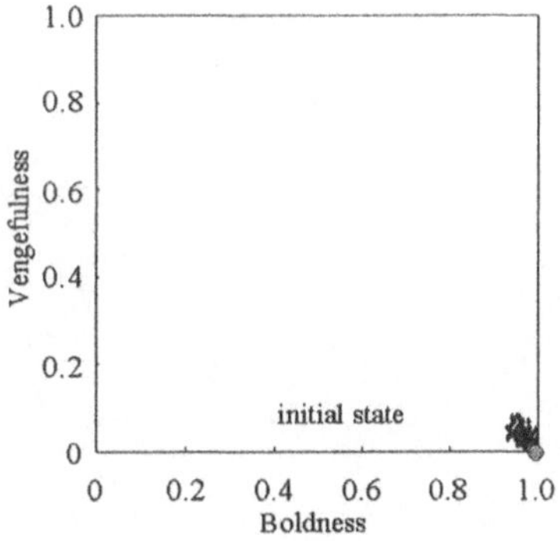

Fig. 3. Result of 500 trials of the N-IPD with unilateral choice: the average boldness $\bar{B}$ and vengefulness $\bar{V}$ of all players in 5,000 generations under the initial condition, $V_i = 1$ and $B_i = 0$ $(\forall i \in N)$

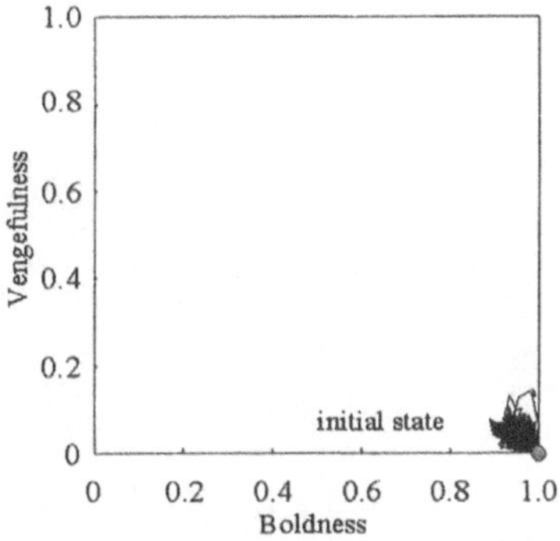

Fig. 4. Example of the transition in the N-IPD with unilateral choice: the transition of the average boldness $\bar{B}$ and vengefulness $\bar{V}$ of all players from 0 to 5,000 generations

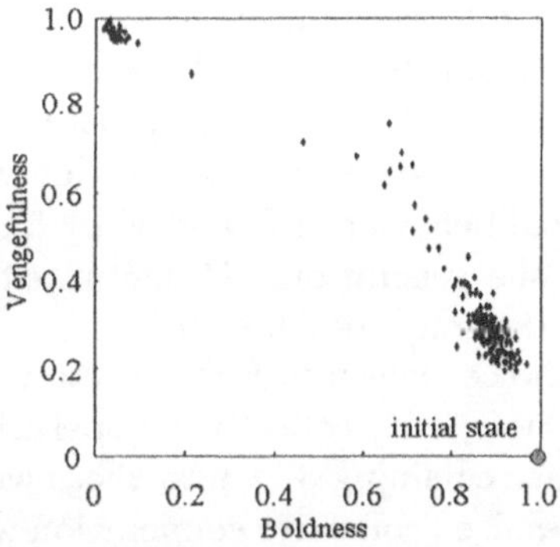

Fig. 5. Result of 500 trials of the N-IPD with mutual choice: the average boldness $\bar{B}$ and vengefulness $\bar{V}$ of all players in 5,000 generations under the initial condition, $V_i = 1$ and $B_i = 0$ $(\forall i \in N)$

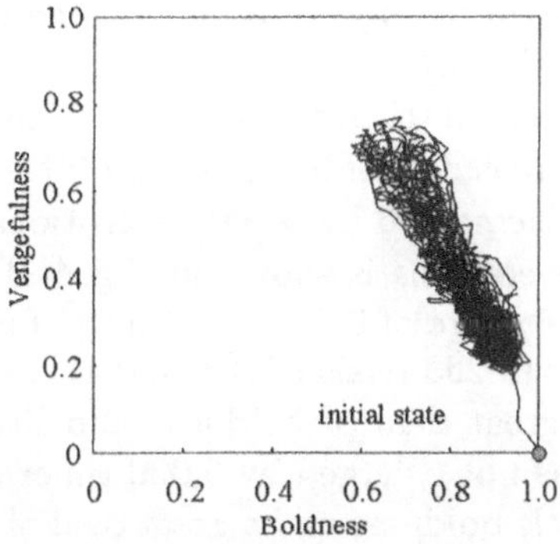

Fig. 6. Example of the transition in the N-IPD with mutual choice: the transition of the average boldness $\bar{B}$ and vengefulness $\bar{V}$ of all players from 0 to 5,000 generations

6 Simulation

The results of 300 trials of each of the N-IPD with unilateral choice, mutual choice and extended mutual choice are shown in Figs. 3 $\sim$ 8. The 300 circles in Figs. 3, 5, 7 indicate the average boldness $(\bar{B})$ and vengefulness $(\bar{V})$ of all players after 5,000 generations. The typical behavior of $\bar{B}$ and $\bar{V}$ in each are

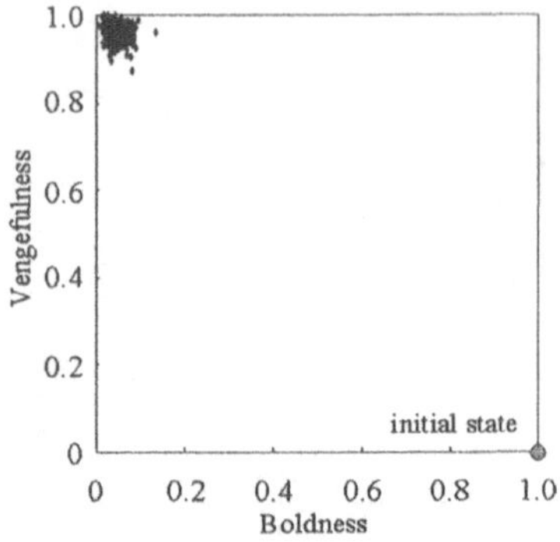

Fig. 7. Result of 500 trials of the N-IPD with extended mutual choice: the average boldness $\bar{B}$ and vengefulness $\bar{V}$ of all players in 5,000 generations under the initial condition, $V_i = 1$ and $B_i = 0$ ($\forall i \in N$)

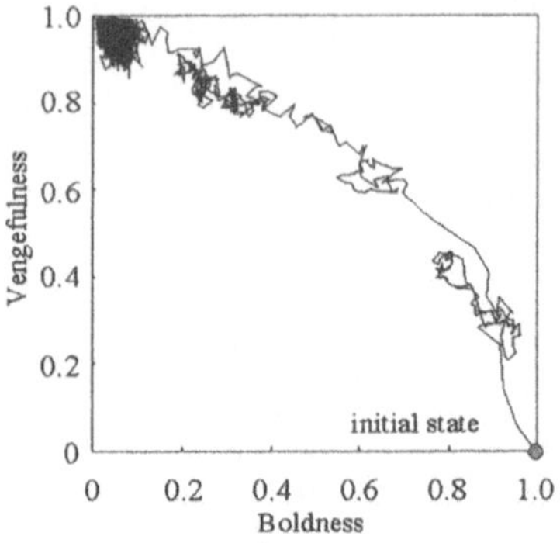

Fig. 8. Example of the transition in the N-IPD with extended mutual choice: the transition of the average boldness $\bar{B}$ and vengefulness $\bar{V}$ of all players from 0 to 5,000 generations

shown in Figs. 4, 6, 8. In these graphs, the x-axis is the average boldness $\bar{B}$ of all players and the y-axis is the average vengefulness $\bar{V}$ of all players.

In all trials of the N-IPD with unilateral choice, shown in Fig. 3, there were a great deal of boldness and little vengefulness, i. e., collective cooperation was not emerged by 5,000 generations. The typical behavior of $\bar{B}$ and $\bar{V}$ by 5,000 generations is shown in Fig. 4. Throughout the generations, $\bar{B}$ fluctuated in the range of 0.9 to 1.0, and $\bar{V}$ fluctuated in the range of 0.0 to 0.1.

In 263 trials of the N-IPD with mutual choice, shown in Fig. 5, there were a great deal of boldness and little vengefulness, i. e., collective cooperation was not emerged by 5,000 generations. In the remaining 37 trials, there were little boldness and a great deal of vengefulness, i. e., collective cooperation was emerged. The typical behavior of $\bar{B}$ and $\bar{V}$ when collective cooperation was not emerged by 5,000 generations is shown in Fig. 6. Usually, $\bar{B}$ remained in the range of 0.8 to 1.0 and $\bar{V}$ remained in the range of 0.2 to 0.4. However, occasionally $\bar{B}$ fell to 0.6 and $\bar{V}$ rose to 0.7, and then $\bar{B}$ and $\bar{V}$ returned to the original state. This fluctuation was repeated throughout the generations. The behavior of the emergence of collective cooperation is similar to that in the following N-IPD with extended mutual choice.

In all trials of the N-IPD with extended mutual choice, shown in Fig. 7, there were little boldness and a great deal of vengefulness, i.e., collective cooperation was emerged. The typical dynamics of $\bar{B}$ and $\bar{V}$ is shown in Fig. 8. Although, at first, $\bar{B}$ fluctuated in the range of 0.6 to 0.8, and $\bar{V}$ fluctuated in the range of 0.4 to 0.8, $\bar{B}$ turned to 0 and $\bar{V}$ turned to 1.0 in the end. Once $\bar{B}$ reached 0 and $\bar{V}$ reached 1.0, this state continued to remain there and did not transfer to another state.

7 Discussion

We discuss the emergence of collective cooperation based on the strategy (B_i and V_i) in the N-IPD with unilateral choice, mutual choice and extended mutual choice. In the following explanation, a player with a high level of boldness is represented as B_{high}, and a player with a low level of boldness as B_{low}. In the same way, players with high and low levels of vengefulness are represented as V_{high} and V_{low}.

7.1 N-IPD with Unilateral Choice

In all trials of the N-IPD with unilateral choice shown in Fig. ??, there were a great deal of boldness and little vengefulness. In these trials, why did the emergence of cooperation fail? A player with boldness decreased by mutation, i. e., a player with B_{low}, cannot acquire a higher payoff than the players with B_{high}. The reason is that the players with B_{low} cannot refuse a game offer from the players with B_{high} in group formation based on unilateral choice and the players with B_{high} have a freeride on the players with B_{low}. Accordingly, the players with B_{low} do not increase in the next generation. Therefore, collective cooperation never becomes emerged.

7.2 N-IPD with Mutual Choice

In 181 trials shown in Fig. 5, a great deal of boldness and little vengefulness were observed. In these trials, why did the emergence of cooperation fail? We analyze below the factors leading to the failure of the emergence of cooperation in the N-IPD with mutual choice.

At first, we assumed that there was only one player with B_{low} and V_{high}. Mutation of player strategies decreases boldness or increases vengefulness because the initial condition is $V_i = 0$ and $B_i = 1$ $(\forall i \in N)$. A player with boldness decreased by mutation, i.e., a player with B_{low}, cannot acquire a higher payoff than the players with B_{high} because these players with B_{high} have a freeride on the players with B_{low}. Accordingly, the players with B_{low} do not increase in the next generation. A player with vengefulness increased by mutation, i.e., a player with V_{high}, cannot acquire a higher payoff than the players with B_{high} because players with V_{high} do not join a group consisting of players with B_{high}. Therefore, one player with B_{low} and V_{high} by mutation cannot acquire a higher payoff than the players with B_{high} and V_{low}. Consequently, the player with B_{low} and V_{high} is not selected in genetic operation, and then perishes.

Next, we assumed that there were plural players with B_{low} and V_{high}. If a group consists of only players with B_{low} and V_{high}, the group refuses the game offers of players with B_{high}. If a group consists of both players with B_{low} and V_{high} and players with B_{low} and V_{low}, it is possible that a player with B_{high} would join this group and have a freeride. The player with B_{high} can join the group because, while the players with B_{low} and V_{high} oppose the acceptance of his/her game offer, the players with B_{low} and V_{low} agree to it. If the players with B_{low} and V_{low} win the majority vote over the players with B_{low} and V_{high}, the player with B_{high} can join the group. In such a group, the players with B_{high} have a free-ride on the players with B_{low}. The players with B_{low} and V_{high} are not selected in genetic operation, and then perish because they cannot acquire higher payoffs than the free-rider. Although there are plural players with B_{low} and V_{high}, the players with B_{high} and the players with B_{low} and V_{low} prevent the emergence of cooperation. The players with B_{high} directly prevent the emergence of cooperation because they have a freeride on the players with B_{low} and V_{high}. On the other hand, the players with B_{low} and V_{low} indirectly prevent because they accept game offers from the players with B_{high} who have a freeride on the players with B_{low}. Therefore, in N-IPD with mutual choice collective cooperation is not often emerged.

In the remaining 16 trials, shown in Fig. 5, there was little boldness and a great deal of vengefulness. Why did the emergence of cooperation succeed? Here, we analyze the factors leading to the emergence of cooperation in N-IPD with mutual choice. The reason for the failure to emerge collective cooperation was that the player with B_{high} can join the group consisting of both players with B_{low} and V_{high} and players with B_{low} and V_{low}. If there are players with B_{low} and V_{high} but no players with B_{low} and V_{low}, the player with B_{high} cannot join the group, and then the players will defect from each other. As a result, the player with B_{high} acquires a lower payoff than the players with B_{low} and V_{high} who cooperate with each other. If the number of the players with B_{low} and V_{high} increases, and the players predominate in the population for a few generations before the number of the players with B_{low} and V_{low} increases by crossover or mutation, collective cooperation becomes emerged. Therefore, since the simulation results (Fig. 5) show that collective cooperation was emerged in 16 out of 300 trials, it is not impossible, but it is difficult, to emerge collective cooperation in the N-IPD with mutual choice.

7.3 N-IPD with Extended Mutual Choice

In all trials of the N-IPD with extended mutual choice shown in Fig. 7, there were little boldness and a great deal of vengefulness. In the N-IPD with extended mutual choice, the emergence of cooperation fails because the players with B_{low} and V_{low} accept the game offer of the players with B_{high}. Here, we analyze the factors leading to the success of the emergence of cooperation in the N-IPD with extended mutual choice.

In the N-IPD with extended mutual choice, if the players with B_{low} and V_{low} agree to accept the game offer of a player with B_{high}, and the group as a whole also accepts it, the players with B_{low} and V_{high} leave the group based on extended mutual choice; they refuse to play the N-PD game with those who play with defectors. The extended mutual choice prevents the player with B_{high} from having a freeride on the players with B_{low} and V_{high}. Because, if the player with B_{high} joins the group, the players with B_{low} and V_{high} leave. As a result, if there are some players with B_{low} and V_{high}, they can form a group without the player with B_{high}. The players with B_{low} and V_{high} can acquire higher payoffs because they cooperate with each other. Throughout this process, the number of players with B_{low} and V_{high} increases and they predominate in the population. Therefore, collective cooperation becomes emerged and continues to be formed.

8 Conclusion

In this paper, a N-IPD with unilateral choice, mutual choice and extended mutual choice were proposed based on an oringnal 2-IPD with mutual choice. Agent-based modeling that an agent considered as a player with bounded rationality was applied for describing the N-IPD with mutual choice. In order to examine how these partner choice mechanisms influence the behaviors of players in N-IPD, the emergence of cooperation was focused on. Agent-based simulations with an evolutionary approach were conducted to confirm the possibility of the emergence of cooperation in the N-IPD with unilateral choice, mutual choice and extended mutual choice. As a result, the followings were confirmed. In the N-IPD with unilateral choice, it is impossible that collective cooperation is emerged. In the N-IPD with mutual choice, it is not impossible, but it is difficult, that collective cooperation is emerged. However, once collective cooperation is emerged, it can be maintained. In the N-IPD with extended mutual choice, extended mutual choice accelerates and maintains the emergence of collective cooperation just as in the result of an oringnal 2-IPD with mutual choice.

References

1. Ashlock, D., Smucker, S., Stanley, A., Tesfatsion, L. (1996): Preferential Partner Selection in an Evolutionary Study of the Prisoner's Dilemma. BioSystems **37(1–2)**, 99–125
2. Axelrod, R. (1984): The Evolution of Cooperation. Basic Books, New York
3. Axelrod, R. (1997): The Complexity of Cooperation. Princeton University Press
4. Axtell, R. (1999): The Emergence of Firms in a Population of Agents: Local Increasing Returns, Unstable Nash Equilibria, and Power Law Size Distributions. The Brookings Institution, CSED Working Paper **3**
5. Batali, J., Kitcher, P. (1995): Evolution of Altruism in Optional and Compulsory Games. Journal of Theoretical Biology **175**, 161–171
6. Banker, R. D., Lee, S. Y. (2000): Mutual Monitoring and Peer Pressure in Teamwork. The University of Texas, Accounting and Information Management Working Paper Series, 2000-11
7. Chen, M. D., Riolo, R. L., Axelrod, R. (1999): The Emergence of Social Organization in the Prisoner's Dilemma: How Context-Preservation and Other Factors Promote Cooperation. Santa Fe Institute Working Paper, 99-01-002

8. Dawes, R. M. (1981): Social Dilemmas. Annual Review of Psychology **31**, 169–193
9. Hauk, E. (1996): Leaving the Prison: A Discussion of the Iterated Prisoner's Dilemma under Preferential Partner Selection. Second International Conference on Computing in Economics and Finance'96
10. Hauk, E. (1999): Multiple Prisoner's Dilemma Games with(out) an Outside Option: An Experimental Study. Universitat Pompeu Fabra, Economics Working Papers
11. Hauk, E., Nagel, R. (2000): Choice of Partners in Multiple Prisoner's Two-person Prisoner's Dilemma Games: An Experimental Study. Universitat Pompeu Fabra, Economics Working Papers
12. Hirshleifer, D., Rasmusen, E. (1989): Cooperation in a Repeated Prisoners' Dilemma with Ostracism. Journal of Economic Behavior and Organization **12**, 87–106
13. Namatame, A. (1999): Multi-agents and Complex Theory. Morikita Shuppan
14. Uno, K., Namatame, A. (1998): An Evolutionary Design of Commitment Networks. In: Proceedings of The Forth International Symposium on Artificial Life and Robotics, Vol. 2, 528–531
15. Schelling, T. C. (1978): Micromotives and Macrobehavior. Norton, New York
16. Shussler, R. (1989): Exit Threats and Cooperation under Anonymity. Journal of Conflict Resolution **33**, 728–749
17. Stanley, E. A., Ashlock, D., Tesfatsion, L. (1994): Iterated Prisoner's Dilemma with Choice and Refusal of Partners. In: Artificial Life III. Santa Fe Institute Studies in the Sciences of Complexity, Vol. XVII, 131–175
18. Tesfatsion, L. (1997): A Trade Network Game with Endogenous Partner Selection. In: Computational Approaches to Economic Problems. Kluwer Academic Publishers, 249–269
19. Yamagishi, T., Hayashi, N. (1996): Selective Play: Social Embeddedness of Social Dilemmas. In: Frontiers in Social Dilemmas Research. Springer, Berlin, Heidelberg, New York, 337–362
20. Yao, X., Darwen, P. (1994): An Experimental Study of N-person Iterated Prisoner's Dilemma games. Infomatica **18(4)**, 375–450.

Randomly-Furcating Stochastic Differential Games

David W. K. Yeung

Department of Finance & Decision Sciences, Hong Kong Baptist University, and Center for Game Theory, St. Petersburg State University
E-mail: wkyeung@hkbu.edu.hk

Summary. This paper presents a class of games – designated as *Randomly Furcating Stochastic Differential Game* – in which random shocks in the stock dynamics and (future) stochastic changes in payoffs are present. Since future payoff are not known with certainty, the term "randomly furcating" is introduced to emphasize that a particularly useful way to analyze such a situation is to assume that payoffs change at any future time instant according to (known) probability distributions defined in terms of multiple-branching stochastic processes. New and significant mathematical results are obtained, under which it becomes possible to characterize the conditions under which previously unsolvable games can be solved. Two illustrations are provided.

Key words: stochastic differential games, randomly furcating payoffs, stochastic control

1 Introduction

A particularly complex and fruitful branch of game theory is dynamic (continuous-time) or differential games, which investigates interactive decision making over time. Classic contributions in this field include [12, 5, 17]. Since institutions like firms, markets, governments, and legal systems are designed by human beings and function through human action, it also follows that "the social world is a differential game". In the existing literature, random considerations are introduced into games played over time through stochastic specification of their state dynamics. This approach leads to stochastic differential games or piecewise deterministic games (see [2, 3, 4]), [8], [15], [14], and [22, 23] for examples of stochastic differential games; and [18, 19], [9] and [21] for examples of piecewise deterministic games).

The new class of "randomly furcating stochastic differential games" presented in this paper attempts to capture and model an essential feature of interactive decision making over time. A necessary property of time – and hence of decision making over time – is that though the individual may, through the expenditure of effort and resources, gather past and present information, the future is inherently unknown and therefore (in the mathematical sense) uncertain. There is no escape from this fact, regardless of what costs the individual is willing to bear in order to obtain data, information, and to forecasting the outcomes of decisions. An empirically meaningful theory of differential games must therefore incorporate the time-uncertainty of decisions and their consequences in an appropriate manner. In the new approach presented in this research, time-uncertainty is modeled in terms of random shocks in the game's stock dynamics and stochastic changes in future payoffs. It therefore follows that to a high degree of truth, "life and the social world are randomly furcating stochastic differential games".

In this paper, a framework is proposed to formalize game-theoretic situations with stochastic dynamics and uncertain environments over time in the following manner. Since the consequences of any action in the game are not known with certainty, the concept of "random furcating" is introduced to describe future payoffs as changing at any instant of time according to (known) probability distributions, which turn are defined in terms of multiple-branching stochastic processes. The introduction of this stochastic specification lead to a novel approach to solve differential games in terms of properties and solution concepts not explored in the literature before. New and significant mathematical results are obtained, under which it becomes possible to characterize the conditions under which previously unsolvable games can be solved.

The paper is organized as follows. Section 2 presents the basic model. Section 3 characterizes the feedback equilibrium of the Basic Model. Section 4 consider the Infinite-horizon problem. Section 5 provides two illustrations and Section 6 concludes the paper.

2 Basic Model of Randomly Furcating Stochastic Differential Games

This section develops a new class of games – designated as *Randomly Furcating Stochastic Differential Game* – in which random shocks in the stock dynamics and (future) stochastic changes in payoffs are present. An essential characteristic of time – and hence decision making over time – is that though the individual may, through the expenditure of resources, gather past and present information, the future is inherently unknown and therefore (in the mathematical sense) uncertain. There is no escape from this fact, regardless of what resources the individual should choose to devote to obtaining data, information, and to forecasting. An empirically meaningful theory of games must therefore incorporate time-uncertainty in an appropriate manner. This development establishes a framework or paradigm for modeling game-theoretic situations with stochastic dynamics and uncertain environments over time.

Definition 1. *A randomly furcating stochastic differential game of prespecified fixed duration involves the following:*

(i) There are N players to be denoted by $P_1, P_2, \ldots, P_N$. Let us further denote the index set $\{1, 2, \ldots, N\}$ by $\mathbf{N}$.

(ii) A time interval $[t_0, T]$ which specifies the duration of the game.

(iii) The control space is an infinite set U^i with some topological structure, for each $i \in \mathbf{N}$. It elements $\{u^i(s), t_0 \leq s \leq T\}$ are control functions of P_i. Furthermore, there exists a set $S^i \subseteq R^{m_i}$ $(i \in \mathbf{N})$ so that, for each fixed $t \in [t_0, T]$, $u^i(s) \in S^i$.

(iv) The state space of the game $X \in R^n$, whose elements $\{x(s), t_0 \leq s \leq T\}$ constitute the permissible state trajectories. The state dynamics is characterized by the vector-valued stochastic differential equations:

$$dx(s) = f[s, x(s), u^1(s), u^2(s), \ldots, u^N(s)]ds + \sigma[s, x(s)]dz(s), x(t_0) = x_0, \quad (1)$$

where $\sigma[s, x(s)]$ is a $n \times \theta$ matrix and $z(s)$ is a θ-dimensional Wiener process and the initial state x_0 is given. Let $\Omega[s, x(s)] = \sigma[s, x(s)]\sigma s[s, x(s)]'$ denote the covariance matrix with its element in row h and column ζ denoted by $\Omega^{h\zeta}[s, x(s)]$.

(v) When the game starts at t_0, the payoff of player i is

$$g^{i(0)}[s, x(s), u^1(s), u^2(s), \ldots, u^N(s)]$$

for $i \in \mathbf{N}$. In future time instants t_k $(k = 1, 2, \ldots, \tau)$, where $t_\tau < T \equiv t_{\tau+1}$, the payoffs are affected by a series of random events θ^k, for $k = 1, 2, \ldots, \tau$, which is a random variable stemming from the branching process as described below.

$\theta^1 = \{\theta^1_1, \theta^1_2, \ldots, \theta^1_{\eta_1}\}$ with corresponding probability $\{\lambda^1_1, \lambda^1_2, \ldots, \lambda^1_{\eta_1}\}$.

Given that $\theta^1_{a_1}$ occurred in time interval $[t_1, t_2)$, for $a_1 = 1, 2, \ldots, \eta_1$

$$\theta^2 = \{\theta_1^{2[(1,a_1)]}, \theta_2^{2[(1,a_1)]}, \ldots, \theta_{\eta_{2[(1,a_1)]}}^{2[(1,a_1)]}\}$$

with probability

$$\{\lambda_1^{2[(1,a_1)]}, \lambda_2^{2[(1,a_1)]}, \ldots, \lambda_{\eta_{2[(1,a_1)]}}^{2[(1,a_1)]}\},$$

Given that $\theta^1_{a_1}$ occurred in time interval $[t_1, t_2)$ and $\theta_{a_2}^{2[(1,a_1)]}$ occurred in time interval $[t_2, t_3)$, for $a_1 = 1, 2, \ldots, \eta_1$ and $a_2 = 1, 2, \ldots, \eta_{2[(1,a_1)]}$,

$$\theta^3 = \{\theta_1^{3[(1,a_1)(2,a_2)]}, \theta_2^{3[(1,a_1)(2,a_2)]}, \ldots, \theta_{\eta_{3[(1,a_1)(2,a_2)]}}^{3[(1,a_1)(2,a_2)]}\}$$

with probability

$$\{\lambda_1^{3[(1,a_1)(2,a_2)]}, \lambda_2^{3[(1,a_1)(2,a_2)]}, \ldots, \lambda_{\eta_{3[(1,a_1)(2,a_2)]}}^{3[(1,a_1)(2,a_2)]}\}.$$

In general, $\theta^1_{a_1}$ occurred in time interval $[t_1, t_2)$, $\theta_{a_2}^{2[(1,a_1)]}$ occurred in time interval $[t_2, t_3)$, ..., and $\theta_{a_{k-1}}^{k-1[(1,a_1)(2,a_2)\ldots(k-2,a_{k-2})]}$ occurred in time interval $[t_{k-1}, t_k)$, for $a_1 = 1, 2, \ldots, \eta_1$, $a_2 = 1, 2, \ldots, \eta_{2[(1,a_1)]}$, ..., $a_{k-1} = 1, 2, \ldots, \eta_{k-1[(1,a_1)(2,a_2)\ldots(k-2,a_{k-2})]}$,

$$\theta^k = \{\theta_1^{k[(1,a_1),(2,a_2)\dots(k-1,a_{k-1})]}, \dots, \theta_{\eta_{k[(1,a_1),(2,a_2)\dots(k-1,a_{k-1})]}}^{k[(1,a_1),(2,a_2)\dots(k-1,a_{k-1})]}\}$$

with probability

$$\{\lambda_1^{k[(1,a_1),(2,a_2)\dots(k-1,a_{k-1})]}, \dots, \lambda_{\eta_{k[(1,a_1),(2,a_2)\dots(k-1,a_{k-1})]}}^{k[(1,a_1),(2,a_2)\dots(k-1,a_{k-1})]}\}$$

for $k = 1, 2, \dots, \tau$.

If the event $\theta_{a_k}^{k[(1,a_1),(2,a_2)\dots(k-1,a_{k-1})]}$ *is realized in the time interval* $[t_k, t_{k+1})$, *the payoff of player* i *in this period of time becomes*

$$g^{[i(k),\theta_{a_k}^{k[(1,a_1),(2,a_2)\dots(k-1,a_{k-1})]}]}[s, x(s), u^1(s)u^2(s), \dots, u^N(s)],$$

for $s \in [t_k, t_{k+1})$, $i \in \mathbf{N}$, $a_k = 1, 2, \dots, \eta_{k[(1,a_1)(2,a_2)\dots(k-1,a_{k-1})]}$, $k = 1, 2, \dots, \tau$.

For clarity in presentation, $g^{[i(k),\theta^k]}[s, x(s), u^1(s)u^2(s), \dots, u^N(s)]$ *is written as* $g^{i(k)}[\theta^k; s, x(s), u^1(s)u^2(s), \dots, u^N(s)]$.

The structure of the payoffs of all players will remain unchanged within the time interval $[t_k, t_{k+1})$ *and are common knowledge to all players within that time interval. Moreover, at time* t_0, *the terminal payoffs at time* T *are known to be:*

$$q^{i(T)}[\theta^T, x(T)], \quad \text{for } i \in \mathbf{N},$$

where θ^T *is a random variable with range* $\{\theta_1^T, \theta_2^T, \dots, \theta_{\eta_T}^T\}$ *and corresponding probability* $\{\lambda_1^T, \lambda_2^T, \dots, \lambda_{\eta_T}^T\}$.

(vi) The payoff to be maximized by player i, *for* $i \in \mathbf{N}$, *can be expressed as:*

$$E_{t_0}\left\{\int_{t_0}^{t_1} g^{i(0)}[s, x(s), u^1(s)u^2(s), \dots, u^N(s)]ds + \right.$$

$$\left. + \sum_{k=1}^{\tau} \int_{t_k}^{t_{k+1}} g^{i(k)}[\theta^k; s, x(s), u^1(s)u^2(s), \dots, u^N(s)]ds + q^{i(T)}[\theta^T, x(T)]\right\} =$$

$$E_{t_0}\left\{\int_{t_0}^{t_1} g^{i(0)}[s, x(s), u^1(s)u^2(s), \dots, u^N(s)]ds + \right.$$

$$+ \sum_{a_1=1}^{\eta_1} \lambda_{a_1}^1 \int_{t_1}^{t_2} g^{i(1)}[\theta_{a_1}^1; s, x(s), u^1(s), u^2(s), \dots, u^N(s)]ds +$$

$$\sum_{a_1=1}^{\eta_1} \lambda_{a_1}^1 \sum_{a_2=1}^{\eta_{2(1,a_1)}} \lambda_{a_2}^{2[(1,a_1)]} \int_{t_2}^{t_3} g^{i(2)}[\theta_{a_2}^{2[(1,a_1)]}; s, x(s), u^1(s), u^2(s), \dots, u^N(s)]ds +$$

$$\dots$$

$$+ \sum_{a_1=1}^{\eta_1} \lambda_{a_1}^1 \sum_{a_2=1}^{\eta_{2(1,a_1)}} \lambda_{a_2}^{2[(1,a_1)]} \dots \sum_{a_k}^{\eta_{k[(1,a_1),\dots(k-1,a_{k-1})]}} \lambda_{a_k}^{k[(1,a_1)\dots(k-1,a_{k-1})]} \times$$

$$\times \int_{t_\tau}^{t_\tau+1} g^{i(\tau)}[\theta_{a_\tau}^{\tau[(1,a_1)\ldots(\tau-1,a_{\tau-1})]}; s, x(s), u^1(s), \ldots, u^N(s)]ds + q^{i(T)}[\theta^T, x(T)] \Bigg\}$$

$$= E_{t_0}\left\{\int_{t_0}^{t_1} g^{i(0)}[s, x(s), u^1(s)u^2(s), \ldots, u^N(s)]ds + \right.$$

$$+\sum_{k=1}^{\tau}\sum_{a_1=1}^{\eta_1}\lambda_{a_1}^1\sum_{a_2=1}^{\eta_{2(1,a_1)}}\lambda_{a_2}^{2[1,a_1]}\ldots\sum_{a_k}^{\eta_{k[(1,a_1)\ldots(k-1,a_{k-1})]}}\lambda_{a_k}^{k[(1,a_1)\ldots(k-1,a_{k-1})]}\times$$

$$\times\int_{t_k}^{t_{k+1}} g^{i(k)}[\theta_{a_k}^{k[(1,a_1)\ldots(k-1,a_{k-1})]}; s, x(s), u^1(s), u^2(s), \ldots, u^N(s)]ds +$$

$$\left. + \sum_{a_T=1}^{\eta T}\lambda_{a_T} q^{i(T)}[\theta_{a_T}^T, x(T)]\right\}, \tag{2}$$

where E_{t_0} denotes the expectation operator performed at time t_0, and equals 1 with probability one.

The game (1)–(2) allows random shocks in the stock dynamics, future stochastic changes in the payoffs. Since future payoff are not known with certainty, the term "randomly furcating" is introduced to emphasize that a particularly useful way to analyze such a situation is to assume that payoffs change at any future time instant according to (known) probability distributions defined in terms of multiple-branching stochastic processes.

This new approach widens the application of differential game theory to problems where future environments are not known with certainty. Important cases abound in finance and economics: in particular, the (real) returns to major asset classes such as equities and bonds over time are subject to significant uncertainty from stochastic shocks in economic activity, government policy, and unanticipated effects of strategic behavior in imperfectly competitive markets. Finally, this approach also represents a new way of modeling dynamic game situations under uncertainty.

3 Characterization of Feedback Nash Equilibrium the Basic Model

Since in an environment with uncertainty, precommitted actions are meaningless, feedback strategies have to be used. A feedback strategy is a decision rule $\phi(t, x)$ such that it is continuous in t and uniformly Lipschitz in x for each t. We denote the set of all feedback strategies for player i in the time interval $[t_k, t_{k+1}]$ by U_k^i, for $i \in \mathbf{N}$ and $k = 0, 1, \ldots, \tau$.

A feedback solution of game (1)–(2) can be characterized as:

Theorem 1. *An N-tuple of feedback strategies* $\{[u_0^{i^*}(t) = \phi_0^{i^*}(t,x)]; i \in \mathbf{N}]\}$ *and a set feedback strategies* $\{[u_{a_k}^{i[(1,a_1)\ldots(k-1,a_{k-1})]^*}(t) = \phi_0^{i[(1,a_1)\ldots(k-1,a_{k-1})]^*}(t,x)]$*;* $i \in \mathbf{N}$*;* $k = 1,2,\ldots,\tau\}$*, contingent upon the event that* $\theta^1 = \theta_{a_1}^1$*,* $\theta^2 = \theta_{a_2}^{2[(1,a_1)]}$*, ...,* $\theta^\tau = \theta_{a_\tau}^{\tau[(1,a_1)(2,a_2)\ldots(\tau-1,a_{\tau-1})]}$*, for* $a_1 = 1,2,\ldots,\eta_1$*,* $a_2 = 1,2,\ldots,\eta_{2[(1,a_1)]}$*, ...,* $a_\tau = 1,2,\ldots,\eta_{\tau[(1,a_1)(2,a_2)\ldots(\tau-1,a_{\tau-1})]}$*, constitutes a Nash equilibrium solution for the game (1)–(2), if there exist functionals* $V^{i[(1,a_1)(2,a_2)\ldots(\tau-1,a_{\tau-1})]a_\tau}(t,x)\colon [t_\tau,T] \times R^n \to R$*, for* $i \in \mathbf{N}$*, which satisfy the following set of partial differential equations:*

$$-V_t^{i[(1,a_1)(2,a_2)\ldots(\tau-1,a_{\tau-1})]a_\tau} - \frac{1}{2}\sum_{h,\zeta=1}^{N} \Omega^{h\zeta}(t,x)V_{x_h x_\zeta}^{i[(1,a_1)(2,a_2)\ldots(\tau-1,a_{\tau-1})]a_\tau}$$

$$= \max_{\phi_{a_\tau}^i \in U^i} \Big\{ g^{i(\tau)}[\theta_{a_\tau}^{\tau[(1,a_1)\ldots(\tau-1,a_{\tau-1})]a_\tau}; t, x(t), \phi_{a_\tau}^{1[(1,a_1)\ldots(\tau-1,a_{\tau-1})]^*}(t,x), \ldots$$
$$\ldots, \phi_{a_\tau}^{i-1[(1,a_1)\ldots(\tau-1,a_{\tau-1})]^*}(t,x), \phi_{a_\tau}^{i}(t,x), \phi_{a_\tau}^{i+1[(1,a_1)\ldots(\tau-1,a_{\tau-1})]^*}(t,x), \ldots$$
$$\ldots, \phi_{a_\tau}^{N[(1,a_1)\ldots(\tau-1,a_{\tau-1})]^*}(t,x)]$$
$$+ V_x^{i[(1,a_1)\ldots(\tau-1,a_{\tau-1})]a_\tau} f[t, x(t), \phi_{a_\tau}^{1[(1,a_1)\ldots(\tau-1,a_{\tau-1})]^*}(t,x), \ldots$$
$$\ldots, \phi_{a_\tau}^{i-1[(1,a_1)\ldots(\tau-1,a_{\tau-1})]^*}(t,x), \phi_{a_\tau}^{i}(t,x), \phi_{a_\tau}^{i+1[(1,a_1)\ldots(\tau-1,a_{\tau-1})]^*}(t,x), \ldots$$
$$\ldots, \phi_{a_\tau}^{N[(1,a_1)\ldots(\tau-1,a_{\tau-1})]^*}(t,x)] \Big\}, \tag{3}$$

$$V^{i[(1,a_1)(2,a_2)\ldots(\tau-1,a_{\tau-1})]a_\tau}(T,x) = \sum_{b=1}^{\eta_T} \lambda_b^T q^{i(T)}[\theta_b^T, x]\,; \tag{4}$$

and functionals $V^{i[(1,a_1)(2,a_2)\ldots(k-1,a_{k-1})]a_k}(t,x)\colon [t_k,t_{k+1}] \times R^n \to R$*, for* $i \in \mathbf{N}$ *and* $k = 1,2,\ldots,\tau-1$*, which satisfy the following set of partial differential equations:*

$$-V_t^{i[(1,a_1)(2,a_2)\ldots(k-1,a_{k-1})]a_k} - \frac{1}{2}\sum_{h,\zeta=1}^{N} \Omega^{h\zeta}(t,x)V_{x_h x_\zeta}^{i[(1,a_1)(2,a_2)\ldots(k-1,a_{k-1})]a_k}$$

$$= \max_{\phi_{a_k}^i \in U^i} \Big\{ g^{i(k)}[\theta_{a_k}^{k[(1,a_1)\ldots(k-1,a_{k-1})}; t, x(t), \phi_{a_k}^{1[(1,a_1)\ldots(k-1,a_{k-1})]^*}(t,x), \ldots$$
$$\ldots, \phi_{a_k}^{i-1[(1,a_1)\ldots(k-1,a_{k-1})]^*}(t,x), \phi_{a_k}^{i}(t,x), \phi_{a_k}^{i+1[(1,a_1)\ldots(k-1,a_{k-1})]^*}(t,x), \ldots$$
$$\ldots, \phi_{a_k}^{N[(1,a_1)\ldots(k-1,a_{k-1})]^*}(t,x)]$$
$$+ V_x^{i[(1,a_1)\ldots(k-1,a_{k-1})]a_k} f[t, x(t), \phi_{a_k}^{1[(1,a_1)\ldots(k-1,a_{k-1})]^*}(t,x), \ldots$$
$$\ldots, \phi_{a_k}^{i-1[(1,a_1)\ldots(k-1,a_{k-1})]^*}(t,x), \phi_{a_k}^{i}(t,x), \phi_{a_k}^{i+1[(1,a_1)\ldots(k-1,a_{k-1})]^*}(t,x), \ldots$$
$$\ldots, \phi_{a_k}^{N[(1,a_1)\ldots(k-1,a_{k-1})]^*}(t,x)] \Big\} \tag{5}$$

$$V^{i[(1,a_1)(2,a_2)\dots(k-1,a_{k-1})]a_k}(T,x)$$

$$= \sum_{a_{k+1}=1}^{\eta_{k+1[(1,a_1)\dots(k,a_k)]}} [\lambda_{a_{k+1}}^{k+1[(1,a_1)\dots(k,a_k)]} V^{i[(1,a_1)\dots(k,a_k)]a_{k+1}}(t_{k+1},x)] \; ; \tag{6}$$

and functionals $V^{i(0)}(t,x)$: $[t_0, t_1] \times R^n \to R$, for $i \in \mathbf{N}$, which satisfy the following set of partial differential equations:

$$-V_t^{i(0)} - \frac{1}{2} \sum_{h,\zeta=1}^{N} \Omega^{h\zeta}(t,x) V_{x_h x_\zeta}^{i(0)} = \max_{\phi_0^i \in U^i} \{ g^{i(0)}[t, x(t), \phi_0^{1^*}(t,x), \dots,$$

$$\phi_0^{i-1^*}(t,x), \phi_0^i(t,x), \phi_0^{i+1^*}(t,x), \dots, \phi_0^{N^*}(t,x)] + V_x^{i(0)} f[t, x(t), \phi_0^{1^*}(t,x), \dots,$$
$$\phi_0^{i-1^*}(t,x), \phi_0^i(t,x), \phi_0^{i+1^*}(t,x), \dots, \phi_0^{N^*}(t,x)] \} \, , \tag{7}$$

$$V^{i(0)}(t_1, x) = \sum_{a_1=1}^{\eta_1} [\lambda_{a_1}^1 V^{1(a_1)}(t_1, x)] \, . \tag{8}$$

Proof. This result follows readily from the optimality conditions in stochastic control problems derived by Fleming [12] and Fleming and Rishel [11], and the definition of Nash equilibrium.

Theorem 1 suggests that a set of sufficiently informed paths, in the shape of strategies contingent upon information regarding the realisation of the $a_h \times h$ random variables

$$\theta_{a_h}^{h[(1,a_1)(2,a_2)\dots(h-1,a_{h-1})]}, \quad a_h = 1, 2, \dots, \eta_{h[(1,a_1)(2,a_2)\dots(h-1,a_{h-1})]} \, ,$$
$$h = 1, 2, \dots, \tau \, ,$$

forms an essential element in the feedback solution of the game (1)–(2). Since for player i the probability distribution of future payoffs in the time interval $[t_k, t_{k+1})$ is given by:

$$g^{i(k)}[\theta^k; t, x(t), u^1(t), \dots, u^N(t)] \text{ for } k = 1, 2, \dots, \tau; \; t \in [t_k, t_{k+1}), \; i \in \mathbf{N} \, ,$$

where θ^k, for $k = 1, 2, \dots, \tau$, is a random variable stemming from the branching process: $\theta^1 = \{\theta_1^1, \theta_2^1, \dots, \theta_{\eta_1}^1\}$ with corresponding probability $\{\lambda_1^1, \lambda_2^1, \dots, \lambda_{\eta_1}^1\}$.

Given that $\theta_{a_1}^1$ is realised in time interval $[t_1, t_2)$, $\theta_{a_2}^{2[(1,a_1)]}$ is realised in time interval $[t_2, t_3), \dots,$ and $\theta_{a_{k-1}}^{k-1[(1,a_1)(2,a_2)\dots(k-2,a_{k-2})]}$ is realised in time interval $[t_{k-1}, t_k)$, for $a_1 = 1, 2, \dots, \eta_1$, $a_2 = 1, 2, \dots, \eta_{2[(1,a_1)]}, \dots, a_{k-1} = 1, 2, \dots, \eta_{k-1[(1,a_1)(2,a_2)\dots(k-2,a_{k-2})]}$,

$$\theta^k = \{\theta_1^{k[(1,a_1),(2,a_2)\dots(k-1,a_{k-1})]}, \dots, \theta_{\eta_{k[(1,a_1),(2,a_2)\dots(k-1,a_{k-1})]}}^{k[(1,a_1),(2,a_2)\dots(k-1,a_{k-1})]}\}$$

would be realised with corresponding probabilities

$$\{\lambda_1^{k[(1,a_1),(2,a_2)\ldots(k-1,a_{k-1})]}, \ldots, \lambda_{\eta_{k[(1,a_1),(2,a_2)\ldots(k-1,a_{k-1})]}}^{k[(1,a_1),(2,a_2)\ldots(k-1,a_{k-1})]}\}$$

for $k = 1, 2, \ldots, \tau$; information is produced along different branches or paths in temporal sequence, in each of which it is sufficient (in the statistical sense) for the execution of feedback strategies within a given interval of time. Payoffs to all players are assumed to change at the end point of each time interval $[t_k, t_{k+1})$, while their values during the interval are common knowledge to all players. This result captures the fact that in markets, news rapidly disseminate over the trading day, until it becomes market information which is commonly known and available to all individuals for the framing of optimal (feedback) investment strategies. At the beginning of the next day more news arrive, after which the process repeats itself.

The informational sufficiency and recursive nature (in the sense of dynamic programming) of this sequential realization of payoffs render the analysis operational. Given any specification of the branching stochastic process, it is possible to simulate a set of such paths, which in turn would supply sufficient informational support for the execution of optimal feedback strategies within the corresponding time interval. If the simulation results are robust (in the statistical sense) to changes in stochastic specifications, a representative solution can be computed as a benchmark for applications. Of course, specific and detailed calibrations would be required before the benchmark can be applied in any given case.

4 Infinite-Horizon Furcating Stochastic Differential Games

In many game situations, the terminal time of the game, T, is either very far in the future or unknown to the players. For example, the value of a publicly listed firm is the present value of its discounted expected future earnings. Nobody knows when the firm will be out of business. As argued by Dockner et al. [11], in this case setting $T = \infty$ may very well be the best approximation for the true game horizon. Even if the firm's management restricts itself to considering profit maximization over the next year, it should value its asset positions at the end of the year by the earning potential of these assets in the years to come. In this section, game situations with infinite horizon and stochastic dynamics and uncertain future payoffs are considered. Important examples of this kind of problems include renewable resources extraction, environmental management, and the pricing of corporate equities. More specifically, randomly furcating stochastic differential games with an infinite horizon are examined.

Consider the game in which player i, for $i \in \mathbf{N} = [1, 2, \ldots, N]$, seeks to maximize:

$$E_{t_0}\left\{\int_{t_0}^{t_1} g^{[i,\theta_{a0}]}[x(s),u^1(s),u^2(s),\ldots,u^N(s)]e^{-r(s-t_0)}ds\right.$$

$$\left.+\sum_{h=1}^{\infty}\int_{t_h}^{t_{h+1}} g^{[i,\theta^h]}[\theta^h;s,x(s),u^1(s)u^2(s),\ldots,u^N(s)]e^{-r(s-t_0)}ds\right\}$$

$$=E_{t_0}\left\{\int_{t_0}^{t_1} g^i[\theta_{a_0};x(s),u^1(s),u^2(s),\ldots,u^N(s)]e^{-r(s-t_0)}ds\right.$$

$$\left.+\sum_{h=1}^{\infty}\int_{t_h}^{t_{h+1}} g^i[\theta^h;s,x(s),u^1(s)u^2(s),\ldots,u^N(s)]e^{-r(s-t_0)}ds\right\}, \tag{9}$$

where $x(s) \in X \subset R^m$ denotes the state variables of game, $u^i \in U^i$ is the control of player i, for $i \in \mathbf{N}$, E_{t_0} denotes the expectation operator performed at time t_0, and θ^h, $h = 1, 2, \ldots$, are independent and identically distributed random variables with range $\{\theta_1, \theta_2, \ldots, \theta_\eta\}$ and corresponding probabilities $\{\lambda_1, \lambda_2, \ldots, \lambda_\eta\}$, which are realized in the time interval $[t_h, t_{h+1})$, with $\theta_{a_0} \in \{\theta_1, \theta_2, \ldots, \theta_\eta\}$ being known at time t_0.

Moreover, the time intervals $[t_h, t_{h+1}]$ are of equal length T for all h. We can interpret the stochastic process generated by θ^h as a steady-state branching process.

The game's state dynamics is characterized by the vector-valued stochastic differential equation:

$$dx(s) = f[x(s),u^1(s),u^2(s),\ldots,u^N(s)]ds+\sigma[x(s)]dz(s), \quad x(t_0) = x_0 \tag{10}$$

where $\sigma[x(t)]$ is a $m \times \Theta$ matrix, and $z(t)$ a Θ-dimensional Wiener process. We let $\Omega[x(t)] = \sigma[x(t)]\sigma[x(t)]'$ denote the covariance matrix, where the element in row h and column ζ is denoted by $\Omega^{h\zeta}[x(t)]$. Uncertainty in future payoffs is introduced through the random variable θ^h and its effects on the functions $g^i[\theta, x(s), u^1(s), u^2(s), \ldots, u^N(s)]$. It should be noted that the functions $g^i[\theta, x(s), u^1(s), u^2(s), \ldots, u^N(s)]$, $i \in \mathbf{N}$, $f[x(s), u^1(s), u^2(s), \ldots, u^N(s)]$ and $\sigma[x(s)]$ do not depend on time s explicitly.

The objective function of player i, for $i \in \mathbf{N}$, in (9) can be expressed alternatively as:

$$E_{t_0}\left\{\int_{t_0}^{t_1} g^i[\theta_{a_0};x(s),u^1(s),u^2(s),\ldots,u^N(s)]e^{-r(s-t_0)}ds\right.$$

$$\left.+\sum_{h=1}^{\infty}\sum_{a_h=1}^{\eta}\lambda_{a_h}\int_{t_h}^{t_{h+1}} g^i[\theta_{a_h};x(s),u^1(s)u^2(s),\ldots,u^N(s)]e^{-r(s-t_0)}ds\right\}. \tag{11}$$

We can view the stochastic differential game (10)–(11) as an infinite-horizon autonomous problem with discounting. In particular, the discount factor is introduced to ensure the convergence of the objective functions. Existing studies in infinite-horizon autonomous stochastic differential games with discounting do not explicitly allow for the effects of the random variable θ^h. Examples can be founded in [8, 20, 15, 14, 22, 23]. Dropping the random variable θ^h, player i's instantaneous payoff is then given by $g^i[x(s), u^1(s), u^2(s), \ldots, u^N(s)]$, $i = 1, 2, \ldots, N$. Applying a result from [11], a Nash equilibrium of the resulting game can be characterized by the familiar condition in [16].

To characterize the solution of infinite-horizon furcating stochastic differential games (10)–(11), consider first the game in which player i, for $i \in \mathbf{N}$, seeks to maximize:

$$J_{k(a_k)} = E_{t_k}\left\{\int_{t_k}^{t_{k+1}} g^i[\theta_{a_k}; x(s), u^1(s), u^2(s), \ldots, u^N(s)]e^{-r(s-t_k)}ds \right.$$
$$\left. + \sum_{h=k+1}^{\infty} \sum_{a_h=1}^{\eta} \lambda_{a_h} \int_{t_h}^{t_{h+1}} g^i[\theta_{a_h}; x(s), u^1(s)u^2(s), \ldots, u^N(s)]e^{-r(s-t_k)}ds\right\}, \quad (12)$$

subject to the stock dynamics:

$$\begin{aligned} &dx(s) = f[x(s), u^1(s), u^2(s), \ldots, u^N(s)]ds + \sigma[x(s)]dz(s)\,, \\ &x(t_k) = x_k \in X\,. \end{aligned} \quad (13)$$

One can readily verify that the problem (12)–(13) satisfies the condition presented in the following remark.

Remark The infinite-horizon autonomous problem (12)–(13) is independent of the choice of t_k and dependent only upon x_k (the state x at time t_k) and the outcome θ_{a_k}. In particular, the problem in which player i, for $i \in \mathbf{N}$, seeks to maximize:

$$J_{\tau(a_\tau)} = E_{t_\tau}\left\{\int_{t_\tau}^{t_{\tau+1}} g^i[\theta_{a_\tau}; x(s), u^1(s), u^2(s), \ldots, u^N(s)]e^{-r(s-t_\tau)}ds \right.$$
$$\left. + \sum_{h=\tau+1}^{\infty} \sum_{a_h=1}^{\eta} \lambda_{a_h} \int_{t_h}^{t_{h+1}} g^i[\theta_{a_h}; x(s), u^1(s)u^2(s), \ldots, u^N(s)]e^{-r(s-t_\tau)}ds\right\}, \quad (14)$$

subject to the stock dynamics:

$$\begin{aligned} &dx(s) = f[x(s), u^1(s), u^2(s), \ldots, u^N(s)]ds + \sigma[x(s)]dz(s)\,, \\ &x(t_\tau) = x_\tau \in X\,, \end{aligned} \quad (15)$$

is identical to the game (12)–(13) if $\theta_{a_k} = \theta_{a_\tau}$ and $x_k = x_\tau$, even for $t_\tau \neq t_k$. We then define:

Definition 2. *A set of feedback strategies* $\{u_{a_\omega}^{i[\omega]^*}(t) = \phi_{a_\omega}^{i[\omega]^*}(t, x); i \in \mathbf{N}\}$ *contingent upon the events* θ_{a_ω}, *for* $t \in [t_\omega, t_{\omega+1})$, $\omega = k, k+1, k+2, \ldots, \infty$, *and* $a_\omega = 1, 2, \ldots, \eta$, *constitutes a Nash equilibrium solution for the game (12)–(13), if there exists functions* $V^{i(k,a_k)}(t, x)$, *for* $t \in [t_k, t_{k+1})$, *satisfying the following relations for each* $i \in \mathbf{N}$:

$$V^{i(k,a_k)}(t, x) = E_{t_k}\left\{\int_{t}^{t_{k+1}} g^i[\theta_{a_k}; x(s), u_{a_k}^{1[k]^*}(s), \ldots, u_{a_k}^{N[k]^*}(s)]e^{-r(s-t_k)}ds\right.$$

$$+\sum_{h=k+1}^{\infty}\sum_{a_h=1}^{\eta}\lambda_{a_h}\int_{t_h}^{t_{h+1}} g^i[\theta_{a_h};x(s),u_{a_k}^{1[k]^*}(s),\ldots,u_{a_k}^{N[k]^*}(s)]e^{-r(s-t_k)}ds\Bigg\}$$

$$=\max_{\left\{u_{a_b}^{i[b]}\right\}_{b=k}^{\infty}} E_{t_k}\Bigg\{\int_t^{t_{k+1}} g^i[\theta_{a_k};x(s),u_{a_k}^{1[k]^*}(s),\ldots$$

$$\ldots,u_{a_k}^{i-1[k]^*},u_{a_k}^{i[k]},u_{a_k}^{i+1[k]^*},\ldots,u_{a_k}^{N[k]^*}(s)]e^{-r(s-t_k)}ds$$

$$+\sum_{h=k+1}^{\infty}\sum_{a_h=1}^{\eta}\lambda_{a_h}\int_{t_h}^{t_{h+1}} g^i[\theta_{a_h};x(s),u_{a_h}^{1[h]^*}(s),u_{a_h}^{2[h]^*}(s),\ldots$$

$$\ldots,u_{a_h}^{i-1[h]^*},u_{a_h}^{i[h]},u_{a_h}^{i+1[h]^*},\ldots,u_{a_h}^{N[h]^*}(s)]e^{-r(s-t_k)}ds\Bigg\}, \quad \textit{for all } i\in\mathbf{N},$$

subject to dynamics (13).

Furthermore, $V^{i(k,a_k)}(t,x)$, for $t\in[t_k,t_{k+1})$, can be expressed as:

$$V^{i(k,a_k)}(t,x)=E_{t_k}\Bigg\{\int_t^{t_{k+1}} g^i[\theta_{a_k};x(s),u_{a_k}^{1[k]^*}(s),\ldots,u_{a_k}^{N[k]^*}(s)]e^{-r(s-t_k)}ds$$

$$+e^{-r(t_{k+1}-t_k)}\sum_{a_{h+1}}^{\eta}\lambda_{a_{k+1}}\Bigg(\int_{t_{k+1}}^{t_{k+2}} g^i[\theta_{a_{k+1}};x(s),u_{a_{k+1}}^{1[k+1]^*}(s),\ldots]e^{-r(s-t_{k+1})}ds$$

$$+\sum_{h=k+2}^{\infty}\sum_{a_h}^{\eta}\lambda_{a_h}\int_{t_h}^{t_{h+1}} g^i[\theta_{a_h};x(s),u_{a_h}^{1[h]^*}(s),\ldots,u_{a_h}^{N[h]^*}(s)]e^{-r(s-t_{k+1})}ds\Bigg)\Bigg\}$$

$$=E_{t_k}\Bigg\{\int_t^{t_{k+1}} g^i[\theta_{a_k};x(s),u_{a_k}^{1[k]^*}(s),u_{a_k}^{2[k]^*}(s),\ldots,u_{a_k}^{N[k]^*}(s)]e^{-r(s-t_k)}ds$$

$$+e^{-r(t_{k+1}-t_k)}\sum_{a_{k+1}=1}^{\eta}\lambda_{a_{k+1}}V^{i(k+1,a_{k+1})}[t_{k+1},x(t_{k+1})]\Bigg\}. \tag{16}$$

The characteristics of a Nash equilibrium solution of the game (12)–(13) are presented below.

Lemma 1. *A set of feedback strategies* $\{u_{a_\omega}^{i[\omega]^*}(t)=\phi_{a_\omega}^{i[\omega]^*}(t,x);i\in\mathbf{N}\}$ *contingent upon the events* θ_{a_ω}*, for* $\omega=k,k+1,k+2,\ldots$ *and* $a_\omega=1,2,\ldots,\eta$*, constitutes a Nash equilibrium solution for the game (12)–(13), if there exist suitably smooth functions* $V^{i(\omega,a_k)}(t,x):[t_\omega,t_{\omega+1}]\times R^n\to R$ *, for* $i\in\mathbf{N}$*,* $\omega=k,k+1,k+2,\ldots$ *and* $a_\omega=1,2,\ldots,\eta$*, which satisfy the following set of partial differential equations:*

$$-V_t^{i(\omega,a_\omega)}-\frac{1}{2}\sum_{h,\zeta=1}^{N}\Omega^{h\zeta}(x)V_{x_h x_\zeta}^{i(\omega,a_\omega)}$$

$$= \max_{\phi^i \in U_k^i} \left\{ g^{i(k)}[\theta_{a_\omega}; x, \phi_{a_\omega}^{1[\omega]^*}(t,x), \ldots, \phi_{a_\omega}^{i-1[\omega]^*}(t,x), \phi^i(t,x), \phi_{a_\omega}^{i+1[\omega]^*}(t,x), \ldots \right.$$

$$\ldots, \phi_{a_\omega}^{N[\omega]^*}(t,x)] e^{-r(t-t_\omega)}$$

$$\left. + V_x^{i(\omega,a_\omega)} f[x, \phi_{a_\omega}^{1[\omega]^*}(t,x), \ldots, \phi^i(t,x), \ldots, \phi_{a_\omega}^{N[\omega]^*}(t,x)] \right\} , \tag{17}$$

and

$$V^{i(\omega,a_k)}(t_{\omega+1}, x) = e^{-r(t_{\omega+1}-t_\omega)} \sum_{a_{\omega+1}}^{\eta} \lambda_{a_{\omega+1}} V^{i(\omega+1,a_{\omega+1})}(t_{\omega+1}, x) , \tag{18}$$

for $i \in \mathbf{N}$, $a_\omega = 1, 2, \ldots, \eta$, *and* $\omega = k, k+1, k+2, \ldots$.

Proof. This result follows directly from the definition of Nash equilibrium, the stochastic control techniques of [11], and the recursive results arising from (16). □

As indicated in (4), the outcome of the infinite-horizon autonomous problem (12)–(13) is independent of the choice of initial time t_k. Therefore $V^{i(k,a_k)}(t,x) = V^{i(k+1,a_k)}(t + t_{k+1}, x)$, and the expression in (16) can be rewritten as:

$$V^{i(k,a_k)}(t,x) = E_{t_k} \left\{ \int_t^{t_{k+1}} g^i[\theta_{a_k}; x(s), u_{a_k}^{1[k]^*}(s), \ldots, u_{a_k}^{N[k]^*}(s)] e^{-r(s-t_k)} ds \right.$$

$$\left. + e^{-r(t_{k+1}-t_k)} \sum_{a_{k+1}=1}^{\eta} \lambda_{a_{k+1}} V^{i(k,a_{k+1})}(t_k, x) \right\} . \tag{19}$$

A feedback Nash solution of the game (12)–(13) can be characterized as follows.

Theorem 2. *A set of feedback strategies* $\{u_{a_k}^{i[k]^*}(t) = \phi_{a_k}^{i[k]^*}(t,x); i \in \mathbf{N}\}$ *contingent upon the events* θ_{a_k}, *for* $a_k = 1, 2, \ldots, \eta$, *constitutes a Nash equilibrium solution for the game (12)–(13), if there exist suitably functions* $V^{i(k,a_k)}(t,x) : [0,T] \times R^n \to R$, *for* $i \in \mathbf{N}$, $a_k = 1, 2, \ldots, \eta$, *which satisfy the following set of partial differential equations:*

$$-V_t^{i(k,a_k)} - \frac{1}{2} \sum_{h,\zeta=1}^{N} \Omega^{h\zeta}(x) V_{x_h x_\zeta}^{i(k,a_k)}$$

$$= \max_{\phi^i \in U_k^i} \left\{ g^i[\theta_{a_k}; x, \phi_{a_k}^{1[k]^*}(t,x), \ldots, \phi^i(t,x), \ldots, \phi_{a_k}^{N[k]^*}(t,x)] e^{-rt} \right.$$

$$\left. + V_x^{i(k,a_k)} f[x, \phi_{a_k}^{1[k]^*}(t,x), \ldots, \phi^i(t,x), \ldots, \phi_{a_k}^{N[k]^*}(t,x)] \right\} , \tag{20}$$

and

$$V^{i(k,a_k)}(T,x) = e^{-rT} \sum_{a_{k+1}=1}^{\eta} \lambda_{a_{k+1}} V^{i(k,a_{k+1})}(0,x) . \tag{21}$$

Proof. Since the game (12)–(13) is independent of the choice of t_k, we select $t_k = 0$, after which (2.78) becomes:

$$V^{i(k,a_k)}(t,x) = E_{t_k}\left\{\int_t^T g^i[\theta_{a_k}; x(s), \phi_{a_k}^{1[k]^*}(s,x), \ldots, \phi_{a_k}^{N[k]^*}(s,x)]e^{-rs}ds\right.$$

$$\left. +e^{-rT}\sum_{a_{k+1}}^{\eta} \lambda_{a_{k+1}} V^{i(k,a_{k+1})}(0,x)\right\}. \tag{22}$$

With (19) and Remark 4, Lemma 1 reduces to (20) and (21). Hence theorem 2 follows. □

5 Illustrations

In this section, two illustrations are given. The first illustration is a finite-time randomly-furcating stochastic differential game and the second one is an infinite-horizon branching stochastic differential game.

5.1 A Finite-Time Randomly-Furcating Stochastic Differential Game of Resource Extraction

Consider an economy which is endowed with a single renewable resource and has $N \geq 2$ resource extractors. Let $u^i(s)$ denote the rate of resource extraction of extractor i at time s, for $i \in \mathbf{N} = \{1,2,\ldots,N\}$. Each extractor controls his rate of extraction. Let U^i be the set of admissible extraction rates and let $x(s)$ be the size of the resource stock at time s. In particular, $U^i \in R^+$ for $x > 0$ and $U^i = \{0\}$ for $x = 0$.

The costs of extraction for extractor $i \in \mathbf{N}$ depend on the quantity of resource extracted $u^i(s)$, the resource stock size $x(s)$ and a cost parameter c. In particular, player i's costs of extraction is specified as follows:

$$C^i = \frac{c}{x(s)^{1/2}}u^i(s). \tag{23}$$

This specification implies that the cost per unit of resource extracted, $cx(s)^{-\frac{1}{2}}$, decreases when $x(s)$ increases. A decreasing unit cost follows from two assumptions: (i) the cost of extraction is proportional to extraction effort, and (ii) the amount of resource extracted, seen as the output of a production function of two inputs (effort and stock level), is increasing in both inputs (cf. [6]).

The market price of the resource depends on the total amount extracted and brought to the market. The price-output relationship at time s is given by the following downward sloping inverse demand curve:

$$P(s) = \theta^k Q(s)^{-1/2}, \quad \text{for } s \in [t_k, t_{k+1}), \tag{24}$$

where $Q(s) = \sum_{i\in N} u^i(s)$ is the total amount of resource harvested and marketed at time s, θ^0 equal to 1 with probability one in the time interval $[t_0, t_1)$ and θ^k, for $k = 1, 2, \ldots, \tau$, is a random variable stemming from the branching process: $\theta^1 = \{\theta_1^1, \theta_2^1, \ldots, \theta_{\eta_1}^1\}$ with corresponding probability $\{\lambda_1^1, \lambda_2^1, \ldots, \lambda_{\eta_1}^1\}$.

Given that $\theta_{a_1}^1$ occurred in time interval $[t_1, t_2)$, for $a_1 = 1, 2, \ldots, \eta_1$,

$$\theta^2 = \{\theta_1^{2[(1,a_1)]}, \theta_2^{2[(1,a_1)]}, \ldots, \theta_{\eta_{2[(1,a_1)]}}^{2[(1,a_1)]}\}$$

with probability

$$\{\lambda_1^{2[(1,a_1)]}, \lambda_2^{2[(1,a_1)]}, \ldots, \lambda_{\eta_{2[(1,a_1)]}}^{2[(1,a_1)]}\}.$$

Given that $\theta_{a_1}^1$ occurred in time interval $[t_1, t_2)$ and $\theta_{a_2}^{2[(1,a_1)]}$ occurred in time interval $[t_2, t_3)$, for $a_1 = 1, 2, \ldots, \eta_1$ and $a_2 = 1, 2, \ldots, \eta_{2[(1,a_1)]}$,

$$\theta^3 = \{\theta_1^{3[(1,a_1)(2,a_2)]}, \theta_2^{3[(1,a_1)(2,a_2)]}, \ldots, \theta_{\eta_{3[(1,a_1)(2,a_2)]}}^{3[(1,a_1)(2,a_2)]}\}$$

with probability

$$\{\lambda_1^{3[(1,a_1)(2,a_2)]}, \lambda_2^{3[(1,a_1)(2,a_2)]}, \ldots, \lambda_{\eta_{3[(1,a_1)(2,a_2)]}}^{3[(1,a_1)(2,a_2)]}\}.$$

In general, $\theta_{a_1}^1$ occurred in time interval $[t_1, t_2)$, $\theta_{a_2}^{2[(1,a_1)]}$ occurred in time interval $[t_2, t_3)$,..., and $\theta_{a_{k-1}}^{k-1[(1,a_1)(2,a_2)\ldots(k-2,a_{k-2})]}$ occurred in time interval $[t_{k-1}, t_k)$, for $a_1 = 1, 2, \ldots, \eta_1$, $a_2 = 1, 2, \ldots, \eta_{2[(1,a_1)]}$, ..., $a_{k-1} = 1, 2, \ldots, \eta_{k-1[(1,a_1)(2,a_2)\ldots(k-2,a_{k-2})]}$,

$$\theta^k = \{\theta_1^{k[(1,a_1),(2,a_2)\ldots(k-1,a_{k-1})]}, \ldots, \theta_{\eta_{k[(1,a_1),(2,a_2)\ldots(k-1,a_{k-1})]}}^{k[(1,a_1),(2,a_2)\ldots(k-1,a_{k-1})]}\}$$

with probability

$$\{\lambda_1^{k[(1,a_1),(2,a_2)\ldots(k-1,a_{k-1})]}, \ldots, \lambda_{\eta_{k[(1,a_1),(2,a_2)\ldots(k-1,a_{k-1})]}}^{k[(1,a_1),(2,a_2)\ldots(k-1,a_{k-1})]}\}$$

for $k = 1, 2, \ldots, \tau$.

The random variables characterize the random elements concerning future demand. Earlier studies have assumed that market price is independent of output. Clark [7] pointed out that the relative simplicity of his pure competition model, which stems from its linearity in the control variables, allows one to derive an optimal policy in feedback form. Dockner et al. [10] used a demand curve with unit elasticity to capture the characteristics of an Eskimoan economy in a differential game model of fishery. The demand curve in (24) was similar to that suggested in [14]. It is elastic and the frivolous outcomes that would occur in monopolistic extraction or joint-maximization under unit elasticity demand are avoided.

The lease for resource extraction starts at time t_0 and ends at time T. When the lease terminates, the resource extractors may be offered a bonus (in present value) $e^{-r(T-t_0)}\theta^T x(T)^{1/2}$, where θ^T is a random variable with range $\{\theta_1^T, \theta_2^T, \ldots, \theta_{\eta_T}^T\}$and corresponding probability $\{\lambda_1^T, \lambda_2^T, \ldots, \lambda_{\eta_T}^T\}$. The dependence of the terminal bonus on the terminal resource stock reflects the leasing body's preference for the extractors to conserve the resource. The random elements in the size of the bonus (denoted by the value of θ^T) is related to the economic environment at time T.

The objective of extractor $i \in \mathbf{N}$ is to maximize his expected profits:

$$E_{t_0}\left\{\sum_{k=0}^{T}\int_{t_k}^{t_{k+1}}\left[\theta^k\Big(\sum_{j=1}^{N}u^j(s)\Big)^{-1/2}u^i(s) - \frac{c}{x(s)^{1/2}}u^i(s)\right]e^{-r(s-t_0)}ds\right.$$

$$\left. + e^{-r(T-t_0)}\theta^T x(T)^{1/2}\right\}, \tag{25}$$

where r is a discount rate which is common to all extractors.

With commercial extraction the resource stock evolves according to:

$$dx(s) = \left[ax(s)^{1/2} - bx(s) - \sum_{j=1}^{N}u^j(s)\right]ds + \sigma x(s)dz(s). \tag{26}$$

Following the analysis in Section 3, we obtain:

Proposition 1. *Given the occurrence of* $\theta_{a_2}^{2[(1,a_1)]}$, *for* $a_1 \in [1,2]$ *and* $a_2 \in [1,2]$, *in time interval* $[t_2, t_3)$, *the value functions of player* i_2 *in the interval* $[t_2, t_3)$, *for* $i_2 \in J_2 = \{1,3,4\}$, *can be obtained as:*

$$V^{i_2[(1,a_1)]a_2}(t_3, x) = e^{-r(t-t_2)}\left[A_{[(1,a_1)]a_2}(t)x^{1/2} + B_{[(1,a_1)]a_2}(t)\right], \tag{27}$$

where $A_{[(1,a_1)]a_2}(t)$ *and* $B_{[(1,a_1)]a_2}(t)$ *satisfy:*

$$\dot{A}_{[(1,a_1)]a_2}(t) = \left[r + \frac{1}{8}\sigma^2 + \frac{b}{2}\right]A_{[(1,a_1)]a_2}(t) - \frac{1}{2[c_{[(1,a_1)(2,a_2)]} + A_{[(1,a_1)]a_2}(t)/2]}$$

$$+\frac{c_{[(1,a_1)(2,a_2)]}}{4[c_{[(1,a_1)(2,a_2)]} + A_{[(1,a_1)]a_2}(t)/2]^2} + \frac{3A_{[(1,a_1)]a_2}(t)}{8[c_{[(1,a_1)(2,a_2)]} + A_{[(1,a_1)]a_2}(t)/2]^2}, \tag{28}$$

$$\dot{B}_{[(1,a_1)]a_2}(t) = rB_{[(1,a_1)]a_2}(t) - \frac{b}{a}A_{[(1,a_1)]a_2}(t), \tag{29}$$

$$A_{[(1,a_1)]a_2}(t_3) = \sum_{b=1}^{2}\hat{\lambda}_b\hat{\theta}_b, \tag{30}$$

$$\text{and } B_{[(1,a_1)]a_2}(t_3) = 0. \tag{31}$$

Proposition 2. *Given the occurrence of* $\theta_{a_1}^1$, *for* $a_1 \in [1,2]$, *in time interval* $[t_1, t_2)$, *the value functions of player* i_1 *in the interval* $[t_1, t_2)$, *for* $i_1 \in J_1 = \{1,3\}$ *and* $a_1 \in [1,2]$, *can be obtained as:*

$$V^{i_1[0]a_1}(t,x) = e^{-r(t-t_1)}\left[A_{(0)a_1}(t)x^{1/2} + B_{(0)a_1}(t)\right], \tag{32}$$

where $A_{(0)a_1}(t)$ and $B_{(0)a_1}(t)$ satisfy:

$$\dot{A}_{(0)a_1}(t) = \left[r + \frac{1}{8}\sigma^2 + \frac{b}{2}\right] A_{(0)a_1}(t) - \frac{1}{2[c_{[(1,a_1)]} + A_{(0)a_1}(t)/2]} +$$

$$+\frac{c_{[(1,a_1)]}}{4[c_{[(1,a_1)]} + A_{(0)a_1}(t)/2]^2} + \frac{2A_{(0)a_1}(t)}{8[c_{[(1,a_1)]} + A_{(0)a_1}(t)/2]^2}, \tag{33}$$

$$\dot{B}_{(0)a_1}(t) = rB_{(0)a_1}(t) - \frac{b}{a}A_{(0)a_1}(t), \tag{34}$$

$$A_{(0)a_1}(t_2) = \sum_{a_2=1}^{2} \lambda_{a_2}^{2[(1,a_1)]} A_{[(1,a_1)]a_2}(t), \tag{35}$$

$$\text{and } B_{(0)a_1}(t_2) = \sum_{a_2=1}^{2} \lambda_{a_2}^{2[(1,a_1)]} B_{[(1,a_1)]a_2}(t_2). \tag{36}$$

Proposition 3. *The value functions of player 1 and player 2 in the interval $[t_0, t_1)$ are respectively:*

$$V^{1[]0}(t,x) = e^{-r(t-t_0)}\left[A_{1[]}(t)x^{1/2} + B_{1[]}(t)\right], \text{ and} \tag{37}$$

$$V^{2[]0}(t,x) = e^{-r(t-t_0)}\left[A_{2[]}(t)x^{1/2} + B_{2[]}(t)\right], \tag{38}$$

where $A_{1[]}(t)$,$B_{1[]}(t)$,$A_{2[]}(t)$ and $B_{2[]}(t)$ satisfy:

$$\dot{A}_{2[]}(t) = \left[r + \frac{1}{8}\sigma^2 + \frac{b}{2}\right] A_{2[]}(t) - \frac{1}{2[c_0 + A_{2[]}(t)/2]} +$$

$$+\frac{c_0}{4[c_0 + A_{2[]}(t)/2]^2} + \frac{A_{2[]}(t)}{8[c_0 + A_{2[]}(t)/2]^2} + \frac{A_{2[]}(t)}{8[c_0 + A_{1[]}(t)/2]^2} \tag{39}$$

$$\dot{B}_{2[]}(t) = rB_{2[]}(t) - \frac{b}{a}A_{2[]}(t), \tag{40}$$

$$A_{2[]}(t_1) = \sum_{b=1}^{2} \hat{\lambda}_b \hat{\theta}_b, \quad \text{and} \quad B_{2[]}(t_1) = 0. \tag{41}$$

$$\dot{A}_{1[]}(t) = \left[r + \frac{1}{8}\sigma^2 + \frac{b}{2}\right] A_{1[]}(t) - \frac{1}{2[c_0 + A_{1[]}(t)/2]} +$$

$$+\frac{c_0}{4[c_0 + A_{1[]}(t)/2]^2} + \frac{A_{1[]}(t)}{8[c_0 + A_{1[]}(t)/2]^2} + \frac{A_{1[]}(t)}{8[c_0 + A_{2[]}(t)/2]^2}, \tag{42}$$

$$\dot{B}_{1[]}(t) = rB_{1[]}(t) - \frac{b}{a}A_{1[]}(t)\,, \tag{43}$$

$$A_{1[]}(t_1) = \sum_{a_1=1}^{2} \lambda_{a_1}^1 A_{(0)a_1}(t_1)\,,$$

$$and \;\; B_{1[]}(t_1) = \sum_{a_1=1}^{2} \lambda_{a_1}^1 B_{(0)a_1}(t_1)\,. \tag{44}$$

5.2 A Infinite-Horizon Randomly-Furcating Stochastic Differential Game of Resource Extraction

Consider an infinite horizon counterpart of the game in Section 5.2. In particular, resource firms are awarded an infinite lease for resource extraction, beginning at time t_0. The objective of firm $i \in \mathbf{N}$ is to maximize expected profits:

$$E_{t_0}\Bigg\{\int_{t_0}^{t_1}\left[\theta_{a_0}\left(\sum_{j=1}^{N}u^j(s)\right)^{-1/2}u^i(s) - \frac{c}{x(s)^{1/2}}u^i(s)\right]e^{-r(s-t_0)}ds$$

$$+\sum_{h=1}^{\infty}\int_{t_h}^{t_{h+1}}\left[\theta^h\left(\sum_{j=1}^{N}u^j(s)\right)^{-1/2}u^i(s) - \frac{c}{x(s)^{1/2}}u^i(s)\right]e^{-r(s-t_0)}ds\Bigg\}\,, \tag{45}$$

where r is the common discount rate. The resource stock evolves according to:

$$dx(s) = \left[ax(s)^{1/2} - bx(s) - \sum_{j=1}^{N}u^j(s)\right]ds + \sigma x(s)dz(s), \quad x(t_0) = x_0 \,, \tag{46}$$

where θ^h is a random variable realizable in the time interval $[t_h, t_{h+1})$, and θ^h, $h = 1, 2, \ldots$, are independent and identical random variables with range $\{\theta_1, \theta_2\}$ and corresponding probabilities $\{\lambda_1, \lambda_2\}$. In particular, $\theta_2 > \theta_1$ and $\theta_{a_0} \in \{\theta_1, \theta_2\}$ is known by all players in the time interval $[t_0, t_1)$. The lengths of the interval $[t_h, t_{h+1}]$ are the same for $h = 0, 1, 2, \ldots$.

Following the analysis in Section 4, we obtain:

Proposition 4. *Given the occurrence of θ_{a_k}, for $a_k \in [1, 2]$, in time interval $[t_k, t_{k+1})$, the value functions of player $i \in \mathbf{N}$ in the interval can be obtained as:*

$$V^{i(k,a_k}(t,x) = e^{-rt}\left[A_{a_k}(t)x^{1/2} + B_{a_k}(t)\right], \quad for \;\; i \in \mathbf{N}\,, \; a_k = 1, 2\,, \tag{47}$$

where $A_{a_k}(t)$ and $B_{a_k}(t)$ satisfy:

$$\dot{A}_{a_k}(t) = \left[r + \frac{1}{8}\sigma^2 + \frac{b}{2}\right]A_{a_k}(t) - \frac{2N-1}{2N^2}\frac{(\theta_{a_k})^2}{\left(c + \frac{A_{a_k}(t)}{2}\right)}$$

$$+\frac{(2N-1)^2}{4N^3}\frac{(\theta_{a_k})^2 c}{\left(c+\frac{A_{a_k}(t)}{2}\right)^2}+\frac{(2N-1)^2(\theta_{a_k})^2 A_{a_k}(t)}{8N^2\left(c+\frac{A_{a_k}(t)}{2}\right)^2}, \tag{48}$$

$$\dot{B}_{a_k}(t)=rB_{a_k}(t)-\frac{a}{2}A_{a_k}(t), \tag{49}$$

$$A_{a_k}(T)=\sum_{a_{k+1}=1}^{2}\lambda_{a_{k+1}}A_{a_{k+1}}(0), \tag{50}$$

$$B_{a_k}(T)=\sum_{a_{k+1}=1}^{2}\lambda_{a_{k+1}}B_{a_{k+1}}(0). \tag{51}$$

6 Concluding Remarks

In this paper, we present and characterize a class of "Randomly Furcating Stochastic Differential Games" to formalize game-theoretic situations with stochastic dynamics and uncertain environments over time in the following manner. The concept of "random furcating" is introduced to describe future payoffs as changing at any instant of time according to (known) probability distributions, which turn are defined in terms of multiple-branching stochastic processes. The introduction of this stochastic specification lead to a novel approach to solve differential games in terms of properties and solution concepts not explored in the literature before. New and significant mathematical results are obtained, under which it becomes possible to characterize the conditions under which previously unsolvable games can be solved. This new approach widens the application of differential game theory to problems where future environments are not known with certainty. Two illustrations are presented.

The analysis can be extended to cover solution techniques for randomly furcating endogenous horizon differential games, infinite-time randomly furcating differential games with uncertain number and types of players, and infinite overlapping generations randomly furcating differential games. The results can also be applied to optimization theory, especially to the solution of problems in infinite-horizon randomly furcating stochastic optimal control and non-autonomous infinite-horizon randomly furcating stochastic optimal control.

References

1. Basar, T. (1977a): Existence of Unique Equilibrium Solutions in Nonzero-Sum Stochastic Differential Games. In: Roxin, E. O., Liu, P. T., Sternberg, R. (Eds.): Differential Games and Control Theory II. Marcel Dekker, Inc., 201–228
2. Basar, T. (1977b): Informationally Nonunique Equilibrium Solutions in Differential Games. SIAM Journal of Control and Optimization **15**, 636–660

3. Basar, T. (1980b): On the Existence and Uniqueness of Closed-Loop Samled-Data Nash Controls in Linear-Quadratic Stochastic Differential Games. In: Iracki, K. et al. (Eds.): Optimization Techniques. Lecture Notes in Control and information Sciences. Springer-Verlag, New York, ch. 22, 193–203
4. Berkovitz, L. D. (1964): A variational approach to differential games. In: Dresher, M., Shapley, L. S., Tucker, A. W. (Eds.): Advances in Game Theory. Princeton, Princeton University Press, NJ, 127–174
5. Clark, C. W. (1976): Mathematical Bioeconomics: The Optimal Management of Renewable Resources. John Wiley, New York
6. Clark, C. W. (1980): Restricted assess to common-property Fishery resources: a game theoretic analysis. In: Liu, P. T. (Ed.): Dynamic Optimization and Mathematical Economics. Plenum, New York, 117–132
7. Clemhout, S., Wan, H. Y. Jr. (1985): Dynamic common-property resources and environmental problems. Journal of Optimization Theory and Applications **46**, 471–481
8. Davis, M. H. A. (1984): Piecewise-deterministic markov processes: a general class of non-diffusion stochastic models. Journal of the Royal Statistical Society (B) **46**, 353–388
9. Dockner, E. J., Feichtinger, G., Mehlmann, A. (1989): Noncooperative solutions for a differential game model of fishery. Journal of Economic Dynamics and Control **13**, 1–20
10. Dockner, E., Jorgensen, S., Long, N. V., Sorger, G. (2000): Differential Games in Economics and Management Science. Cambridge University Press, Cambridge
11. Fleming, W. H. (1969): Optimal continuous-parameter stochastic control. SIAM Review **11**, 470–509
12. Fleming, W. H., Rishel, R. W. (1975): Deterministic and Stochastic Optimal Control. Springer-Verlag, Berlin
13. Isaacs, R. (1965): Differential Games. Wiley, New York
14. Jorgensen, S., Yeung, D. (1996): Stochastic differential game model of a common property fishery. Journal of Optimization Theory and Applications **90**, 391–403
15. Kaitala, V. (1993): Equilibria in a stochastic resource management game under imperfect information. European Journal of Operational Research **71**, 439–453
16. Kamien, Schwartz (1991): Identification of classes of differential games for which the open-loop is a degenerate feedback Nash equilibrium. Journal of Optimization Theory and Applications **55**, 217–231
17. Leitmann, G., Mon, G. (1967): Some geometric aspects of differential games. Journal of Astronaut and Science **14**, 56
18. Rishel, R. (1975a): Control of systems with jump Markov disturbances. IEEE Transactions on Automatic Control **20**, 241–244
19. Rishel, R. (1975b): Dynamic programming and minimum principles for systems with jump Markov disturbances. SIAM Journal of Control **13**, 338–371
20. Sorger, G. (1989): Competitive dynamic advertising: a modification of the Case game. Journal of Economic Dynamics and Control **13**, 55–80
21. Vermes, D. (1985): Optimal control of piecewise-deterministic Markov processes. Stockastics **14**, 165–208
22. Yeung, D. W. K. (1998): A Class of Differential Games Which Admits a Feedback Solution with Linear Value Functions. European Journal of Operational Research **107 (3)**, 737–754

23. Yeung, D. W. K. (1999): A stochastic differential game model of institutional investor speculation. Journal of Optimization Theory and Applications **102**, 463–477

On Modulo 2 Game

Chizhong Zhou

Computer and Information Department, Yueyang Normal University,
Yueyang, Hunan 414000, PR China
E-mail: `chizhongz@163.com`

Summary. In this paper we deal with a type of extended combinatorial game,the modulo 2 game. In such a game a tie position may appear. We introduce a system of logical functions to display the properties of the game. Necessary and sufficient conditions for the existence of a winning strategy for the first player or second player are determined. Algorithms for implementing the optimal strategy and its complexity are then presented.

Key words: modulo 2 game, combinatorial game, extensive game, winning strategy, algorithmic game

1 Introduction

A combinatorial game typically involves two players, alternating play in well-defined moves. It is a perfect-information game, for which winning strategies can be stated explicitly, or at least proved to exist. A common assumption is that the game terminates after a finite number of moves, and the result is a unique winner. And in the combinatorial-game setting, it is usual to define the winner as the last player who is able to move, or to define the winner as the first player who cannot move. Of course, there are exceptions: some games can be drawn out forever, and some games define ties in certain cases [2, 3].

In the paper of Z. Huang and S. Huang [6] there is an unusual and interesting combinatorial game. Give a take-away game where two players alternately remove a positive number of tokens from a single heap with n (n is odd) tokens. The number of tokens that may be taken by each player at any move is restricted to one of $1, 2, \ldots, s$ ($s \geq 1$). As well as the heap is empty the total number of tokens removed by each player is counted separately. The winner is the player whose total number of tokens is odd.

In the above paper it is proved: if s is odd the first player to move can be the winner iff $n \neq s+2 \pmod{2s+2}$; if s is even the first player to move can be the winner iff $n \neq s+1 \pmod{s+2}$. A winning strategy for the first player is also given in the paper.

The essential difference between the game and the usual combinatorial game is that the standard to determine the winner is neither the last player being able to move nor the first player being unable to move. Instead, the standard is the total numbers of tokens, removed by each player separately, modulo 2. This is of interests. A natural problem is whether the game can be extended to a more general form. In other words, we want to consider whether the condition of odd n can be abolished and whether the restrict condition of the number of tokens taken at any move can be widened. In this paper we give a positive answer. First we state the following definitions.

Give a take-away game where two players alternately remove a positive number of tokens from a single heap with n tokens. The number of tokens that may be taken by each player at any move is restricted to belong to the set

$$\mathbf{S} = \{a_1,; a_2, \cdots, a_s\} \quad (0 < a_1 < a_2 < \cdots < a_s)\,. \tag{1}$$

As well as the number of tokens in the heap is less than a_1 the game is ended and the total number of tokens removed by each player is counted separately.

1. Define the player whose total number is even to be winner if the two total numbers have different pairities, otherwise define the position to be tie. Under the definition we call the game a *even-type modulo 2 game* with parameters n and $\mathbf{S}$ denoted by $\mathbf{M_0}(n,\ \mathbf{S})$.
2. Define the player whose total number is odd to be winner if the two total numbers have different pairities, otherwise define the position to be tie. Under the definition we call the game a *odd-type modulo 2 game* with parameters n and $\mathbf{S}$ denoted by $\mathbf{M_1}(n,\ \mathbf{S})$.

Even-type modulo 2 game and odd-type modulo 2 game, with parameters n and $\mathbf{S}$, are said to belong to *modulo 2 game* with parameters n and $\mathbf{S}$ denoted by $\mathbf{M}(n,\ \mathbf{S})$. In $\mathbf{M}(n,\ \mathbf{S})$ when the game is ended the total number of tokens moved by each player is called his *score*.

The above extended game is more difficult and complicated than the initial game. Even if n is odd, for some $\mathbf{S}$, one of the games $\mathbf{M_0}(n,\ \mathbf{S})$ and $\mathbf{M_1}(n,\ \mathbf{S})$ or both of them may have a tie position. For example, for $\mathbf{S} = 5,\ 8,\ 11,\ 13$, all $\mathbf{M_0}(55,\ \mathbf{S})$ and $\mathbf{M_1}(55,\ \mathbf{S})$ can have tie positions. Moreover, for example in $\mathbf{M_0}(n,\ \mathbf{S})$, a player who wants to win must make great efforts to ensure not only his own score to be even but also his opponent's score to be odd. It seems not easy to do so. Hence, it is difficult to solve the problem by the initial method of paper [6] or by other traditional methods. In Sec. 2 we will introduce a system of logical functions to expose the properties of modulo 2 game. Furthermore, we discover and prove the sufficient and necessary conditions that a player has a winning strategy and that there can be a tie position. In Sect. 3 the algorithms of optimal strategies of modulo 2 games are given. It is proved that the time complexity for each of these algorithm are $\mathcal{O}(n)$. In Sect. 4 two further problems are presented.

2 Logical Test Functions and Winning Strategy

First we give additional definitions: In a modulo 2 game, if a player has a strategy that ensures his own score is even (or odd) and makes his opponent's score is odd (or even, correspondingly) as well we call it a *strong even strategy* (or *strong odd strategy*, correspondingly) of him; If he has a strategy that ensures his own score is even (or odd) and also makes his opponent's score is even (or odd, correspondingly) as well we call it a *balanced even strategy* (or *balanced odd strategy*, correspondingly) of him. If the opponent of a player has a strategy that makes the player's score is even (or odd) we call the player *constrained to even status* (or *constrained to odd status*, correspondingly). It is clear that the first player to move (To be simple, in the sequent discussions we will delete "to move") has a winning strategy in $\mathbf{M_0}(n,\ \mathbf{S})$ (or $\mathbf{M_1}(n,\ \mathbf{S})$) iff he has a strong even (or odd, correspondingly) strategy. To answer whether the winning strategy exists we introduce a system of logical functions. A number of authors, e. g. [4, 5] had applied the methods of logic to analyze games. However, our method is different. In this paper we always assume that $\mathbf{S}$ is defined by (1). For integer x define

$$m(x) = \begin{cases} 1, & \text{for even x,} \\ 0, & \text{otherwise.} \end{cases} \tag{2}$$

Define logical functions $P_0(n) = P_0(n,\ \mathbf{S})$, $P_1(n) = P_1(n,\ \mathbf{S})$, $p_0(n) = p_0(n,\ \mathbf{S})$, $p_1(n) = p_1(n,\ \mathbf{S})$, $f_0(n) = f_0(n,\ \mathbf{S})$, $f_1(n) = f_1(n,\ \mathbf{S})$, $t_0(n) = t_0(n,\ \mathbf{S})$ and $t_1(n) = t_1(n,\ \mathbf{S})$, recursively, as below:
For $0 \le n < a_1$

$$P_0(n) = p_0(n) = t_0(n) = 1 \text{ and } P_1(n) = p_1(n) = f_0(n) = f_1(n) = t_1(n) = 0\,; \tag{3}$$

For $n \ge a_1$

$$P_0(n) = \bigvee_{a_i \le n} P_0(n - a_i)\,, \tag{4}$$

$$P_1(n) = \bigvee_{a_i \le n} P_1(n - a_i)\,. \tag{5}$$

$$p_0(n) = \bigwedge_{a_i \le n} (m(a_i) \wedge P_0(n - a_i) \vee \neg m(a_i) \wedge P_1(n - a_i))\,, \tag{6}$$

$$p_1(n) = \bigvee_{a_i \le n} (m(a_i) \wedge P_1(n - a_i) \vee \neg m(a_i) \wedge P_0(n - a_i))\,, \tag{7}$$

$$f_0(n) = p_1(n - a_i) \bigvee_{a_i \le n} \bigwedge_{a_j \le n - a_i} (u_{00}(a_i,\ a_j) \vee u_{01}(a_i,\ a_j) \vee$$

$$\vee u_{10}(a_i,\ a_j) \vee u_{11}(a_i,\ a_j))\ , \tag{8}$$

where

$u_{00}(a_i,\ a_j) = m(a_i) \wedge m(a_j) \wedge f_0(n - a_i - a_j)$,
$u_{01}(a_i,\ a_j) = m(a_i) \wedge \neg m(a_j) \wedge t_0(n - a_i - a_j)$,
$u_{10}(a_i,\ a_j) = \neg m(a_i) \wedge m(a_j) \wedge t_1(n - a_i - a_j)$,
$u_{11}(a_i,\ a_j) = \neg m(a_i) \wedge \neg m(a_j) \wedge f_1(n - a_i - a_j))$,
$f_1(n) = \neg m(a_1)$ for $a_1 \le n < 2 * a_1$,
and for $n \ge 2 * a_1$

$$f_1(n) = p_0(n - a_i) \wedge \bigvee_{a_i \le n} \bigwedge_{a_j \le n - a_i} (v_{00}(a_i,\ a_j) \vee v_{01}(a_i,\ a_j) \vee$$

$$\vee v_{10}(a_i,\ a_j) \vee v_{11}(a_i,\ a_j))\ , \tag{9}$$

where

$v_{00}(a_i,\ a_j) = m(a_i) \wedge m(a_j) \wedge f_1(n - a_i - a_j)$,
$v_{01}(a_i,\ a_j) = m(a_i) \wedge \neg m(a_j) \wedge t_1(n - a_i - a_j)$,
$v_{10}(a_i,\ a_j) = \neg m(a_i) \wedge m(a_j) \wedge t_0(n - a_i - a_j)$,
$v_{11}(a_i,\ a_j) = \neg m(a_i) \wedge \neg m(a_j) \wedge f_0(n - a_i - a_j)$,
$t_0(n) = m(a_1)$ for $a_1 \le n < 2 * a_1$,
and for $n \ge 2 * a_1$

$$t_0(n) = p_0(n - a_i) \bigvee_{a_i \le n} \bigwedge_{a_j \le n - a_i} (w_{00}(a_i,\ a_j) \vee w_{01}(a_i,\ a_j) \vee$$

$$\vee w_{10}(a_i,\ a_j) \vee w_{11}(a_i,\ a_j))\ , \tag{10}$$

where

$w_{00}(a_i,\ a_j) = m(a_i) \wedge m(a_j) \wedge t_0(n - a_i - a_j)$,
$w_{01}(a_i,\ a_j) = m(a_i) \wedge \neg m(a_j) \wedge f_0(n - a_i - a_j)$,
$w_{10}(a_i,\ a_j) = \neg m(a_i) \wedge m(a_j) \wedge f_1(n - a_i - a_j)$,
$w_{11}(a_i,\ a_j) = \neg m(a_i) \wedge \neg m(a_j) \wedge t_1(n - a_i - a_j)$,

$$t_1(n) = p_1(n - a_i) \bigvee_{a_i \le n} \bigwedge_{a_j \le n - a_i} (x_{00}(a_i,\ a_j) \vee x_{01}(a_i,\ a_j) \vee$$

$$\vee x_{10}(a_i,\ a_j) \vee x_{11}(a_i,\ a_j))\ , \tag{11}$$

where

$x_{00}(a_i,\ a_j) = m(a_i) \wedge m(a_j) \wedge t_1(n - a_i - a_j)$,
$x_{01}(a_i,\ a_j) = m(a_i) \wedge \neg m(a_j) \wedge f_1(n - a_i - a_j)$,
$x_{10}(a_i,\ a_j) = \neg m(a_i) \wedge m(a_j) \wedge f_0(n - a_i - a_j)$,
$x_{11}(a_i,\ a_j) = \neg m(a_i) \wedge \neg m(a_j) \wedge t_0(n - a_i - a_j)$.

Additionally, for $n \ge a_1$ define functions $g_0(n)$ and $g_1(n)$ as below:

$g_0(n) = \bigwedge_{a_i \le n} (m(a_i) \wedge f_0(n - a_i) \vee \neg m(a_i) \wedge t_0(n - a_i))$;
$g_1(n) = \bigwedge_{a_i \le n} (m(a_i) \wedge f_1(n - a_i) \vee \neg m(a_i) \wedge t_1(n - a_i))$.

We call each of the system of the above functions a *test function* of $\mathbf{M}(n, \mathbf{S})$. The effects of test functions will be observed from the following lemmas and theorems.

Lemma 1. *In* $\mathbf{M}(n, \mathbf{S})$

1. The first player is even constrained iff $p_0(n) = 1$.
2. The first player is odd constrained iff $p_1(n) = 1$.

Proof. We only prove the conclusion 1 by induction. It will be seen that the conclusion 2 can be proved in the same way. For $n < a_1$, from (3) the conclusion is obvious. Assume that the conclusion is true for all integer less than n $(n \ge a_1)$. We are to prove that it is also true for n. Suppose $p_0(n) = 1$. From (6), $m(a_i) \wedge P_0(n-a_i) \vee \neg m(a_i) \wedge P_1(n-a_i) = 1$ holds for all $a_i (a_i \le n)$. For even a_i it implies $P_0(n - a_i) = 1$. From (4) there exists a_j $(a_j \le n - a_i)$ such that $p_0(n - a_i - a_j) = 1$. By the induction hypothesis the first player is even constrained in $\mathbf{M}(n - a_i - a_j, \mathbf{S})$. That is, in $\mathbf{M}(n, \mathbf{S})$, whenever the first player removes a_i tokens the second player can remove some a_j tokens to force the first player's score to be even. Thus the first player is even constrained. The case of odd a_i can be proved in the same way. Conversely, suppose that the first player is even constrained in $\mathbf{M}(n, \mathbf{S})$. Following the reverse process of the above it is easy to prove $p_0(n) = 1$. Hence conclusion 1 is proved. □

Lemma 2. *In* $\mathbf{M}(n, \mathbf{S})$

1. The first player has a strong even strategy iff $f_0(n) = 1$.
2. The first player has a strong odd strategy iff $f_1(n) = 1$.
3. The first player has a balanced even strategy iff $t_0(n) = 1$.
4. The first player has a balanced odd strategy iff $t_1(n) = 1$.

Proof. We prove these conclusions by induction uniformly. For $n < a_1$, from (3) the conclusions are obvious. Assume that all the conclusions are true for any integer less than n $(n \ge a_1)$. We are to prove that these are also true for n.

First we prove the sufficiency for the conclusion 1. Suppose $f_0(n) = 1$. From (8), there exists an a_i $(a_i \le n)$ such that $p_1(n-a_i) = 1$ and the equality

$$u_{00}(a_i, a_j) \vee u_{01}(a_i, a_j) \vee u_{10}(a_i, a_j) \vee u_{11}(a_i, a_j) = 1\,, \quad \text{for all } a_j \quad (a_j \le n - a_i) \tag{12}$$

holds. For convenience, we call the first player in $\mathbf{M}(n, \mathbf{S})$ Left and call his opponent Right. We consider the case of even a_i first. From (8)–(2), if a_j is even then $f_0(n-a_i-a_j) = 1$. By the induction hypothesis in $\mathbf{M}(n-a_i-a_j, \mathbf{S})$ the first player has a strong even strategy to ensure his own score to be even and to force the second player's score to be odd as well. It is to say that, in

$\mathbf{M}(n, \mathbf{S})$, when the game begins the Left can remove tokens of even number a_i such that whenever the second player removes any tokens of even number a_j he has sequential steps to make his own total score even and to make the total score of Right odd as well. If a_j is odd then $t_0(n - a_i - a_j) = 1$. By the induction hypothesis in game $\mathbf{M}(n - a_i - a_j, \mathbf{S})$ the first player has a balanced even strategy to make all his own score and the opponent's score to be even. Whence for the Left removing a_i tokens whenever the Right removes any tokens of even or odd number a_j the Left has sequential steps to make his own total score even and to make the total score of right odd as well. That is in game $\mathbf{M}(n, \mathbf{S})$ the first player has a strong even strategy. For the case of odd a_1 we can derive the same conclusion. This proves the sufficiency for the conclusion 1. The sufficiencies for the rest conclusion can be proved in the same way.

Conversely, assume that in game $\mathbf{M}(n, \mathbf{S})$ the Left has a strong even strategy. At the beginning, after the Left removes tokens of any number a_i the Right is changed to the first player in game $\mathbf{M}(n - a_i, \mathbf{S})$. However, the Left must select and can select such a_i that the right is odd constrained in game $\mathbf{M}(n - a_i, \mathbf{S})$. Whence by the induction hypothesis $p_1(n - a_i) = 1$ holds. Now suppose that the Right removes tokens of number a_j. There are four possibilities: even a_i and even a_j, even a_i and odd a_j, odd a_i and even a_j, odd a_i and odd a_j. We only consider one of them for the sample. For example, odd a_i and even a_j. At this time the Left is changed to the first player in game $\mathbf{M}(n - a_i - a_j, \mathbf{S})$, and so to be the winner he must have a balanced odd strategy in game $\mathbf{M}(n - a_i - a_j, \mathbf{S})$. Also by the induction hypothesis $t_1(n - a_i - a_j) = 1$ holds. That is, (12) holds. Summarizing $p_1(n - a_i) = 1$ the equality $f_0(n) = 1$ follows from (8). Thus the necessity for conclusion 1 is proved. The necessities for the rest conclusions of the lemma can be proved in the same way. □

Lemma 3. *In* $\mathbf{M}(n, \mathbf{S})$, *for* $n \geq a_1$

1. *The second player has a strong even strategy iff* $g_0(n) = 1$.
2. *The second player has a strong odd strategy iff* $g_1(n) = 1$.

Proof. Let the terminologies 'Left' and 'Right' have the same meanings as in the proof of lemma 2. We only prove the conclusion 1. We first prove the sufficiency. Suppose $g_0(n) = 1$. Then, from (2), for all a_i $(a_i \leq n)$ even a_i implies $f_0(n - a_i) = 1$ and odd a_i implies $t_0(n - a_i) = 1$. By lemma 2 the Right, as the first player in $\mathbf{M}(n - a_i, \mathbf{S})$, has a strong even strategy in $\mathbf{M}(n - a_i, \mathbf{S})$ for even a_i and has a balanced even strategy in $\mathbf{M}(n - a_i, \mathbf{S})$ for odd a_i. Whence we can easily observe that the Right has a strong even strategy. Following the inverse process we can prove the necessity. □

From Lemma 2 and Lemma 3 we can obtain the following theorem immediately.

Theorem 1. *For* $n \geq a_1$, *In* $\mathbf{M}_0(n, \mathbf{S})$

1. *The first player has a winning strategy iff* $f_0(n) = 1$*;*
2. *The second player has a winning strategy iff* $g_0(n) = 1$*;*
3. *There can be a tie position iff* $f_0(n) = 0$ *and* $g_0(n) = 0$*.*

Theorem 2. *For* $n \geq a_1$*, In* $\mathbf{M_1}(n,\ \mathbf{S})$

1. *The first player has a winning strategy iff* $f_1(n) = 1$*;*
2. *The second player has a winning strategy iff* $g_1(n) = 1$*;*
3. *There can be a tie position iff* $f_1(n) = 0$ *and* $g_1(n) = 0$*.*

Due to Theorem 1 and Theorem 2 we call, especially, the function $f_0(n)$ and $g_0(n)$ the *first test function* and the *second test function* of $\mathbf{M_0}(n,\ \mathbf{S})$; and call the function $f_1(n)$ and $g_1(n)$ the *first test function* and the *second test function* of $\mathbf{M_1}(n,\ \mathbf{S})$. For example, according to the values of the test functions, for $\mathbf{S} = \{5,\ 8,\ 11,\ 13\}$ and for n from 5 to 100 we can get the following data results.

The cases that the first player can be the winner in $\mathbf{M_0}(n,\ \mathbf{S})$ are
n= 13, 14, 15, 23, 24, 25, 29, 30, 31, 32, 33;
the cases that the first player can be the winner in $\mathbf{M_1}(n,\ \mathbf{S})$ are
n= 5, 6, 7, 8, 9, 11, 12, 13, 14, 15, 16, 17, 31, 41, 47, 49;
the cases that the second player can be the winner in $\mathbf{M_0}(n,\ \mathbf{S})$ are
$n = 5$, 6, 7.

There is only one case that the second player can be the winner in $\mathbf{M_1}(n,\ \mathbf{S})$. That is
n= 23.

The cases that the game can be tied in $\mathbf{M_0}(n,\ \mathbf{S})$ are
n= 8, 9, 10, 11, 12, 16, 17, 18, 19, 20, 21, 22, 26, 27, 28, 34, 35, 36, 37, 38, 39, 40, 41, 42, 43, 44, 45, 46, 47, 48, 49, 50, 51, 52, 53, 54, 55, 56, 57, 58, 59, 60, 61, 62, 63, 64, 65, 66, 67, 68, 69, 70, 71, 72, 73, 74, 75, 76, 77, 78, 79, 80, 81, 82, 83, 84, 85, 86, 87, 88, 89, 90, 91, 92, 93, 94, 95, 96, 97, 98, 99, 100;
the cases that the game can be tied in $\mathbf{M_1}(n,\ \mathbf{S})$ are
n= 10, 18, 19, 20, 21, 22, 24, 25, 26, 27, 28, 29, 30, 32, 33, 34, 35, 36, 37, 38, 39, 40, 42, 43, 44, 45, 46, 48, 50, 51, 52, 53, 54, 55, 56, 57, 58, 59, 60, 61, 62, 63, 64, 65, 66, 67, 68, 69, 70, 71, 72, 73, 74, 75, 76, 77, 78, 79, 80, 81, 82, 83, 84, 85, 86, 87, 88, 89, 90, 91, 92, 93, 94, 95, 96, 97, 98, 99, 100.

3 Algorithms for Optimal Strategies and Complexities

Our purpose is that, to find out the winning strategies if there is a unique winner, otherwise to find out the strategies that each player will not lost, in $\mathbf{M}(n,\ \mathbf{S})$. To realize our purpose for $n \geq a_1$ we define a system of sets $\mathbf{F}_0(n)$, $\mathbf{F}_1(n)$, $\mathbf{T}_0(n)$, $\mathbf{T}_1(n)$ as below:

$a_i \in \mathbf{F}_0(n)$ iff

$p_1(n - a_i) = 1$

and for all a_j $(a_j \leq n - a_i)$

$$u_{00}(a_i,\ a_j) \vee u_{01}(a_i,\ a_j) \vee u_{10}(a_i,\ a_j) \vee u_{11}(a_i,\ a_j) = 1;$$

$a_i \in \mathbf{F}_1(n)$ iff $n < 2a_1$ and a_1 is odd, or

$$p_0(n - a_i) = 1$$

and for all a_j $(a_j \leq n - a_i)$

$$v_{00}(a_i,\ a_j) \vee v_{01}(a_i,\ a_j) \vee v_{10}(a_i,\ a_j) \vee v_{11}(a_i,\ a_j) = 1;$$

$a_i \in \mathbf{T}_0(n)$ iff $n < 2a_1$ and a_1 is even, or

$$p_0(n - a_i) = 1$$

and for all a_j $(a_j \leq n - a_i)$

$$w_{00}(a_i,\ a_j) \vee w_{01}(a_i,\ a_j) \vee w_{10}(a_i,\ a_j) \vee w_{11}(a_i,\ a_j) = 1;$$

$a_i \in \mathbf{T}_1(n)$ iff

$$p_1(n - a_i) = 1$$

and for all a_j $(a_j \leq n - a_i)$

$$x_{00}(a_i,\ a_j) \vee x_{01}(a_i,\ a_j) \vee x_{10}(a_i,\ a_j) \vee x_{11}(a_i,\ a_j) = 1.$$

The following lemma is clear.

Lemma 4. *In* $\mathbf{M}(n,\ \mathbf{S})$, *for the first player,*

1. *If* $f_0(n) = 1$, *he can realize his strong even strategy iff he removes tokens of number in* $\mathbf{F}_0(n)$;
2. *If* $f_1(n) = 1$, *he can realize his strong odd strategy iff he removes tokens of number in* $\mathbf{F}_1(n)$;
3. *If* $t_0(n) = 1$, *he can realize his balanced even strategy iff he removes tokens of number in* $\mathbf{T}_0(n)$;
4. *If* $t_1(n) = 1$, *to realize his balanced odd strategy iff he removes tokens of number in* $\mathbf{T}_1(n)$.

Now we state the algorithms for optimal strategies. We only state the algorithms for even-type modulo 2 game. The algorithms for odd-type game are similar. We prescribe that in the sequential discussion denote the number of tokens removed by the first player (Left) and the second player (Right) at each return remove, by L and R, respectively.

Theorem 3. *In* $\mathbf{M}_0(n,\ \mathbf{S})$, $n \geq a_1$, *if* $f_0(n) = 1$ *then the first player can win by the following*

Algorithm M_{01}:

1. $k := 0$;
2. *If* $k = i$ $(i = 0,\ 1)$ *then* $L \in \mathbf{F}_i(n)$; *If* $k = j$ $(j = 3,\ 4)$ *then* $L \in \mathbf{T}_{j-1}(n)$;
3. $n := n - L$; *If* $n < a_1$ *then goto 9;*

4. Read(R);
5. $n := n - R$; *If* $n < a_1$ *then goto 9;*
6. $u := m(L) \wedge m(R)$; $v = m(L) \wedge \neg m(R)$; $w = \neg m(L) \wedge m(R)$; $x = \neg m(L) \wedge \neg m(R)$;
7. case k of
0: If v *then* $k := 2$; *If* w *then* $k := 3$; *If* x *then* $k := 1$;
1: If v *then* $k := 3$; *If* w *then* $k := 2$; *If* x *then* $k := 0$;
2: If v *then* $k := 0$; *If* w *then* $k := 1$; *If* x *then* $k := 3$;
0: If v *then* $k := 1$; *If* w *then* $k := 0$; *If* x *then* $k := 2$
end;
8. goto 2;
9. end.

Proof. The label $k = 0$, 1, 2 and 3 corresponds to status $f_0(n) = 1$, $f_1(n) = 1$, $t_0(n) = 1$ and $t_1(n) = 1$, respectively. By lemma 4, corresponding to each of the above statuses the first player must remove tokens of number L in $\mathbf{F}_0(n)$, $\mathbf{F}_1(n)$, $\mathbf{T}_0(n)$ and $\mathbf{T}_1(n)$, respectively. It is capable to do this since all these sets are nonempty in the above statuses. From (8)–(11) we observe that for the status $f_0(n) = 1$ after the first player removes a_i tokens and the opponent removes a_j tokens there are four possible transfers of status

- For even a_i and even a_j it is transferred to the status $f_0(n - a_i) = 1$ (after executing $n := n - a_i$ it becomes the initial status);
- For even a_i and odd a_j it is transferred to the status $t_0(n - a_i) = 1$;
- For odd a_i and even a_j it is transferred to the status $t_1(n - a_i) = 1$;
- For odd a_i and odd a_j it is transferred to the status $f_1(n - a_i) = 1$.

The tranofero of thc rest statuses are similar. Thus the correctness of the algorithm is proved. □

Theorem 4. *In* $\mathbf{M}_0(n, \mathbf{S})$, $n \geq a_1$, *if* $g_0(n) = 1$ *then the second player can win by the following*

Algorithm M_{02}*:*

1. Read(L);
2. $n := n - L$; *if* $n < a_1$ *goto 11;*
3. If $m(L)$ *then* $k := 0$ *else* $k := 2$;
4. If $k = i$ $(i = 0, 1)$ *then* $R \in \mathbf{F}_i(n)$; *If* $k = j$ $(j = 3, 4)$ *then* $R \in \mathbf{T}_{j-1}(n)$;
5. $n := n - R$; *If* $n < a_1$ *then goto 11;*
6. Read(L);
7. $n := n - L$; *If* $n < a_1$ *then goto 11;*
8. $u := m(R) \wedge m(L)$; $v = m(R) \wedge \neg m(L)$; $w = \neg m(R) \wedge m(L)$; $x = \neg m(R) \wedge \neg m(L)$;
9. case k of
0: If v *then* $k := 2$; *If* w *then* $k := 3$; *If* x *then* $k := 1$;
1: If v *then* $k := 3$; *If* w *then* $k := 2$; *If* x *then* $k := 0$;
2: If v *then* $k := 0$; *If* w *then* $k := 1$; *If* x *then* $k := 3$;

0: If v then $k := 1$; If w then $k := 0$; If x then $k := 2$
end {case};
10. goto 4;
11. end.

Proof. At the beginning after executing "Read(L); $n := n - L$;" if $n \geq a_1$ the second player (Right) becomes the first player in the new $\mathbf{M_0}(n,\ \mathbf{S})$ at the status $f_0(n) = 1$ for even L or at the status $t_0(n) = 1$ for odd L. Thus the rest steps of the algorithm is the same as Algorithm M_{01} with L and R exchanged. This completes the proof. □

Theorem 5. *In $\mathbf{M_0}(n,\ \mathbf{S})$, $n \geq a_1$, if $f_0(n) = g_0(n) = 0$ then there can be a tie position if both players stand simultaneously by the following Algorithm M_{00}:*

1. $k := 0$;
2. case k of
0: For $i := 1$ to s if $a_i \leq n$ then
begin
if $m(a_i) \wedge \neg f_0(n - a_i)$ then $L := a_i$;
if $\neg m(a_i) \wedge \neg t_0(n - a_i)$ then begin $L := a_i$; $k := 1$ end
end;
1: For $i := 1$ to s if $a_i \leq n$ then
begin
if $m(a_i) \wedge \neg t_1(n - a_i)$ then begin $L := a_i$; $k := 2$ end;
if $\neg m(a_i) \wedge \neg f_1(n - a_i)$ then begin $L := a_i$; $k := 3$ end
end;
2: For $i := 1$ to s if $a_i \leq n$ then
begin
if $m(a_i) \wedge \neg t_0(n - a_i)$ then begin $L := a_i$; $k := 1$ end;
if $\neg m(a_i) \wedge \neg f_0(n - a_i)$ then begin $L := a_i$; $k := 0$ end
end;
3: For $i := 1$ to s if $a_i \leq n$ then
begin
if $m(a_i) \wedge \neg f_1(n - a_i)$ then $L := a_i$;
if $\neg m(a_i) \wedge \neg t_1(n - a_i)$ then begin $L := a_i$; $k := 2$ end
end
end;
3. $n := n - L$; If $n < a_1$ then goto 7;
4. case k of
0: For $i := 1$ to s if $a_i \leq n$ then
begin
if $m(a_i) \wedge \neg f_0(n - a_i)$ then $R := a_i$;
if $\neg m(a_i) \wedge \neg t_0(n - a_i)$ then begin $R := a_i$; $k := 1$ end
end;
1: For $i := 1$ to s if $a_i \leq n$ then

begin
if $m(a_i) \wedge \neg t_1(n-a_i)$ *then begin* $R := a_i$; $k := 2$ *end;*
if $\neg m(a_i) \wedge \neg f_1(n-a_i)$ *then begin* $R := a_i$; $k := 3$ *end*
end;
2: For $i := 1$ *to* s *if* $a_i \leq n$ *then*
begin
if $m(a_i) \wedge \neg t_0(n-a_i)$ *then begin* $R := a_i$; $k := 1$ *end;*
if $\neg m(a_i) \wedge \neg f_0(n-a_i)$ *then begin* $R := a_i$; $k := 0$ *end*
end;
3: For $i := 1$ *to* s *if* $a_i \leq n$ *then*
begin
if $m(a_i) \wedge \neg f_1(n-a_i)$ *then* $R := a_i$;
if $\neg m(a_i) \wedge \neg t_1(n-a_i)$ *then begin* $R := a_i$; $k := 2$ *end*
end
end;
5. $n := n - R$; *If* $n < a_1$ *then goto 7;*
6. goto 2;
7. end.

Proof. Denote the accumulative sum of tokens moved by the first player and the second player, after a series of steps, by s_1 and s_2, respectively. The label k indicates four status: $k = 0$ for even s_1 and even s_2; $k = 1$ for even s_1 and odd s_2; $k = 2$ for odd s_1 and even s_2; $k = 3$ for odd s_1 and odd s_2. For $k = 0$ let the first player (Left) remove $L = a_i$ tokens. There are two problems: how the status to be transferred and how to select a_i. First we consider the first problem. If a_i is even the status keep to $k = 0$. If a_i is odd it is transferred to odd s_1 and even s_2 which should have $k = 3$. However, relative to the Right he will be the first player in the new game $\mathbf{M}(n-a_i,\ \mathbf{S})$, which becomes $\mathbf{M}(n,\ \mathbf{S})$ after executing "$n := n - L$". Hence for odd a_i the status the Right facing to is $k = 1$. Now we consider the second problem. If a_i is even, $f_0(n-a_i) = 0$ must hold. Suppose that $f_0(n-a_i) = 1$. Then the Right have a strong even strategy in the new game. This implies that the Right can be a winner and contradicts to the given condition. By the same reason if a_i is odd, $t_0(n-a_i) = 0$ must hold. Summarizing the above we can conclude that for $k = 0$ the number a_i should be selected by the following rule: a_i is even and $f_0(n-a_i) = 0$, the status is kept to $k = 0$; a_i is odd and $t_0(n-a_i) = 0$, the status is transferred to $k = 1$. For $k = 1,\ 2,\ 3$ we can get similar conclusions in the same way. Thus the proof is finished. □

Theorem 6. *In game* $\mathbf{M}(n,\ \mathbf{S})$ *all algorithm for computing optimal strategy can be implemented in* $\mathcal{O}(n)$ *time.*

Proof. From above theorems we observe that to implement any algorithm for computing optimal strategy it is sufficient to compute all test functions. Assume that for compute all values of all test functions for parameter values $a_1,\ a_1+1,\ \cdots\ n$ it is needed to do at most $q(n)$ times of operations. Then from the expressions of these test functions we can easily see that to compute the values of all test functions continuously for parameter values $n+1,\ n+2,\ \cdots,\ n+a_1$ it is needed to do at most a constant c times of operations. Whence we have

$$q(n+a_1)=q(n)+c.$$

Let $n=ka_1+r\ (0\leq r<a_1)$. We can solve the difference equation to get

$$q(n)=ck+d=c(n-r)/a_1+d,\ \ d=constant.$$

Hence the proof is finished. □

4 Further Problems

The first problem is: Are there any more special $\mathbf{S}$ like $\{1,\ 2,\ \cdots,\ s\}$ such that in $\mathbf{M}(n,\ \mathbf{S})$ all the test functions have closed forms? Classify them if there are.

The existence of such $\mathbf{S}$ is certain. We have searched several examples and will introduce them in another paper. However, to classify them maybe more difficult.

The second problem is that for fixed $\mathbf{S}$ and n. Among the system of games $\mathbf{M}_i(j,\ \mathbf{S})$ $(i=0,\ 1;\ j=1,\ 2,\ \cdots\ n)$ find the distribution of each total number of games: 1) the first player can be the winner; 2) the second player can be the winner; 3) there can be a tie position.

It seems that, by the data results, the number of case 1 and the number of case 2 are much less than the number of case 3.

References

1. Back, R. J. R., von Wright (1995): Games and Winning Strategies. Inform. Process. Lett. **53**, 165–172
2. Berlekamp, E., Conway, J., Guy, R. (2001): Winning Ways, second edition. A. K. Peters Ltd, Vol. I
3. Bonanno, G. (1993): The Logical Representation of Extensive Games. Intenational Journal of Games Theory **22(2)**, 153–169
4. Canning, D. (1992): Rationality, Computabolity and Nash Equilibrium. Economica **60**, 877–888
5. Huang, Z., Huang, S. (1990): Solution to a Combinatorial Game. Journal of National University of Defense Technology **12(3)**, 15–20

Bargaining in Dynamic Games*

Leon A. Petrosyan

St. Petersburg State University,
Faculty of Applied Mathematics and Control Processes,
Center of Game Theory,
Universitetskiy pr., 35, Petrodvorets, St. Petersburg, 198504, Russia
E-mail: lp@isdgrus.ru

Summary. A regularization method is proposed, which enables time-consistent bargaining procedures to be constructed.

Key words: dynamic game, bargaining, time-consistency

1 Bargaining Problem

Consider n-person differential game $\Gamma(x_0, T-t_0)$ with prescribed duration $T-t_0$:

$$\dot{x} = f(x, u_1, \ldots, u_n), \quad x \in R^n, \; u_i \in U_i, \; i \in N; \tag{1}$$

$$x(t_0) = x_0; \tag{2}$$

with payoffs

$$K_i(x_0, T-t_0; u_1, \cdots, u_n) = \int_{t_0}^{T} h_i(x(t))\,dt + H_i(x(T)), \quad i \in N, \qquad H_i, \; h_i > 0, \quad i \in N. \tag{3}$$

Define in $\Gamma(x_0, T-t_0)$ the set $\kappa(x_0, T-t_0)$ of all possible payoff-vectors

$$K(x_0, T-t_0) = \{K_i(x_0, T-t_0)\} \tag{4}$$

for all different open-loop controls $u(t) = \{u_i(t)\}$, $t \in [t_0, T]$. Imbedding the game $\Gamma(x_0, T-t_0)$ in a family of games $\Gamma(x, T-t)$ for all possible initial conditions $x \in R^n$ and $t \in [t_0, T]$ we get the family of sets $\kappa(x, T-t)$ each corresponding to the game $\Gamma(x, T-t)$.

For each $\Gamma(x, T-t)$ define a vector $\overline{K}(x, T-t) \in \kappa(x, T-t)$ as "status-quo" point in $\Gamma(x, T-t)$. The status-quo point can be defined in different ways: $\overline{K}_i(x, T-t)$ can be defined as maximal guaranteed payoff of player $i \in N$

* This work is financially supported by scientific program "Universities of Russia".

in the zero-sum differential game $\Gamma_i(x, T-t)$ played between the players $i \in N$ and $N \setminus \{i\}$ with the payoff of $i \in N$ defined as $\{\widetilde{K}_i(x, T-t)\}$, or as vector of players' payoffs in some given Nash equilibrium, there are also other approaches on how to define the "status-quo" point $\overline{K}_i(x, T-t)$ in the n-person game $\Gamma(x, T-t)$.

Given the set $\kappa(x, T-t)$ and the "status-quo" point $\overline{K}(x, T-t)$ one can define the bargaining solution as mapping B which to every pair

$$\kappa(x, T-t), \quad \overline{K}(x, T-t)$$

corresponds the point $\widehat{K}(x, T-t) \in \kappa(x, T-t)$:

$$B[\kappa(x, T-t); \overline{K}(x, T-t)] = \widehat{K}(x, T-t) . \tag{5}$$

Any n-tuple of open-loop controls $\hat{u}(t) = \{\hat{u}_i(t)\}$ and corresponding trajectory $[\hat{x}(\tau; x, T-t]$ $(\hat{x}(t) = x,\ \tau \in [t, T])$ in the game $\Gamma(x, T-t)$ for which

$$\begin{aligned} \widehat{K}(x, T-t) &= K(x, T-t; \hat{u}_1, \ldots, \hat{u}_n) \\ &= \int_{t_0}^{T} h_i(\hat{x}(\tau))\, dt + H_i(\hat{x}(T)) , \quad i \in N , \end{aligned} \tag{6}$$

is called *optimal control* and corresponding *optimal trajectory*.

Let us fix now the initial conditions in the game and consider one special game $\Gamma(x_0, T-t_0)$, and let $[\hat{x}(\tau), x_0, T-t_0]$ be any optimal trajectory in $\Gamma(x_0, T-t_0)$.

Consider the family of games $\Gamma(\hat{x}(t), T-t)$, $t \in [t_0, T]$, along $[\hat{x}(t), x_0, T-t_0]$. For each $t \in [t_0, T]$ one can define the bargaining solution by the rule

$$B[\kappa(\hat{x}(t), T-t); \overline{K}(\hat{x}(t), T-t)] = \widehat{K}(\hat{x}(t), T-t) , \tag{7}$$

where $\kappa(\hat{x}(t), T-t)$ is the set of all possible payoff vectors in $\Gamma(\hat{x}(t), T-t)$ and $\overline{K}(\hat{x}(t), T-t)$ is a "status-quo" point in $\Gamma(\hat{x}(t), T-t)$.

Definition 1. *The bargaining solution $\widehat{K}(x_0, T-t_0)$ is called time-consistent if following condition holds*

$$\widehat{K}_i(x_0, T-t_0) = \int_{t_0}^{t} h_i(\hat{x}(t))\, dt + \widehat{K}(\hat{x}(t), T-t) , \quad t \in [t_0, T] . \tag{8}$$

It is clear that this is a strong condition and it holds, for a very trivial classes of bargaining solutions (see [3, 2]).

2 Regularization of Bargaining Solution

Definition 2. *The vector-function*

$$\beta(t) = \{\beta_i(t)\} \tag{9}$$

is called bargaining payoff distribution procedure (BPDP) if

$$\widehat{K}_i(x_0, T-t_0) = \int_{t_0}^{T} \beta_i(t)\, dt + H_i(\hat{x}(T))\,. \tag{10}$$

It is clear that there exist a large class of BPDP functions.

Now by changing the payoff along the optimal trajectory $[\hat{x}(t), x_0, T-t_0]$ in $\Gamma(x_0, T-t_0)$ we shall try to fulfill the time-consistency condition (8).

Define $\beta(t) = (\beta_1(t), \dots, \beta_n(t))$ as

$$\int_{t}^{T} \beta_i(\tau) d\tau + H_i(\hat{x}(T)) = \widehat{K}_i(\hat{x}(t), T-t) \tag{11}$$

(which does not contradict (10) in the definition 2).

Then if $\widehat{K}(x, T-t)$ is differentiable we get the following formula for $\beta(t)$:

$$\beta_i(t) = -\frac{d}{dt}\widehat{K}_i(\hat{x}(t), T-t)\,, \quad t \in [t_0, T]\,,\ i \in N\,. \tag{12}$$

¿From (11) we get

$$\widehat{K}_i(x_0, T-t_0) = \int_{t_0}^{t} \beta_i(t)\, dt + \widehat{K}_i(\hat{x}(t), T-t)\,. \tag{13}$$

Regularization.

If we now change the payoffs of players along optimal trajectory as following

$$\widehat{K}(x_0, T-t_0) = \int_{t_0}^{T} \beta_i(t), dt + H_i(\hat{x}(T)) - \int_{t_0}^{T} h_i(\hat{x}(t))\, dt \mid H_i(\hat{x}(T))$$

(which is actually the changing of instantaneous payoff of player i from $h_i(\hat{x}(t))$ to $\beta_i(t)$), the bargaining solution $\widehat{K}_i(x_0, T-t_0)$ will be time-consistent as it follows from (13).

The BPDP can be interpreted as a changed "instantaneous" payoff (compared with the "natural" instantaneous payoff $h_i\,(\hat{x}(t))$) of the players along the optimal trajectory (bargaining trajectory) defined to guarantee the time-consistency of the bargaining process in dynamic games. It is important to mention that the total payoff of the player $i \in N$ along the optimal trajectory is not changing, as it is seen from (10).

$$\hat{K}_i(x_0, T-t_0) = \int_{t_0}^{T} h_i\,(\hat{x}(t))\, dt + H_i\,(\hat{x}(T)) = \int_{t_0}^{T} \beta_i(t) dt + H_i\,(\hat{x}(T))\,. \tag{14}$$

Consider a special case when $h_i(x) \equiv 0$, $i \in N$. This is the case of the game with terminal payoff, then from (14) we'll have

$$\int_{t_0}^{T} \beta_i(t) dt = 0, \qquad i \in N \tag{15}$$

which means that except trivial cases ($\beta(t) \equiv 0$) BPDP in this case $\beta(t)$ on the time interval $[t_0, T]$ has to take positive and negative values. Consider for this special case (terminal payoffs) the Nash bargaining solution. For this one has to maximize the Nash product

$$\max_{\substack{K \geq \overline{K}(x_0, T-t_0) \\ K \in \kappa(x_0, T-t_0)}} \prod_{i=1}^{n} (K_i - \overline{K}_i(x_0, T-t_0)) =$$

$$= \prod_{i=1}^{n} \left(\hat{K}_i(x_0, T-t_0) - \overline{K}_i(x_0, T-t_0) \right). \tag{16}$$

Now considering the equation (15) along the optimal trajectory $\hat{x}(t)$ we get

$$\max_{\substack{K \geq \overline{K}(\hat{x}(t), T-t) \\ K \in \kappa(\hat{x}(t), T-t)}} \prod_{i=1}^{n} (K_i - \overline{K}_i(\hat{x}(t), T-t)) =$$

$$= \prod_{i=1}^{n} \left(\hat{K}_i(\hat{x}(t), T-t) - \overline{K}_i(\hat{x}(t), T-t) \right). \tag{17}$$

And since the values of $\overline{K}_i(\hat{x}(t), T-t)$ change in time when $t \in [t_0, T]$ and the set $\kappa(\hat{x}(t), T-t)$ is monotone decreasing in t, the bargaining solution $\hat{K}_i(\hat{x}(t), T-t)$, $i \in N$, for the subgame $\Gamma(\hat{x}(t), T-t)$ can not remain constant over the time interval $[t_0, T]$ except some very trivial cases, which would imply BPDP $\beta \equiv 0$ on the time interval $[t_0, T]$ as seen from (12). Thus in the case of Nash bargaining solutions for games with terminal payoffs we get positive and negative values for BPDP for $t \in [t_0, T]$. The existence of negative values for BPDP can initiate some criticism of proposed regularization method since it demands that on some time intervals the players give up their earnings (the earnings h_i, H_i are positive). For this cases other regularization method was constructed based upon the derivation of a special motion equation for the "status-quo" point along the optimal trajectory (see [2]).

From (8) we get that Nash bargaining solution for games with terminal payoff is in general time-inconsistent. This statement is true also for the games with the payoffs of the form (6) and other bargaining solutions. This implies the need of regularization for all such bargaining solutions.

3 Implementation of Regularization Schemes

Consider now the general bargaining problem for the game $\Gamma(x_0, T-t_0)$ with the payoff functions defined as in (3).

Along the optimal trajectory $\hat{x}(t)$ we have the following equation for the "natural" and regularized payoffs

$$\int_{t_0}^{T} h_i(\hat{x}(t))dt + H_i(\hat{x}(T)) = \int_{t_0}^{T} \beta_i(t)dt + H_i(\hat{x}(T)), \tag{18}$$

but only in very special cases we can be sure that $h_i(\hat{x}(t)) \equiv \beta_i(t)$. Thus if we suppose the continuity of the above mentioned functions on some open time intervals we shall have $h_i(\hat{x}(t)) > \beta_i(t)$ and on some $h_i(\hat{x}(t)) < \beta_i(t)$.

In the first case when $h_i > \beta_i$ the regularization is admissible since the players "natural" instantaneous income h_i exceeds the regularized payment β_i. And β_i can be realized on this time interval. In the second case some additional sources are necessary to guarantee the regularized payment β_i to the player since it is less than his "natural" income.

There can be different ways to overcome this difficulty. But under some conditions the savings can be made for the player, which can be used to cover the additional regularization costs in the second case.

To illustrate the idea consideer the following simple case for a fixed player $i \in N$. Suppose that there exists such a sequence $t_0, t_1, \ldots, t_{l-1}, t_l = T$ (suppose l is odd)

$$h_i(\hat{x}(t)) > \beta_i(t), \qquad t \in (t_0, t_1)$$
$$h_i(\hat{x}(t)) < \beta_i(t), \qquad t \in (t_1, t_2)$$
$$\ldots$$
$$h_i(\hat{x}(t)) > \beta_i(t), \qquad t \in (t_{k-1}, t_k)$$
$$h_i(\hat{x}(t)) < \beta_i(t), \qquad t \in (t_k, t_{k+1})$$
$$\ldots$$
$$h_i(\hat{x}(t)) > \beta_i(t), \qquad t \in (t_{l-1}, t_l)$$

Then if the following condition is satisfied for all $m = 1, \ldots, \frac{l-1}{2}$

$$\sum_{k=0}^{m} \int_{t_{2k}}^{t_{2k+1}} [h_i(\hat{x}(t)) - \beta_i(t)]dt > \sum_{k=0}^{m} \int_{t_{2k+1}}^{t_{2k+2}} [\beta_i(t) - h_i(\hat{x}(t))]dt \tag{19}$$

then the positive savings $h_i - \beta_i$ can be accumulated on the time intervals of the first type to cover the additional regularization costs on the time intervals of the second type.

If l is even the corresponding result is similar.

References

1. Petrosyan, L. A. (1993): Strongly Time-Consistent Optimality Principles in Multicriterial Problems of Optimal Control. Technical Cybernetics **1**, 165–174 (in Russian)
2. Petrosyan, L. A. (1995): The Regularization of NB-Scheme in Differential Games. Dynamic and Control **5**, 31–31

Part II

Equilibrium Properties

Extensions of Hart and Mas-Colell's Consistency to Efficient, Linear, and Symmetric Values for TU-Games*

Theo Driessen[1] and Tadeusz Radzik[2]

[1] Department of Applied Mathematics, University of Twente,
P. O. Box 217, 7500 AE Enschede, The Netherlands
E-mail: `t.s.h.driessen@math.utwente.nl`
(author in charge of any correspondence)

[2] Institute of Mathematics, Technical University of Wroclaw,
50–370 Wroclaw, Wyspianskiego 27, Poland
E-mail: `tradzik@pwr.jgora.pl`

Summary. By Hart and Mas-Colell's axiomatization, it is known that the Shapley value for TU-games is fully characterized by its 1–standardness for two-person games and its consistency property with respect to a particular reduced game. In the framework of TU-games, this paper establishes a similar axiomatization (with reference to some kind of consistency and standardness for two-person games) for values that verify efficiency, linearity, and symmetry. The fundamental idea in this unified consistency approach involves the introduction of a new type of reduced game. The construction of this game takes into account, besides the value itself, the probabilities of two events that a removed player joins or does not join a proposed coalition. Although the reduced game varies whenever the efficient, linear, and symmetric value varies, an operational criterion is presented to determine the appropriate reduced game by solving an associated system of linear equations recursively. Finally, the impact of the unified consistency approach is illustrated in the context of several known values, in particular the least square values and the Shapley value.

Key words: cooperative TU-game, value, consistency, reduced game
1991 Mathematics Subject Classifications: 90D12, 90D40

---

* The research for this paper was partially done during a stay (September 1996) of the second author at the Department of Applied Mathematics, University of Twente, Enschede, The Netherlands.

1 Introduction and Game Theoretic Notions

This paper concerns the problem of allocating some overall profits among a number of agents who have undertaken a joint venture. Such an allocation problem may be solved in a variety of ways, but an allocation rule which prescribes somehow a solution for the allocation problem should be justifiable on the basis of some generally accepted principles.

For instance, given any real-valued function $v\colon 2^N \to \mathbb{R}$ defined on the power set of a finite set of agents $N := \{1, 2, \ldots, n\}$, consider the following optimization problem of which the objective function is given by the weighted sum of the square of the "excesses" of all subsets.

Minimize

$$\sum_{\substack{S \subseteq N, \\ S \neq N, S \neq \emptyset}} m_{n,s} \cdot \left[v(S) - \sum_{j \in S} x_j \right]^2 \text{ subject to } (x_j)_{j \in N} \in \mathbb{R}^n, \sum_{j \in N} x_j = v(N), \quad (1)$$

where s denotes the cardinality of set $S \subseteq N$ and $m_{n,s}$, $s \in \{1, 2, \ldots, n-1\}$, is a (yet unspecified) collection of non-negative constants. It is known that this least square problem (1) has a unique optimal solution and in fact, the unique optimal solution of (1) may be interpreted as a function of the input, the variable function v. Moreover, the optimal solution turns out to meet several principles, like efficiency (by definition); symmetry (the optimal solution behaves natural under any renumbering of the agents); covariance (the optimal solution behaves natural under any affine transformation on v); and linearity (in v).

In the game theoretic framework, any function $v\colon 2^N \to \mathbb{R}$ is known as a game (with player set N) and the optimal solution of the least square problem may be regarded as a one-point solution concept called value (in fact, known as a least square value). Besides the common principles of efficiency, linearity, symmetry, and covariance, there exists a particular principle with firm roots in the game theoretic literature, known as the consistency property (cf. [1], see Chapters 13 and 18; [3]). Usually, the impact of the consistency property is exploited for the sake of an axiomatization of the solution concept. The consistency approach we are going to develop throughout this paper has to be classified as a unified one because the mathematical treatment focuses on an appropriately chosen class of solution concepts including the least square values. The fundamental result claims that any solution from this class can be fully characterized by two axioms, of which the main one refers to the consistency axiom. As an adjunct, we rediscover the axiomatic characterization of the well-known Shapley value presented by Hart and Mas-Colell (cf. [6]); the single result so far that inspired us to search for a unified consistency approach in the context of TU-games.

Formally, a *transferable utility game* (or cooperative game or coalitional game with side payments) is a pair $\langle N, v \rangle$, where N is a finite set of *players* and $v : 2^N \to \mathbb{R}$ is a *characteristic function* satisfying $v(\emptyset) = 0$. An element of N (notation: $i \in N$) and a nonempty subset S of N (notation: $S \subseteq N$ or $S \in 2^N$ with $S \neq \emptyset$) is called a *player* and *coalition* respectively, and the associated real number $v(S)$ is called the *worth* of coalition S. The size (cardinality) of coalition S is denoted by $|S|$ or, if no ambiguity is possible, by s. Particularly, n denotes the size of the player set N. Given a (transferable utility) game $\langle N, v\rangle$ and a coalition S, we write $\langle S, v\rangle$ for the *subgame* obtained by restricting v to subsets of S only (i.e., to 2^S).

The former allocation problem will be replaced by the game theoretic problem of how to divide the overall earnings the amount of $v(N)$ among the players in the TU-game $\langle N, v\rangle$. In conformity with the traditional solution part, with every TU-game there is associated a single allocation called the value of the TU-game. Let $\mathcal{G}$ denote the universal game space consisting of all TU-games with a finite (but variable) player set. Formally, a *value* on $\mathcal{G}$ is a function Φ that assigns a single payoff vector $\Phi(N,v) = (\Phi_i(N,v))_{i\in N} \in \mathbb{R}^N$ to every TU-game $\langle N, v\rangle$. The so-called value $\Phi_i(N,v)$ of player i in the game $\langle N, v\rangle$ represents an assessment by i of his gains from participating in the game. In the framework of values for TU-games, we review several known properties that will be treated in the Sections 3 and 4.

Definition 1. *A value Φ on the universal game space $\mathcal{G}$ is said to be*

(i) efficient, if $\sum_{i\in N} \Phi_i(N,v) = v(N)$ for all games $\langle N, v\rangle$;

(ii) linear, if $\Phi(N, \alpha\cdot v + \beta\cdot w) = \alpha\cdot\Phi(N,v) + \beta\cdot\Phi(N,w)$ for all games $\langle N, v\rangle$, $\langle N, w\rangle$, and all $\alpha\in\mathbb{R}$, $\beta\in\mathbb{R}$. Here the game $\langle N, \alpha\cdot v+\beta\cdot w\rangle$ is defined by $(\alpha\cdot v+\beta\cdot w)(S) := \alpha\cdot v(S)+\beta\cdot w(S)$ for all $S\subseteq N$;

(iii) symmetric, if $\Phi_{\pi(i)}(N,\pi v) = \Phi_i(N,v)$ for all games $\langle N, v\rangle$, all $i\in N$, and every permutation π on N. Here the game $\langle N, \pi v\rangle$ is defined by $(\pi v)(\pi S) := v(S)$ for all $S\subseteq N$;

(iv) covariant, if $\Phi(N, \alpha\cdot v + \mathbf{a}) = \alpha\cdot\Phi(N,v) + \mathbf{a}$ for all games $\langle N, v\rangle$, all $\alpha\geq 0$, and all $\mathbf{a} = (a_i)_{i\in N}\in\mathbb{R}^N$. Here the game $\langle N, \alpha\cdot v+\mathbf{a}\rangle$ is defined by $(\alpha\cdot v+\mathbf{a})(S) := \alpha\cdot v(S) + \sum_{j\in S} a_j$ for all $S\subseteq N$;

(v) coalitionally monotonic, if the following holds: for all games $\langle N, v\rangle$,$\langle N, w\rangle$, such that $v(S) > w(S)$ for some coalition $S\subseteq N$ and $v(T) = w(T)$ for all $T\subseteq N$, $T\neq S$, it holds that $\Phi_i(N,v)\geq\Phi_i(N,w)$ for all $i\in S$;

(vi) λ-standard for two-person games (where $\lambda\in\mathbb{R}$) if, for every two-person game $\langle\{i,j\}, v\rangle$ and every player $k\in\{i,j\}$, it holds that

$$\Phi_k(\{i,j\},v) = \lambda\cdot v(\{k\}) + \frac{1}{2}\cdot\Big[v(\{i,j\}) - \lambda\cdot v(\{i\}) - \lambda\cdot v(\{j\})\Big].$$

The λ-standardness for a value with reference to two-person games expresses that the value allocates the *surplus* $v(\{i,j\}) - \lambda\cdot v(\{i\}) - \lambda\cdot v(\{j\})$ equally to the two players i and j after each player k is conceded to get his *weighted individual worth* $\lambda\cdot v(\{k\})$. In the context of λ-standardness for two-person games, it is supposed that the proportionality rate λ is non-negative in order to avoid that the player with the bigger individual worth receives the smaller payoff. The 1-standardness for two-person games is a very common notion for solution concepts and it is satisfied, among many values, by the Shapley value.

2 The Reduced Game and the Consistency Property for Values

Informally, the notion of a reduced game and the corresponding reduced game property may be put forward as follows. Fix a value Φ that associates a payoff with every player in every (transferable utility) game. A game is always described by a finite player set as well as a real-valued characteristic function on the power set of the player set. A so-called reduced game is deducible from a given game by removing one (or more) player(s) on the understanding that the removed player(s) will be paid according to the value under consideration. The remaining players form the player set of the reduced game; the characteristic function of which is composed somehow of the original characteristic function, the value under consideration and some constants. With this informal description of a reduced game at hand, the given value Φ is called consistent with respect to the reduced game in question, or possessing the reduced game property, if $\Phi_j(N\backslash\{i\}, v^{\Phi,i}) = \Phi_j(N,v)$, where $\langle N,v\rangle$ is a game, $i \in N$, $j \in N\backslash\{i\}$, and $\langle N\backslash\{i\}, v^{\Phi,i}\rangle$ denotes the reduced game in question. The consistency (or reduced game) property for the value expresses that, if a player is supposed to be paid according to the value, then there is no inconsistency in what each of the players of the reduced game will get, in either the original game or the reduced game.

In the framework of values for TU-games, let us continue with the introduction of the following type of reduced game. Fix a value Φ, defined over the set of all games, and remove one player, say $i \in N$, from a game $\langle N,v\rangle$ with at least three players. To determine the worth of coalition S in the reduced game with player set $N\backslash\{i\}$, one assumes on the one hand that the remaining players in $(N\backslash\{i\})\backslash S$ are not present, on the other that the members of S approach player i for joining or not joining their coalition (in order to try to increase their earnings). If player i decides not to join, the members of S are left with the subgame $\langle S,v\rangle$ to benefit by its worth $v(S)$. If player i decides to join, he does so with the convention that he accepts the offer made by S to receive a reward the amount of his value payoff $\Phi_i(S\cup\{i\},v)$ in the resulting subgame $\langle S\cup\{i\},v\rangle$ with player set $S\cup\{i\}$. Consequently, the members of S are conceded to benefit by what is left of the worth $v(S\cup\{i\})$ in the resulting subgame $\langle S\cup\{i\},v\rangle$.

Generally speaking, a player will not be indifferent between joining or not joining a proposed coalition. Because we are driving at a unified approach, we assign arbitrary probabilities to both events, provided that these probabilities are symmetric with respect to the size of the player set and the size of the coalition as well. Particularly, one assumes that, for (not) joining a coalition, a player is indifferent between coalitions of the same size (given the size of the player set). If, for every $S \subseteq N\backslash\{i\}$, $S \neq \emptyset$, let us say of size s, the expressions $A_{n-1,s}$ and $B_{n-1,s}$ are seen as the probabilities that player i joins or does not join coalition S respectively, then the worth of coalition S in the reduced game with player set $N\backslash\{i\}$ is simply interpreted as the expected worth $A_{n-1,s}\cdot[v(S \cup \{i\}) - \Phi_i(S \cup \{i\}, v)] + B_{n-1,s}\cdot v(S)$. In the next definition we formalize our previous thoughts.

For notation' sake, let $\mathcal{A}$ and $\mathcal{B}$ represent two arbitrary collections of real numbers written as $\mathcal{A} := \{A_{k,s} \mid k \in \mathbb{N}\backslash\{1\}, s \in \{1,2,\ldots,k\}\}$ and $\mathcal{B} := \{B_{k,s} \mid k \in \mathbb{N}\backslash\{1\}, s \in \{1,2,\ldots,k\}\}$ respectively. Here $\mathbb{N} := \{1,2,3,\ldots\}$ denotes the set of all natural numbers.

Definition 2. *Let Φ be a value on $\mathcal{G}$ and $\mathcal{A},\mathcal{B}$ two collections of constants. Given a game $\langle N,v\rangle$ with $n \geq 3$ and a player $i \in N$, the corresponding reduced game $\langle N\backslash\{i\}, v^{\Phi,i,\mathcal{A},\mathcal{B}}\rangle$ associated with Φ and $\mathcal{A},\mathcal{B}$ is defined to be as follows: for all $S \subseteq N\backslash\{i\}$, $S \neq \emptyset$,*

$$v^{\Phi,i,\mathcal{A},\mathcal{B}}(S) := A_{n-1,s} \cdot \Big[v(S \cup \{i\}) - \Phi_i(S \cup \{i\}, v)\Big] + B_{n-1,s} \cdot v(S). \quad (2)$$

We say Φ possesses the reduced game property (RGP) or is consistent with respect to the reduced game of the form (2) if it satisfies the following condition: for every game $\langle N,v\rangle$ with $n \geq 3$ and every player $i \in N$,

$$\textbf{(RGP)} \qquad \Phi_j(N\backslash\{i\}, v^{\Phi,i,\mathcal{A},\mathcal{B}}) = \Phi_j(N,v), \quad \textit{for all } j \in N \setminus \{i\}. \quad (3)$$

Due to Hart and Mas-Colell's result (cf. [6], see Proposition 4.5, p. 597), it is known that the Shapley value Sh on the universal game space is consistent with respect to the reduced game of the form $v^{Sh,i,\mathcal{A},\mathcal{B}}(S) = v(S \cup \{i\}) - Sh_i(S \cup \{i\}, v)$ for all $S \subseteq N\backslash\{i\}$, $S \neq \emptyset$. As far as to the present knowledge of the authors, it is not yet known whether similar consistency properties hold uniformly on a suitably chosen class of values for TU-games.

In the main Section 4 we investigate this open problem for the class of values that verify efficiency, linearity, and symmetry. In fact, we present necessary and sufficient conditions on the coefficients that describe the reduced game of the form (2), so that a given efficient, linear, and symmetric value is consistent (see theorem 4). Moreover, we complete the axiomatization of these values in terms of its associated consistency property and a minor property referring to some kind of standardness for two-person games. In other words, the uniqueness part as treated in the axiomatization of these values is provided by theorem 2. In the remaining part of Section 3 we pay attention to the properties of efficiency, (weak versions of) linearity, and symmetry in the setting of consistent values. Section 4 concludes with a discussion of the developed consistency theory applied to several examples of values, including the least square values. The appendix contains the technical proof of the fundamental theorem 4.

3 Properties for and Uniqueness of Consistent Values

In this section we show that consistency, together with some kind of standardness with reference to two-person games, yields efficiency as well as weak versions of both linearity and symmetry (weak in the sense that if the relevant equality holds with respect to one player, then similar equalities do hold with respect to all players). Secondly, for the sake of the axiomatization treated in the next section, we establish the uniqueness of a consistent value (if it exists), provided that some kind of standardness for two-person games holds as well as some elementary property concerning the equal division of incremental gains for the grand coalition in a game (cf. theorem 2). Throughout this section let $\mathcal{A}$ and $\mathcal{B}$ be two arbitrary collections of constants with $A(k,k) := 1$ and $B(k,k) := 0$ for all $k \in \mathbb{N}\backslash\{1\}$.

Lemma 1. *Suppose a value Φ on $\mathcal{G}$ verifies the λ–standardness for two-person games as well as the consistency property with respect to the reduced game of the form (2) associated with Φ and $\mathcal{A}, \mathcal{B}$. Then the value Φ verifies efficiency.*

Proof. We prove by induction on the size n of the player set N that $\sum_{j \in N} \Phi_j(N,v) = v(N)$ for all games $\langle N,v \rangle$ with $n \geq 2$. The case $n = 2$ is a direct consequence of the λ–standardness for Φ with reference to two-person games.

From now on fix a game $\langle N,v \rangle$ with $n \geq 3$ and, by the induction hypothesis, it holds that $\sum_{j \in M} \Phi_j(M,w) = w(M)$ for all games $\langle M,w \rangle$ with $2 \leq |M| \leq n-1$. Choose an arbitrary player $i \in N$. Since Φ is consistent with respect to the reduced game of the form (2) (associated with Φ and $\mathcal{A}, \mathcal{B}$), it holds that $\Phi_j(N\backslash\{i\}, v^{\Phi,i,\mathcal{A},\mathcal{B}}) = \Phi_j(N,v)$ for all $j \in N\backslash\{i\}$. From this and the induction hypothesis applied to the reduced game $\langle N\backslash\{i\}, v^{\Phi,i,\mathcal{A},\mathcal{B}} \rangle$ respectively, we deduce the following chain of equalities:

$$\sum_{j \in N\backslash\{i\}} \Phi_j(N,v) = \sum_{j \in N\backslash\{i\}} \Phi_j(N\backslash\{i\}, v^{\Phi,i,\mathcal{A},\mathcal{B}}) = v^{\Phi,i,\mathcal{A},\mathcal{B}}(N\backslash\{i\})$$
$$= v(N) - \Phi_i(N,v) \, .$$

We conclude that $\sum_{j \in N} \Phi_j(N,v) = v(N)$ for all games $\langle N,v \rangle$ with $n \geq 3$. This completes the inductive proof of the efficiency property for the value Φ (under the given circumstances). □

Theorem 1. *Suppose a value Φ on $\mathcal{G}$ verifies the λ–standardness for two-person games as well as the consistency property with respect to the reduced game of the form (2) associated with Φ and $\mathcal{A}, \mathcal{B}$. Then the following two equivalences hold.*

(i) The value Φ verifies linearity on $\mathcal{G}$ $\Longleftrightarrow$ $\Phi_i(N, \alpha \cdot v + \beta \cdot w) = \alpha \cdot \Phi_i(N,v) + \beta \cdot \Phi_i(N,w)$ for all games $\langle N,v \rangle$, $\langle N,w \rangle$, and all $\alpha \in \mathbb{R}$, $\beta \in \mathbb{R}$, and some $i \in N$ (instead of all $i \in N$).

(*ii*) *The value* Φ *verifies symmetry on* $\mathcal{G} \iff \Phi_{\pi(i)}(N,\pi v) = \Phi_i(N,v)$ *for all games* $\langle N,v\rangle$, *every permutation* π *on* N, *and some* $i \in N$ *(instead of all* $i \in N$*).*

Proof. (*i*) Suppose the expression at the right hand of the equivalence holds. The proof of the linearity for Φ proceeds by induction on the size n of the player set N. The case $n = 2$ is a direct consequence of the efficiency for Φ (arising from Lemma 1). From now let $\langle N,v\rangle$ and $\langle N,w\rangle$ be two games with $n \geq 3$ and $\alpha \in \mathbb{R}$, $\beta \in \mathbb{R}$. By assumption, there exists some $i \in N$ such that $\Phi_i(N,\alpha\cdot v+\beta\cdot w) = \alpha\cdot\Phi_i(N,v)+\beta\cdot\Phi_i(N,w)$. From this and the induction hypothesis, we obtain that $\Phi_i(S\cup\{i\},\alpha\cdot v+\beta\cdot w) = \alpha\cdot\Phi_i(S\cup\{i\},v)+\beta\cdot\Phi_i(S\cup\{i\},w)$ for all $S \subseteq N\backslash\{i\}$ and consequently, by (2), $(\alpha\cdot v+\beta\cdot w)^{\Phi,i,\mathcal{A},\mathcal{B}} = \alpha\cdot v^{\Phi,i,\mathcal{A},\mathcal{B}}+\beta\cdot w^{\Phi,i,\mathcal{A},\mathcal{B}}$. From this, together with the consistency property for Φ and the induction hypothesis, we conclude that the following chain of equalities holds:

$$\begin{aligned}\Phi_j(N,\alpha\cdot v+\beta\cdot w) &\overset{(3)}{=} \Phi_j(N\backslash\{i\},(\alpha\cdot v+\beta\cdot w)^{\Phi,i,\mathcal{A},\mathcal{B}})\\ &= \Phi_j(N\backslash\{i\},\alpha\cdot v^{\Phi,i,\mathcal{A},\mathcal{B}}+\beta\cdot w^{\Phi,i,\mathcal{A},\mathcal{B}})\\ &= \alpha\cdot\Phi_j(N\backslash\{i\},v^{\Phi,i,\mathcal{A},\mathcal{B}})+\beta\cdot\Phi_j(N\backslash\{i\},w^{\Phi,i,\mathcal{A},\mathcal{B}})\\ &\overset{(3)}{=} \alpha\cdot\Phi_j(N,v)+\beta\cdot\Phi_j(N,w) \quad \text{for all } j \in N\backslash\{i\}.\end{aligned}$$

(*ii*) Suppose the expression at the right hand of the equivalence holds. The proof of the symmetry property for Φ proceeds by induction on the size n of the player set N. The case $n = 2$ is a direct consequence of the efficiency for Φ (arising from Lemma 1). From now let $\langle N,v\rangle$ be a game with $n \geq 3$ and π a permutation on N. By assumption, there exists some $i \in N$ such that $\Phi_{\pi(i)}(N,\pi v) = \Phi_i(N,v)$. From this and the induction hypothesis, we obtain that $\Phi_{\pi(i)}(\pi S\cup\{\pi(i)\},\pi v) = \Phi_i(S\cup\{i\},v)$ for all $S \subseteq N\backslash\{i\}$ and consequently, by (2), $(\pi v)^{\Phi,\pi(i),\mathcal{A},\mathcal{B}} = \pi(v^{\Phi,i,\mathcal{A},\mathcal{B}})$. From this, together with the consistency property for Φ and the induction hypothesis, we conclude that the following chain of equalities holds:

$$\begin{aligned}\Phi_{\pi(j)}(N,\pi v) &= \Phi_{\pi(j)}(N\backslash\{\pi(i)\},(\pi v)^{\Phi,\pi(i),\mathcal{A},\mathcal{B}})\\ &= \Phi_{\pi(j)}(N\backslash\{\pi(i)\},\pi(v^{\Phi,i,\mathcal{A},\mathcal{B}}))\\ &= \Phi_j(N\backslash\{i\},v^{\Phi,i,\mathcal{A},\mathcal{B}})\\ &\overset{(3)}{=} \Phi_j(N,v) \quad \text{for all } j \in N\backslash\{i\}.\end{aligned}$$

□

Theorem 2. *There exists at most one value* Φ *on* $\mathcal{G}$ *that verifies the next three properties.*

(*i*) λ*-standardness for two-person games (where* $\lambda \in \mathbb{R}$ *is fixed).*

(*ii*) *Equal split of incremental gains, that is* $\Phi(N,w) = \Phi(N,v)+\frac{w(N)-v(N)]}{n}\cdot(1,1,\ldots,1)$ *for all games* $\langle N,v\rangle$ *and* $\langle N,w\rangle$ *with* $w(S) = v(S)$ *for all* $S \subsetneq N$.

(iii) *The consistency property with respect to the reduced game of the form (2) associated with Φ and $\mathcal{A}, \mathcal{B}$.*

Proof. Suppose that, besides Φ, another value σ on $\mathcal{G}$ verifies the λ–standardness for two-person games, the equal split property, as well as the consistency property with respect to the reduced game of the form (2) associated with σ and $\mathcal{A}, \mathcal{B}$ (cf. (3) in terms of σ).

We prove by induction on the size n of the player set N that $\sigma(N,v) = \Phi(N,v)$ for all games $\langle N,v\rangle$ with $n \geq 2$. The case $n = 2$ is a direct consequence of the λ–standardness for both values σ and Φ with reference to two-person games.

From now on fix a game $\langle N,v\rangle$ with $n \geq 3$ and, by the induction hypothesis, it holds that $\sigma(M,w) = \Phi(M,w)$ for all games $\langle M,w\rangle$ with $2 \leq |M| \leq n-1$. Let $i \in N$. We claim the two reduced games $\langle N\backslash\{i\}, v^{\sigma,i,\mathcal{A},\mathcal{B}}\rangle$ and $\langle N\backslash\{i\}, v^{\Phi,i,\mathcal{A},\mathcal{B}}\rangle$ are related in such a way that

$$v^{\sigma,i,\mathcal{A},\mathcal{B}}(S) = v^{\Phi,i,\mathcal{A},\mathcal{B}}(S) \quad \text{for all } S \subsetneq N\backslash\{i\} \qquad \text{and} \tag{4}$$

$$v^{\sigma,i,\mathcal{A},\mathcal{B}}(N\backslash\{i\}) = v^{\Phi,i,\mathcal{A},\mathcal{B}}(N\backslash\{i\}) + \Phi_i(N,v) - \sigma_i(N,v)\,. \tag{5}$$

In order to prove (4), let $S \subsetneq N\backslash\{i\}$. The induction hypothesis applied to the subgame $\langle S \cup \{i\}, v\rangle$ yields $\sigma_i(S \cup \{i\}, v) = \Phi_i(S \cup \{i\}, v)$ and hence, by (2), $v^{\sigma,i,\mathcal{A},\mathcal{B}}(S) = v^{\Phi,i,\mathcal{A},\mathcal{B}}(S)$. This proves (4). Further, from (2) together with $A_{n-1,n-1} := 1$, $B_{n-1,n-1} := 0$, we derive that $v^{\sigma,i,\mathcal{A},\mathcal{B}}(N\backslash\{i\}) = v(N) - \sigma_i(N,v)$ as well as $v^{\Phi,i,\mathcal{A},\mathcal{B}}(N\backslash\{i\}) = v(N) - \Phi_i(N,v)$, and therefore, (5) holds. From the equal split property for Φ applied to the two reduced games, together with the two resulting formulae (4)–(5), we derive that for all $j \in N\backslash\{i\}$

$$\Phi_j(N\backslash\{i\}, v^{\sigma,i,\mathcal{A},\mathcal{B}}) = \Phi_j(N\backslash\{i\}, v^{\Phi,i,\mathcal{A},\mathcal{B}}) + \frac{\Phi_i(N,v) - \sigma_i(N,v)}{n-1}\,. \tag{6}$$

From (6), the induction hypothesis applied to the game $\langle N\backslash\{i\}, v^{\sigma,i,\mathcal{A},\mathcal{B}}\rangle$ and the consistency properties for the two values σ and Φ (associated with the value itself and $\mathcal{A}, \mathcal{B}$), we deduce the following chain of equalities:

$$\begin{aligned}
\sigma_j(N,v) &\overset{(3)}{=} \sigma_j(N\backslash\{i\}, v^{\sigma,i,\mathcal{A},\mathcal{B}}) \\
&= \Phi_j(N\backslash\{i\}, v^{\sigma,i,\mathcal{A},\mathcal{B}}) \\
&\overset{(6)}{=} \Phi_j(N\backslash\{i\}, v^{\Phi,i,\mathcal{A},\mathcal{B}}) + \frac{\Phi_i(N,v) - \sigma_i(N,v)}{n-1} \\
&\overset{(3)}{=} \Phi_j(N,v) + \frac{\Phi_i(N,v) - \sigma_i(N,v)}{n-1} \qquad \text{for all } j \in N\backslash\{i\}\,.
\end{aligned}$$

Thus, so far we arrive at the following conclusion:

$$\sigma_j(N,v) - \Phi_j(N,v) = \frac{\Phi_i(N,v) - \sigma_i(N,v)}{n-1}\,,$$

for all $i \in N$ and all $j \in N \setminus \{i\}$. By interchanging the roles of players i and j, the latter result yields

$$\sigma_i(N,v) - \Phi_i(N,v) = \frac{\sigma_i(N,v) - \Phi_i(N,v)}{(n-1)^2}, \quad \text{for all } i \in N.$$

Since $n \geq 3$, we conclude that $\sigma_i(N,v) = \Phi_i(N,v)$ for all $i \in N$. Thus, $\sigma(N,v) = \Phi(N,v)$ for all games $\langle N,v \rangle$ as was to be shown. □

4 An Axiomatization of Consistent Values

For arbitrary values without specific properties, the development of a unified consistency approach seems to be out of the question. As far as to the current view of the authors on a unified consistency treatment, the best one can do is to look at values that are fully characterized by explicit formulae or equivalently, by properties widely studied in the game theoretic literature. Throughout this section we focus on the largest possible class of values, determined by the efficiency, linearity, and symmetry axioms. For values of this particular type our aim is to present necessary and sufficient conditions to fulfil the consistency property with reference to the reduced game of (2). Note that the class of efficient, linear, and symmetric values contains the Least Square family and the well-known Shapley value in particular.

Theorem 3. *The following three statements for a value Φ on $\mathcal{G}$ are equivalent.*

(i) Φ verifies efficiency, linearity, and symmetry.

(ii) There exists a (unique) collection of constants $\{\rho_{n,s} \mid n \in \mathbb{N}\setminus\{1\}, s \in \{1,2,\ldots,n-1\}\}$ so that, for every n-person game $\langle N,v \rangle$ with at least two players, the value payoff vector $(\Phi_i(N,v))_{i\in N} \in \mathbb{R}^N$ is of the following form (cf. [10], Lemma 9, p. 117):

$$\Phi_i(N,v) = \frac{v(N)}{n} + \sum_{\substack{S \subsetneq N, \\ S \ni i}} \frac{\rho_{n,s}}{s} \cdot v(S) - \sum_{\substack{S \subseteq N, \\ S \not\ni i}} \frac{\rho_{n,s}}{n-s} \cdot v(S) \quad \textit{for all } i \in N. \tag{7}$$

(iii) There exists a (unique) collection of constants $\{b_{n,s} \mid n \in \mathbb{N}\setminus\{1\}, s \in \{1,2,\ldots,n\}\}$, with $b_{n,n} := 1$, so that, for every n-person game $\langle N,v \rangle$ with at least two players, the value payoff vector $(\Phi_i(N,v))_{i\in N} \in \mathbb{R}^N$ is of the following form:

$$\Phi_i(N,v) = \sum_{S \subseteq N\setminus\{i\}} \frac{1}{n \cdot \binom{n-1}{s}} \cdot \left[b_{n,s+1} \cdot v(S \cup \{i\}) - b_{n,s} \cdot v(S) \right], \tag{8}$$

$$\textit{for all } i \in N.$$

It is left to the reader to verify, by straightforward computations, that the expression at the right hand of (7) is equivalent to the one at the right hand of (8) by choosing $b_{n,s} = \binom{n}{s} \cdot \rho_{n,s}$ for all $s \in \{1, 2, \dots, n-1\}$. We remark that the associated collection of constants $\rho_{n,s}$ or $b_{n,s}$ is non-negative if and only if the value Φ, as given by (7), verifies the coalitional monotonicity (cf. Definition 1(v)). Throughout this section and the remaining proofs we prefer to make use of formula (8) because of the next interpretations.

Whenever $b_{n,s} := 1$ for all $s \in \{1, 2, \dots, n\}$, the expression at the right hand of (8) is well-known as the *Shapley value* $Sh_i(N, v)$ of player i in the n-person game $\langle N, v \rangle$ (cf. [11], [8]). In words, the Shapley value payoff of a player in a game equals some kind of expected outcome of his marginal contributions in the game itself. Generally speaking, the value payoff $\Phi_i(N, v)$ of player i in a game $\langle N, v \rangle$, as given by the right hand of (8), may be interpreted as player i's Shapley value in a disrupted game arising from the game $\langle N, v \rangle$. For every coalition, its worth in the disrupted game equals a given fraction of its worth in the original game, where the fraction depends upon the size of the coalition (assuming symmetry with respect to the size of the player set). To be exact, in the setting of the expression (8) and the Shapley value, it holds that $\Phi_i(N, v) = Sh_i(N, v^d)$ where $v^d\colon 2^N \to \mathbb{R}$ is defined to be $v^d(S) := b_{n,s} \cdot v(S)$ for all $S \subseteq N$. The fundamental result of this paper is the next theorem concerning the validity of the consistency property for the values under consideration as well as the corresponding axiomatization of these values.

Theorem 4. *Let $\mathcal{A}$ and $\mathcal{B}$ be two (unknown) collections of constants satisfying $A_{k,k} = 1$ and $B_{k,k} = 0$ for all $k \in \mathbb{N}\backslash\{1\}$. Suppose Φ is an efficient, linear, and symmetric value on $\mathcal{G}$, as given by (8) associated with the collection of constants $b_{n,s}$.*

For every $k \in \mathbb{N}\backslash\{1\}$, let the sequence of real numbers $\{x_{k,s} \mid s \in \{0, 1, \dots, k\}\}$ be the unique solution of the following system of equations: $x_{k,k} = k + 1$ and

$$x_{k,s} = (s+1) \cdot \sum_{\ell=s+1}^{k} \frac{b_{\ell+1,s+1}}{\ell \cdot (\ell+1)} \cdot x_{k,\ell} \quad \textit{for all } s \in \{0, 1, \dots, k-1\}\,. \tag{9}$$

Then the value Φ is consistent with respect to the reduced game of the form (2) associated with Φ and $\mathcal{A}, \mathcal{B}$ if and only if the two collections $\mathcal{A}, \mathcal{B}$ satisfy the following system of equations:

$$b_{k,s} \cdot A_{k,s} = x_{k,s} \quad \textit{and} \tag{10}$$

$$b_{k,s} \cdot B_{k,s} = b_{k+1,s} - x_{k,s-1}\,, \tag{11}$$

for all $k \in \mathbb{N}\backslash\{1\}$ and all $s \in \{1, 2, \dots, k-1\}$. In addition, Φ is the unique value σ on $\mathcal{G}$ that is $b_{2,1}$–standard for two-person games and consistent with respect to the reduced game of the form (2) associated with σ and $\mathcal{A}, \mathcal{B}$ (cf. the reduced game property (3) in terms of the value σ).

The uniqueness part of theorem 4 follows immediately from theorem 2 because any value Φ, as given by either (8) or (7), is $b_{2,1}$–standard (or $2 \cdot \rho_{2,1}$–standard) for two-person games and it clearly verifies the equal split property with reference to incremental gains for the grand coalition in the game.

For expositional convenience, the technical and lengthy proof of the consistency part of theorem 4 is postponed till the appendix. Practically, the techniques used in the consistency proof are not relevant for the theory to be developed further. In short, the consistency proof consists of tedious computations arising from substituting the reduced game of the form (2) in the expression (8) involving the value. In the remainder of this section we aim to clarify the contents and impact of theorem 4.

Remark In the setting of theorem 4, let us discuss the solvability of the system of equations (10)–(11) for the two unknown collections of constants $\mathcal{A}, \mathcal{B}$ that determine the reduced game of the form (2). Clearly, the relevant system of equations is (uniquely) solvable whenever the underlying collection of constants $b_{k,s}$ is supposed to be non-zero, i. e., $b_{k,s} \neq 0$ for all $k \in \mathbb{N} \backslash \{1\}$ and all $s \in \{1, 2, \dots, k-1\}$. However, if some of the constants $b_{k,s}$, $s \in \{1, 2, \dots, k-1\}$, happen to be zero, then the relevant system of equations may fail to be solvable (for instance, if $b_{k,s} = 0$ for all $s \in \{1, 2, \dots, k-2\}$ and $b_{k,k-1} \neq 0$ for all $k \in \mathbb{N} \backslash \{1\}$) and hence, the corresponding efficient, linear, and symmetric value is not consistent with respect to any reduced game of the form (2).

By theorem 4, the determination of the coefficients $A_{k,s}$, $B_{k,s}$ that describe the reduced game of the form (2), is fully based on the computation of the (unique) solution of the system of equations (9). For every $k \in \mathbb{N} \backslash \{1\}$, the relevant solution has to be computed recursively (in a backwards manner) and generally speaking, it is impossible to achieve an explicit formula for the solution $x_{k,s}$, $s \in \{0, 1, \dots, k-1\}$. Under certain general circumstances, however, we are able to deduce an explicit formula for the solution of the system of equations (9). For that purpose we say the underlying collection of constants $b_{n,s}$ is *separable* if it results from a product of two independent sequences, of which one sequence refers to the size n of the player set and the other sequence to the size s of the coalition. See the forthcoming examples 1–3.

Corollary 1. *Suppose a value Φ on $\mathcal{G}$ is efficient, linear, and symmetric, as given by (8) associated with the collection of non-zero constants $\{b_{n,s} \mid n \in \mathbb{N} \backslash \{1\}, s \in \{1, 2, \dots, n\}\}$, where $b_{n,n} := 1$ and $b_{n,s} = \alpha_n \cdot \beta_s$ for all $n \in \mathbb{N} \backslash \{1\}$ and all $s \in \{1, 2, \dots, n-1\}$ (separability condition). Under this non-zero separability assumption, the following holds.*

(i) The unique solution of the system of equations (9) is given by

$$x_{k,s} = \alpha_{k+1} \cdot \beta_{s+1} \cdot \prod_{\ell=s+2}^{k} [1 - \gamma_\ell], \quad \textit{for all } s \in \{0, 1, \dots, k-1\}, \tag{12}$$

with the convention that $\gamma_\ell := \frac{1}{\ell} \cdot [1 - \alpha_\ell \cdot \beta_\ell]$ for all $\ell \in \mathbb{N} \backslash \{1\}$ and $\prod_{\ell=t}^{k} y_\ell := 1$ if $t > k$.

(*ii*) Φ *is consistent with respect to the reduced game of the form (2) such that, for all* $k \in \mathbb{N}\backslash\{1\}$ *and all* $s \in \{1, 2, \ldots, k-1\}$, *the coefficients* $A_{k,s}$ *and* $B_{k,s}$ *are given by*

$$A_{k,s} = \frac{\alpha_{k+1}}{\alpha_k} \cdot \frac{\beta_{s+1}}{\beta_s} \cdot \prod_{\ell=s+2}^{k} [1-\gamma_\ell] \quad \textit{and} \quad B_{k,s} = \frac{\alpha_{k+1}}{\alpha_k} \cdot \left[1 - \prod_{\ell=s+1}^{k} [1-\gamma_\ell]\right]$$

(*iii*) Φ *is the unique value on* $\mathcal{G}$ *that is* $\alpha_2 \cdot \beta_1$*–standard for two-person games and consistent with respect to this particular type of a reduced game.*

Proof. Fix $k \in \mathbb{N}\backslash\{1\}$. Under the non-zero separability assumption, the system of equations (9) reduces as follows:

$$\frac{x_{k,s}}{(s+1)\cdot\beta_{s+1}} = \sum_{\ell=s+1}^{k} \frac{\alpha_{\ell+1}}{\ell\cdot(\ell+1)} \cdot x_{k,\ell} \quad \text{for all } s \in \{0, 1, \ldots, k-1\}.$$

Hence $\dfrac{x_{k,s-1}}{s\cdot\beta_s} - \dfrac{x_{k,s}}{(s+1)\cdot\beta_{s+1}} = \dfrac{\alpha_{s+1}}{s\cdot(s+1)} \cdot x_{k,s}$ or equivalently,

$$x_{k,s-1} = \frac{[s+\alpha_{s+1}\cdot\beta_{s+1}]}{s+1} \cdot \frac{\beta_s}{\beta_{s+1}} \cdot x_{k,s}\,, \quad \forall\, s \in \{1, 2, \ldots, k-1\}\,.$$

From this, together with $x_{k,k-1} = b_{k+1,k} = \alpha_{k+1}\cdot\beta_k$, we conclude that the unique solution of the system of equations (9) is given by the expression at the right hand of (12). With part (i) at hand, the remaining statements mentioned in parts (*ii*)–(*iii*) follow immediately from theorem 4 and the formulae (10)–(11) in particular. □

Example 1. For every $\delta \in \mathbb{R}\backslash\{0\}$, the *$\delta$-discounted Shapley value* Sh^δ is obtained from the right hand of (8) with corresponding constants $b_{n,s} := \delta^{n-s}$ for all $s \in \{1, 2, \ldots, n\}$. That is, the worth of a coalition in an n-person game is weakly discounted whenever the size s of the coalition is relatively large (or strongly discounted, if the size of the coalition is relatively small) in comparison with the size n of the player set.

Clearly, the separability applies (so that $\alpha_n := \delta^n$, $\beta_s := \delta^{-s}$, and $\gamma_\ell = 0$). Thus, by Corollary 1, for every $\delta \in \mathbb{R}\backslash\{0\}$, the δ-discounted Shapley value Sh^δ is consistent with respect to the reduced game of the form $v^{Sh^\delta,i,\mathcal{A},\mathcal{B}}(S) = v(S\cup\{i\}) - Sh_i^\delta(S\cup\{i\}, v)$ for all $S \subseteq N\backslash\{i\}$, $S \neq \emptyset$. That is, without regard to the non-zero discount coefficient δ, the characteristic function of the reduced game with player set $N\backslash\{i\}$ arises from the belief that, with probability one, the removed player i joins every coalition S, on the understanding that player i receives the amount of his δ-discounted Shapley value payoff in the resulting subgame $\langle S\cup\{i\}, v\rangle$ with player set $S\cup\{i\}$. Together with the δ–standardness for two-person games, this particular type of a reduced game property fully characterizes the δ-discounted Shapley value Sh^δ. Our unified result applied to the special case in which $\delta = 1$ is in accordance with Hart and Mas-Colell's axiomatization of the Shapley value (cf. [6], see Proposition 4.5, p. 597).

Example 2. By its definition (cf. [7], see theorem, p. 45), the *solidarity value* ψ is an efficient, linear and symmetric value on $\mathcal{G}$, hence the solidarity value is of the form (8) associated with the constants $b_{n,s} := 1/(s+1)$ for all $s \in \{1,2,\dots,n-1\}$. Clearly, the separability applies (so that $\alpha_n := 1$, $\beta_s := 1/(s+1)$, and $\gamma_\ell = 1/(\ell+1)$). Thus, by Corollary 1, the solidarity value ψ is consistent with respect to the reduced game of the form

$$v^{\psi,i,\mathcal{A},\mathcal{B}}(S) = \frac{s+1}{n} \cdot \Big[v(S \cup \{i\}) - \psi_i(S \cup \{i\}, v)\Big] + \frac{n-1-s}{n} \cdot v(S)$$

for all $S \subseteq N \backslash \{i\}$, $S \neq \emptyset$. That is, the characteristic function of the reduced game with player set $N \backslash \{i\}$ arises from the belief that the probability that the removed player i joins a coalition of size s is proportional to the size $s+1$ of the resulting coalition, i. e., $A_{n-1,s} = (s+1)/n$. The larger the size of the coalition, the higher the probability for the player to join (where the probability increases to one). Together with the 1/2–standardness for two-person games, this particular type of a reduced game property fully characterizes the solidarity value ψ.

Example 3. The *per-capita Shapley value* Sh^{pc} is obtained from the right hand of (8) with corresponding constants $b_{n,s} := \frac{1}{s}$ for all $s \in \{1,2,\dots,n-1\}$. In words, the per-capita Shapley value payoff of a player in a game equals the expected outcome of his marginal contributions in the per-capita game (instead of the game itself).

Clearly, the separability applies (so that $\alpha_n := 1$, $\beta_s := \frac{1}{s}$, and $\gamma_\ell = \frac{\ell-1}{\ell^2}$). Thus, by Corollary 1, the per-capita Shapley value Sh^{pc} is consistent with respect to the reduced game of the form (2) such that the coefficients $A_{n-1,s}$ and $B_{n-1,s}$ are given by

$$A_{n-1,s} = \frac{s}{s+1} \cdot \prod_{\ell=s+2}^{n-1} \left[1 - \frac{(\ell-1)}{\ell^2}\right] \quad \text{and} \quad B_{n-1,s} = 1 - \prod_{\ell=s+1}^{n-1} \left[1 - \frac{(\ell-1)}{\ell^2}\right].$$

Together with the 1–standardness for two-person games, this particular type of a reduced game property fully characterizes the per-capita Shapley value. Note that $A_{n-1,s} + B_{n-1,s} = 1 - \frac{1}{s\cdot(s+1)} \cdot A_{n-1,s}$ for all $s \in \{1,2,\dots,n-2\}$. In words, the *probabilistic* interpretation of the characteristic function of the reduced game, as presented in the paragraph preceding Definition 2, fails to hold in the context of the per-capita Shapley value. From the mathematical viewpoint, this example illustrates that there exist reduced games of the form (2) that are interesting on their own, but without a probabilistic interpretation. For this very reason, we excluded the probabilistic condition $A_{n-1,s} + B_{n-1,s} = 1$ for all $s \in \{1,2,\dots,n-1\}$ from Definition 2.

This paper was devoted to the consistency theory developed on the class of efficient, linear, and symmetric values for TU-games. We conclude with a short discussion on the consistency theory restricted to the class of least square values.

Example 4. With every collection of nonnegative constants $\mathcal{M} = \{m_{n,s} \mid n \in \mathbb{N}\backslash\{1\}, s \in \{1,2,\ldots,n-1\}\}$, there is associated the *least square value* $LS^{\mathcal{M}}$ on $\mathcal{G}$ to be defined as the unique optimal solution of the least square problem (1). It was shown in [10] (cf. theorem 8, p. 116) that the Least Square family is fully characterized by five properties: efficiency, linearity, symmetry, covariance and coalitional monotonicity (cf. Definition 1). As a matter of fact (cf. [10], p. 114), every least square value $LS^{\mathcal{M}}$ is of the form (7) associated with the constants $\rho_{n,s} = \frac{s\cdot(n-s)}{n} \cdot \frac{m_{n,s}}{S_{\mathcal{M}}(n)}$ or equivalently, of the form (8) associated with the constants $b_{n,s} = s\cdot\binom{n-1}{s}\cdot\frac{m_{n,s}}{S_{\mathcal{M}}(n)}$ for all $n \in \mathbb{N}\backslash\{1\}$ and all $s \in \{1,2,\ldots,n-1\}$, where $S_{\mathcal{M}}(n) := \sum_{s=1}^{n-1}\binom{n-2}{s-1}\cdot m_{n,s}$.

By its definition (cf. [9], see theorem 5, p. 116), the *least square prenucleolus* is given by the least square value $LS^{\mathcal{M}}$ whenever $m_{n,s} := 1$ for all $s \in \{1,2,\ldots,n-1\}$. Further, it turns out that the Shapley value agrees with the least square value $LS^{\mathcal{M}}$ whenever $m_{n,s} := s^{-1}\cdot\binom{n-1}{s}^{-1}$ for all $s \in \{1,2,\ldots,n-1\}$. The solidarity value, however, does not agree with any least square value since the solidarity value fails to meet the 1–standardness for two-person games.

Concerning the class of all least square values, neither the corresponding collection of constants $b_{n,s}$ is necessarily separable nor the corresponding system of equations (10)–(11) is necessarily solvable. Thus, it happens that a least square value is not consistent with respect to any reduced game of the form (2). This lack of any consistency property only occurs if some of the underlying nonnegative constants $m_{n,s}$ are degenerated to zero. If none of the nonnegative constants $m_{n,s}$ equals zero, then the induced least square value is consistent with respect to some reduced game of the form (2). For every $k \in \mathbb{N}\backslash\{1\}$, by substituting $y_{k,\ell} := x_{k,\ell}/(\ell+1)$, the fundamental system of equations (9) reduces to $y_{k,k} = 1$ and

$$y_{k,s} = \sum_{\ell=s+1}^{k} \binom{\ell-1}{s} \cdot \frac{m_{\ell+1,s+1}}{S_{\mathcal{M}}(\ell+1)} \cdot y_{k,\ell} \quad \text{for all } s \in \{0,1,\ldots,k-1\}\,.$$

Generally speaking, it is impossible to achieve an explicit formula for its solution $y_{k,s}$. Even in the context of the least square prenucleolus (that is, $m_{n,s} := 1$ and $S_{\mathcal{M}}(n) = 2^{n-2}$), an explicit formula for its solution $y_{k,s}$ is not available. Of course, the relevant system is explicitly solvable in the context of a suitably chosen least square value, like the Shapley value (in which case $x_{k,k} = k+1$ and $x_{k,s} = 1$ otherwise).

Remark For the sake of further research, it seems worthwhile to investigate whether a similar consistency theory can be developed in the framework of non-efficient values for TU-games (cf. [5]) or a suitably defined class of values for NTU-games. Our first goal is to develop a somehow related consistency theory for efficient, linear, and symmetric values for TU-games, where we deviate from the reduced game à la Hart and Mas-Colell, but adopt the reduced game à la Sobolev in which case the reduced game refers to an arbitrary allocation instead of the value itself (cf. [12], [3], [14], [15]).

We recall that Hart and Mas-Colell [6] established that the Shapley value is the unique efficient value that admits a so-called potential representation (described as the discrete gradient of a potential function). An extension of Hart and Mas-Colell's potential approach to an appropriately determined subclass of efficient, linear and symmetric values was treated by Driessen and Radzik [4]. It is beyond the scope of this paper to notice that the so-called weighted pseudo-potential representation of a certain type of values is an additional tool to establish the obtained consistency results for such subclass of values.

Appendix: Proof of Theorem 4

Let Φ be an efficient, linear, and symmetric value, as given by (8) associated with the collection of constants $\{b_{k,s} \mid k \in \mathbb{N}\backslash\{1\}, s \in \{1,2,\ldots,k\}\}$, where $b_{k,k} := 1$ for all $k \in \mathbb{N}\backslash\{1\}$. Until further notice we impose no constraints upon two collections of constants $\mathcal{A} = \{A_{k,s} \mid k \in \mathbb{N}\backslash\{1\}, s \in \{1,2,\ldots,k\}\}$, $\mathcal{B} = \{B_{k,s} \mid k \in \mathbb{N}\backslash\{1\}, s \in \{1,2,\ldots,k\}\}$, except for $A_{k,k} := 1$ and $B_{k,k} := 0$ for all $k \in \mathbb{N}\backslash\{1\}$.

Fix an n-person game $\langle N, v\rangle$ with $n \geq 3$ and two arbitrary players $i \in N$, $j \in N\backslash\{i\}$. In order to investigate the reduced game property for Φ, player j's value payoff $\Phi_j(N,v)$, as given by (8), has to be compared with a similar expresssion to be constructed for $\Phi_j(N\backslash\{i\}, v^{\Phi,i,\mathcal{A},\mathcal{B}})$. To be exact, we aim to determine the coefficient of the terms $v(S)$, $S \subseteq N$, in the resulting expression for $\Phi_j(N\backslash\{i\}, v^{\Phi,i,\mathcal{A},\mathcal{B}})$ and for that purpose we distinguish four types of coalitions in the following manner:

$$\Phi_j(N\backslash\{i\}, v^{\Phi,i,\mathcal{A},\mathcal{B}}) = \sum_{T\subseteq N\backslash\{i,j\}} \Big[\alpha_{|T|}^{+i,+j} \cdot v(T\cup\{i,j\}) + \alpha_{|T|}^{+i,-j} \cdot v(T\cup\{i\}) + \alpha_{|T|}^{-i,+j} \cdot v(T\cup\{j\}) + \alpha_{|T|}^{-i,-j} \cdot v(T)\Big]. \tag{13}$$

For expositional convenience we write

$$\begin{aligned} p_{k,s} &:= k^{-1} \cdot \binom{k-1}{s}^{-1} && \text{for all } k \in \mathbb{N} \text{ and all } s \in \{0,1,\ldots,k-1\}; \\ bA_{k,s} &:= b_{k,s} \cdot A_{k,s} \\ bB_{k,s} &:= b_{k,s} \cdot B_{k,s} \\ b_{k,0} &:= A_{k,0} = B_{k,0} = 0 && \text{for all } k \in \mathbb{N}\backslash\{1\} \text{ and all } s \in \{0,1,\ldots,k\}. \end{aligned}$$

We apply (8) to the reduced game $\langle N\backslash\{i\}, v^{\Phi,i,\mathcal{A},\mathcal{B}}\rangle$, as given by (2), and thus we obtain

$$\begin{aligned}
&\Phi_j(N\backslash\{i\}, v^{\Phi,i,\mathcal{A},\mathcal{B}}) \\
&\overset{(8)}{=} \sum_{S\subseteq N\backslash\{i,j\}} p_{n-1,s}\cdot\Big[b_{n-1,s+1}\cdot v^{\Phi,i,\mathcal{A},\mathcal{B}}(S\cup\{j\}) - b_{n-1,s}\cdot v^{\Phi,i,\mathcal{A},\mathcal{B}}(S)\Big] \\
&\overset{(2)}{=} \sum_{S\subseteq N\backslash\{i,j\}} p_{n-1,s}\cdot b_{n-1,s+1}\cdot\Big[A_{n-1,s+1}\cdot v(S\cup\{i,j\}) + B_{n-1,s+1}\cdot v(S\cup\{j\})\Big] \\
&\quad - \sum_{S\subseteq N\backslash\{i,j\}} p_{n-1,s}\cdot b_{n-1,s}\cdot\Big[A_{n-1,s}\cdot v(S\cup\{i\}) + B_{n-1,s}\cdot v(S)\Big] \\
&\quad + \sum_{S\subseteq N\backslash\{i,j\}} p_{n-1,s}\cdot b_{n-1,s}\cdot A_{n-1,s}\cdot\Phi_i(S\cup\{i\}, v) \qquad (14) \\
&\quad - \sum_{S\subseteq N\backslash\{i,j\}} p_{n-1,s}\cdot b_{n-1,s+1}\cdot A_{n-1,s+1}\cdot\Phi_i(S\cup\{i,j\}, v)\,. \qquad (15)
\end{aligned}$$

Next we determine the subexpressions (14)–(15) by applying (8) to the subgames $\langle S\cup\{i\}, v\rangle$ and $\langle S\cup\{i,j\}, v\rangle$ respectively. Secondly, we obtain

$$\begin{aligned}
&\sum_{S\subseteq N\backslash\{i,j\}} p_{n-1,s}\cdot bA_{n-1,s}\cdot\Phi_i(S\cup\{i\}, v) \\
&\overset{(8)}{=} \sum_{S\subseteq N\backslash\{i,j\}} p_{n-1,s}\cdot bA_{n-1,s}\cdot\sum_{T\subseteq S} p_{s+1,t}\cdot b_{s+1,t+1}\cdot v(T\cup\{i\}) \\
&\quad - \sum_{S\subseteq N\backslash\{i,j\}} p_{n-1,s}\cdot bA_{n-1,s}\cdot\sum_{T\subseteq S} p_{s+1,t}\cdot b_{s+1,t}\cdot v(T) \\
&= \sum_{R\subseteq N\backslash\{i,j\}}\Big[\sum_{s=r}^{n-2}\tbinom{n-2-r}{s-r}\cdot p_{n-1,s}\cdot p_{s+1,r}\cdot bA_{n-1,s}\cdot b_{s+1,r+1}\Big]\cdot v(R\cup\{i\}) \\
&\quad - \sum_{R\subseteq N\backslash\{i,j\}}\Big[\sum_{s=r}^{n-2}\tbinom{n-2-r}{s-r}\cdot p_{n-1,s}\cdot p_{s+1,r}\cdot bA_{n-1,s}\cdot b_{s+1,r}\Big]\cdot v(R)\,,
\end{aligned}$$

where the latter equality follows from a careful, but straightforward determination of the former two double sums. Using similar reasonings, we also obtain

$$\sum_{S\subseteq N\setminus\{i,j\}} p_{n-1,s}\cdot bA_{n-1,s+1}\cdot \Phi_i(S\cup\{i,j\},v)$$

$$\stackrel{(8)}{=}\sum_{S\subseteq N\setminus\{i,j\}} p_{n-1,s}\cdot bA_{n-1,s+1}\cdot \sum_{T\subseteq S\cup\{j\}} p_{s+2,t}\cdot b_{s+2,t+1}\cdot v(T\cup\{i\})$$

$$-\sum_{S\subseteq N\setminus\{i,j\}} p_{n-1,s}\cdot bA_{n-1,s+1}\cdot \sum_{T\subseteq S\cup\{j\}} p_{s+2,t}\cdot b_{s+2,t}\cdot v(T)$$

$$=\sum_{R\subseteq N\setminus\{i,j\}}\left[\sum_{s=r}^{n-2}\binom{n-2-r}{s-r}\cdot p_{n-1,s}\cdot p_{s+2,r+1}\cdot bA_{n-1,s+1}\cdot b_{s+2,r+2}\right]\cdot v(R\cup\{i,j\})$$

$$+\sum_{R\subseteq N\setminus\{i,j\}}\left[\sum_{s=r}^{n-2}\binom{n-2-r}{s-r}\cdot p_{n-1,s}\cdot p_{s+2,r}\cdot bA_{n-1,s+1}\cdot b_{s+2,r+1}\right]\cdot v(R\cup\{i\})$$

$$-\sum_{R\subseteq N\setminus\{i,j\}}\left[\sum_{s=r}^{n-2}\binom{n-2-r}{s-r}\cdot p_{n-1,s}\cdot p_{s+2,r+1}\cdot bA_{n-1,s+1}\cdot b_{s+2,r+1}\right]\cdot v(R\cup\{j\})$$

$$-\sum_{R\subseteq N\setminus\{i,j\}}\left[\sum_{s=r}^{n-2}\binom{n-2-r}{s-r}\cdot p_{n-1,s}\cdot p_{s+2,r}\cdot bA_{n-1,s+1}\cdot b_{s+2,r}\right]\cdot v(R)\,.$$

Simplifying and adding all subexpressions results in the expression at the right hand of (13), with the convention that for all $t\in\{0,1,\ldots,n-2\}$

$$\alpha_t^{+i,+j} := p_{n-1,t}\cdot bA_{n-1,t+1} - n\cdot p_{n,t+1}\cdot\sum_{s=t}^{n-2}\frac{bA_{n-1,s+1}\cdot b_{s+2,t+2}}{(s+1)\cdot(s+2)}\,,$$

$$\alpha_t^{+i,-j} := -p_{n-1,t}\cdot bA_{n-1,t} + p_{n-1,t}\cdot\sum_{s=t}^{n-2}\frac{bA_{n-1,s}\cdot b_{s+1,t+1}}{s+1}$$

$$-p_{n-1,t}\cdot\sum_{s=t}^{n-2}\frac{(s+1-t)\cdot bA_{n-1,s+1}\cdot b_{s+2,t+1}}{(s+1)\cdot(s+2)}\,,$$

$$\alpha_t^{-i,+j} := p_{n-1,t}\cdot bB_{n-1,t+1} + n\cdot p_{n,t+1}\cdot\sum_{s=t}^{n-2}\frac{bA_{n-1,s+1}\cdot b_{s+2,t+1}}{(s+1)\cdot(s+2)}\,,$$

$$\alpha_t^{-i,-j} := -p_{n-1,t}\cdot bB_{n-1,t} - p_{n-1,t}\cdot\sum_{s=t}^{n-2}\frac{bA_{n-1,s}\cdot b_{s+1,t}}{s+1}$$

$$+p_{n-1,t}\cdot\sum_{s=t}^{n-2}\frac{(s+1-t)\cdot bA_{n-1,s+1}\cdot b_{s+2,t}}{(s+1)\cdot(s+2)}\,.$$

So far we conclude that the reduced game property for Φ holds if and only if the two unknown collections of constants $\mathcal{A},\mathcal{B}$ satisfy the following system of equations:

$$\alpha_t^{+i,+j} = p_{n,t+1} \cdot b_{n,t+2}\,, \tag{16}$$
$$\alpha_t^{+i,-j} = -p_{n,t+1} \cdot b_{n,t+1}\,, \tag{17}$$
$$\alpha_t^{-i,+j} = p_{n,t} \cdot b_{n,t+1}\,, \tag{18}$$
$$\alpha_t^{-i,-j} = -p_{n,t} \cdot b_{n,t}\,, \quad \text{for all } t \in \{0,1,\dots,n-2\}. \tag{19}$$

It turns out that, for $t = n-2$, both (16) and (18) hold trivially (because of $b_{k,k} = A_{k,k} = 1$ and $B_{k,k} = 0$). Further, (17) holds trivially for $t = 0$, whereas the case $t = 0$ has to be excluded from (19) beforehand. It is straightforward to verify that (16) and (17) are fully identical and similarly, (18) and (19) are fully identical. In summary, the reduced game property for Φ holds if and only if the two unknown collections $\mathcal{A}, \mathcal{B}$ satisfy the following (simplified) system of equations: for all $t \in \{1,2,\dots,n-2\}$

$$\frac{bA_{n-1,t}}{t+1} = \frac{b_{n,t+1}}{n-1} + \sum_{s=t+1}^{n-2} \frac{bA_{n-1,s} \cdot b_{s+1,t+1}}{s \cdot (s+1)}\,, \tag{20}$$

$$bB_{n-1,t} = \frac{(n-t) \cdot b_{n,t}}{n} - t \cdot \sum_{s=t}^{n-1} \frac{bA_{n-1,s} \cdot b_{s+1,t}}{s \cdot (s+1)}\,. \tag{21}$$

Given (20), it turns out that, for all $t \in \{2,3,\dots,n-2\}$, (21) is equivalent with

$$bB_{n-1,t} = b_{n,t} - bA_{n-1,t-1}\,. \tag{22}$$

Finally we remark that, for all $t \in \{1,2,\dots,n-2\}$, (20) involving $bA_{n-1,t}$ fully agrees with the expression for $x_{k,s}$, as given by (9), where $s \in \{1,2,\dots, k-1\}$. In addition, for all $t \in \{2,3,\dots,n-2\}$, (22) involving $bB_{n-1,t}$ fully agrees with (11), where $s \in \{2,3,\dots,k-1\}$. Moreover, (21) applied to the remaining case $t = 1$ fully agrees with (11) applied to $s = 1$. This completes the proof of the consistency part as stated in theorem 4. □

References

1. Aumann, R. J., Hart, S. (Eds.) (1992): Handbook of Game Theory with Economic Applications. In: Handbooks in Economics. Elsevier Science Publishers, Amsterdam, The Netherlands, No. 11, Vol. 1
2. Driessen, T. S. H. (1988): Cooperative Games, Solutions, and Applications. Kluwer Academic Publishers, Dordrecht, The Netherlands
3. Driessen, T. S. H. (1991): A Survey of Consistency Properties in Cooperative Game Theory. SIAM Review **33**, 43–59
4. Driessen, T. S. H., Radzik, T. (2002): A Weighted Pseudo-Potential Approach to Values for TU Games. International Transactions in Operational Research (ITOR) **9**, 1–18
5. Dubey, P., Neyman, A., Weber, R. J. (1981): Value Theory without Efficiency. Mathematics of Operations Research **6**, 122–128

6. Hart, S., Mas-Colell, A. (1989), Potential, Value, and Consistency. Econometrica **57**, 589–614. Also in [8], 127–137
7. Nowak, A. S., Radzik, T. (1994): A Solidarity Value for n-Person Transferable Utility Games. International Journal of Game Theory **23**, 43–48
8. Roth, A. E. (Ed.) (1988): The Shapley Value: Essays in Honor of Lloyd S. Shapley. Cambridge University Press, Cambridge, USA
9. Ruiz, L. M., Valenciano, F., Zarzuelo, J. M. (1996): The Least Square Prenucleolus and the Least Square Nucleolus: two Values for TU-Games Based on the Excess Vector. International Journal of Game Theory **25**, 113–134
10. Ruiz, L. M., Valenciano, F., Zarzuelo, J. M. (1998): The Family of Least Square Values for Transferable Utility Games. Games and Economic Behavior **24**, 109–130
11. Shapley, L. S. (1953): A Value for n-Person Games. Annals of Mathematics Study **28**, 307–317 (Princeton University Press). Also in [8], 31–40
12. Sobolev, A. I. (1973): The Functional Equations that Give the Payoffs of the Players in an n-Person Game. In: Vilkas, E. (Ed.): Advances in Game Theory. Izdat. "Mintis", Vilnius, 151–153 (in Russian)
13. Weber, R. J. (1988): Probabilistic Values for Games. In: [8], 101–119
14. Yanovskaya, E. (1996): Axiomatic Characterizations of Reduced Games. Working Paper, St. Petersburg Institute for Economics and Mathematics, Russian Academy of Sciences, St. Petersburg, Russia
15. Yanovskaya, E., Driessen, T. S. H. (2002): Note: On Linear Consistency of Anonymous Values for TU-Games. International Journal of Game Theory (to appear)

On the Stability of Cooperation Structures

Guillaume Haeringer

University of Warwick, Department of Economics,
CV4 7AL Coventry, UK
E-mail: g.haeringer@warwick.ac.uk

Summary. This paper studies the payoff structure of stable cooperation structures in link formation games. Players choose non-cooperatively with whom they want to form a link, and the payoffs are given by the Myerson value of the cooperation structure obtained. We characterize the class of TU-games that ensure the stability of the full cooperation structure, which turns out to be much larger than the class of superadditive TU-games. We then provide an exact characterization of the Moderer and Shapley potential of the link formation game, and establish its equivalence with the potential as defined by Hart and Mas-Colell [13]. We use this result to show that stable but Pareto dominated graphs can emerge under simple best-response dynamics.

Key words: cooperation structure, graph, Myerson value, stability, potential

1 Introduction

In the last few years, the economic theory literature witnessed a growing interest on networks, with the purpose to understand how networks emerge, and to study their stability and efficiency properties.[1] Here, networks, also called cooperation structures, represent situations in which players may cooperate, or communicate only with a subset of the population. An interesting feature of these situations is that two players having a relationship may not have a relationship with the same subset of players. Cooperation structures are usually described with (non-directed) graphs, whose vertices represent players and edges (or links) represent the relationships players have between each other.

Among all the contributions on that topic, the work of Aumann and Myerson [1] has received particular attention. In their seminal paper, Aumann and Myerson proposed a model of endogenous cooperation structure, that was associating in a subtle way both cooperative and non-cooperative games. That

[1] See [14] for a recent contribution about such issues and the references therein.

is, the cooperation structure is modeled by means of a cooperative game, but it is formed in a non-cooperative way. The building block of Aumann and Myerson's model, which became common to most of the so-called *link formation games*,[2] is that a link between two players is formed only if they both want it. In such games a player's strategy is the set of players with whom she wants to form a link, and players' payoffs are given by the Myerson value [19] in the cooperation structure thus obtained. Since Aumann and Myerson's game is an extensive form game, a major drawback of their approach is that the order of play has to bet set exogenously. Later, Qin [20] reformulated Aumann and Myerson's game as a normal-form game. The main results obtained by Qin' are that *(i)* the game admits a potential – in the sense of Monderer and Shapley [18]; and *(ii)* if the underlying TU-game is superadditive, then the full cooperation structure (each player cooperates with all the other players) is stable in the sense that no player can profitably deviate by severing her links or any part thereof.

The purpose of this paper is to scrutinize the relationship between the stability of cooperation structures and the payoff structure of the game of link formation. Since this latter is built upon a cooperative TU-game, the properties we look for simply consists of designing the class of cooperative games that ensure that some specific cooperation structure is stable. Among all the possible cooperation structures there is one on which several papers (including this one) have brought specific attention: the full cooperation structure, which is the structure where all players are linked between each other. The motivation for studying this structure is twofold. First, in many situations, the full cooperation structure is the efficient one, and as such, one would like it to be the stable one, too. Second, it has an interesting appealing property, that of making a bridge between networks and coalitions. Indeed, from the network point of view, a coalition can be seen as a network where all members are linked to each other. Recently, both models on network formation and coalition formation have received a great deal of attention.[3] However, little as been done to study how coalitions and networks are inter-related. This paper is aimed at providing new insights in that respect as well.

A first result we propose is a complete characterization of the class of TU-games that admit a stable full cooperation structure. It turns out that this class is much larger than the class of superadditive games. This is to be contrasted with other papers (e. g., [20] and [9]) that focused on superadditive environments only. More precisely, we show that the full cooperation structure is stable if and only if the (Shapley) value of each player is individually rational and the TU-game satisfies a mild condition on average marginal

[2] See among others Jackson and Wolinsky [16], Dutta and Mutuswami [8], or Jackson and Watts [15] who used the same principle in their models.

[3] See for instance [3] or [5], and [16] or [2] for recent contributions on coalition formation and on network formation respectively. The interested reader will find further references in these papers. See also [7].

contributions. In particular, this condition implies that if, for some TU-game (N, v) and two players i and j, we have $v(\{i,j\}) < v(\{i\}) + v(\{j\})$, then the full cooperation structure is not stable. This result suggests that "small coalitions" have a greater role in the stability of the full cooperation structure than bigger ones.[4] Nonetheless, this result also shows that the underlying TU-game does not need to be superadditive to ensure the stability of the full cooperation structure. This is, we believe, an interesting property given that non-superadditive economic environments abound. Such environments are for instance when there are congestion effects in the production of public goods. Another case that can generate non-superadditivity is when there are inefficiencies in decision making in large organization. In large firms or large organizations, increasing return to scale can be dominated by the cost of coordination (see [10, 11, 12]).

A second set of results presented in this paper concerns the potential(s) of the link formation game. This approach follows that of Qin, who showed that this game admits a potential – in the sense of Monderer and Shapley [18], henceforth MS-potential – if and only if the allocation rule used in the underlying TU-game is the Myerson value. In this paper, we propose a formula for the potential, and use it to derive further results about the stability of cooperation structures. The existence of an MS-potential is quite appealing,[5] since this latter has strong learning properties. That is, best-response and fictitious play learning processes converge to equilibria of the game, and these equilibria are those maximizing the MS-potential.[6] A second result concerning the MS-potential is its equivalence with the potential of Hart and Mas-Colell [13] – henceforth HM-potential – which, contrary to that of Monderer and Shapley, is a solution concept for cooperative games. At first sight, this result is quite surprising, since the two potentials are apparently two different mathematical objects. Nevertheless, the structure of the link formation game makes this result quite predictable. First, both concepts are built upon marginal variation of the payoffs. Second, Hart and Mas-Colell showed that their potential is closely related to the Shapley value, which is precisely the allocation rule used to compute players' payoffs in the link formation game, and thus the MS-potential. Moreover, it is worth to point out that a similar result has already been observed by Monderer and Shapley [18, theorems 6.1 and 6.2], and Ui [21].

The paper is organized as follows. In section 2 we present the framework. In section 3 we define the game of endogenous formation of cooperation structure. Our first characterization result is presented in section 4. Section 5 is devoted

[4] In fact, all coalitions do matter since whether or not the Shapley value of each player is individually rational depends on the worth of *all* coalitions.

[5] Not all games admit an MS-potential.

[6] See [4] and [17, 18] for results on the learning properties of the potential, and [15] for a study of dynamic network formation with myopic players.

to the analysis of the potentials of the game. We study an example in section 6 and conclude in section 7.

2 Definitions

A transferable utility game (or a TU-game) is defined by a couple (N, v) where $N = \{1, \ldots, n\}$ is the set of players, and $v\colon 2^N \to \mathbb{R}$ is the characteristic function, with the convention that $v(\varnothing) \equiv 0$). The number $v(S)$ is the worth of coalition $S \in 2^N$. Given a game (N, v) and a subset of players S, (S, v) is the sub-game obtained by restricting v to subsets of S only. If N, R, S, T, ... are coalitions, then n, r, s, t, ... denote their respective size. The space of all TU-games with the set of players N is denoted by Γ_N.

We assume that players are able to establish meaningful relationships with a subset of players, and we call the set of all private relationships a *cooperation structure*. A useful way to represent cooperation structures is by a non-directed graph, whose vertices represent players and whose edges represent the relationships that players have between each other.

We define a graph, g, as a set of links between players. We write $i{:}j \in g$ when i and j are linked in g. Two players i and j are indirectly linked in g if there exists a sequence $i_1, \ldots, i_k$ in N, with $k \geq 3$, such that $i = i_1$, $i_{t-1}{:}i_t$ for all $t = 2, \ldots, k$, and $i_k = j$. The sequence $i_1, \ldots, i_k$ is called a path. Two players are said to be *connected* if they are (indirectly) linked. A coalition S is said to be connected if any pair of players $i, j \in S$ is connected by a path involving only players in S. We denote by GR_N the set of all graphs, by g_N the complete graph, i. e. the graph in which each player is linked with all other players,

$$g_N = \{i{:}j \mid i \in N\ ,\ j \in N\ ,\ i \neq j\}\ , \tag{1}$$

$$GR_N = \{g \mid g \subseteq g_N\}\ , \tag{2}$$

and $g_\varnothing$ the empty graph, i. e., there is no pair $i, j \in N$ such that $i{:}j \in g_\varnothing$.

For any graph g, $g \backslash \{i{:}j\}$ denotes the graph in which the link $i{:}j$ has been deleted, and $g \cup \{i{:}j\}$ is the graph to which the link $i{:}j$ has been added. We denote the graph g restricted to the set of vertices S by $g(S)$,

$$g(S) = \{i{:}j \mid i, j \in S,\ i{:}j \in g\}\ .$$

For any subset of players S, there is a unique partition of players that groups together players connected by $g(S)$. Such a partition is denoted by S/g. In other words, for any $i \in S$, the element of the partition S/g containing i also contains all other players that are connected to i by $g(S)$. A formal definition of this partition is defined as follows,

$$S/g = \{\{i \in S \mid i \text{ and } j \text{ are connected in } S \text{ by } g\} \mid j \in S\}\ .$$

3 The Game

Qin's model works as follows. Given a TU-game (N, v), we allow players to choose who they want to cooperate, or communicate with. Hence, each player may have private relationships with a subset of the grand coalition only. More formally, we define a non-cooperative game

$$G_h^{(N,v)} = \langle N, (\Sigma_i)_{i\in N}, (h_i)_{i\in N} \rangle ,$$

where the TU-game (N, v) is called the *underlying game* of $G_h^{(N,v)}$, and $G_h^{(N,v)}$ is called a *link formation game*. As in (N, v), the set of players is N. The (finite) strategy set of player i is Σ_i, and $h = (h_i)_{i\in N}$ is the payoff function, $h : \times_{i\in N} \Sigma_i \to \mathbb{R}^N$.

3.1 Strategies

A strategy σ_i for player i is defined as a n-dimensional vector whose coordinates are either 0 or 1. When $\sigma_i(j) = 0$ player i does not wish to form a link with player j, while he would be glad to create such a link when $\sigma_i(j) = 1$. The ith coordinate of σ_i is assumed to be equal to zero, $\sigma_i(i) \equiv 0$. In other words, no player can make a link with herself. A link between players i and j is formed only if both i and j wish to create it: $\sigma_i(j) = \sigma_j(i) = 1$. We denote the n-tuple of strategies by $\sigma = (\sigma_1, \ldots, \sigma_n)$. Each $\sigma \in (\Sigma_i)_{i\in N}$ gives a unique cooperation structure $g(\sigma)$,

$$g(\sigma) = \{i{:}j \mid \sigma_i(j) = \sigma_j(i) = 1\} .$$

For each cooperation structure g, players are rewarded using the Myerson value of the game (N, v) on g, denoted by the n-dimensional vector $\mu(v, g)$.

We design by σ^S the strategy profile such that *(i)* for all $i, j \in S$, $\sigma_i(j) = 1$ and $\sigma_j(i) = 1$; and *(ii)* for all $i \in S$ and $j \notin S$, $\sigma_i(j) = 0$ or $\sigma_j(i) = 0$. That is, $g(\sigma^S)$ is a graph such that all players in S are connected to each other and players not in S have no links.

3.2 The Payoffs

Payoffs in the link formation game defined by Qin are given by the Myerson value [19], which is an allocation rule for TU-games with graphs. In order to compute the Myerson value for some game (N, v) and some cooperation structure g, we first need to define a characteristic function embedded on graphs,

$$\forall\, S \subseteq N, \quad (v/g)(S) = \sum_{T \in S/g} v(T) . \tag{3}$$

The Myerson value is the unique allocation μ rule that satisfies the *Component Efficiency* axiom, $\forall S \in N/g$, $g \in GR_N$, $\sum_{i\in S} \mu_i(v, g) = v(S)$, and the *Fairness* axiom, $\forall\ g \in GR_N,\ i{:}j \in g,\ \mu_i(v,g) - \mu_i(v,g\backslash i{:}j) = \mu_j(v,g) - \mu_j(v,g\backslash i{:}j)$.

Myerson proved that his value is closely related to the Shapley value. Indeed, we have

$$\forall\,(N,v) \in \Gamma_N\,,\ \forall\, g \in GR_N\,, \qquad \mu(v,g) = \varphi(v/g)\,, \tag{4}$$

where φ is the Shapley value. It is straightforward to see that for the full cooperation structure, $v/g_N = v$, and hence $\mu(v,g_N) = \varphi(v)$. Moreover, if some player i is not connected to anyone in the cooperation structure g, then $\mu_i(v,g) = v(\{i\})$. Note that it may happen that for some game (N,v) and some g, we have $\mu_i(v,g) < v(\{i\})$.

Thus, the payoff of a player in the game $G_\mu^{(N,v)}$ under the strategy profile σ is her Myerson value of the corresponding cooperation structure,

$$h_i(\sigma) \equiv \mu_i(v,g(\sigma))\,. \tag{5}$$

4 The Stability of Cooperation Structures

Consider now a cooperation structure g such that players i and j are not linked, i. e., $i{:}j \notin g$. Then, player i desires to link up with player j if and only if

$$\mu_i(v, g \cup \{i{:}j\}) \geq \mu_i(v,g)\,. \tag{6}$$

Given that the Myerson value satisfies the Fairness axiom, it is readily verified that if (6) holds true for player i, then it also holds true for player j.[7]

A cooperation structure g is *stable* in Qin's sense if there is a strategy profiles σ such that $g(\sigma) = g$ and σ is a Nash equilibria of the game $G_\mu^{(N,v)}$. This result was also obtained in a similar framework by Dutta, van den Nouweland, and Tijs [9].[8] We now show that this result can be generalized to a larger class of games.

Proposition 1. *Let (N,v) be a TU-game and μ the Myerson value. Then the full cooperation structure is stable if and only if $\mu_i(v,g_N) \geq v(\{i\})$ for all $i \in N$, and for all $S \subseteq N$, for all $i \in S$,*

$$\sum_{T \subseteq S, T \ni i} \rho_T(v(T) - v(T\backslash' i) - v(\{i\})) \geq 0\,, \tag{7}$$

where $\rho_S = (n-1)!(n-s)!/n!$.

[7] Reciprocity may not hold if player i wants to link up with several players at a time.

[8] Dutta et al. employed several equilibrium concepts, (undominated Nash equilibrium, coalition-proof Nash equilibrium and Strong Nash equilibrium) and did not use a specific solution concept to analyze to equilibria of the game.

Proof. See the Appendix. □

Proposition 1 says that if the Shapley value of each player in the game (N, v) is individually rational then g_N is stable if and only if the game (N, v) is in the average "superadditive". It is worth to point out that if the game is zero normalized, then condition (7) is nothing but requiring that in the average marginal contributions are non-negative. Hence, the two conditions – individual rationality and (7) – fully characterize the class of TU-games that ensure the stability of the full cooperation structure.

Qin showed that if the underlying TU-game is superadditive, then for any partition P of N, there is a Nash equilibrium σ game such that $N/g(\sigma) = P$ – see [20, remark 2]. We now show that Qin's result can be extended.[9]

Proposition 2. *Let (N, v) be a superadditive TU-game. For any $g \subseteq g_N$, there exist a strategy profile σ such that $g = (\sigma) = g$ and σ is a Nash equilibrium of $G_\mu^{(N,v)}$.*

Proof. See the Appendix. □

5 The Potential

In his paper, Qin used Monderer and Shapley's potential to study the stability of cooperation structures. He showed that a link formation game with an underlying TU-game admits an MS-potential if and only if the allocation rule that gives players' payoffs is the Myerson value. In this section, we propose a formula for the MS-potential, and show its equivalence with the potential of Hart and Mas-Colell.

Definition 1. *A potential in the sense of Monderer and Shapley, or MS-potential, for $G_\mu^{(N,v)}$ is a function $P\colon \Sigma \to \mathbb{R}$ such that for any $i \in N$, $\sigma \in \Sigma$, and $\hat{\sigma}_i \in \Sigma_i$,*

$$h_i(\hat{\sigma}_i, \sigma_{-i}) - h_i(\sigma) = P(\hat{\sigma}_i, \sigma_{-i}) - P(\sigma)\,. \tag{8}$$

A non-cooperative game $\gamma = (N, \Sigma, h)$, with player set N, Strategy space Σ and payoff functions $h = (h_i)_{i \in N}$ is a *potential game* if it admits an MS-potential.

Since the MS-potential is defined up to a constant, we can assign a value for the empty graph, and then recursively deduce its value for all graphs.

Proposition 3. *Let σ be any strategy profile, and let $g = g(\sigma)$.*

$$P(\sigma) = \sum_{S \subseteq N} \frac{(n-s)!(s-1)!}{n!}(v/g)(S)\,. \tag{9}$$

[9] Proposition 2 is actually a corollary of Proposition 1 in [9].

Clearly, the formula of the MS-potential is closely related to the Shapley value of the game. Nonetheless, it is easy to see that the potential in (9) is also the potential as defined by Hart and Mas-Colell. Their potential, defined upon cooperative games is defined as follows.

Definition 2. *A function $P^{HM}\colon \Gamma_N \to \mathbb{R}$ is a potential in the sense of Hart and Mas-Colell, or HM-potential, if it satisfies the following condition*

$$\sum_{i\in N} D_i P(N,v) = v(N)\,, \tag{10}$$

where $D_i P(N,v) = P(N,v) - P(N\backslash\{i\},v)$.

Hart and Mas-Colell [13] proved that the potential P^{HM} can be characterized (up to a constant) by the following equality:

$$P^{HM}(N,v) = \sum_{S\subseteq N} \frac{(n-s)!(s-1)!}{n!} v(S)\,. \tag{11}$$

We can then show the following result.

Proposition 4. *Let σ be any strategy profile, and $g = g(\sigma)$.*

$$P^{HM}(N,v/g) = P_v(g)\,.$$

Proof. Observe that the potential P^{HM} is defined on games with full cooperation structures. Hence, to prove the proposition, it suffices to obtain the potential P^{HM} for any graph g. For any graph g, $(N,v/g) = (N,(v/g)/g_N)$. Thus,

$$P^{HM}(N,v/g) = \sum_{S\subseteq N} \frac{(n-s)!(s-1)!}{n!} (v/g)(S)\,. \tag{12}$$

Clearly, (9) and (12) do coincide, which proves the result. □

A similar result has been already observed by Monderer and Shapley and Ui. Their proofs work in opposite directions. Monderer and Shapley proved that for any cooperative game we can construct a non-cooperative game (called *participation game*) such that it admits an MS-potential if and only if the allocation rule used in the original cooperative game (which serves to compute players' payoffs in the participation game) is the Shapley value. Conversely, Ui started from a non-cooperative game, and showed that we can construct an associated cooperative game with payoffs being given by the Shapley value if and only if the original game admits an MS-potential. Here, we took a different route, simply showing that the formula for the HM-potential and the MS-potential do coincide.

6 An Example

As the following example shows, superadditivity of the game (N,v) is not necessary for the stability of the full cooperation structure.

Consider the following TU game (N,v) with $N = \{a,b,c,d\}$, and v defined by

$$v(S) = \begin{cases} 0, & \text{if } |S| = 1\,, \\ 180, & \text{if } |S| = 2\,, \\ 168, & \text{if } |S| = 3\,, \\ 144, & \text{if } |S| = 4\,. \end{cases}$$

The Myerson value of each player for the full cooperation structure is $\mu_i(v, g_N) = 36$, for $i = a,b,c,d$, which is obviously individually rational. Moreover, condition (7) is satisfied. Hence, Proposition 1 implies that the full cooperation structure is stable. Yet, it is obvious that the full cooperation structure is Pareto dominated, i. e. there is another cooperation structure (for instance $g' = \{a{:}b,\ c{:}d\}$) that yields a value of 90 to each player, which is strictly higher than the value when the full cooperation structure is formed.[10]

It turns out that this example has more to say. Consider the potential of the game. Since the game is symmetric, all graphs (and thus all the values of the potential) can be characterized by the number of links, and for the case when there are two, three or four links on the number of link per player. In table 1 when a graph has for distribution of links per player the value 3, 2, 2, 1 it means that there is one player with three links, two players with two links and one player with one link. This yields the following values for the potential.

Table 1. Values of the potential

♯ of links	link distribution	P	♯ of links	link distribution	P
0	0,0,0,0	0	3	2,2,2,0	146
1	1,1,0,0	90	3	3,1,1,1	123
2	1,1,1,1	180	4	2,2,2,2	152
2	2,1,1,0	116	4	3,2,2,1	153
3	2,2,1,1	139	5	3,3,2,2	167
			6	3,3,3,3	182

This example is quite interesting since it shows that the complete graph does maximize the potential, although the game is not superadditive (it is

[10] A similar result was obtained by Jackson and Wolinsky [16], who showed that some games $G_\mu^{(N,v)}$ may have a stable cooperation structures that are not efficient, i. e. structures that do not maximize the worth of coalition(s).

even not monotonic). Yet we can see that the potential is maximum when the cooperation structure is complete.[11]

Further scrutiny shows that the complete graph can be obtained by fictitious play. For, consider the sequence of graphs $\bar{g} = \{g_1, g_2, \ldots, g_7\}$, where $g_1 = g_\varnothing$ and $g_7 = g_N$, with the following distribution of links.

Table 2. Distribution of links

Graph	♯ of links	link distribution	Potential
g_1	0	0,0,0,0	0
g_2	1	1,1,0,0	90
g_3	2	2,1,1,0	116
g_4	3	2,2,1,1	139
g_5	4	2,2,2,2	152
g_6	5	3,3,2,2	167
g_7	6	3,3,3,3	182

It is easy to see that for any $h = 1, \ldots 6$, the graph g_{h+1} is obtained just by adding an additional link to g_h. Indeed, it is easy to construct an initial strategy profile such that each step is obtained from the previous one by just *one* deviation.[12] Clearly, for all $h = 1, \ldots 6$ $P(g_h) < P(g_{h+1})$. Hence, according to Monderer and Shapley's terminology, the sequence $\bar{g}$ in an *finite improvement path*, which reflects a myopic learning process.

7 Conclusion

We first showed that superadditivity (like monotonicity) are not necessary to obtained the stability of the full cooperation structure. In fact, the class of TU-games that admit full cooperation structure turns out to be very large. One relevent point is that small coalitions seem to have a particular role. It is often argued that the Nash equilibrium concept is not appropriate for this game since it admits too many stable cooperation structures. It it worth to point out that our result still holds if one uses Jackon and Wolinsky's *pairwise stability* concept.[13] Our result is even more striking when we look

[11] See [16] for a related result showing that stable graphs are not necessarily those maximizing the total value.

[12] Recall that to make a link between i and j we must have $\sigma_i(j) = \sigma_j(i) = 1$. To see that the sequence can be obtained by a sequence of 6 best-response deviations, consider for instance that players a and b make the first link (the unique link in g_2). Then, we can start with the strategy profile σ such that $\sigma_a(b) = 0$ and $\sigma_b(a) = 1$. The deviating player is a, and doing so her payoff in g_2 is strictly higher than her payoff in g_1.

[13] When we look only at the full cooperation structure the pairwise stability concept and the Nash stability used in this paper do coincide.

at the potential of the game. As our example shows, neither monotonicity nor superadditivity are necessary to obtain the full cooperation as the stable cooperation structure when considering fictitious play. This amounts to say that simple best-response dynamics can yield players to build inefficient cooperation structure.

An other result we obtained is the equivalence of the potential of Monderer and Shapley [18], and that of Hart and Mas-Colell. Since these two concepts are designed for two different frameworks (non-cooperative games for the former and cooperative games for the latter), it has been usually thought that the only common aspect they share was their name. The model analyzed in this paper allows us to study the two potentials proposed in the literature. To see this, it is important to have in mind how the link formation game is built. Recall that we start first with a *cooperative* TU-game. Hence, this game has an HM-potential. The second step consists of building a *non-cooperative* game of link formation, which allows us to study the MS-potential. It is worth pointing out that not all non-cooperative games admit a potential. However, Qin proved that this game *does* admit an MS-potential (if and only if the allocation rule is the Myerson value). In this paper, we showed that these two potential are identical (for this game). Though being quite surprising at first sight, this result can be expected. Indeed, both concepts are defined upon marginal differences in terms of payoffs. Hart and Mas-Colell showed that their potential is closely related to the Shapley value of the TU-game, and the Myerson value, which defines the payoffs in the link formation game, is known as being the Shapley value of some specific TU-game.

Appendix

In order to prove proposition 1, we introduce more definitions. For any cooperation structure g and players i and j such that $i{:}j \in g$, define the set

$$\mathbf{S}(g, i{:}j) = \{S \subseteq N \mid S/g = S \text{ and } S/(g \backslash \{i{:}j\}) \neq S\} .$$

In other words, $\mathbf{S}(g, i{:}j)$ is the set of coalitions S such that S is connected in $g(S)$ but is non-connected in $g(S)\backslash\{i{:}j\}$. When no confusion is possible, we write $\mathbf{S}(i{:}j)$ instead of $\mathbf{S}(g, i{:}j)$. In other words, $\mathbf{S}(i{:}j)$ represents the set of coalitions S that are connected thanks to $i{:}j$ when the graph is restricted to S.[14] Thus, for any $S \in \mathbf{S}(i{:}j)$, if $i{:}j$ is deleted, $S/(g\backslash\{i{:}j\} = \{S_i, S_j\})$, with S_i and S_j being the sets in $S/(g\backslash\{i{:}j\})$ that contain i and j respectively.

Proof of Proposition 1.

Observe that there is a unique strategy profile σ^* such that $g(\sigma^*) = g_N$. Hence, σ^* is a Nash equilibrium if and only if no player breaks one or more

[14] In graph terminology, we say that $\mathbf{S}(g, i{:}j)$ is the set of all subgraphs $\hat{g}$ of g such that the link $i{:}j$ is *critical* in $\hat{g}$ – see [6].

links. Hence, it suffices to compare, for any player i, her payoffs under the strategy profile σ and under the strategy profile $(\sigma_i', \sigma_{-i}^*)$, where $\sigma_i'(j) = 0$ for at least one $j \in N\backslash\{i\}$.

Clearly, if the value of some player is not individually rational, then that player has an incentive to break up all her links (or any part thereof), and then the full cooperation structure is not stable. In other words, individual rationality is a necessary condition for the full cooperation structure to be stable. We now show that if the value of each player is individual rational, then stability is equivalent to condition (7).

We claim that for any graph g, if a player contemplates breaking a link, yielding g', the change of her payoff (i. e., her Myerson value) will only depend on the worth of the coalitions that are connected in g and not connected in g'. To see this, consider the value of player i for a cooperation structure g and $g\backslash\{i{:}j\}$:

$$\mu_i(v, g) = \sum_{S \ni i} \rho_S \left((v/g)(S) - (v/g)(S\backslash i)\right) ,$$

$$\mu_i(v, g\backslash\{i{:}j\}) = \sum_{S \ni i} \rho_S \left((v/g(\backslash\{i{:}j\}))(S) - (v/g(\backslash\{i{:}j\}))(S\backslash i)\right) .$$

Notice that

(i) $g/(S\backslash\{i\}) = (g\backslash\{i{:}j\})/(S\backslash\{i\})$,

(ii) $(g\backslash\{i{:}j\})/S \neq g/S$ when $i{:}j$ connects S in g, i. e., when $S \in \mathbf{S}(i{:}j)$,

(iii) $(g\backslash\{i{:}j\})/S = g/S$ when $i{:}j$ does not connects S in g.

Using equation (3) we deduce

(a) *(i)* $\Rightarrow \forall\, S \ni i,\ (v/g)(S\backslash\{i\}) = (v/(g\backslash\{i{:}j\}))(S\backslash\{i\})$,

(b) *(ii)* $\Rightarrow$ when $S \in \mathbf{S}(i{:}j)$, $(v/g\backslash\{i{:}j\})(S)$ and $(v/g)(S)$ may differ

(c) Let S such that $S/g = S$. Then *(iii)* $\Rightarrow (v/g\backslash\{i{:}j\})(S) = (v/g)(S)$ when $S \notin \mathbf{S}(i{:}j)$.

If $S/g \neq S$, then there is one element of S/g that contains i. Let $\hat{S}$ be this element. Clearly, for all $T \in S/g$ such that $T \neq \hat{S}$, we have $(v/g\backslash\{i{:}j\})(T) = (v/g)(T)$. Hence, when doing $(v/g)(S) - (v/g(\backslash\{i{:}j\}))(S)$ we can restrict to $\hat{S}$, and either we are in case *(b)* or in the first part of case *(c)*.

Combining these observations we obtain

$$\mu_i(g) - \mu_i(g\backslash\{i{:}j\}) \geq 0 \iff \sum_{S \in \mathbf{S}(i:j)} \rho_S(v(S) - v(S_i) - v(S_j)) \geq 0\,, \quad (13)$$

which proves the claim.

Observe that if $g = g_N$, then $\mathbf{S}(i{:}j) = \{\{i, j\}\}$. In other words, in the full cooperation structure, if a player i breaks a link with player j, the only coalition that becomes non-connected is $\{i, j\}$.

More generally, suppose that some player i severs her links in g_N with players $j \in T$, where T is some subset of $N\backslash\{i\}$. Then we have $\{\mathbf{S}(i{:}j)\}_{j \in N} = \{R \cup \{i\}\colon R \subseteq T\backslash\{i\}\}$. This implies that i severs her links with players $j \in S \subseteq N\backslash\{i\}$ if, and only if

$$\sum_{T\subseteq S, T\ni i} \rho_T(v(T) - v(T\backslash' i) - v(\{i\})) < 0 \,. \tag{14}$$

As (14) holds for any $S \subseteq N\backslash\{i\}$, for any $i \in N$, the result follows. □

Proof of Proposition 2. Consider any superadditive game (N, v) and any cooperation structure g. Hence, (13) always holds and we can deduce that no player wishes to break a link. Consider the following strategy profile. For each player $i \in N$, let $\sigma_i(j) = 0$ if $i{:}j \notin g$ and $\sigma_i^*(j) = 1$ if $i{:}j \in g$. Because players do not sever links, if a player i deviates with the strategy $\hat{\sigma}$, it should be the case that for some j we have $\hat{\sigma}_i(j) = 1$ and $\sigma_i^*(j) = 0$. But deviations are individuals, which implies that $\sigma_j^*(i) = 0$. Hence, $g(\hat{\sigma}_i, \sigma_{-i}^*) = g$, and player's i payoff remains unchanged. Thus, i has no incentives to deviates. □

Lemma 1. *Let g be a graph such that there exists a set of players* **S** *and a player $i \notin \hat{S}$ such that for all $j \in \hat{S}$, i and j are not connected. Then*

$$\mu_i(v, g, N) = \mu_i(v, g, N\backslash\mathbf{S}) \,. \tag{15}$$

Proof. First, observe that if i is not connected to any player in $\hat{S}$, then it is also the case for any player that is connected to i.

We prove the lemma when $\hat{S}$ contains only one player, say j. When $\hat{S}$ contains more than one player, it suffices to repeat the argument presented below.

If we withdraw player j from the set of players, we have a new set of player N' of size $n' = n - 1$. With this new player set, we have (the right-hand side of equation (15))

$$\mu_i(v, g, N') = \sum_{R\ni i} \frac{(n'-r)!(r-1)!}{n'!} \left[(v/g)(R) - (v/g)(R\backslash\{i\})\right] \,. \tag{16}$$

We now compute the left-hand side of equation (15). Observe that for any $T \ni j$, $(v/g)(T) = (v/g)(T\backslash\{j\}) + (v)(\{j\})$.[15] Thus, we have for any $i \in N\backslash\hat{S}$

$$\begin{aligned}\mu_i(v, g, N) &= \sum_{\substack{R\ni i\\ j\notin R}} \rho(r) \left[(v/g)(R) - (v/g)(R\backslash\{i\})\right] \\ &\qquad - \sum_{\substack{R\ni i\\ j\in R}} \rho(r) \left[(v/g)(R\backslash\{j\}) - (v/g)(R\backslash\{i,j\})\right] \\ &= \sum_{\substack{R\ni i\\ j\notin R}} (\rho(r) + \rho(r+1)) \left[(v/g)(R) - (v/g)(R\backslash\{i\})\right] \,. \end{aligned} \tag{17}$$

Straightforward computation shows that

[15] If $\hat{S}$ contains more than one player, then this equality becomes: for any $T \supset \hat{S}$, $(v/g)(T) = (v/g)(T\backslash\hat{S}) + (v)(\hat{S})$.

$$\rho(r) + \rho(r+1) = \frac{(n-1-r)!(r-1)!}{(n-1)!} = \frac{(n'-r)!(r-1)!}{n'!} \,. \tag{18}$$

Combining combining (17) and (18) we obtain (16). □

Lemma 2. *Assume that $P(\varnothing) \equiv 0$. Let S be a nonempty subset of N and let $g = g(\sigma^S)$. Then*

$$P(\sigma^S) = \sum_{i \in S} -\frac{s-1}{s} v(\{i\}) + \sum_{T \subseteq S \,:\, t \geq 2} \frac{(s-t)!(t-1)!}{s!} v(T) \,. \tag{19}$$

Proof. We prove the lemma by induction on the size of S. Assume first that $S = \{i, j\}$. Consider the empty graph and the graph $g = g(\sigma^S)$. Since players i and j are the only players whose payoff changes between g and the empty graph, we have $P(\sigma^S) - P(\sigma^\varnothing) = \mu_i(v, g) - \mu_i(v, \varnothing)$. By lemma 1, we know that $\mu_h(v, g, N) = \mu_h(v, g, \{i, j\})$ for $h = i, j$. Thus, with the convention that $P(\varnothing) \equiv 0$, we obtain

$$\begin{aligned} P(\sigma^S) &= \mu_i(v, g) - \mu_i(v, \varnothing) = \varphi_i(v/g) - \varphi_i(v/\varnothing) \\ &= \frac{1}{2} v(\{i\}) + \frac{1}{2} [v(\{i, j\}) - v(\{j\})] - v(\{i\}) \\ &= -\frac{1}{2} [v(\{i\}) + v(\{j\})] + \frac{1}{2} v(\{i, j\}) \,, \end{aligned}$$

which completes the proof for the case $s = 2$.

Let S be any set of players of size three or more, and let $g = g(\sigma^S)$. By the induction hypothesis, the potential for the graph g is given by (19). Let $i \notin S$, and let $g' = g(\sigma^{S \cup \{i\}})$. We write S' for $S \cup \{i\}$. We have

$$\begin{aligned} P(\sigma^{S'}) - P(\sigma^S) &= \mu_i(v, g') - \mu_i(v, g) = \varphi_i(v/g') - \varphi_i(v/g) \\ \iff P(\sigma^{S'}) &= P(\sigma^S) + \varphi_i(v/g') - \varphi_i(v/g) \,. \end{aligned} \tag{20}$$

Recall that

$$\begin{aligned} \varphi_i(v, g') &= \sum_{T \ni i} \frac{(n-t)!(t-1)!}{n!} [(v/g)(T) - (v/g)(T \backslash \{i\})] \,, \\ &= \sum_{T \subseteq S' \,:\, T \ni i} \frac{(s'-t)!(t-1)!}{s'!} [v(T) - v(T \backslash \{i\})] \,, \end{aligned} \tag{A}$$

$$\varphi_i(v, g) = v(\{i\}) \,, \tag{B}$$

$$P(\sigma^S) = \sum_{j \in S} -\frac{s-1}{s} v(\{i\}) + \sum_{T \subseteq S \,:\, t \geq 2} \frac{(s-t)!(t-1)!}{s!} v(T) \,. \tag{C}$$

where (A) is obtained using lemma 1. Thus,

$$P(\sigma^{S'}) = (A) - (B) + (C) \,. \tag{21}$$

First, observe that $v(\{i\})$ is only present in (A) (for $T = \{i\}$) and (B). Thus, in $P(\sigma^{S'})$, the coefficient of $v(\{i\})$ is given by

$$\frac{(s'-t)!(t-1)!}{s'!} - 1 = \frac{(s'-1)!}{s'!} - 1 = -\frac{s'-1}{s},$$

where the first equality comes from the fact that $t = 1$.

We now compute the coefficient for $v(\{j\})$, $\forall j \in S$. For each $v(\{j\})$, it is given by

$$-\frac{(s'-t)!(t-1)!}{s'!} - \frac{s-1}{s} = -\frac{(s'-2)!}{s'!} - \frac{s'-2}{s'-1} = -\frac{s'-1}{s'}.$$

In order to compute the coefficient for coalitions $R \subseteq S$, we must take into account that when a coalition R appears in (A), $t = r + 1$, whereas in (C) we have $t = r$. Hence, the coefficient is given by

$$\begin{aligned}-\frac{(s'-t-1)!t!}{s'!} + \frac{(s-t)!(t-1)!}{s!} &= -\frac{(s'-t-1)!t!}{s'!} + \frac{(s'-t-1)!(t-1)!}{(s'-1)!}\\ &= -\frac{(s'-t)!(t-1)!}{s'!}.\end{aligned}$$

It remains to compute the coefficient for the coalitions T that include i *and* some players in S. It is easy to see that such coalitions only appear in (A). Thus we have,

$$P(\sigma^{S'}) = \sum_{j\in S'} -\frac{s'-1}{s'} v(\{j\}) + \sum_{T\subseteq S'\,:\,t\geq 2} \frac{(s'-t)!(t-1)!}{s'!} v(T), \tag{22}$$

which completes the proof. □

Lemma 3. *Assume that $P(\emptyset) \equiv 0$. Let S be a nonempty subset of N and let $g = g(\sigma^S)$. Then*

$$P(\sigma^{S}) = \sum_{i\in N} -\frac{n-1}{n} v(\{i\}) + \sum_{T\subseteq N\,:\,t\geq 2} \frac{(n-t)!(t-1)!}{n!} (v/g)(T). \tag{23}$$

The difference between this lemma and lemma 2 is in the second summation. In lemma 2 the summation is done over $T \subseteq N$ and in lemma 3 the summation is done over $T \subseteq N$.

Proof. We prove the lemma by induction on the size of S. Assume first that $S = \{i, j\}$. From the proof of lemma 2, we know that $P(\sigma^S) = -\frac{1}{2}[v(\{i\}) + v(\{j\})] + \frac{1}{2}v(\{i,j\})$. Using lemma 1 it is straightforward to see that in this case the following equality holds true

$$\begin{aligned}&\sum_{h\in S} -\frac{s-1}{s} v(\{h\}) + \sum_{T\subseteq S\,:\,t\geq 2} \frac{(s-t)!(t-1)!}{s!} v(T)\\ &= \sum_{i\in N} -\frac{n-1}{n} v(\{i\}) + \sum_{S\subseteq N\,:\,s\geq 2} \frac{(n-s)!(s-1)!}{n!} (v/g)(S).\end{aligned}$$

Let S be any set of players of size three or more and lower than n (if $S = N$ then lemmata 2 and 3 are identical). Denote $P^S(\sigma)$ and $P^N(\sigma)$ the potential when the summation is done over S and N respectively (equations (19) and (23)).

By the induction hypothesis, $P^N(\sigma^S) = P^S(\sigma^S)$. Consider some $i \notin S$, and let $S' = S \cup \{i\}$, $g' = g(\sigma^{S'})$. We claim that $P^N(\sigma^{S'}) = P^{S'}(\sigma^{S'})$. We have

$$
\begin{aligned}
P^N(\sigma^{S'}) &\equiv P^N(\sigma^S) + \varphi_i(v/g', N) - \varphi_i(v/g, N) \\
&= P^S(\sigma^S) + \varphi_i(v/g', S') - \varphi_i(v/g,') = P^{S'}(\sigma^{S'}) \, ,
\end{aligned}
$$

The first equality is simply the definition of the potential. The second and third equalities are respectively obtained using lemma 1,[16] and equation (20). □

Let σ be any strategy profile, and let $g = g(\sigma)$. Denote by $P_v(\sigma)$ and $P_{v/g}(\sigma^N)$ the potential for the games $G_\mu^{(N,v)}$ and $G_\mu^{(N,v/g)}$ respectively.

Lemma 4. $P_v(\sigma) = P_{v/g}(\sigma^N)$.

Proof. Since the potential is deduced (up to a constant) from the payoffs of the game, it suffices to show that $\mu(v,g) = \mu(v/g, g_N)$, for all $g \subseteq g_N$. Recall that $\mu(v,g) = \varphi(v/g)$, and that $\mu(v, g_N) = \varphi(v)$. Hence, $\mu(v/g, g_N) = \varphi(v/g)$, which yields the desired equality. □

Proof of Proposition 3. From lemmata 4 and 3, it is easy to deduce that

$$
P(\sigma) = \sum_{i \in N} -\frac{n-1}{n} v(\{i\}) + \sum_{S \subseteq N \,:\, s \geq 2} \frac{(n-s)!(s-1)!}{n!} (v/g)(S) \, .
$$

Since the potential is defined up to a constant, it suffices to add $\sum_{i \in N} \frac{n-1}{n} v(\{i\})$ to the potential of any graph and the result follows. □

References

1. Aumann, R. J., Myerson, R. B. (1988): Endogenous Formation of Links between Players and of Coalitions: An Application of the Shapley Value. In: Roth, A. (Ed.): The Shapley value. Cambridge Univ. Press, Cambridge, 175–191
2. Bala, V., Goyal, S. (2000): A Noncooperative Model of Network Formation. Econometrica **68**, 1181–1229
3. Banerjee, S., Konishi, H., Sönmez, T. (2001): Core in a Simple Coalition Formation Game. Social Choice and Welfare **18**, 135–153
4. Blume, L. E. (1993): The Statistical Mechanics of Strategic Interaction. Games and Economic Behavior **5**, 387–424

[16] No player in S' is linked to a player in $N \backslash S'$.

5. Bogomolnaia, A., Jackson, M. (2001): The Stability of Hedonic Coalition Structures. Games and Economic Behavior, forthcoming, October 2001
6. Bollobás, B. (1998): Modern Graph Theory. Springer–Verlag, New York
7. Dutta, B., Jackson, M. (2001): On the Formation of Networks and Groups. In: Dutta, B., Jackson, M. O. (Eds.): The Models of the Strategic Formation of Networks and Groups. Springer–Verlag, Heidelberg
8. Dutta, B., Mutuswami, S. (1997) Stable Networks. Journal of Economic Theory **76**, 322–344
9. Dutta, B., van den Nouweland, A., Tijs, S. (1998): Link Formation in Cooperative Situations. International Journal of Game Theory bf 27, 245–256
10. Guesnerie, R., Oddou, C. (1979): On Economic Games that are not Necessarily Superadditive. Economics Letters **3**, 301–306
11. Guesnerie, R., Oddou, C. (1981): Second Best Taxation as a Game. Journal of Economic Theory **25**, 67–91
12. Guesnerie, R., Oddou, C. (1988): Increasing Returns to Size and their Limits. Scandinavian Journal of Economics **90**, 259–273
13. Hart, S., Mas-Colell, A. (1989): Potential, Value, and Consistency. Econometrica **57**, 589–614
14. Jackson, M. O., van den Nouweland, A. (2001): Strongly Stable Networks. CalTech (mimeo)
15. Jackson, M. O., Watts, A. (1999): The Evolution of Social and Economic Network. Journal of Economic Theory (to appear)
16. Jackson, M. O., Wolinsky, A. (1996): A Strategic Model of Social and Economic Networks. Journal of Economic Theory **71**, 44–74
17. Monderer, D., Shapley, L. S. (1996a): Fictitious Play Property for Games with Identical Interests. Journal of Economic Theory **68**, 258–265
18. Monderer, D., Shapley, L. S. (1996b): Potential Games. Games and Economic Behavior **14**, 124–143
19. Myerson, R. B. (1977): Graphs and Cooperation in Games. Mathematics of Operation Research **2**, 225–229
20. Qin, C. Z. (1996): Endogeneous Formation of Cooperation Structures. Journal of Economic Theory **69**, 218–226
21. Ui, T. (2000): A Shapley Value Representation of Potential Games. Games and Economic Behavior **31**, 121–135

Time-Consistent Imputation Distribution Procedure for Multistage Game*

Victor V. Zakharov[1] and Maria B. Dementieva[2]

[1] St. Petersburg State University,
Faculty of Applied Mathematics and Control Processes,
Bibliotechnaya pl. 2, St. Petersburg, 198504, Russia
E-mail: mcvictor@icape.nw.ru

[2] St. Petersburg State University,
Faculty of Applied Mathematics and Control Processes,
Bibliotechnaya pl. 2, St. Petersburg, 198504, Russia
E-mail: ppdanhom@peterlink.ru

Summary. In this paper the existence of time-consistent selectors of the subcore is considered. Necessary and sufficient conditions for time-consistency of an imputation from the subcore in multistage TU-cooperative games are determined. The imputation distribution procedure and corresponding algorithm, which lead to time-consistency of the solution of a multistage cooperative game, are presented.

Key words: multistage TU-cooperative games, core, subcore, time-consistency

1 Introduction

Under the framework of classical TU-cooperative game theory many optimality concepts are treated. One of the important problems in a dynamic cooperative game is the problem of time-consistency. When we try to use optimality concepts from the static games in dynamic games the problem of time-consistent realization of the solution along optimal trajectory crops up. In this paper we treat time-consistency of subcore selectors. When we consider continuous dynamic an imputation distribution procedure (IDP) may be proposed (see [1]). It provides time-consistency of some imputations from subcore and non negative payoffs to the players at each moment of time along optimal trajectory of a balanced game (see [2]). Sometimes it is not appropriate to use the same IDP when we analyze multistage cooperative game. Using the proof of Theorem 1, which formulates necessary and sufficient conditions of the time-consistency of an imputation from subcore in mu! ltistage

* This work is financially supported by scientific program "Universities of Russia".

TU-cooperative game, we propose an algorithm for IDP and introduce a notion of time-consistent grand subcore (TCGSC). This algorithm constructs a procedure of non negative payoffs to the players in a balanced multistage TU-cooperative games.

2 Preliminaries and Main Definitions

Let us denote a division of time period $t_0 < t_1 < \cdots < t_m$ by $\mathbf{T} = \{t_r\}_{r=\overline{0,m}}$, $m \in \mathbb{N}$. We call a pair (N, v) a multistage cooperative game. Here N is a finite set of players and $v\colon 2^N \times \mathbf{T} \mapsto I\!R$ is a characteristic function of the game, $v(\emptyset, t) = 0$ for all $t \in \mathbf{T}$, $v(S, t_m) = 0$ for all $S \subset N$. The sign $(N, v(t^*))$ means the cooperative game at moment $t^* \in \mathbf{T}$. We assume that $v(N, t)$ is the decreasing function with respect to t.

Definition 1. *A vector $\xi = (\xi_1, \ldots, \xi_n)$ is called an imputation in a cooperative game $(N, v(t))$ if its components satisfy the following conditions*

$$1)\xi_i \geq v(i,t)\,, \quad \forall i \in N\,, \tag{1}$$

$$2)\sum_{i \in N} \xi_i = v(N,t)\,. \tag{2}$$

Definition 2. *A subset $C(N, v(t))$ of imputation set is called a core of a cooperative game $(N, v(t))$ if its elements satisfy the inequalities*

$$\sum_{i \in S} \xi_i \geq v(S,t)\,, \quad \forall S \subset N\,. \tag{3}$$

We denote by $Con(\xi^0)$ a cone described by inequality $\xi \geq \xi^0$.

Let us define $X^0(t)$ as a set of solutions of the following linear programming problem

$$\min \sum_{i \in N} \xi_i\,, \tag{4}$$

$$\text{subject to} \sum_{i \in S} \xi_i \geq v(S,t)\,, \quad S \subset N,\ S \neq N\,, \tag{5}$$

and $Y^0(t)$ as the union of cones

$$Y^0(t) = \bigcup_{\xi^0 \in X^0(t)} Con(\xi^0)\,. \tag{6}$$

Now let us redefine some solution concepts and known propositions (see [3], [4]) for the multistage cooperative game (N, v) using $Y^0(t)$.

Definition 3. *We call a set*

$$SC(v(t),\xi^0(t)) = \{\xi \in Con(\xi^0(t)) \mid \sum_{i\in N} \xi_i = v(N,t)\}$$

a subcore of a game $(N, v(t))$ *with respect to* $\xi^0(t)$ *from* $X^0(t)$.

Definition 4. *We call a set*

$$GSC(N,v(t)) = \bigcup_{\xi^0(t)\in X^0(t)} SC(v(t),\xi^0(t)) = \{\xi \in Y^0(t) \mid \sum_{i\in N} \xi_i = v(N,t)\}$$

a grand subcore of a game $(N, v(t))$.

It is clear that subcore and grand subcore are subsets of the core in the balanced TU-games.

Definition 5. *The solution concept* $\phi(t)$ *of a multistage cooperative game* (N, v) *is called time-consistent if for every* $\xi \in \phi(t)$ *and for all* $t^* \in \mathbf{T}$ *there exists a vector* $\alpha(t^*) \geq 0$, *such that* $\xi - \alpha(t^*) \in \phi(t^*)$.

Let us now introduce a subset of grand subcore, which we denote by $TCGSC$.

Definition 6. *We call a set*

$$TCGSC(N,v(t_k)) = \{\xi \in \bigcap_{r=\overline{k,m}} Y^0(t_r) \mid \sum_{i\in N} \xi_i = v(N,t_k)\}$$

a time-consistent grand subcore of a game $(N, v(t_k))$, $t_k \in \mathbf{T}$.

Definition 7. *We call a multistage cooperative game* (N, v) *a balanced game if subgame* $(N, v(t_k))$ *is a balanced game for every* $t_k \in \mathbf{T}$.

We assume that $\alpha(t)$ is a payoff vector to the players by the moment t. In the paper we consider non negative non decreasing vector-functions $\alpha(t)$, which satisfy a natural acceptance condition from the following definition.

Definition 8. *A non decreasing vector* $\alpha(t)$, *is called acceptable in a game* $(N, v(t))$ *if its elements satisfy* $\alpha_i(t) \geq 0$ *for all* $i \in N$ *and* $\sum_{i\in N} \alpha_i(t) \leq (v(N,t_0) - v(N,t))$.

Theorem 1. *In a balanced multistage game* (N, v), $t \in \mathbf{T}$, *a vector* $\xi(t_0) \in GSC(N, v(t_0))$ *is time-consistent iff* $\xi(t_0) \in Y^0(t)$ *for all* $t \in \mathbf{T}$.

Proof. The condition of balancedness of (N, v) is equivalent to the nonemptyness of $GSC(N, v(t))$ for every $t \in \mathbf{T}$. That is the following arguments make sense.

Necessity. Suppose that a vector $\xi(t_0)$ is a time-consistent imputation from $GSC(N, v(t_0))$. We show that there exists such a vector $\xi^0(t) \in X^0(t)$ that $\xi(t_0) \geq \xi^0(t)$.

By the time-consistency of $\xi(t_0)$ we have $\xi(t_0) - \alpha(t) \in GSC(N, v(t))$. It implies that there exists $\xi^0(t) \in X^0(t)$ which satisfies

$$\xi(t_0) - \alpha(t) \geq \xi^0(t) . \tag{7}$$

Taking into account the non negativity of $\alpha_i(t)$, we obtain the required inequality $\xi(t_0) \geq \xi^0(t)$ for all $t \in \mathbf{T}$. That is $\xi(t_0) \in Y^0(t)$.

Sufficiency. Suppose that for every $t \in [t_0, T]$ there exists such a vector $\xi^0(t) \in X^0(t)$ that $\xi(t_0) \geq \xi^0(t)$. To prove the time-consistency we have to construct an acceptable payoff vector $\alpha(t)$. Let us find $\alpha(t)$ in the form $\alpha(t) = \beta(t)[v(N, t_0) - v(N, t)]$. Here $\beta(t)$ is a vector which satisfies the following conditions:

(i) $\beta(t) \geq \mathbf{0}$,

(ii) $\sum_{i \in N} \beta_i(t) = 1$.

It is obvious that the vector $\beta(t) = \frac{\xi(t_0) - \xi^0(t)}{v(N,t_0) - \sum_{i \in N} \xi_i^0(t)}$ satisfies (i) and (ii). Indeed, non negativity of $\beta_i(t)$ is evident for all $i \in N$ and

$$\sum_{i \in N} \beta_i(t) = \frac{\sum_{i \in N} \xi_i(t_0) - \sum_{i \in N} \xi_i^0(t)}{v(N, t_0) - \sum_{i \in N} \xi_i^0(t)} = \frac{v(N, t_0) - \sum_{i \in N} \xi_i^0(t)}{v(N, t_0) - \sum_{i \in N} \xi_i^0(t)} = 1 .$$

Thus, we have found the payoff vector $\alpha(t)$ providing (7)

$$\begin{aligned}\xi(t_0) - \alpha(t) = \xi(t_0) - \frac{\xi(t_0) - \xi^0(t)}{v(N, t_0) - \sum_{i \in N} \xi_i^0(t)}(v(V, t_0) - v(N, t)) \geq \\ \xi(t_0) - (\xi(t_0) - \xi^0(t)) \cdot 1 = \xi^0(t) .\end{aligned}$$

It completes the proof. □

Corollary 1. *An imputation $\xi(t_0) \in GSC(N, v(t_0))$ is time-consistent in a balanced multistage game (N, v) iff it belongs to $TCGSC(N, v(t_0))$ (or, as equivalent, $\xi(t_0) \in \bigcap_{k=\overline{0,m}} Y^0(t_k)$).*

Corollary 2. *In the conditions of Theorem 1 the nonemptiness of $TCGSC(N, v(t_k))$, $k = \overline{0,m}$, and the decreasing of $v(N, t)$ with respect to t imply that there exists a sequence $\{\xi(t_k)\}_{k=\overline{0,m}}$, such that $\xi(t_0) \geq \xi(t_1) \geq \cdots \geq \xi(t_{m-1}) \geq 0$. Here $\xi(t_k) \in TCGSC(N, v(t_k))$.*

Proof. We have to show that for every vector $\xi(t_k) \in TCGSC(N, v(t_k))$ there exists a vector $\zeta \in TCGSC(N, v(t_{k+1}))$, such that $\zeta \leq \xi(t_k)$.

Indeed, by $\xi(t_k) \in TCGSC(N, v(t_k))$ we have

$$\xi(t_k) \in \bigcap_{r=\overline{k,m}} Y^0(t_r) .$$

Thus

$$\xi(t_k) \in \bigcap_{r=\overline{k+1,m}} Y^0(t_r)\,,$$

and there exists a non empty subset

$$M(t_{k+1}) \subset \bigcap_{r=\overline{k+1,m}} Y^0(t_r),$$

such that a vector $\zeta^* \in M(t_{k+1})$ satisfies $\zeta^* \leq \xi(t_k)$. Since $TCGSC(N, v(t_k + 1)) \neq \emptyset$ and $v(N, t_k) > v(N, t_{k+1})$ we can choose a vector $\zeta \in M(t_{k+1})$ with $\sum_{i\in N} \zeta_i = v(N, t_{k+1})$. Our reasoning is true for all $k = \overline{0, m-1}$. It completes the proof. □

Now by Corollary 2 we have the following IDP providing the non negativity of payoffs to every player $i \in N$. In the condition of Corollary 2 we choose a vector $\xi(t_k) \in TCGSC(N, v(t_k))$ for every $k = \overline{1, m}$, such that $\xi(t_k) \leq \xi(t_{k-1})$. After that we set the payoff vectors at every moment $t_k \in \mathbf{T}$ as

$$\alpha([t_{k-1}, t_k]) = \xi(t_{k-1}) - \xi(t_k), \ k = \overline{1, m}\,.$$

However, sometimes $TCGSC(N, v(t_k)) = \emptyset$ at some moment $t_k \in \mathbf{T}$. In such a case we can delay a part of difference $v(N, t_{k-1}) - v(N, t_k)$ to provide time-consistency.

3 Imputation Distribution Procedure

Let us consider now an algorithm based on the maximization of total payoffs at every step. Assume that the conditions of Theorem 1 is fulfilled for a multistage cooperative game (N, v).

Algorithm

Let us introduce new signs. We denote a vector at a period $[t_k, t_m]$ by $\alpha(t_k)$, a payoff vector at a moment t_k by $\alpha([t_{k-1}, t_k])$ and a new guaranteed gain at a period $[t_k, t_m]$ by $\tilde{v}(N, t_k)$.

We define a set $Z^0(t_k)$ as solution set of the following minimization problem

$$\min \sum_{i\in N} \omega_i\,,$$
$$\text{subject to} \, \omega \in \bigcap_{r=\overline{k,m}} Y^0(t_r)\,. \tag{8}$$

Remark, that for all $t_k \in \mathbf{T}$ there exists a solution of the problem 8. It ensues by construction of $Y^0(t_k)$.

The initial step. We choose a vector $\xi(t_0) \in TCGSC(N, v(t_0))$. This imputation the players will receive by the end of the game (the moment t_m). And we set

$$\alpha(t_0) := \xi(t_0),\ \tilde{v}(N, t_0) := v(N, t_0)\,.$$

The step number k. At this step we find a non negative payoff vector to the players at moment t_k with respect to the vector $\alpha(t_{k-1})$ and $\tilde{v}(N, t_{k-1})$ from the previous step.

Consider a set $Z^0(t_k)$. If for a vector $\omega \in Z^0(t_k)$ we have $\sum_{i \in N} \omega_i \leq v(N, t_k)$, then we set

$$\alpha(t_k) := \xi(t_k),\ \text{ and } \tilde{v}(N, t_k) := v(N, t_k)\,.$$

Here $\xi(t_k) \in TCGSC(N, v(t_k))$ and $\xi(t_k) \leq \xi(t_{k-1})$.

Otherwise we set

$$\alpha(t_k) := \omega,\ \text{ and } \tilde{v}(N, t_k) := \sum_{i \in N} \omega_i\,.$$

Here ω is a vector from $Z^0(t_k)$. At last we set

$$\alpha([t_{k-1}, t_k]) := \alpha(t_{k-1}) - \alpha(t_k)\,.$$

The last step. By the definition of a multistage game we have

$$\alpha(t_m) := (0, 0),\ \tilde{v}(N, t_m) := 0\ ,\ \text{ and } \alpha([t_{m-1}, t_m]) = \alpha(t_{m-1})\,.$$

Let us note that existence of applicable $\alpha(t_{k+1})$ with respect to $\alpha(t_k)$ is proved by analogy with Corollary 2. Indeed, it is sufficient to displace $\xi(t_k)$ by $\alpha(t_k)$, ζ by $\alpha(t_{k+1})$ and $v(N, t)$ by $\tilde{v}(N, t)$ in the proof.

As the result of the algorithm we have the payoff sequence

$$\{\alpha([t_{k-1}, t_k])\}_{k=\overline{1,m}}\ ,$$

which guarantees the time-consistency of $\xi(t_0)$.

Let us apply the algorithm to an example.

Example 1. Consider the multistage cooperative game (N, v) with $N = \{1, 2\}$ and $m = 5$. The values of the characteristic function is presented in Table 1.

The initial step. We choose a vector $\xi(t_0) = (12, 13) \in TCGSC(N, v(t_0))$. And we set

$$\alpha(t_0) := \xi(t_0) = (12, 13),\ \tilde{v}(N, t_0) := v(N, t_0) = 25\,.$$

The 1st step. Here we find the allocation vector $\alpha([t_0, t_1])$. The solution vector of the problem (8) at the moment t_1 is $(10, 12)$. Since $10 + 12 < 23$ we set

Table 1.

t	$v(\{1\},t)$	$v(\{2\},t)$	$v(N,t)$	$X^0(t)$
t_0	9	12	25	(9,12)
t_1	10	10	23	(10,10)
t_2	8	2	15	(8,2)
t_3	5	8	14	(5,8)
t_4	3	3	8	(3,3)
t_5	0	0	0	(0, 0)

$$\alpha(t_1) := (11.5; 11.5), \text{ and } \tilde{v}(N, t_1) := v(N, t_1) = 23 .$$

Thus, we have

$$\alpha([t_0, t_1]) := \alpha(t_0) - \alpha(t_1) = (12, 13) - (11.5; 11.5) = (0.5; 1.5) .$$

The 2nd step. Here we find the payoff vector $\alpha([t_1, t_2])$. The solution vector of the problem (8) at the moment t_2 is $(8, 8)$. Since $8 + 8 > 15$ we set

$$\alpha(t_2) := (8, 8), \text{ and } \tilde{v}(N, t_2) := 16 .$$

Thus, we have

$$\alpha([t_1, t_2]) := \alpha(t_1) - \alpha(t_2) = (11.5; 11.5) - (8, 8) = (3.5; 3.5) .$$

The 3rd step. Here we find the payoff vector $\alpha([t_2, t_3])$. The solution vector of the problem (8) at the moment t_3 is $(5, 8)$. Since $5 + 8 < 14$ we set

$$\alpha(t_3) := (6, 8), \text{ and } \tilde{v}(N, t_3) := v(N, t_3) = 14 .$$

Thus, we have

$$\alpha([t_2, t_3]) := \alpha(t_2) - \alpha(t_3) = (8, 8) - (6, 8) = (2, 0) .$$

The 4th step. Here we find the payoff vector $\alpha([t_3, t_4])$. The solution vector of the problem (8) at the moment t_4 is $(3, 3)$. Since $3 + 3 < 8$ we set

$$\alpha(t_4) := (5, 3), \text{ and } \tilde{v}(N, t_4) := v(N, t_4) = 8 .$$

Thus, we have

$$\alpha([t_3, t_4]) := \alpha(t_3) - \alpha(t_4) = (6, 8) - (5, 3) = (1, 5) .$$

The last step. Here we have

$$\alpha([t_4, t_5]) := \alpha(t_4) = (5, 3) .$$

We can see that the proposed algorithm constructs a time-consistent IDP with non negative payoffs to the players at each period.

4 Conclusion

In the paper we formulated necessary and sufficient conditions of time consistency of an imputation from the subcore in a balanced multistage TU-cooperative games. It was shown that time consistent IDP without any delay of total payments to the players may be constructed if the game is balanced.

References

1. Petrosjan, L. (1995): The Shapley Value in Differential Games. In: van Olsder, G. (Ed.): Annals of the International Society of Dynamic Games, Vol. 3, 409–420
2. Villiger, R., Petrosjan, L. (2001): Construction of Time-Consistent Imputations in Differential Games. In: Proceedings of the 2nd International Conference Logic, Game Theory and Social Choice. St. Petersburg, 235–240
3. Zakharov, V. (1996): About Selectors of the Core in Dynamic Games. In: Proceedings of the 7th ISDG symposium on Dynamic Game and Applications. Kanagawa, Japan
4. Zakharov, V., Kwon, O-H. (1999): Selectors of the Core and Consistency Properties. Game Theory and Applications 4, 237–250

Rationality of Final Decisions Leads to Sequential Equilibrium*

Ryuichiro Ishikawa[1] and Takashi Matsuhisa[2]

[1] Graduate School of Economics, Hitotsubashi University,
Naka 2-1, Kunitachi-shi, Tokyo 186-8601, Japan
E-mail: MXB02762@nifty.ne.jp

[2] Department of Liberal Arts and Sciences,
Ibaraki National College of Technology,
866 Nakane, Hitachinaka-shi, Ibaraki 312-8508, Japan
E-mail: mathisa@ge.ibaraki-ct.ac.jp

Summary. This article investigateS epistemic conditions for a sequential equilibrium in an extensive form game with imperfect information: If players mutually know that all players maximize own expected payoffs for any information sets in their final decisions, then their behaviors with belief yield a sequential equilibrium. This result is an extension of [2] in a perfect information game. In this article, we propose the notion of μ-rationality, by which we mean that player knows that he maximizes his own payoff according to the common-belief μ. We also introduce the notion of μ-consistency in imperfect information game. Our main theorem states that mutual knowledge of both μ-rationality and μ-consistency induces a sequential equilibrium outcome in an extensive form game.

Key words: knowledge, rationality, epistemic conditions, backward induction, sequential equilibrium

1 Introduction

This paper investigates what epistemic conditions induce a sequential equilibrium, that is, what each player should know in order to achieve the sequential equilibrium in a given game. There are many equilibrium solutions in an extensive form game, however it is not clear how players achieve these solutions. This paper aims to fill this gap for sequential equilibrium in an extensive form game with imperfect information.

---

* This is submitted for possible presentation in the International Congress of Mathematicians(ICM2002), Satellite Conference on Game Theory and Applications, August, 2002. The paper is preliminary and the final form will be published elsewhere.

In a normal form game, Aumann and Brandenburger [3] gives epistemic conditions for leading to Nash equilibrium: Suppose that the players have a common prior, that their payoff functions and their rationality are mutually known, and that their conjectures for the opponents' actions are commonly known. Then the conjectures form Nash equilibrium.

In an extensive form game there is a contradictory relationship between players' rationality and solution concepts. The contradiction is informally presented by Rosenthal [9] and is formally investigated by Reny [8] and Ben-Porath [5]. They show that players' rationality at the root in the extensive form game does not always lead to the backward induction outcome by examining the centipede game.

On the other hand Aumann [2] establishes the theorem that players' rationality at every node in a perfect information game can lead to the backward induction outcome.

In this paper we investigate in the same line of Aumann. We extend his result into in an imperfect information game as follows.

Main Theorem. *The mutual knowledge of μ-rationality in players' final decisions leads to a sequential equilibrium of an extensive form game with imperfect information.*

Precisely, if everybody knows that each maximizes his own expected payoff according to the common belief μ at each information set in their final decisions, then the assignment associated with μ induces a sequential equilibrium.

This paper is organized as follows: in Section 2 we recall an extensive form game and the sequential equilibrium based on [6]. In addition, we introduce knowledge of players and μ-rationality. In section 3 we formally state the main theorem and give examples to illustrate it. In section 4 we give the proof of the main theorem, and in the final section we conclude some remarks.

2 Game and Knowledge

2.1 Extensive Form Games

We consider a finite extensive form game. By this we mean a structure $G = \langle (T, \prec), N, (\mathcal{I}_i)_{i \in N}, (A_i)_{i \in N}, (u_i)_{i \in N} \rangle$ consisting of as follows: T is the finite set of nodes that is divided into the set of players' decision nodes X and the set of the terminal nodes Z. We assume there is no chance moves for simplicity.[3] $(T, \prec)$ forms a tree with the unique root: The relation $\prec$ is a totally order on the predecessors $P(x)$ of each member x in T and $p(x)$ is the immediate predecessor of x. N is a set of finitely many players. For each $i \in N$, X_i is the subset of X that consists of i's decision nodes and thus X is the disjoint

[3] We restrict our attention into the case that the number of the initial node is just one for simplicity.

union of all the sets of X_i for $i \in N$. We denote by $\iota(x)$ the player making his decision at $x \in X$.

The information that player i possesses is represented by i's *information partition* $\mathcal{I}_i$ on X_i consisting of components I_i called *i's information set*. When a set $I_i \in \mathcal{I}_i$ contains a node $x \in X_i$, we denote it by $I_i(x)$ (or simply by $I(x)$). Each information set is identified with the set of all the decision nodes among which the player can not distinguish. In addition $\mathcal{I}$ denotes the disjoint union of all $\mathcal{I}_i$ for $i \in N$.

Each player i has a feasible action set $A_i(I)$ at every $I \in \mathcal{I}_i$. Since each of i's information sets is the set of nodes that she can not distinguish, the feasible action sets $A_i(x)$, $A_i(x')$ at x, $x' \in I$ are identified with each other, which denotes $A_i(I)$. We denote by A_i the set of all profiles of i's feasible-actions; that is, $A_i \equiv \times_{I \in \mathcal{I}_i} A_i(I)$.

In this paper we focus on games with *perfect recall.*[4] An extensive form game G is said to be with perfect recall if the following conditions are satisfied.

- For any two nodes in a same information set, it is impossible that one node is the predecessor of the other one.
- For any three nodes $x, x', x'' \in X_i$ with $x' \in I(x'')$ and $x \in P(x')$, there exist $\hat{x} \in I(x) \cap P(x'')$ and $a \in A_i(I(x))$ such that if a respectively reaches x' and x'' then it is played at both x and $\hat{x}$.

The assumption of perfect recall plays a crucial role in the main theorem. Let $u_i \colon Z \to \mathbf{R}$ be i's payoff function on the terminal nodes of T.

A *local strategy* at $I \in \mathcal{I}_i$ for player i is a probability distribution b_i^I on $A_i(I)$, and i's *behavior strategy* b_i is the profile $(b_i^I)_{I \in \mathcal{I}_i}$. A behavior strategy b_i is called i's *pure strategy* if each component of b_i assigns the probability one to the specific action of $A_i(I)$ at each information set I. Let $\mathcal{B}_i(I)$ denote the set of all local strategies at I and $\mathcal{B}_i$ the set of all behavior strategies for player i. We set $\mathcal{B} = \times_{i \in N} \mathcal{B}_i$ the set of all profiles of behavior strategies.

Each strategy $b \in \mathcal{B}$ induces the probability distribution $\mathbf{P}^b$ on T defined as follows: for $x \in T$,

$$\mathbf{P}^b(x) := \prod_{a \in \pi(x)} b(a) , \tag{1}$$

where $\pi(x)$ is the set of all actions reaching x from the root. The formula (1) represents the probability to reach x from the root calculated by the strategies on $P(x)$. i's *expected utility* U_i induced from $\mathbf{P}$ on $\mathcal{B}$ is defined by

$$U_i(b) := \sum_{z \in Z} \mathbf{P}^b(z) u_i(z) . \tag{2}$$

By Kuhn's theorem in [7] we restrict our attention to behavior strategies; hereafter behavior strategies are simply called *strategies* in this paper.

[4] Kuhn [7].

2.2 Sequential Equilibrium[5]

A *system of beliefs* is the class of probability distributions μ on each information set $I \in \mathcal{I}$; hence $\sum_{x \in I} \mu(x) = 1$ for each $I \in \mathcal{I}$. Let $\mu(x)$ interpret as a belief assigned by $\iota(x)$ to $x \in I$ if an information set I is reached. Let $\mathcal{M}$ denote the set of beliefs. Each member of $\mathcal{B} \times \mathcal{M}$ is called an *assessment*. Given an assessment $(b, \mu) \in \mathcal{B} \times \mathcal{M}$, we define the *conditional* probability $\mathbf{P}^{b,\mu}(\cdot \mid I)$ over Z by

$$\mathbf{P}^{b,\mu}(z \mid I) = \begin{cases} 0, & \text{if } x \notin P(z) \cap I, \\ \mu(x) \prod_{a \in \pi(x,z)} b(a), & \text{if } x \in P(z) \cap I, \end{cases} \tag{3}$$

where $\pi(x, z)$ is the set of actions taken to reach z from $x \in I$. This formula represents the probability of player's assessment of reaching each terminal node when she is at an information set I. Then we define the conditional expectation U_i^μ under i's information set I by

$$U_i^\mu(b \mid I) := \sum_{z \in Z} \mathbf{P}^{b,\mu}(z \mid I) u_i(z) . \tag{4}$$

Let $\mathcal{B}^+$ denote the set of strategies $b \in \mathcal{B}$ such that $b(a) \gneqq 0$ for any $a \in A$, and $\mathcal{M}^+$ the subset of $\mathcal{M}$ which consists of $\mu \in \mathcal{M}$ such that $\mu(x) \gneqq 0$ at each $x \in X$. For given $b \in \mathcal{B}^+$, we say that the belief μ is *associated* with b if it is defined by the Bayes' rule:

$$\mu(x \mid b) = \mathbf{P}^b(x) \Big/ \sum_{\hat{x} \in I} \mathbf{P}^b(\hat{x}) . \tag{5}$$

We can now define the sequential equilibria of G as follows.

Definition 1. *Let G be an extensive form game. We denote by $\mathcal{SE}(G \mid I)$ the set of all the assessments (b^*, μ^*) satisfying both the conditions* ($\mathbf{C_I}$) *and* ($\mathbf{SR_I}$) *at an information set I.*

($\mathbf{C_I}$) *An assessment (b^*, μ^*) is consistent at the information set I. That is, there exists a sequence $\{(b^n, \mu(\cdot \mid b^n))\} \subsetneqq \mathcal{B}^+ \times \mathcal{M}^+$ such that for all $x \in I$ and all $a \in A_{\iota(I)}(I)$,*

$$\lim_{n \to \infty} (b^n(a), \mu(x \mid b^n)) = (b^*(a), \mu^*(x)) .$$

($\mathbf{SR_I}$) *An assessment (b^*, μ^*) is sequential rational at the information set I. That is, for the information set I and for any alternative strategy profile $b_i' \in \mathcal{B}_i$,*

$$U_i^\mu(b^* \mid I) \geqq U_i^\mu(b_i', b_{-i}^* \mid I) ,$$

where $i = \iota(I)$ and $b_{-i}^* = (b_j^*)_{j \in N \setminus \{i\}}$.

Let $\mathcal{SE}(G)$ denote the intersection of $\mathcal{SE}(G|I)$ over $I \in \mathcal{I}$. We call $(b^*, \mu^*) \in \mathcal{SE}(G)$ a *sequential equilibrium* of the game G.

[5] Kreps and Wilson [6].

2.3 Knowledge Structure on G

Aumann [2] introduces the partition model of knowledge on extensive form games. He shows that the backward induction outcome is reached by the common knowledge of rationality in perfect information games. We will extend the model of knowledge on perfect information games into that on imperfect information games.

A *knowledge structure* on an extensive form game G is a quadruple

$$\langle \Omega, (\Pi_i)_{i\in N}, (K_i)_{i\in N}, \mathbf{b} \rangle$$

consisting of the following structures and interpretations: Ω is a non-empty set, each element ω is called a *state*, a subset E of Ω an *event*, and Π_i is a mapping of Ω into 2^Ω such that the image makes a partition on Ω consisting of components $\Pi_i(\omega)$ for $\omega \in \Omega$. To avoid confusion we call Π_i i's *knowledge partition*. Intuitively a component $\Pi_i(\omega)$ of i's knowledge partition is interpreted as the event consisting of all the states that player i cannot distinguish from ω. i's *knowledge operator* K_i on 2^Ω is defined by

$$K_i E = \{\omega \in \Omega \mid \Pi_i(\omega) \subseteqq E\} \quad \text{for } E \subseteqq \Omega\,. \tag{6}$$

A mapping $\mathbf{b}$ from Ω to $\mathcal{B}$ assigns a $|N|$-tuple $(\mathbf{b}_i(\omega))_{i\in N}$ of players' strategies to each $\omega \in \Omega$.

We record the properties of the knowledge operator.[6] For any $E, F \subseteqq \Omega$,

(N) $K_i\Omega = \Omega$;
(M) if $E \subseteqq F$, then $K_iE \subseteqq K_iF$;
(K) $K_i(E \cap F) = K_iE \cap K_iF$;
(T) $K_iE \subseteqq E$;
(4) $K_iE \subseteqq K_i(K_iE)$;
(5) $\Omega \setminus K_iE \subseteqq K_i(\Omega \setminus K_iE)$.

The *mutual knowledge* operator K_E on Ω is defined by $K_EF = \bigcap_{i\in N} K_iF$. The event K_EF is interpreted as that 'every player knows F. An event F is called *publically known* among the players if $F \subseteqq K_EF$. The *common-knowledge* operator K_C is defined by

$$K_CE := \bigcap_{k=1,2,\dots} \; \bigcap_{\{i_1,i_2,\dots,i_k\}\subseteq N} K_{i_1}K_{i_2}\cdots K_{i_k}E\,.$$

The event K_CE is interpreted as that all players know that all players know that $\cdots$ that all players knows E.'

Now, if ϕ is a function on Ω and v is its value then $[\phi = v]$ (or simply $[v]$) denotes the event $\{\omega \in \Omega \mid \phi(\omega) = v\}$. Therefore for any $b_i \in \mathcal{B}_i$, we denote by $[b_i]$ the set $\{\omega \in \Omega \mid \mathbf{b}_i(\omega) = b_i\}$. We assume that every behavior strategy is publically known among the players; that is,

[6] Bacharach [4].

(PK) $[b_i] \subseteqq K_E[b_i]$ for every $b_i \in \mathcal{B}_i$.

This is interpreted as that everybody knows every behavior strategy for each player. In view of the assumption **(PK)** we can observe that each strategies of player i is Π_i-measurable, and thus $K_i[b_i] = [b_i]$ by **(T)**.

Example 1. Let G be an extensive form game $\langle (T, \prec), N, (\mathcal{I}_i)_{i\in N}, (A_i)_{i\in N}, (u_i)_{i\in N} \rangle$. Let $\Omega = T \setminus Z$ and Π_i the function from Ω to 2^Ω defined by

$$\Pi_i(\omega) = \begin{cases} I_i(\omega), & \text{if } \omega \in X_i; \\ \Omega \setminus X_i, & \text{otherwise.} \end{cases} \tag{7}$$

Let $\mathbf{b}_i^I : \Omega \to \mathcal{B}_i(I)$ be defined as follows: We let take $a \in A_i(I)$ and $b_i \in \mathcal{B}_i(I)$ for each $i \in N$, and we define

$$\mathbf{b}_i^I(\omega) = \begin{cases} b_i(a), & \text{if } \omega \in I \in \mathcal{I}_i, \\ 0, & \text{otherwise.} \end{cases}$$

We set $\mathbf{b}_i = (\mathbf{b}_i^I)_{I\in\mathcal{I}_i}$ as the profile of $\mathbf{b}_i^I$ over i's information sets, and set the mapping $\mathbf{b}$ on Ω into $\mathcal{B}$ by $\mathbf{b} = (\mathbf{b}_i)_{i\in N}$. Let us define the knowledge operator K_j for player j as follows:

$$K_j[b_i^I] = \begin{cases} I, & \text{if } i = j; \\ \emptyset, & \text{if } i \neq j, \end{cases}$$

Then for any $b_i \in \mathcal{B}_i$ and $b_i^I \in \mathcal{B}_i(I)$, it can be observed that $[b_i] = \sqcup_{I\in\mathcal{I}_i}[b_i^I] \subseteqq \cup K_E[b_i^I] = \sqcup_{I\in\mathcal{I}_i} I$ by **(M)**, where the symbol $\sqcup$ denotes the disjoint union operator.

2.4 Rationality and Consistency

We introduce the notion of rationality which is an extension of rationality in [2]. Let μ be a system of beliefs. An player i is said to be *μ-rational* at $I \in \mathcal{I}_i$ if each strategy that i does not know never yields his expected utility value according to μ at $I \in \mathcal{I}_i$ greater than the actual expected utility value at I. If he is rational at any $I \in \mathcal{I}_i$, then i is said to be μ-rational. Formally, let $\mathcal{R}_i^\mu(I)$ denote the event:

$$R_i^\mu(I) := \bigcap_{b_i' \in \mathcal{B}_i} \sim K_i\left[U_i^\mu(b_i', \mathbf{b}_{-i}) \mid I) \gneqq U_i^\mu(\mathbf{b} \mid I)\right], \tag{8}$$

where $\sim$ denotes the complementation and $\mathbf{b}_{-i}$ denotes $(\mathbf{b}_j)_{j\in N\setminus\{i\}}$. This is interpreted as the event that 'player i is μ-rational at $I \in \mathcal{I}_i$.' We denote

$$R_i^\mu = \bigcap_{I\in\mathcal{I}_i} R_i^\mu(I) \qquad \text{and} \qquad R^\mu = \bigcap_{i\in N} R_i^\mu .$$

The former event is interpreted as that player i is μ-rational and the latter as that all players are μ-rational. Furthermore we define the notion of μ-*consistency*. For given $\mu \in \mathcal{M}$, the event of μ-consistency C^μ is the set of all the states ω such that there exists a sequence $\{(b^n, \mu(\cdot \mid b^n))\} \subseteqq \mathcal{B}^+ \times \mathcal{M}^+$ with $\lim_{n\to\infty} (b^n, \mu(\cdot \mid b^n)) = (\mathbf{b}(\omega), \mu)$.

It is well end this section in a remark: rationality in perfect information game is clearly equivalent to μ-rationality when the belief μ is the constant function 1. That is, the rationality in [2] is the 1-rationality R^1 for all players in our sense. One of the purposes in this paper is to extend the result of Aumann [2] in the case of μ-rationality.

3 The Result

Let G be an extensive form game and $\mu \in \mathcal{M}$. We denote by $SE^\mu(G)$ the event consisting of the states $\omega \in \Omega$ such that the assessment $(\mathbf{b}(\omega), \mu) \in \mathcal{B} \times \mathcal{M}$ constitutes a sequential equilibriums in G; that is,

$$SE^\mu(G) = \{\omega \in \Omega \mid (\mathbf{b}(\omega), \mu) \in \mathcal{SE}(G)\} . \tag{9}$$

Furthermore, $SE^\mu(G \mid I)$ denotes the event consisting of the states $\omega \in \Omega$ such that $(\mathbf{b}(\omega), \mu)$ is a member of $\mathcal{SE}(G \mid I)$ for each information set I.

We denote by $\mathcal{I}_F$ the subset of $\mathcal{I}$ consisting of all the information sets in which each player finally decides in the game G. By *final decisions* of player i we mean the set of all the nodes in his information sets where he can reach some terminal node without any other his decisions. Let R^μ_F be the event of μ-rationality over $\mathcal{I}_F$, that is, $R^\mu_F = \bigcap_{h \in \mathcal{I}_F} R^\mu_i(h)$.

The main theorem states that if μ-rationality at the information sets in final decisions for each players under μ-consistency for some $\mu \in \mathcal{M}$ is mutually known then the sequential equilibrium is achieved in the given game G. We can now state the main theorem formally as follows. Notations and assumptions being the same as above,

Theorem 1. $K_E(R^\mu_F \cap C^\mu) = SE^\mu(G)$.

Before proceeding with the proof we shall give two examples to illustrate the theorem.

Example 2. The first example is an extensive game with perfect information. N is $\{1, 2\}$. First player 1 chooses either L or R as his action, and player 2 chooses a pair of a, b, c, d contingently. The backward induction strategy is uniquely determined: the player 1's is R and the player 2's is (b, d). Figure 1 shows this situation.

Now let $\langle \Omega, (\Pi_i)_{i\in N}, (K_i)_{i\in N}, \mathbf{b}\rangle$ be a knowledge structure on G consisting of as follows: Ω is the set of all action profiles:

$$\Omega = \left\{ \begin{array}{l} (L,(a,c)),\ (R,(a,c)) \\ (L,(a,d)),\ (R,(a,d)) \\ (L,(b,c)),\ (R,(b,c)) \\ (L,(b,d)),\ (R,(b,d)) \end{array} \right\};$$

the knowledge partition Π_i for $i \in N$ is defined by $\Pi_i(\omega) = \{\omega\}$ for every $\omega \in \Omega$; the knowledge operator K_i is given by (6); and finally the map $\mathbf{b} : \Omega \to \mathcal{B}$ is the identity map. It is easily seen that $SE^1(G) = \{(R,(b,d))\}$ because the sequential equilibrium implies the backward induction strategy, and it can be also observed that $R^1 \cap C^1 = \{(R,(b,d))\}$. Therefore $K_E(R^1 \cap C^1) = SE^1(G)$.

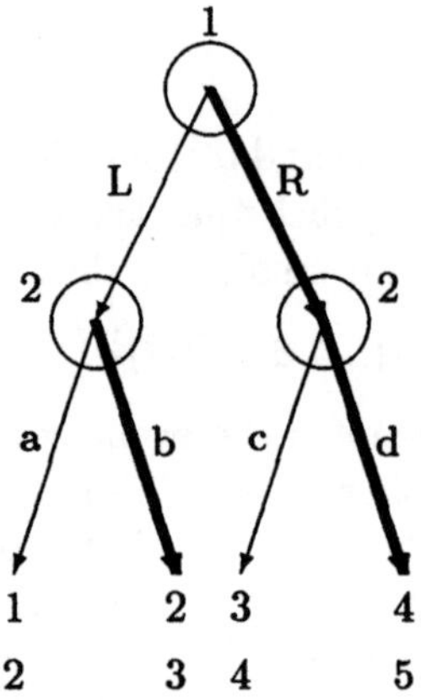

Fig. 1. Case of Perfect Information

Example 3. The next example is an extensive game with imperfect information. N is $\{1,2\}$. First player 1 chooses either L or R and player 2 chooses either a or b. The sequential equilibrium uniquely determined: the player 1's is R and the player 2's is b when $\mu(R) = 1$. Figure 2 shows this situation.

Now let $\langle \Omega, (\Pi_i)_{i \in N}, (K_i)_{i \in N}, \mathbf{b} \rangle$ be the knowledge structure on G consisting of as follows: Ω is the set of all behavior strategies on A; that is, $\Omega = \mathcal{B}$. The knowledge partition Π_i for player $i \in N$ is defined by $\Pi_i(\omega) = \{\omega\}$; the knowledge operator K_i is given by (6); and finally the map $\mathbf{b} : \Omega \to \mathcal{B}$ is the identity map. It is easily seen that $SE^\mu(G) = \{(R,b)\}$ with $\mu(R) = 1$ and it can be also observed that $R^\mu \cap C^\mu = \{R,b\}$. Therefore $K_E(R^\mu \cap C^\mu) = SE^\mu(G)$.

4 Proof of Theorem

On noting the assumption (**PK**) it can be plainly observed that $SE^\mu(G) \subsetneqq K_E(R^\mu_F \cap C^\mu)$. The converse will be shown by induction as follows. It may

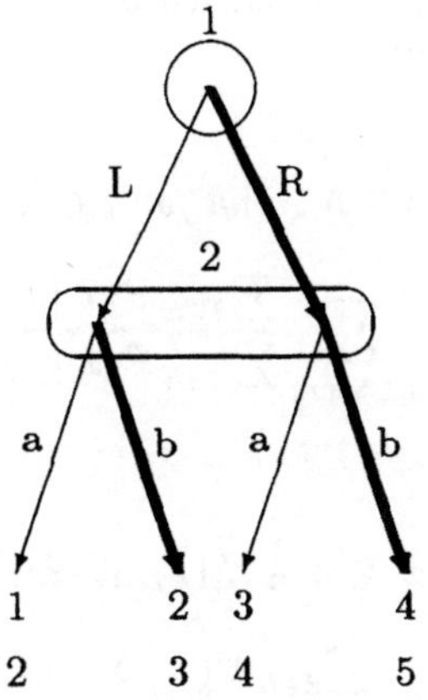

Fig. 2. Case of Imperfect Information

assumed that $K_E(R_F^\mu \cap C^\mu) \neq \emptyset$. For each information set $I \in \mathcal{I}_i$, let $S_i(I)$ be the subset of $\mathcal{I}_i$ consisting of i's information sets next after i decides at I. Let $\mathcal{I}^{\prec}(I)$ denote the set of all the information sets at which $\iota(I)$ decides after I. We shall show the two points: First that for each $i \in N$ and any $h \in \mathcal{I}_F \cap \mathcal{I}_i$, $K_i(R_F^\mu \cap C^\mu) \subseteqq SE^\mu(G \mid h)$, and secondly that $K_i(R_F^\mu \cap C^\mu) \subseteqq SE^\mu(G \mid I)$ if $K_i(R_F^\mu \cap C^\mu) \subseteqq SE^\mu(G \mid h)$ at any $h \in \mathcal{I}^{\prec}(I)$.

We shall verify the first point: For each player $i \in N$, it follows that

$$K_E(R_F^\mu \cap C^\mu) \subseteqq \bigcap_{b_i' \in \mathcal{B}_i} \sim K_i\Big[U_i^\mu(b_i', \mathbf{b}_{-i} \mid h) \gneqq U_i^\mu(\mathbf{b} \mid h)\Big] \cap C^\mu .$$

We note that for any $\omega \in K_E(R_F^\mu \cap C^\mu)$ and for any $b_i' \in \mathcal{B}_i$,

$$\begin{aligned} \omega \notin K_i\Big[U_i^\mu(b_i', \mathbf{b}_{-i} \mid I) \gneqq U_i^\mu(\mathbf{b} \mid I)\Big] \\ \Leftrightarrow \exists\, \xi \in \Pi_i(\omega),\ \xi \notin \Big[U_i^\mu(b_i', \mathbf{b}_{-i} \mid I) \gneqq U_i^\mu(\mathbf{b} \mid I)\Big] \\ \Leftrightarrow \exists\, \xi \in \Pi_i(\omega),\ U_i^\mu(\mathbf{b}(\xi) \mid I) \geqq U_i^\mu(b_i', \mathbf{b}_{-i}(\xi) \mid I) . \end{aligned}$$

Furthermore, it is observed that $\mathbf{b}_i(\omega) = b_i(\xi)$ for any $\xi \in \Pi_i(\omega)$ by (**PK**), and thus it can be plainly obtained that for any $\omega \in K_E(R_F^\mu \cap C^\mu)$ and for any $b_i' \in \mathcal{B}_i$,

$$U_i^\mu(\mathbf{b}(\omega) \mid I) \geqq U_i^\mu(b_i', \mathbf{b}_{-i}(\omega) \mid I) .$$

Therefore we have shown that for each $\omega \in K_i(R_F^\mu \cap C^\mu)$, the assessment $(\mathbf{b}(\omega), \mu)$ is μ-rational on any $h \in \mathcal{I}_F \cap \mathcal{I}_i$, and it is easily observed that the assessment is μ-consistent. It follows that $K_i(R_F^\mu \cap C^\mu) \subseteqq SE^\mu(G \mid h)$.

The following lemma is needed to verify the second point. For $I \in \mathcal{I}_i$ let $S_i(I)$ be the set of all nodes x in X_i such that there is a node y in $I \cap P(x)$

with the property: If $z \in X_i \cap P(x)$ with $y \prec z \prec x$ then $z = y$. We denote $\mathbf{P}(x \mid b) := \mathbf{P}^b(x)$ for simplicity.

Lemma 1. *For $b \in \mathcal{B}$, each $i \in N$, and for $I \in \mathcal{I}_i$ such that $S_i(I) \neq \emptyset$,*

$$U_i^\mu(b \mid I) = \sum_{h \in S_i(I)} \frac{\sum_{\tilde{x} \in h} \mathbf{P}(\tilde{x} \mid b)}{\sum_{\hat{x} \in h} \mathbf{P}(\hat{x} \mid b)} U_i^\mu(b \mid h) .$$

Proof. For $b \in \mathcal{B}, x \in I$ and $x' \in h \in S_i(I)$, it can be observed that

$$\mu(x \mid b^n) = \frac{\sum_{\tilde{x} \in h} \mathbf{P}(\tilde{x} \mid b^n)\, \mu(x' \mid b^n)}{\sum_{\hat{x} \in h} \mathbf{P}(\hat{x} \mid b^n)\, \prod_{a \in \pi(x,x')} b^n(a)} .$$

Therefore it follows that

$$\begin{aligned} U_i^\mu(b \mid I) &= \lim_{n \to \infty} \sum_{x \in I} \mu(x \mid b^n) \prod_{a \in \pi(x,z)} b^n(a) u_i(z) \\ &= \lim_{n \to \infty} \sum_{x \in I} \frac{\sum_{\tilde{x} \in h} \mathbf{P}(\tilde{x} \mid b^n)}{\sum_{\hat{x} \in I} \mathbf{P}(\hat{x} \mid b^n)} \mu(x' \mid b^n) \prod_{a \in \pi(x',z)} b(a) u_i(z) \\ &= \lim_{n \to \infty} \sum_{h \in S_i(I)} \frac{\sum_{\tilde{x} \in h} \mathbf{P}(\tilde{x} \mid b^n)}{\sum_{\hat{x} \in I} \mathbf{P}(\hat{x} \mid b^n)} \sum_{x' \in h} \mu(x' \mid b^n) \prod_{a \in \pi(x',z)} b(a) u_i(z) \\ &= \sum_{h \in S_i(I)} \frac{\sum_{\tilde{x} \in h} \mathbf{P}(\tilde{x} \mid b)}{\sum_{\hat{x} \in I} \mathbf{P}(\hat{x} \mid b)} U_i^\mu(b \mid h) , \end{aligned}$$

in completing the proof of the lemma. □

Let us return to the proof of theorem, and we shall verify the second point. Assume now that $K_i(R_F^\mu \cap C^\mu) \subseteqq SE^\mu(G \mid h)$ for each $i \in N$ and each information set $h \in \mathcal{I}_i$ with $h \subseteqq S_i(I)$. We shall show that $K_i(R_F^\mu \cap C^\mu) \subseteqq SE^\mu(G \mid I)$. Suppose to the contrary that there exists $\tilde{b}_i \in \mathcal{B}_i$ such that at $\omega \in K_i(R_F^\mu \cap C^\mu)$,

$$U_i^\mu(\tilde{b}_i, \mathbf{b}_{-i}(\omega) \mid I) \gneqq U_i^\mu(\mathbf{b}(\omega) \mid I) . \tag{10}$$

It suffices to prove that for any $\bar{b}_i \in \mathcal{B}_i$

$$U_i^\mu(\mathbf{b}(\omega) \mid I) \geqq U_i^\mu(\bar{b}_i, \mathbf{b}_{-i}(\omega) \mid I) . \tag{11}$$

In fact, it can be easily observed that the inequality (11) is in contradiction to (10), completing the proof of the theorem.

We shall verify the inequality (11) as follows: it follows from the above lemma that

$$\begin{aligned}
U_i^{\mu}(\mathbf{b}(\omega) \mid I) &= \sum_{h\in I} \frac{\sum_{\tilde{x}\in h} \mathbf{P}(\tilde{x} \mid \mathbf{b}(\omega))}{\sum_{\hat{x}\in I} \mathbf{P}(\hat{x} \mid \mathbf{b}(\omega))} U_i^{\mu}(\mathbf{b}(\omega) \mid h) \\
&\geqq \sum_{h\in S_i(I)\setminus\{h'\}} \frac{\sum_{\tilde{x}\in h} \mathbf{P}(\tilde{x} \mid \mathbf{b}(\omega))}{\sum_{\hat{x}\in I} \mathbf{P}(\hat{x} \mid \mathbf{b}(\omega))} U_i^{\mu}(\mathbf{b}(\omega) \mid h) \\
&\quad + \frac{\sum_{\tilde{x}\in h} \mathbf{P}(\tilde{x} \mid \mathbf{b}(\omega))}{\sum_{\hat{x}\in I} \mathbf{P}(\hat{x} \mid \mathbf{b}(\omega))} U_i^{\mu}(\bar{b}_i, \mathbf{b}_{-i}(\omega) \mid h') \\
&= \lim_{n\to\infty} \sum_{h\in S_i(I)\setminus\{h'\}} \frac{\sum_{\tilde{x}\in h} \mathbf{P}(\tilde{x} \mid b^n)}{\sum_{\hat{x}\in I} \mathbf{P}(\hat{x} \mid b^n)} U_i^{\mu(\cdot|b^n)}(b^n \mid h) \\
&\quad + \lim_{n\to\infty} \frac{\sum_{\tilde{x}\in h} \mathbf{P}(x' \mid b^n)}{\sum_{\hat{x}\in I} \mathbf{P}(\hat{x} \mid b^n)} U_i^{\mu(\cdot|b^n)}(\bar{b}_i^n, b_{-i}^n \mid h') \\
&= U_i^{\mu}(\bar{b}_i, \mathbf{b}_{-i}(\omega) \mid I) \,.
\end{aligned}$$

This completes the proof of the inequality (11) and so does the proof of the theorem. □

5 Concluding Remarks

This paper examines what epistemic conditions about players' rationality lead to the outcomes induced by a sequential equilibrium. Aumann [2] shows that if players act on the rational behavior in a perfect information game then they can obtain the outcome by the backward induction solution. In this paper we extend this result into the case for sequential equilibrium. We require here only the mutual knowledge of rationality for all players instead of common knowledge of it in [2]. Therefore it is sufficient only to know rationality at the information sets in final decisions for each player.

Some related works (e. g. [8, 5]) lead to the different results from Aumann's. In [2] and this paper rationality on information sets is required, however they do only beliefs about players' rationality at the beginning of a game. Since players have the Bayesian rationality in [8, 5] players can revise their own beliefs about their opponents' behaviors or their present nodes through moving plays. These are the different views in examining extensive form games. Aumann regards rationality of players as an representation of the equilibrium, while Reny and Ben-Porath capture it as playability in a given game. We would like to examine the relationship between the two views in the further research.

References

1. Aumann, R. J. (1995): Backward Induction and Common Knowledge of Rationality. Games and Economic Behavior **8**, 6–19

2. Aumann, R. J., Brandenburger, A. (1995): Epistemic Conditions for Nash Equilibrium. Econometrica **63(5)**, 1161–1180
3. Bacharach, M. (1985): Some Extensions of a Claim of Aumann in an Axiomatic Model of Knowledge. Journal of Economic Theory **37**, 167–190
4. Ben-Porath, E. (1997): Rationality, Nash equilibrium and Backwards Induction in Perfect-information Games. Review of Economic Studies **64**, 23–46
5. Kreps, D. M., Wilson, R. (1982): Sequential Equilibria, Econometrica **50(4)**, 863–894
6. Kuhn, H. W. (1953): Extensive Games and the Problem of Information. In: Kuhn, H. W., Tucker, A. W. (Eds.): Contributions to the Theory of Games. Annals of Mathematical Studies. Princeton University Press, Princeton, No. 28, Vol. II, 193–216
7. Reny, P. J. (1993): Common Belief and the Theory of Games with Perfect Information. Journal of Economic Theory **59**, 257–274
8. Rosenthal, R. W. (1981): Games of Perfect Information, Predatory Pricing and the Chain-Store Paradox. Journal of Economic Theory **25**, 92–100

The Core in the Presence of Externalities*

László Á. Kóczy

Katholieke Universiteit Leuven, Centre for Economic Studies
Naamsestraat 69, B-3000 Leuven, Belgium
E-mail: laszlo.koczy@econ.kuleuven.ac.be

Summary. We generalise the coalition structure core to partition function games. Our definition relies only on a single crucial assumption, namely that there is some internal consistency in the game. Residuals of the deviation play a game similar to the initial one, and – whenever this is possible – they come to a residual core outcome. Deviating players form their optimistic or pessimistic expectations with this in mind. This leads to a recursive definition of the core. Compared to existing approaches, our core concept displays a reduced sensitivity to behavioural assumptions. We consider the core of an economy with a common pool resource defined by Funaki and Yamato [4], and show that for a number of numerical examples our core concept resolves the puzzle arising when more naive approaches were used. We outline possibilities for further extensions.

Key words: core, externalities, partition function, behavioural assumptions, equilibrium binding agreements

1 Introduction

Equilibrium binding agreements (EBA's) of [4] introduced a fundamentally new approach to deal with for games where externalities are present. Perhaps theirs is the first solution concept that is originally defined for such games. Funaki and Yamato [4] consider a two-stage game: first a partition is formed then the players play a game with cooperation within, and without cooperation across coalitions. This game is superadditive by construction[2]. Moreover, if a game has several equilibrium binding agreements then the *coarsest* ones

* Discussions with Luc Lauwers, Johan Eyckmans, Carmen Bevía and Henry Tulkens lead to considerable improvement of the paper. The joint support (SOR-H/99/23, SOR-H/00/01 and COE/01/02) of the Katholieke Universiteit Leuven and the Soros Foundation is gratefully acknowledged.

[2] We return to this in Sect. 4.4.

are selected as *the* set of EBA's. Both of these indicate the strong reliance on superadditivity and the bias towards the grand coalition.

We return to the more standard ideas of deviation and the core. Indeed as a starting point we take a game in partition function form and the coalition structure core (in the sequel simply *core*) as defined [5]. Here we do not only consider payoff vectors that are feasible under the grand coalition, but *outcomes* that consist of a payoff vector and a partition that make it feasible.

In partition function form games the payoff of a coalition dependson the entire coalition structure, and in general there are no restrictions as regards to how a change in the coalition structure will affect the payoff of a coalition. externalities the standard definitions cannot be used. Conversely we may say that a deviation will likely change outsiders' payoffs, so the deviating coalition should expect a reaction from the other players. This reaction, however, cannot, in general, be predicted with certainty. Since the payoff for the deviating coalition depends on the entire coalition structure, also its payoff is uncertain; the deviation happens under certain expectations.

In the past rather explicit assumptions have been made to resolve this problem. These assumptions can be of two kinds. In the first group the partition of the residual players is given explicitly: it can be a partition of singletons, as in the case of the γ-core of Tulkens and Chander [7], or the pre-deviation *status quo* partition, typically used in the literature for cartels, for instance by d'Aspremont et al. [6]. In the other approach the reaction is regarded from the deviators' point of view: residual players maximise or minimise the deviating coalition's payoff, such us by Funaki and Yamato [4]. We prefer not to make such assumptions, but instead approach the problem from the side of the residual players, themselves, and treat them as rational payoff-maximising players. After the deviation we assume that the deviating coalitions decided to go alone, and that we may assume that will not change their partition. Then the rest of the players, the *residuals* can play a *residual* game, a partition function game with payoff values influenced by the externalities of the deviation. As the residual players are also rational they will clearly discard outcomes that are very unfavourable to them. If, in their bargaining process they come across an outcome that is not dominated, they will not be able to discard this one and they will eventually select one of the solutions of the residual game provided that such solution exists. In order to remain consistent, we assume that the solution concept used in the residual game is the same that we intend to define for the initial game, namely the core. This will, of course lead to a recursive definition.

The residual core might be empty, in which case there is no reason to prefer one residual outcome to another, and we should assume that any of them can arise. Also, the coalition structure core may contain outcomes with different partitions. Due to these reasons the deviating coalition still has to consider different scenarios. It may be optimistic or pessimistic. A pessimistic coalition deviates only if *all* possibilities represent at least a weak improvement. An optimistic coalition is satisfied with *only one* possibility that represents a (weak) improvement. These two scenarios lead to different definitions of deviation and to different cores. We show that these two concepts behave well when compared to each other, which enables us to use them together, as the two extremes of the same solution concept.

2 Preliminaries

We introduce our basic definitions. We begin with some simple notation then we define games. First we introduce characteristic function games. Strictly speaking these definitions are not needed for our results, but are used in other concepts we will refer to. Finally we introduce partition function form games the type of games that are used for most of this paper.

2.1 Mathematical Notation

We use some abbreviations and relations that need to be defined.A set subscript to a vector of real numbers means a summation: For a vector $x \in I\!R^T$ indexed over the elements of a set T, for any $S \subseteq T$ we have $x_S = \sum_{i \in S} x_i$.

Set subscripts to relation symbols between real vectors mean a restriction of the relation. Let $x \in I\!R^{T_1}, y \in I\!R^{T_2}$ and $S \subseteq T_1$ and $S \subseteq T_2$. Then we write $x >_S y$ if $x_i \geq y_i$ for all $i \in S$ and $x_S > y_S$. Similarly, we write $x =_S y$ if $x_i = y_i$ for all $i \in S$, and $x \leq_S y$ if $x_i \leq y_i$ for all $i \in S$.

2.2 Characteristic Function Games

Characteristic function games are the standard form of coalition formation games where the payoff of a coalition depends only on the coalition itself.

Definition 1 (Characteristic Function Game). *Let $N = \{1, \ldots, n\}$ be a set of players. Nonempty subsets of N are called* coalitions. *A characteristic function $v\colon 2^N \setminus \{\emptyset\} \longrightarrow I\!R$ assigns a real value to each coalition. We allow payoff transfers within coalitions. Then the pair (N, v) is said to be a* transferable utility game in characteristic function form.

Cooperation is often regarded as the ultimate good, and the aim of the game is often to maintain cooperation, to maintain the *grand coalition.* We believe that if cooperation has benefits, then these benefits are incorporated in the characteristic function, and do not make the additional assumption that the grand coalition is formed. We use a more general form of imputations:

Definition 2 (Partition). *A* partition *is a set of pairwise disjoint coalitions, whose union is N. The set of partitions of a set N is denoted by $\Pi(N)$.*

Definition 3 (Outcome). *An* outcome *of a game* (N,v) *is a pair* $(x,\mathcal{P})$, *with* x *in* $I\!R^N$ *and* $\mathcal{P}$ *a partition of* N. *The vector* $x=(x_1,x_2,\ldots,x_n)$ *satisfies*

$$\forall i \in N \;:\; x_i \geq v(\{i\}) \quad \text{and} \quad \forall C \in \mathcal{P} \;:\; x_C = v(C)\;,$$

The first condition is known as individual rationality: player i will only cooperate to form a coalition if it gets at least as much as it would get on his own. The second condition combines feasibility with the myopic behaviour of the players: the members of each coalition in $\mathcal{P}$ *can only distribute the coalition's worth among themselves, and that they always distribute the entire amount. Outcomes generalise imputations. Given this we can define a core-concept.*

Definition 4 (Domination & Deviation). *Outcome* $(x,\mathcal{P})$ dominates $(y,\mathcal{Q})$ *if there exists a coalition* C *in* $\mathcal{P}$ *such that* $x >_C y$. *Then we say that* $(y,\mathcal{Q})$ *is dominated* by $(x,\mathcal{P})$ via C.

A deviation *is the formation of a new coalition* C*, possibly, but not necessarily from members of different coalitions. This deviation is* profitable *if* $v(C)$ *exceeds the pre-deviation payoff* x_C. *As this excess can be divided in infinitely many ways among the members of the coalition each of them giving new outcomes it is often easier not to make a reference to these outcomes. It is clear that if there is a profitable deviation from* $(y,\mathcal{Q})$ *then there is also an outcome dominating it.*

Definition 5 (Coalition Structure Core). *[5] Let* (N,v) *be a game. The* coalition structure core $C(N,v)$ *is the set of undominated outcomes. Equivalently, a pair* $(x,\mathcal{P})$ *is in the coalition structure core if and only if it satisfies:*

- *feasibility: for each coalition* C *in* $\mathcal{P}$ *we have* $x_C \leq v(C)$, *and*
- *coalitional rationality: for each coalition* $S \subseteq N$ *we have* $x_S \geq v(S)$.

The coalition structure core collects all undominated outcomes.

2.3 Partition Function Games

Much of the definitions carries over from characteristic function games, which we do not redefine here.

Definition 6 (Partition Function). *Given a set of players* N *the* partition function

$$\begin{aligned} V:\; \Pi(N) &\longrightarrow (2^N \longrightarrow I\!R)\;, \\ \mathcal{P} &\longmapsto V_{\mathcal{P}}\;, \\ V_{\mathcal{P}}:\; C &\longmapsto V_{\mathcal{P}}(C) \end{aligned}$$

assigns a real valued function to each partition. This payoff function *then assigns a real value to each coalition. This function gives 0 to coalitions that do not belong to the partition. For clarity we write* $V(C,\mathcal{P})$ *instead of* $V_{\mathcal{P}}(C)$.

The partition function assigns a different characteristic function to each partition. The payoffs of the same coalition are in general different in different partitions, since they are not even determined by the same function! On the other hand the partition function can be distilled to a characteristic function if all coalitions get the same payoff regardless of the partition.

Definition 7 (Game). *The pair (N, V) is a* transferable utility game in partition function form, *in short, a game.*

Given the partition function we redefine outcomes for partition function form games.

Definition 8 (Outcome). *Given a game (N, V) an* outcome $(x, \mathcal{P})$ *is a pair with x in $I\!R^N$ and $\mathcal{P}$ a partition of N. The vector $x = (x_1, x_2, \ldots, x_n)$ lists the payoffs of each player and has to satisfy*

$$\forall i \in N \ : \ x_i \geq 0 \quad \textit{and} \quad \forall \ C \in \mathcal{P} \ : \ x_C = V(C, \mathcal{P})\ .$$

The first condition, participation rationality *expresses that players should get a non-negative benefit from participating in the game, and as such replaces individual rationality*[3] *that is not well-defined for partition function games. Let $\Omega(N, V)$ denote the set of outcomes.*

In partition function games the payoff of a coalition depends on the entire coalition structure. So the payoff this coalition experiences after the deviation is not a well-defined number.

Definition 9 (Worth). *The worth of a deviating coalition or coalitions is the set of payoff-vectors that it or they expect as a post-deviation payoff. The worth of a coalition is denoted by W.*

Note that we do not assume the uniqueness of post-deviation scenarios, nor that only one coalition participates. Thus, in general the worth of a deviation is a set of payoff vectors in $I\!R^k$, where k is the number of deviating coalitions. If we only allow single coalition deviations, then we have sets of numbers. Moreover, in the case of some existing models only a single number.

Domination and the core will be redefined in sect. 3.2, for the moment we can say that they implement the same ideas except for partition function games. A remarkable difference is that we allow *multi-coalition deviations*, too. For characteristic function games having single or multiple deviating coalitions makes no difference: a multi-coalition deviation can be reproduced by a sequence of simple deviations. The same property does not hold for partition function games as the payoff of a coalition may change as a result of other deviations. In particular, while the terminal payoff (if reached) is desirable

[3] Our suggestion might seem rather ad hoc. This is only used here to have a lower bound on the payoffs, and the choice for this lower bound does not play a significant role in the definition of the core.

for one of the deviating coalitions, it is possible that when it should deviate the immediate residual response is unfavourable, making the deviation non-profitable. We show an example of such a game in Sect. 4.5 Without this extension the Pareto-efficiency of undominated outcomes would not hold.

3 The Core Concept

Unlike in the characteristic function form, in the partition function form a deviation affects the payoffs of outsider players, even if the composition of their coalition is unaltered. In general we may assume that they try to minimise negative effects or enhance positive effects by reacting to this new situation. This reaction, on the other hand, influences the payoff of the deviating players. Therefore players can only deviate having some expectations about the reaction and hence about their post-deviation payoff. Such expectations have been modelled differently in the past. We will discuss the existing approaches in Sect. 4

We assume that the remaining set of players, which we call residuals and that include both outsiders and players who are in fragments of seceding coalitions, reacts in a payoff maximising way disregarding the effect of their moves on the deviating coalition. After some players made this first move (namely the deviation), the rest begins a bargaining *given* this coalition and argue until no further objections are possible. In order to be consistent we assume that they implement the same core concept that we are about to define for the original game. As in characteristic function games, the core can be empty here as well. In this case our argument does not select particular outcomes, so we allow the entire set of *residual* outcomes. Since our concept generalises the coalition structure core in either case we may have to consider more than possible residual partition giving more than one possible post-deviation payoff for the deviating coalitions. In this case optimistic and pessimistic cases are considered.

In the following we formalise the treatment of residual players.

After a deviation the residual players engage in a game similar to the initial one. The set of players is given; the residual partition function has to be consistent with the partition function of the initial game, and, at the same time, take account of the deviation. Then the residual game is defined as follows:

Definition 10 (Residual Game). *Let (N, V) be a game. Let S be a coalition and R be its complement in N. Let $\mathcal{S}$ be a multi-coalition deviation, a partition of S. Given the deviation $\mathcal{S}$ the residual game $(R, V_{\mathcal{S}})$ is the partition function form game over the player set R and with the partition function*

$$V_{\mathcal{S}}: \Pi(R) \longrightarrow (2^R \longrightarrow IR),$$
$$\mathcal{R} \longmapsto V_{\mathcal{S}}(\mathcal{R}),$$
$$V_{\mathcal{S}}(\mathcal{R}): C \longmapsto \begin{cases} V(C, \mathcal{R} \cup \mathcal{S}), & \textit{if } C \in \mathcal{R}, \\ 0, & \textit{otherwise}. \end{cases}$$

The residual game is a partition function form game on its own it can be solved independently of this deviation, or the initial game. A very trivial, but crucial property of this game is that it has fewer players than the initial one. The residual game is solved with the deviation taken as fixed, in fact independently of this deviation, or the initial game. A very trivial, but crucial property of this game is that it has fewer players than the initial one.

In order to be able to determine the payoff for the deviating coalitions, we have to know which outcomes, or in fact which partitions the residual players will form. We may say that just like in our example, an outcome giving 0 to everyone will be discarded if there are outcomes that give positive payoffs to all players. Can we say more in general? If an outcome is undominated, this means that no coalition can profit from forming a new outcome. In a sense the outcome is at least as good than anything else. If such core outcomes exists we may assume that the residual players prefer these, and discard others. On the other hand if such outcomes do not exist we would need extra assumptions to remove outcomes. Such assumptions could further improve the solution concept, but would also be arbitrary and to remain general we prefer not to make them. Therefore if the residual core is not empty deviating players only consider residual core outcomes. If it is empty, they consider all residual outcomes.

3.1 The Worth of Deviations

Once we know this preferred set of the residual players (the core or if the core is empty, the set of all outcomes), we can give the post-deviation payoffs for the deviating coalitions. This payoff is nothing but the payoff of the deviating coalitions in the partition completed with the preferred residual partitions. Then a deviation is only profitable if it is profitable for *all* deviating coalitions.

In general the preferred set contains outcomes with different partitions, thus giving different values for the coalitions, some of which may be attractive for the deviating coalitions, some may be very bad. It is clear that if the deviation is profitable under *all* options then the deviating coalitions will deviate. On the other hand if we place such a strong restriction on deviation, the set of undominated outcomes will be large. All outcomes that are outside this set are dominated, but, should the players be a bit more optimistic, also some elements of the set may be dominated. In order to find the set of outcomes that are definitely undominated, we consider optimistic players. Optimistic players deviate if there is *at least one* possibility that makes the deviation profitable. Solving the problem with both optimistic and pessimistic players we get a pair of concepts representing the two extremes. We suggest to use them as a set interval (for inclusion).

3.2 The Core

This section is devoted to the definition of our new concepts. Our definition is inductive and is done if four steps. For a trivial single-player game we can give the core explicitly. Given the definition for all at most $k-1$ player games we can give our definition of dominance for k player games. Once dominance is defined, we may define the core. First we give the definition for the pessimistic case, and then a slightly modified version gives the optimistic core.

Pessimistic case.

Definition 11 (Core – Pessimistic Case). *The definition consists of four steps.*

Step 1. The Core of a Trivial Game.
Let (N, V) be a game. The core *of a game with $N = \{1\}$ is the efficient outcome with the trivial partition:*

$$C_{-}(\{1\}, V) = \{(V(1, (1)), (1))\} .$$

Step 2. Inductive Assumption.
Given the definition of the core for every game with at most $k-1$ players we can define dominance for a game of k players.

Step 3. Dominance.
The outcome $(x, \mathcal{P})$ is dominated via a set of coalitions $\mathcal{S}$ if either

1) the residual core is empty *and* for all *residual outcomes the deviation $\mathcal{S}$ is profitable. Formally: for all partitions $\mathcal{Q}$ containing $\mathcal{S}$ there exists an outcome $(y, \mathcal{Q})$ such that $y >_S x$, or*
2) the residual core is not empty *and* for all *residual* core *outcomes the deviation $\mathcal{S}$ is profitable. Formally: for all residual core outcomes $(y_R, \mathcal{R})$ there exists an outcome $(y, \mathcal{S} \cup \mathcal{R})$ with a payoff vector $y =_S y_R$ and $y >_S x$.*

The outcome $(x, \mathcal{P})$ is dominated if it is dominated via a set of coalitions.

Step 4. The Core.
The core *of a game of k players is the set of undominated outcomes and we denote it by $C_{-}(N, V)$.*

Optimistic Case.

Now we define the optimistic case. It only differs in the definition of dominance.

Definition 12 (Core – Optimistic Case). *The definition consists of four steps.*

Step 1. The Core of a Trivial Game.
Let (N, V) be a game. The core of a game with $N = \{1\}$ is the efficient outcome with the trivial partition:

$$C_+(\{1\}, V) = \{(V(1,(1)),(1))\} .$$

Step 2. Inductive Assumption.
Given the definition of the core for every game with at most $k-1$ players we can define dominance for a game of k players.

Step 3. Dominance.
The outcome $(y, \mathcal{Q})$ is dominated via a set of coalitions S if either

1) *the residual core is* empty *and* there exists *a residual outcome that makes the deviation S profitable. Formally: there exists an outcome $(y, \mathcal{Q})$, such that $\mathcal{Q}$ contains S and $y >_S x$,*
2) *the residual core is* not empty *and* there exists *a residual* core *outcome that makes the deviation S profitable. Formally: there exists a residual core n outcome outcome $(y_R, \mathcal{R})$, such that there exists an outcome $(y, S \cup \mathcal{R})$, such that $y =_S y_R$ and $y >_S x$.*

Step 4. The Core.
The core *of a game of k players is the set of undominated outcomes and we denote it by $C_+(N, V)$.*

3.3 The Relation of the Two Cases

In order to be able to refer to the core as an interval for inclusion, we need to prove the following theorem:

Theorem 1. *Given a game (N, V) the pessimistic core contains the optimistic core, $C_+(N, V) \subseteq C_-(N, V)$.*

Proof. There are two key ideas in the proof. Firstly, if deviating players are less hesitant (optimistic) they deviate even if they would not otherwise (being pessimistic) making the core smaller. Secondly, more options make hesitant players even more hesitant.

The proof is by induction on the player set N.

For a single-player game we have $C_+(\{1\}, V) = C_-(\{1\}, V)$ and so the result holds.

Assuming that $C_+(N_{k-1}, V) \subseteq C_-(N_{k-1}, V)$ is satisfied for all games with $|N_{k-1}| \leq k-1$, we consider a deviation S from an outcome $(x, \mathcal{P})$ in a game of k players. As the deviation includes at least one player, the residual game consists of at most $k-1$ players. Since the residual game is a game in partition function form with a player set $R = N \setminus \bigcup_{S \in \mathcal{S}} S$, and a partition function $V_{\mathcal{S}}$, and $|R| \leq k-1$, we have $C_+(R, V_{\mathcal{S}}) \subseteq C_-(R, V_{\mathcal{S}})$.

We discuss four cases, depending on the emptiness of the residual cores. In each of these cases we consider residual outcomes that make the deviation profitable or not, and show that profitability in the pessimistic case implies profitability in the optimistic case.

1. Both Residual Cores Are Non-Empty

Deviating players form their expectations about their post-deviation payoffs with respect to the residual cores $C_+(R, V_{\mathcal{S}})$ and $C_-(R, V_{\mathcal{S}})$. If under pessimistic assumptions the deviation is profitable, it is profitable for all outcomes in the set $C_-(R, V_{\mathcal{S}})$. By our assumption $C_+(R, V_{\mathcal{S}}) \subseteq C_-(R, V_{\mathcal{S}})$ and is non-empty, so elements in $C_+(R, V_{\mathcal{S}})$ all make the deviation profitable, so it will be profitable in the best case as well. On the other hand profitability in the best case has no implications on the worst case.

2. Both Residual Cores Are Empty

Deviating players form their expectations with respect to the entire residual outcome set $\Omega(R, V_{\mathcal{S}})$. If in the worst case the deviation is profitable, it is profitable for all residual outcomes in $\Omega(R, V_{\mathcal{S}})$, and therefore in the best case as well. Again, profitability in the best case has no implication on the worst case.

3. The Optimistic Residual Core is Empty, The Pessimistic Residual Core is Non-Empty

Deviating players form their expectations with respect to the entire residual outcome set $\Omega(R, V_{\mathcal{S}})$ in the optimistic approach and $C_-(R, V_{\mathcal{S}})$ in the pessimistic approach. If in the worst case the deviation is profitable, it is profitable in the entire set $C_-(R, V_{\mathcal{S}})$. Since $C_-(R, V_{\mathcal{S}})$ is contained in $\Omega(R, V_{\mathcal{S}})$, the set $\Omega(R, V_{\mathcal{S}})$ contains residual behaviours for which the deviation is profitable, and hence it is profitable in the best case.

4. The Optimistic Residual Core is Non-Empty, The Pessimistic Residual Core is Empty

Formally we have $\emptyset = C_-(R, V_{\mathcal{S}}) \subset C_+(R, V_{\mathcal{S}})$, which contradicts to our inductive assumption, and hence this case does not arise.

If a deviation is profitable in the pessimistic case, it is profitable in the optimistic one as well. Hence if an outcome does not belong to the pessimistic core $C_-(N, V)$ it does not belong to the optimistic core $C_+(N, V)$ either. □

4 Overview of Existing Concepts

In this section we overview six existing solution concepts or methods that are used to solve games with externalities. The first two are the more naive optimistic and pessimistic approaches. We show that our proposal is less sensitive to behavioural assumptions. In the next two approaches the uncertainty about the residual behaviour is resolved by forcing a specific partition on the residual players. Then we look at the r-core of Huang and Sjostrom [3], which is not so much a new concept, but a method to convert the problem to the characteristic function form. Finally we look at the equilibrium binding agreements of Ray and Vohra [8].

4.1 If Only the Effect Counts

In these two approaches the residual behaviour is only examined from the point of view of the deviators.

The *naive pessimistic* approach corresponds to the α-theory originally defined by Aumann and Peleg [2] for nontransferable utility games. Their β-theory, where the deviating coalition could select its strategy *given* the residual strategy is more optimistic, but still to a lesser extent than the *naive optimistic*, which we denote by ω, being the other extreme. Variations on these approaches are summarised by Cornet [4].

Definition 13 (Worths in the α- and ω-Approaches). *Let $W_\alpha(C)$ and $W_\omega(C)$ respectively denote the worth of a coalition C for the α and ω approaches. Then we have*

$$W_\alpha(C) = \min_{\mathcal{R}\in\Pi(N\setminus C)} \{V(C,\{C\}\cup\mathcal{R})\} ,$$
$$W_\omega(C) = \max_{\mathcal{R}\in\Pi(N\setminus C)} \{V(C,\{C\}\cup\mathcal{R})\} .$$

Definition 14 (Deviation and α- and ω-Core). *Coalition C deviates from the outcome $(x,\mathcal{Q})$ if $W_\alpha(C)\geq x_C$ and $W_\omega(C)\geq x_C$ are respectively satisfied. An outcome is undominated if there exist no deviations from it. The set of undominated outcomes is the core, denoted by $C_\alpha(N,V)$ and $C_\omega(N,V)$, respectively.*

Sometimes the two approaches are used together, but often separately. The two make different assumptions about the players' type. Since the players are probably neither totally optimistic, nor totally pessimistic, the two approaches should be used together to avoid errors of different kinds. If an outcome is in the optimistic core, we can be sure that with the "real" type of the players it is still undominated. Finding that the optimistic core is not empty is a strong result. On the other hand, if an outcome does not belong to the pessimistic core, it is definitely dominated regardless of the players' type. Here emptiness is the stronger result. The two different approaches often, as in the case of [4] lead to different or even contradictory conclusions.

Here our major criticism is that these approaches discuss the reaction solely from the deviators' point of view, and make no attempt to justify them as rational moves by the residual players. Unrealistic beliefs about residual reactions may distort the expectations and make the difference between the two cores large. Our concept is a direct answer to this problem. In the following we show that the new concepts reduce the sensitivity to behavioural assumptions.

The inclusion of the ω-core in the α-core, that is, $C_\omega(N,V) \subseteq C_\alpha(N,V)$ is easy to see, but the recursive definition makes the comparison of the optimistic and the pessimistic core to the α- and ω-cores less plausible. We prove this lemma first.

Lemma 1. *The optimistic contains the ω-core, that is, $C_\omega(N,V) \subseteq C_+(N,V)$. The pessimistic core is contained in the α-core, that is, $C_-(N,V) \subseteq C_\alpha(N,V)$.*

We prove the part for the optimistic approaches; the corresponding result for the pessimistic approaches is proven in a like manner.

Proof. In the given game, consider an outcome and a deviation from it. We want to check the profitability of this deviation. The two concepts, the ω and the optimistic approach look at two different residual outcome sets. The ω-approach considers the entire residual outcome set $\Omega(R, V_S)$, our optimistic approach considers $C_+(R, V_S)$ only, *provided* this set is not empty.

If the residual core is empty the two approaches both look at $\Omega(R, V_S)$ and expect the same post-deviation payoff. In particular, if the deviation is profitable in the optimistic approach it is also profitable for the ω-approach.

Now we look at the case when the residual core is not empty. Since $C_+(R, V_S)$ is contained in $\Omega(R, V_S)$ the best case in the ω approach is at least as good as in our optimistic approach looking only at residual core outcomes. Therefore if the deviation is profitable in the optimistic approach, it is also profitable in our ω-approach.

If an outcome is not in the optimistic core $C_+(N,V)$, there exists a profitable deviation from it in the optimistic approach. As the same deviation is also profitable in the ω-approach the outcome is neither in the $C_\omega(N,V)$.

Corollary 1. *The proposed new concepts, as a pair are a refinement of the α-core, ω-core pair.* $C_\omega(N,V) \subseteq C_+(N,V) \subseteq C_-(N,V) \subseteq C_\alpha(N,V)$.

4.2 Approaches with Explicitly Given Residual Partition

Two approaches give the residual partition explicitly. The first assumes that residuals do not react, the second assumes that they fall apart to singletons.

The *status quo*, or δ approach is often used in cartel games for instance by d'Aspremont et al. [6]. In this approach deviating players simply assume that outsiders do not react. This model completely overlooks the externalities arising from a deviation. The worth does not only depend on the deviating coalition, but also the initial partition. Formally:

Definition 15 (Worth in the Status Quo Approach). *Let $W_{sq}(C, \mathcal{Q})$ denote the worth of a coalition C that deviated from partition $\mathcal{Q} = \{Q_1, Q_2, \ldots, Q_k\}$. Then we have*

$$W_{sq}(C, \mathcal{Q}) = V(C, \{C, Q_1 \setminus C, Q_2 \setminus C, \ldots, Q_k \setminus C\}) .^4$$

[4] If $Q_i \subseteq C$, then $Q_i \setminus C$ is empty and is not listed in the partition. On the other hand, if $Q_i \cap C = \emptyset$, then $Q_i \setminus C = Q_i$.

Definition 16 (Deviation and δ-Core). *Coalition C deviates from the outcome (x, Q) if $W_{sq}(C, Q) \geq x_C$. An outcome is undominated if there exist no deviations from it. The set of undominated outcomes is the core, denoted by $C_{sq}(N, V)$.*

Observe that the set $C_{sq}(N, V)$ is basically the set of strong Nash equilibria.

Studying the pessimistic and optimistic approaches Chander and Tulkens [3] find that it is not reasonable to assume that the residual players act to hurt or help deviating players and not to maximise their own payoffs. Therefore they introduce an approach that is "individually reasonable" in the sense that in the two-stage game they define the residual players choose their strategies according to a Nash behaviour, essentially breaking up to singletons. This approach is known now [7] as the γ-approach. It is, in a sense, the opposite of the *status quo* approach, where we assumed that the smallest number of "links" break up between players. The core based on this approach is the γ-core.

Definition 17 (Worth in the γ-Approach). *Let $W_\gamma(C)$ denote the worth of a coalition C. Then we have*

$$W_\gamma(C) = V(C, \{C, S_1, S_2, \ldots, S_{n-|C|}\}) ,$$

where $|S_i| = 1$ for all $1 \leq i \leq n - |C|$.

Definition 18 (Deviation and γ-Core). *Coalition C deviates from the outcome (x, Q) if $W_\gamma(C) \geq x_C$. An outcome is undominated if there exist no deviations from it. The set of undominated outcomes is the core, denoted by $C_\gamma(N, V)$.*

In a partition function form game where strategies correspond to coalitions the γ-approach leaves the residual players defenseless. Our solution is more general than that. As a special case residual players can fall apart to singletons as in the γ-approach, but they can form any other partition as well if they find it more attractive.

4.3 The r-Theory [3]

The r-theory is a recursive approach similar to ours (though independent) that defines a characteristic function from a normal form game. Once this function is created, the game can be solved using the usual techniques for characteristic function games.

Consider a game in normal form $\Gamma = (N, (A_i, u_i)_{i \in N})$, where N is the set of players, A_i is the strategy set of player i and $u_i : \prod_{i \in N} A_i \longrightarrow I\!R$ is the payoff function of player i. For any coalition $S \subseteq N$, $A_S = \prod_{i \in S} A_i$ and $A_{-S} = \prod_{i \notin S} A_i$. For any strategy vector $a \equiv (a_i)_{i \in N} \in A$ we use a_S to denote $(a_i)_{i \in S}$ and a_{-S} to denote $(a_i)_{i \notin S}$. Similarly $(u_i(a))_{i \in S}$ is denoted by $u_S(a)$.

The game has two stages: first a partition is formed cooperatively then a game is played *given* this coalition structure with cooperation within the coalitions, but without cooperative interaction between coalitions. Thus each partition $\mathcal{P} \in \Pi(N)$ induces a normal form game with the coalitions as composite players. Let $\beta(\mathcal{P})$ denote the set of Nash-equilibria of this game. Since transfers within coalitions are allowed we have $x_S = u_S(a)$ for all coalitions S and strategy-vectors a. The aim of a coalition is to maximise the total payoff of its members.

The idea is to create a characteristic function from this normal form game in a recursive way. If a coalition deviates, it is taken as given, and the rest of the players play a game. Given a subset of the players R the set of strategies for a coalition C in R conditional on the deviations $\mathcal{P} \in \Pi(N \setminus R)$ is denoted by $A(C\,|\mathcal{P})$.

In the recursive definition first the trivial cases are introduced. $|R| = 1$. Then the strategy space of R conditional on the existing deviations described as a partition is the same as the set of Nash equilibria of the corresponding noncooperative game with composite players; Formally:

$$A(R\,|\mathcal{P}(N \setminus R)) = \beta(R \cup \mathcal{P}(N \setminus R)),$$

where $\mathcal{P}(N \setminus R) \in \Pi(N \setminus R)$.

Now assuming that the set of strategies $A(R\,|\mathcal{P}(N \setminus R))$ has been defined for all R, such that $1 \leq |R| \leq r-1$, we define worth in the case of $|R| = r$ and denote it by $W(R\,|\mathcal{P})$ to indicate reliance on the deviations $\mathcal{P} \in \Pi(N \setminus R)$.

$$W(R\,|\mathcal{P}(N \setminus R)) = \min\{u_R(a)\,|a \in \beta(\{R\} \cup \mathcal{P}(N \setminus R))\}$$

and in general, for any $C \subset R$:

$$W(C\,|\mathcal{P}(N \setminus R)) = \min\{u_C(a)\,|a \in A(R \setminus C\,|\{C\} \cup \mathcal{P}(N \setminus R))\}.$$

For simplicity we write $W(S)$ instead of $W(S\,|\mathcal{P}(\emptyset))$.

These are of course the pessimistic cases. The corresponding optimistic cases are similar, with the minima replaced by maxima. The characteristic function for R is defined as

$$v : C \longmapsto W(C\,|\mathcal{P}(N \setminus R)),$$

for all $C \subseteq R$. Using this characteristic function we can determine the core. The set of strategies leading to a core outcome is

$$A_C(R\,|\mathcal{P}(N \setminus R)) = \{a \mid \exists x : \ x_R = u_R(a), x \in C(N, v)\}.$$

If $R = N$ then we have the solution. Otherwise the induction continues; the set of strategies for R are defined as:

$$A(R\,|\mathcal{P}(N \setminus R)) = \left\{a \,\middle|\, \begin{array}{l} a \in \beta(\mathcal{P}(R) \cup \mathcal{P}(N \setminus R)), \mathcal{P}(R) \in \Pi(R) \\ a \in A_C(R\,|\mathcal{P}(N \setminus R)) \end{array} \right\}.$$

Having $A_C(N)$ we know which strategies are played to reach a core outcome, so we know the coalitional payoffs in the core outcomes. Then the r-core is defined as follows:

$$C_r\left(N,(A_i,u_i)_{i\in N}\right)=\left\{(x,\mathcal{P})\left|\begin{array}{l}\exists a:\ \forall C\in\mathcal{P}:\ x_C=u_C(a)\\ \forall S\subseteq N\ x_S\geq W(S)\end{array}\right.\right\}.$$

Due to the different form of the games the similarities with the core are not apparent. The r-theory has the advantage over our approach that it is defined in the rather general and intuitive normal form, and its argumentation can be extended to other forms as well. It uses well-established concepts and results.

However, much of this applies to our concept as well. The core we propose can also be extended to more general forms. Moreover the normal form has also some limitations: it is by definition superadditive, while arguments against superadditivity are well-known. Both superadditivity and the characteristic function used can be accounted for the use of single-coalition deviations only. In the core we propose we have no problem using multi-coalition deviations that are needed for efficiency in partition function form games. The significance of this shortcoming prevails from the example we present in Sect. 4.5.

Finally, a small, but significant difference makes the r-theory very limited in use. Here the characteristic function is only defined if *all* residual cores are non-empty. For larger games this condition becomes -at least mathematically-very strong. One would have difficulties interpreting the three possible results one can get computing the r-core: it can be a nice non-empty core, it can be empty, and what is even worse: undefined. As soon as we allow for sub-additivity in a game we can give characteristic function games with a nonempty core, where the appropriately defined r-core is undefined. Hence the r-core is *not* an extension of the core of a characteristic function form game, while our core concept is. In our approach this existence problem is resolved by looking at the "second best": if the residual core is empty, we take all possible outcomes. In the aforementioned characteristic function form game the residual behaviour does not influence the deviating payoffs, so even if some residual cores can be empty the core can be defined and can even be non-empty.

4.4 Equilibrium Binding Agreements [8]

Equilibrium binding agreements became very popular as a solution concept for games with externalities. Wherever possible we use the same notation as for the r-theory. Here, instead of outcomes, we use strategy-vector and partition pairs. We consider here the same two stage game. The equilibrium binding agreements are Nash-equilibria of this game that are also immune to "credible" defections by a subcoalition.

For a partition $\mathcal{P}$, let $\mathcal{R}(\mathcal{P})$ denote its refinements. The coalitions of a refinement $\hat{\mathcal{P}}$ of $\mathcal{P}$ are subsets of coalitions in $\mathcal{P}$. The deviating coalitions in $\hat{\mathcal{P}}$ that enforced $\hat{\mathcal{P}}$ from $\mathcal{P}$ are called *perpetrators* while the rest are the residuals. As a coalition may break into several, say k subcoalitions, then $k-1$ of these have to be labelled as perpetrators. A *re-merging* is a coalition structure formed by the merger of perpetrators with their respective residuals.

Definition 19 (Equilibrium Binding Agreements for a Given Partition). *The definition is recursive. Let $\mathcal{B}(\mathcal{P})$ denote the set of equilibrium binding agreements for a given partition $\mathcal{P}$.*

1. *For the trivial partition, $\mathcal{P}^*$ of singleton coalitions as no further deviations are allowed, $\mathcal{B}(\mathcal{P}^*) = \beta(\mathcal{P}^*)$.*
2. *Now consider partitions $\mathcal{P}$ with $\mathcal{P}^*$ as the only possible refinement. For any $a \in \beta(\mathcal{P})$ we say that $(a^*, \mathcal{P}^*)$ blocks $(a, \mathcal{P})$ if $a^* \in \mathcal{B}(\mathcal{P})$ and there exists a perpetrator S such that $u(a^*) >_S u(a)$.*
3. *Assume that for some $\mathcal{P}$ the set $\mathcal{B}(\mathcal{P}')$ has been defined for all $\mathcal{P}' \in \mathcal{R}(\mathcal{P})$ and that for each $a' \in \beta(\mathcal{P}')$ the set of outcomes $(u(a''), \mathcal{P}'')$ blocking $(u(a'), \mathcal{P}')$ has been defined.*
4. *Let $a \in \beta(\mathcal{P})$. Then $(a, \mathcal{P})$ is blocked by $(a', \mathcal{P}')$ if $\mathcal{P}' \in \mathcal{R}(\mathcal{P})$ and there exists a collection of perpetrators and residuals in the move from $\mathcal{P}$ to $\mathcal{P}'$ such that*
   a) *a' is a binding agreement for $\mathcal{P}'$,*
   b) *there is a* leading perpetrator *S, which gains from the move, that is, $u(a') >_S u(a)$,*
   c) *any re-merging, $\hat{\mathcal{P}}$ of the* other *perpetrators is blocked by $(a', \mathcal{P}')$ as well, with one of these perpetrators as a leading perpetrator. Formally, let $\mathcal{S}$ be the set of perpetrators other than S in the move from $\mathcal{P}$ to $\mathcal{P}'$. Then $\mathcal{B}(\hat{\mathcal{P}}) = \emptyset$ and there exists a strategy profile $\hat{a} \in \beta(\hat{\mathcal{P}})$ and $S' \in \mathcal{S}$, such that $(\hat{a}, \hat{\mathcal{P}})$ is blocked by $(a', \mathcal{P}')$ with S' as the leading perpetrator.*
5. *A strategy profile a is an* equilibrium binding agreement for *$\mathcal{P}$ if $a \in \beta(\mathcal{P})$ and there is no $(a', \mathcal{P}')$ that blocks $(a, \mathcal{P})$.*

There are obvious similarities between EBA's and the core, beginning with the recursive definitions. Deviations and blocks correspond to each other: in our approach deviating players "leave the room" and then we look at the reaction in the residual game.

However, while for equilibrium binding agreements only refining deviations are allowed and thus residual players are not allowed to coalesce with players from other coalitions this being one of the weaknesses of the concept, in our completely cooperative approach no such restrictions apply. Similarly, binding agreements are only safe against dismantling deviations the core has to be safe against deviations of all sorts, even against deviations that are subject to further deviations. These forces go in the opposite directions making the relation between the core and of the coarsest EBA's somewhat unclear.

In general a game cannot be represented in both forms, making a comparison even more difficult. Only normal form games with unique Nash-equilibria[5] for all partitions can be converted into the partition function form. On the other hand in this normal form game merging coalitions can keep their strategies, and get the same total payoff, hence all such games are superadditive by construction, which is not a restriction we have in our approach. So in this respect our approach is more general.

In the following we give some results about the relation of the core and the EBA's. Since EBA's are defined in an optimistic form, while a corresponding pessimistic form is easy to deduce we will also use the optimistic core for comparison.

Lemma 2. *Consider a game that exists both in normal and in partition function form. If an outcome in the grand coalition is blocked in the definition of Ray and Vohra [8], all outcomes with the grand coalition as partition are dominated in the corresponding partition function form game.*

Proof. Our proof is by induction.

1. The result is trivial for single-player games.
2. Assume that the lemma holds for all games with at most $n-1$ players.
3. Prove the lemma for a game with n-players.
   If the outcome is blocked, then there exists a perpetrator that gets better off *given* the EBA in the residual game.
   a) If the residual core of the corresponding deviation is empty, under the optimistic beliefs the deviating coalition expects at least this much, thus it, too, will deviate.
   b) If the residual core is not empty and it consists of outcomes with the grand coalition as partition, then – using our inductive assumption – the grand coalition is the coarsest residual EBA, hence the perpetrators have the same beliefs as the deviating coalition(s), and so deviation and blocking happens at the same time.
   c) If the residual core is not empty and contains also outcomes that have other partitions[6], then by the same argument the coarsest residual EBA is under the grand coalition. Then the beliefs formed by the deviating coalitions are at least as optimistic as those of the perpetrators.

We have checked all possibilities, and the proposition held in all cases. □

Lemma 3. *For normal form games with a well defined partition function form and only inefficient equilibrium binding agreements the (optimistic) core of the corresponding partition function game is empty.*

[5] This is to be understood to the extent of coalitional payoffs.

[6] By the superadditivity of the partition function, the core *will* contain outcomes with the grand coalition as partition.

Proof. players. All EBA's are inefficient if the binding agreements with the grand coalition as partition are blocked. Then, by Lemma 2 in the partition function form game all outcomes with the grand coalition as partition are also dominated. Then the outcomes of the grand coalition are not in the core. Recall that the partition function form is superadditive by construction. This implies that the core, if nonempty, contains the grand coalition. Hence the core is empty. □

This result, however, does not imply that the set of the coarsest EBA's would include the core. If the core is not empty, by Lemma 3 there exist efficient EBA's that is, there exists an EBA for the grand coalition, too, thus this is *the* coarsest EBA. Thus if the core contains outcomes with different partitions they will not be in the solution of Ray and Vohra [8].

4.5 An Example

We begin with a simple 4-player example with the following partition function, V:

$$\begin{aligned} V(123,4) &= (7,0) , \\ V(12,34) &= (4.4,4.4) , \\ V(12,3,4) &= (3,3,3) , \\ V(1,2,34) &= (3,3,3) , \\ V(1,2,3,4) &= (2,2,2,2) . \end{aligned}$$

We use an abbreviated notation: $(12,3,4)$ is a partition with three coalitions, $\{1,2\}$, $\{3\}$ and $\{4\}$. $V(12,3,4)$ is, in fact, a function, but we write out its values in the same order as the coalitions are arranged in the partition. Payoffs not indicated here are all 0. While the above game has been created to illustrate certain properties of the different approaches, it is not hard to imagine it as a model for a game with companies as players: Monopolies are prohibited by law so the grand coalition gets nothing; forming an alliance that controls the market brings large profits for the alliance and hurts the remaining players, otherwise alliances can get about the same profit. Certain coalitions are inhibited for instance due to geographical location or cultural differences.

Solutions with the Former Approaches

First we solve the game using the existing concepts. The worth of the deviations in the different approaches are summarised in the following table with the obvious notation. Of course we are only concerned with those deviations that have a positive payoff in at least one partition.

Inspecting Table 1 we find that the different approaches give, indeed very different expectations.

Table 1. Worths for the α-, ω-, r-, and γ-approaches

	W_α	W_ω	W_γ	W_r
{1}	0	3	2	2
{2}	0	3	2	2
{3}	0	3	2	2
{4}	0	3	2	0
{12}	3	4.4	3	3
{34}	3	4.4	3	3
{123}	7	7	7	7

The α-core contains outcome with partition $(12,3)$, where player 1 gets at least 2. In the γ- and r-cores we also have the assumption that player 2 gets at least 2, too.

In the case of the status quo approach we have to check each outcome (in fact, only each partition) and each deviation from it. Table 2 summarises these worths. Since any outcome with 0 payoffs is dominated via a deviation by (123), they are not listed. A dash indicates that the coalition is not a deviation for that partition.

Table 2. Worths for the status quo approach

W_{sq}	$(123,4)$	$(12,34)$	$(12,3,4)$	$(1,2,34)$	$(1,2,3,4)$
{1}	0	3	2	–	–
{2}	0	3	2	–	–
{3}	3	3	–	2	–
{4}	–	3	–	2	–
{12}	3	–	–	4.4	3
{34}	4.4	–	4.4	–	3
{123}	–	7	7	7	7

Finally we look at the EBA's. While the above partition function is not superadditive, and so it cannot be a representation of the Nash-equilibria of the normal form games with different coalition structures, it is still possible to apply the same technique, and get meaningful results. We find that the partition $(1,2,3,4)$ is obviously an EBA, moreover with certain payoff configurations the partitions $(12,34)$ and $(123,4)$ are not blocked.

The solutions under different outcomes are summarised in Table 3. Each row represents a partition. The outcomes belonging to a certain concept are the outcomes of the form $(x,\mathcal{P})$, where $\mathcal{P}$ is this partition and the payoff vector x satisfies the conditions listed in the respective column plus the conditions listed in the column "conditions". The only exception is the r-core: this is a core in the characteristic function game created from the original one. Here,

Table 3. Comparison of the existing solutions

partition	conditions	α-	ω-	sq-	γ- cores	r-	coarsest EBA's
$(123,4)$	$x_1+x_2+x_3=7$ $x_4=0$	$x_3 \geq 3$ $x_3 \leq 4$	$\emptyset$	$\emptyset$	$(2;2;3;0)$	$(2;2;3;0)$	$x_1,x_2,x_3 \geq 2$
$(12,34)$	$x_1+x_2=4.4$ $x_3+x_4=4.4$	$x_3 \geq 3$	$\emptyset$	$\emptyset$	$\emptyset$	$x_1,x_2 \geq 2$ $x_3 \geq 2.6$	$x_1,x_2 \geq 2$ $x_3,x_4 \geq 2$
$(1,2,34)$	$x_1=x_2=3$ $x_3+x_4=3$	$x_3 \geq 1$	$\emptyset$	$\emptyset$	$\emptyset$	$x_3 \geq 2$	

as a coalition's payoff cannot vary across partitions, it may differ from that of the original game. This is an example where this happens even for core outcomes. At places the payoff vector x is explicitly given. The sign $\emptyset$ means that the concept does not contain outcomes with that partition. We evaluate this table when we have our result as well.

Our New Approach.

Now we solve the game with the approach we propose. First we compute worths, then test outcomes against possible deviations. The outcomes that survive these tests are the core outcomes.

Computing Worths.

Since we allow multi-coalition deviations, they are really partitions, and so we use the same notation for them. For a four-player game there are a few dozen possible deviations. In this simple example, however only (1), (2), (3), (12), (34), $(1,2)$, $(1,3)$, $(1,4)$, $(2,3)$, $(2,4)$, $(3,4)$, (123), $(1,34)$, $(2,34)$, $(12,3)$, $(12,4)$, $(12,34)$, $(1,2,34)$, $(12,3,4)$, and $(1,2,3,4)$ may be profitable, as these are the only ones with an embedding partition (a partition containing them as subpartition) where all deviating coalitions have a positive payoff.

Since the payoff for a coalition depends on the entire coalition structure it is not, in general, possible to determine the post-deviation payoffs without reference to the remaining players. The only exception is when all players deviate. Here $W(C,\mathcal{P}) = V(C,\mathcal{P})$. So we have $W(12,34) = (4.9,4.9)$, $W(12,3,4) = (3,3,3) = W(1,2,34) = (3,3,3)$, $W(1,2,3,4) = (2,2,2,2)$.

If there is only one remaining player, it has only one possible "partition". We only need to add this last player as an additional coalition, and the result is the post-deviation partition. For instance, in the case of deviation (123) the remaining players' only possible partition is (4). We add this as an additional coalition to find partition $(123,4)$. Here the coalition (123) gets (7), so this is the post-deviation payoff. In a similar manner we get the following worths: $W(123) = (7)$, $W(1,34) = W(2,34) = W(12,3) = W(12,4) = (3,3)$.

Finally, for the rest of the deviations we have to look at the *residual game.* The partition function is defined conditional on the deviation. We do this by reconstructing the original partition (for each partition we add the deviating coalition(s)) and select the payoffs of the residual coalitions from among the payoffs for this partition.

As an example we consider deviation (34). The residual game has two players, 1 and 2. The payoff-vector for $(1, 2)$ and (12) are deduced from the payoff vectors of $(1, 2, 34)$ and $(12, 34)$, respectively. We have $V_3(12) = (4.4)$ and $V_3(1, 2) = (3, 3)$. From this example it is clear that given a deviation by (34), the remaining two players will not form a coalition/stay together, since this gives them 4.4, while otherwise they could get 3 each, which is strictly better for at least one of them. So, although (34) *could* get 4.4 by deviating, this expectation is unrealistic, and we can be certain that it will only get $W(34) = (3)$. Similar arguments help us to determine the worth of other deviations. For the deviation (4) we have to consider a 3-player residual game. The solution of a 3-player game is not so trivial: we have to define the partition function and solve it, just like the initial game. Eventually the results we get are: $W(1) = W(2) = W(3) = (2)$, $W(4) = (0)$, $W(12) = W(34) = (3)$, and $W(1, 2) = W(1, 3) = W(1, 4) = W(2, 3) = W(2, 4) = W(3, 4) = (2, 2)$.

Testing Deviations Against Outcome Sets.

Having the post-deviation payoffs for all possible deviations, we can test these against the outcomes. For instance the coalition $\{1, 2\}$ in $(12, 3, 4)$ can only divide a payoff of 3 between players 1 and 2, who could get 2 each by deviating. So all outcomes with partition $(12, 3, 4)$ (and any feasible payoff-configuration) are dominated. For simplicity we will just say that the *partition* $(12, 3, 4)$ is dominated. Then we can say that all partitions, having 0 as payoff for all of the coalitions, are dominated. The partition $(1, 2, 34)$ is dominated by (3) or (4). The partition $(1, 2, 3, 4)$ is dominated by $(12, 34)$ via $(12, 34)$, since $2 + 2 < 4.4$. Finally an outcome $(x, (123, 4))$ is dominated by $(12, 34)$ unless $x_1 + x_2 \geq 4.4$ or $x_3 + x_4 \geq 4.4$. Since $x_4 = 0$ the latter implies $x_3 \geq 4.4$. Since $x_1 + x_2 + x_3 = 7$ these imply that either $x_1 + x_2 \leq 2.6$ or $x_3 \leq 2.6$, and hence such an outcome is dominated by (12) or (34), each giving 3 to its members. Finally an outcome $(x, (12, 34))$ is not dominated iff $x_1, x_2, x_4 \geq 2$ and $x_3 \geq 2.6$. This is the set of core outcomes.

Comparison of Results.

Due to the simplicity of the game the core is independent of optimistic or pessimistic assumptions. While this does not hold in general, this is a large step from the often very different α- and ω- cores. Indeed, we find that while the ω-core is empty, the α-core is rather large, in fact, it contains most of the other core solutions.

The fact that the status quo-core is also empty implies that deviations without a reaction are likely to be profitable, and comparing it with other cores we may deduce that a residual reaction may harm deviating players in this game. Overlooking these makes deviating players too optimistic.

The r-core takes a similar approach to ours, but as i works via a characteristic function form, it does not, in fact cannot allow multi-coalition deviations. Hence the r-core is not tested against such deviations.

Finally the coarsest EBA's give a very similar equilibrium as ours, except that EBA's are not protected against deviations involving players from different coalitions, and hence the set of the coarsest EBA's is larger than our core.

5 A Core of an Economy with a Common Pool Resource

This section is devoted to our motivating application, a common pool resource game introduced by Funaki and Yamato [4]. Recall that the authors found contradictory results when using the cores with optimistic and pessimistic assumptions. Our aim is to reexamine this game using our more sophisticated concepts, hoping that our results will be conclusive. Our results are not general, indeed we find that the results depend on a number of parameters used, so we only give limited numerical calculations.

5.1 Introduction

In this game a society of fishermen share a lake. Each of them chooses his labour input (effort in fishing) in order to maximise his profit (the value of fish caught minus his effort investment). If we have decreasing returns to labour, in the lack of cooperation the lake becomes overfished and the tragedy of commons occurs while cooperatives or coalitions would choose a more efficient level. Given a coalition structure or partition of the players, we assume a noncooperative play among the coalitions.

Theorem 1 of [4] proves that there is a unique equilibrium, under which all cooperatives invest the same amount of labour. This amount determines the profits for the coalitions and hence a partition function can be defined. From this partition function they generate a characteristic function using optimistic and pessimistic expectations that correspond to the ω and the α approaches.

Funaki and Yamato [4] find that under optimistic assumptions the core is empty, while under pessimistic assumptions the tragedy of the commons can be avoided for sufficiently large games. They attribute this difference to the different behavioural assumptions. It is right to ask, if the different assumptions lead to different results, which one is right? In this section we apply our less sensitive concepts and form definitive conclusions for a number of cases. Due to the recursive definitions the conclusions depend on the results for small games, which are more problematic [4, p. 139], and hence only numeric examples are possible.

5.2 The Model

We overview the main elements of the model of Funaki and Yamato [4].

We consider $n \geq 2$ fishermen, $N = \{1, 2, \dots, n\}$. The amount of labour (effort) exerted by fisherman j to catch fish is denoted by $x_j \geq 0$. The production function f specifies the total amount of fish caught given the total effort $x_N = \sum_{j \in N} x_j$. We assume that $f(0) = 0$, $f'(x_N) > 0$, $f''(x_N) < 0$ and $\lim_{x_N \to \infty} f'(x_N) = 0$. The price of a fish is normalised to 1, and the personal cost of labour is q, such that $0 < q < f'(0)$. The income of a fisherman j is given by

$$m_j(x_1, x_2, \dots, x_n) = \frac{x_j}{x_N} f(x_N) - q x_j \ ,$$

where $m_j(0, 0, \dots, 0) = 0$.

Lemma 4 (Funaki and Yamato [4, theorem 1]). *For any partition $\mathcal{P} = \{C_1, C_2, \dots, C_k\}$, there exists a unique equilibrium $\left(x^*_{C_1}, x^*_{C_2}, \dots, x^*_{C_k}\right)$ given by*

$$f'(x^*_N) + (k-1) f(x^*_N)/x^*_N = kq \ , \tag{1}$$

$$x^*_{C_i} = x^*_N / k \quad \text{for all } i = 1, \dots, k, \text{ and} \tag{2}$$

$$x^*_{C_i} > 0 \quad \text{for all } i = 1, \dots, k \ , \tag{3}$$

where $x^*_N = \sum_{i=1}^{k} x^*_i$.

Thus given a coalition structure the equilibrium effort and hence the equilibrium payoffs are unique. This implies that we can construct a partition function and it is well defined.

Moreover, given the coalition structure the payoff distribution is also determined, the payoff of an individual player is given by the *size* of the coalition it belongs to and the *sizes* of other coalitions. For this reason in the sequel coalitions will be denoted by their size.

5.3 Simulations

We examine three simple types of production functions, all of which satisfy the conditions above. Depending on the choice for q and a parameter γ (where applicable) the tragedy of the commons can be avoided for some games. The three functions considered are the following:

$$f_1(x) = x^\gamma \ , \tag{4}$$

$$f_2(x) = 1 - e^{-x} \ , \tag{5}$$

$$f_3(x) = \frac{1}{\gamma}(x+1)^\gamma - \frac{1}{\gamma} \ . \tag{6}$$

The Function $f_1(x) = x^\gamma$.

This is a function that also Funaki and Yamato [4] use to illustrate that different γ values can lead to different results for 3-player games. We show the calculation of the core for a specific example and give a summary of the results.

The game we look at first is a 4-player game with $\gamma = 0.2$ and $q = 0.5$. As x tends to 0, $f_1'(x)$ tends to infinity so our condition on q is satisfied. In this game the maximum number of coalitions is 4, so we have to solve 1 for $k = 1, 2, 3$ and 4. The respective x_N^* values are 0.318, 1.256, 1.614, and 1.799. Table 4 summarises the payoffs.

Table 4. Payoffs for a 4-player game with $f_1(x)$

Partition	coalitional payoff	coalition size	per-member payoff
(4)	0.636	4	0.159
(3,1)	0.209	3	0.070
(3,1)	0.209	1	0.209
(2,2)	0.209	2	0.105
(2,1,1)	0.098	2	0.049
(2,1,1)	0.098	1	0.098
(1,1,1,1)	0.056	1	0.056

We are interested in the stability of the grand coalition, so we test the membership of only this coalition structure in the core. In the grand coalition each player gets 0.159. The cooperation is threatened only by deviations that give more to the deviating players. In our example there is only one such possibility, the singleton in the partition $(3, 1)$ gets 0.209. Under the overly optimistic ω-assumption he would hope to get this amount, but is this belief realistic? In Table 5 we look at payoffs in the residual game consisting of the remaining 3 players:

Table 5. Payoffs for a 3-player residual game

Coalition size	partition by size	value	per-member value
3	(3)	0.209	0.070
2	(2,1)	0.098	0.049
1	(2,1)	0.098	0.098
1	(1,1,1)	0.056	0.056

Now we check the membership of the (residual) grand coalition in the residual core. Should the grand coalition form, each player would get 0.070. There is only one, possibly profitable deviation, namely by a singleton.

This deviation is profitable if and only if the residual players – now only two of them – stay together and thus get 0.049 each, while breaking up would give them 0.056. Clearly they will break up, and hence the deviation by the singleton is not profitable. Hence, the grand coalition is in the core of the 3-player residual game. Moreover, implies that a deviation by a singleton in the original, 4-player game is also profitable. Hence the grand coalition is not in the core, and the tragedy of the commons cannot be avoided[7].

A similar argument gives the same conclusion for other values of γ and q for all four or five-player games.

The Function $f_2(x) = 1 - e^{-x}$.

In this case we have looked at 3, 4 and 5-player games, and found that for 5-player games the grand coalition does not belong to the core. For 3 and 4 player games there exists cutoff values so that if $q < \widehat{q}$ then the grand coalition belongs to the core, while if $q > \widehat{q}$ it does not. We found $\widehat{q}_3 = 0.213$ and $\widehat{q}_4 = 0.475$.

The Function $f_3(x) = \frac{1}{\gamma}(x+1)^{\gamma} - \frac{1}{\gamma}$.

In this case we have an extra parameter, γ and we find that depending on γ a similar cutoff value may exist, although here it works in the other direction: for a given γ if $q > \widehat{q}$ then the grand coalition belongs to the core, while if $q < \widehat{q}$ it does not. In particular, where the cutoff-value is 0 the grand coalition will be stable for all values of q. Table 6 and Fig. 1 summarise our findings.

Table 6. Critical values for q

γ	$\widehat{q}_3$	$\widehat{q}_4$	$\widehat{q}_5$
0.1	0.191	0.617	0.243
0.3	0.175	0.645	0.243
0.5	0.121	0.687	0.240
0.584	0	0.711	0.223
0.7	0	0.754	0.082
0.704	0	0.756	0
0.9	0	0.881	0

[7] We must also note, however, that after the separation of a single player the rest of the players stay together, and hence the overfishing is only to the extent of about 30%. This structure does of course exhibit the usual instability of a cartel system.

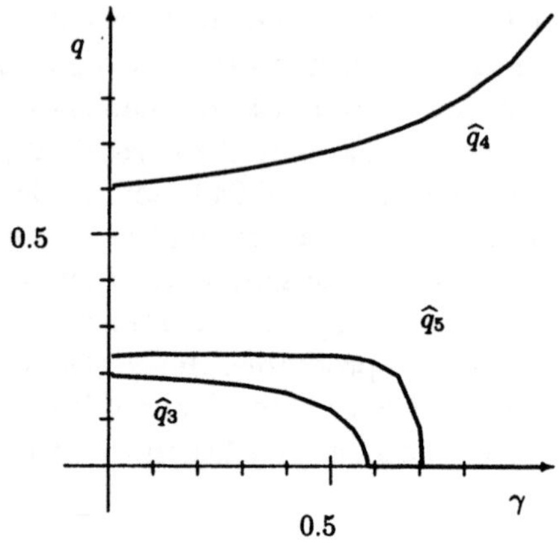

Fig. 1. Critical values for q as a function of γ.

5.4 Evaluation of the Results

We have found that unlike with the α and the ω approaches, the stability of the grand coalition depends highly on the function considered and the parameters used. While our results are far less general than that of Funaki and Yamato, in these examples our approach proved to be not only less sensitive, but insensitive to the approach taken. This means that instead of beliefs, we have now certainty, which stands in strong contrast with the inconclusive results when using the α and ω approaches.

6 Extensions

Its bias towards to grand coalition makes the concept of Ray and Vohra [8] less suitable for extension to non-superadditive games. On the other hand we see no difficulty to extend our approach to games in normal form or even to more general forms that capture the advantages of both forms. Since such an extension is straightforward intuitively, but cumbersome notationally, we prefer to give an outline only.

Firstly note that our approach fails when there are multiple Nash-equilibria in the non-cooperative part of the game, as this inhibits us from defining the partition function. So skip this step and go straight to the solution concept. First redefine deviations. What we had so far was that a deviating coalition gathered all possible future scenarios, and compared these with the current one. If one of them, or all of them (in the optimistic, and pessimistic cases respectively) were better than the current one, the outcome is dominated. There is no reason to restrict ourselves to include only one scenario for each partition. Indeed, if there are multiple Nash-equilibria, all should be included in the set of possible scenarios.

We can go even further. If we fancy to use other than Nash-equilibria to solve our non-cooperative game, we may do so. Should there be no equilibria, we may even include all possible combinations of strategies by the same argument that we used when the residual core is empty.

Finally, we propose a way to combine the strengths of the two approaches. Arguments against superadditivity are well known and the present normal form game is superadditive by construction. Just as the original definition of the partition function assigns a *characteristic function* to each partition, we may assign an *normal form game* to each partition with the coalitions as composite players.

References

1. d'Aspremont, C., Jacquemin, A., Gabszewicz, J. J., Weymark, J. A. (1983): The Stability of Collusive Prive Leadership. Canadian J. Economics **XVI**, 17–25
2. Aumann, R. J., Peleg, B. (1960): Von Neumann-Morgenstern Solutions to Cooperative Games without Side Payments. Bull. Am. Math. Soc. **66**, 173–179
3. Chander, P., Tulkens, H. (1994): The Core of and Economy with Multilateral Environmental Externalities. Nota di Lavoro **69.94**. Fondazione Eni Enrico Mattei, Milano
4. Cornet, M. F. (1998): Game Theoretic Models of Bargaining and Externalities. Tinbergen Inst. Res. Ser: 176. Thesis Publishers, Amsterdam
5. Funaki, Y., Yamato, T. (1999): The Core of an Economy with a Common Pool Resource: A Partition Function Form Approach. Int. J. Game Theory **28(2)**, 157–171
6. Greenberg, J. (1994): Coalition Structures. In: Aumann, R. J., Hart, S. (Eds.): Handbook of Game theory. Elsevier Science Publications B. V., 1305–1337.
7. Huang, C. Y., Sjostrom, T. (2001): Consistent Solutions for Cooperative Games with Externalities (mimeo)
8. Ray, D., Vohra, R. (1997): Equilibrium Binding Agreements. J. Econ. Theory **73**, 30–78
9. Tulkens, H., Chander, P. (1997): The Core of and Economy with Multilateral Environmental Externalities. Int. J. Game Theory **26(3)** 379–401

Network Topology and the Efficiency of Equilibrium

Igal Milchtaich

Department of Economics, Bar-Ilan University,
52900 Ramat-Gan, Israel
E-mail: milchti@mail.biu.ac.il

Summary. Different networks, such as transportation, communication, computer, and supply networks, are susceptible to similar kinds of inefficiencies. These arise when congestion externalities render each user's cost dependent on the other users' choice of routes. If each user chooses the least expensive (e. g., fastest) route from the users' common point of origin to the common destination, the result may be inefficient in the sense that there is an alternative choice of routes that reduces the costs for all users. However, this may happen only for certain kinds of network topologies. This paper gives several alternative characterizations of networks in which inefficiencies may occur. In particular, a necessary and sufficient condition for inefficiency is that specific simple network is embedded in the network.

Key words: Congestion, externalities, network topology, Braess's paradox, transportation networks, Wardrop equilibrium

1 Introduction

In transportation and other kinds of networks, congestion externalities are a potential source of inefficiency. A remarkable example of this, known as Braess's paradox (see, for example, [4]), is shown in Figure 1. Cars arrive at a constant rate at vertex o of the depicted network and leave it at vertex d. The network consists of three fast roads (e_1, e_4, and e_5) and two slow detours (e_2 and e_3). The travel time on each road is an increasing function of the flow through it, or the average number of vehicles passing a fixed point in the road per unit of time. (This is a reasonable assumption if the density of vehicles on the road is relatively low, so that the flow is well below the road's capacity. See Chapter 13 and Figure 1.8 in [23].) However, regardless of the flow, the travel time on the route consisting of the three fast roads is shorter than on any of the alternative routes. Therefore, in equilibrium, all vehicles use this route. The travel time on the network is then 21 minutes. Suppose, however, that the transverse road, e_5, is closed, or its physical condition deteriorates to the point at which the travel time on it becomes similar to that on each of the two detours. The road's new cost curve is higher than the old one: the travel time corresponding to every flow through e_5 is longer than before. As a result of the change in costs, the old equilibrium is replaced by a new one, in which the transverse road is not used at all: half the vehicles go through the left route (e_1 and e_3), and half through the right route (e_2 and e_4). Paradoxically, the new travel time is shorter than before, 20 minutes. The reason for this is that the motorists' choice of routes is only guided by concern for their own good; they do not consider others' welfare. This selfish attitude results in an overuse of the fast roads, and consequently an inefficient equilibrium. As Newell [19] and Sheffi [23] point out, traffic engineers have known for a long time that restrictions of travel choices and reductions in capacity may improve the flow on the network as a whole. This, for instance, is the underlying principle behind many traffic control schemes, such as ramp metering on freeway entrances [23, p. 77].[1]

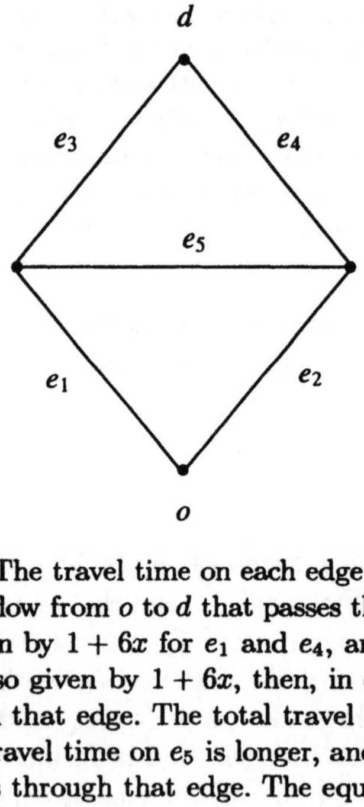

Fig. 1. Braess's paradox. The travel time on each edge is an increasing function of the fraction x of the total flow from o to d that passes through the edge. The travel times, in minutes, are given by $1 + 6x$ for e_1 and e_4, and $15 + 2x$ for e_2 and e_3. If the travel time on e_5 is also given by $1 + 6x$, then, in equilibrium, the entire flow from o to d passes through that edge. The total travel time from o to d is then 21 minutes. If, however, the travel time on e_5 is longer, and given by $15 + 2x$, then, in equilibrium, no flow passes through that edge. The equilibrium travel time is then *shorter*, 20 minutes

[1] When the cost curves are *linear*, as they are in Figure 1, there is a limit to how much can be gained from raising them. Specifically, Roughgarden and Tardos [22] show that, in the linear case, the maximum decrease in equilibrium cost that can be achieved this way does not exceed 25 %.

Braess's paradox is not limited to transportation networks only. There is by now a moderately large literature showing that this or similar paradoxes may also occur in such diverse networks as computer and telecommunication networks, electric circuits, and mechanical systems. Remarkably, much of this literature (e. g., [2, 8, 9]) is concerned with the same network shown in Figure 1, the Wheatstone network. It turns out, that there is a good reason for this. As this paper shows, the Wheatstone network is essentially the *only* one in which Braess's paradox can occur. More precisely, a necessary and sufficient condition for the existence of *some* cost function for which the paradox occurs is that the network has an embedded Wheatstone network. In networks without this property, so-called series-parallel networks, Braess's paradox cannot occur. Several alternative characterizations of series-parallel networks are given below.

The emphasis in this paper is on network topologies for which some cost functions giving rise to inefficiencies exist. Other papers, by contrast, put the emphasis on the cost functions themselves. For example, Steinberg and Zangwill [24] and Dafermos and Nagurney [10] derive formulas yielding, under certain conditions, the change in users' costs induced by the creation of additional routes. These formulas can, in principle, be used to determine whether a form of Braess's paradox occurs in the network. They are, however, rather complicated. Calvert and Keady [7] consider the total power loss in a network in which the potential drop across the two endpoints of each edge is an increasing function of the quantity obtained by dividing the edge flow by some edge-specific conductivity factor. They show that if the functional relation between this quantity and the potential drop is given by a power function, which is the same for all edges, then, when one or more of the conductivity factors is increased, with the total flow through the network kept constant, the total power loss either decreases or remains the same. Conversely, if the functions are *not* all equal to some power function, and there are at least six edges, it is possible to arrange the edges so that if one of the conductivity factors is increased, the total power loss also increases. Thus, a version of Braess's paradox occurs in the network. Calvert and Keady [7] also give a topological result (Theorem 11), which maintains that, in a series-parallel network, this phenomenon cannot occur. In such a network, the total power loss weakly decreases when any of the conductivity factors increases.

Braess's paradox is not the only kind of inefficiency caused by congestion externalities. Consider, for example, the series-parallel network in Figure 2(a), which represents the alternatives faced by weekend visitors to a certain seaside town, where the only attractions are the two nearby beaches. The two edges joining o and ν represent the alternatives of going to the North Beach (e_1) or the South Beach (e_2) on Saturday. The two edges joining ν and d represent the same two alternatives on Sunday. The South Beach is more remote, and so the cost of getting there is two units higher than for the North Beach. On the other hand, it is a longer beach, and therefore does not get crowded as fast. However, the additional pleasure of spending the day on an uncrowded beach never exceeds the difference in travel costs. Therefore, in equilibrium, all the visitors go to the North Beach, both on Saturday and on Sunday. Crowding then costs each person four units of pleasure. However, if people were taking turns in going to the South Beach, half of them going there on Saturday and the other half on Sunday, then the cost for *all* individuals would be lower, 3.5. Thus, this arrangement, which is not an equilibrium, represents a Pareto improvement over the equilibrium. The difference between this example and the one above is that, in the case of Braess's paradox, Pareto improvement results from increasing the costs of certain facilities (e. g., increasing the travel time on the transverse road in Figure 1), thereby creating a new equilibrium that is better for everyone. In the present example, by contrast, it is not possible to make everybody better off simply by increasing the costs (e. g., charging congestion-dependent entry fees to beaches). Since the networks in Figure 2 are series-parallel, Braess's paradox cannot occur, and therefore any Pareto improvement must involve non-equilibrium behavior.

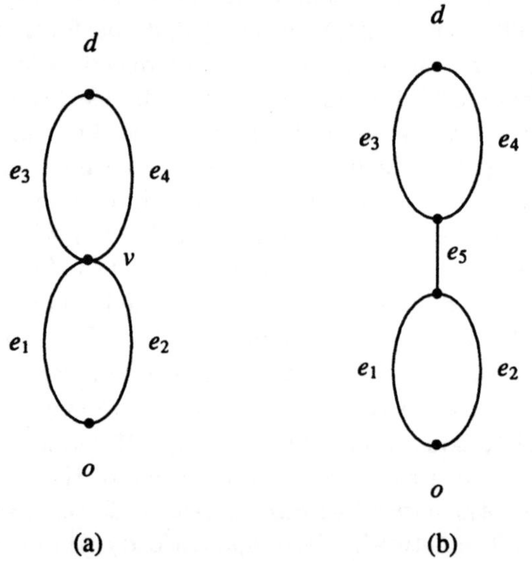

Fig. 2. Another kind of inefficiency caused by congestion externalities. The cost of each edge in network (a) is an increasing function of the fraction x of the total flow from o to d that passes through the edge. For e_1 and e_3, the cost is given by $2x$. For e_2 and e_4, it is $2+x$. In equilibrium, only e_1 and e_3 are used, and the equilibrium cost is 4. However, this outcome is inefficient. Splitting the flow, so that half of it goes through e_1 and e_4 and half through e_2 and e_3, would reduce the cost to 3.5. A similar phenomenon occurs in network (b). Indeed, since all the routes from o to d pass through the middle edge e_5, the cost of this edge is immaterial

One of the main results of this paper is that the three graphs in Figures 1 and 2 essentially represent the *only* kinds of network topologies in which congestion externalities may lead to Pareto inefficient equilibria. For example, inefficiencies never arise in a network like that in Figure 3. The crucial difference between this network and the previously mentioned ones is that routes in

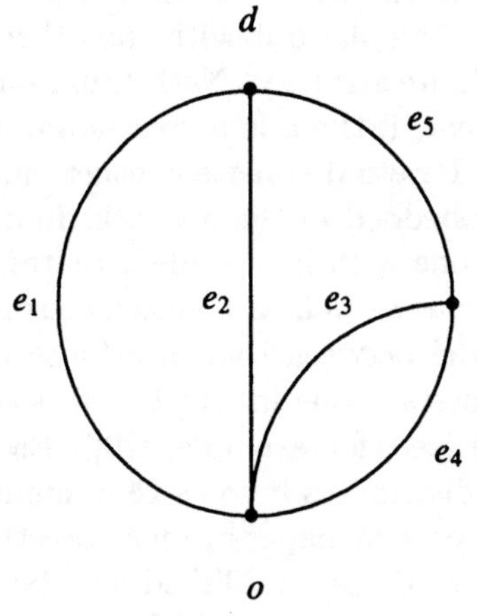

Fig. 3. A network with independent routes

this network are independent in the sense that each route contains at least one edge that is not part of any other route. The first paper to show a connection between the independence of the routes in a network (or, rather, a property equivalent to it) and the efficiency of the equilibria is [11]. Holzman's and Law-Yone's paper considers, in fact, a larger class of strategy spaces, of which routes in networks constitute a sub-class. An explicit treatment of transportation networks can be found in [14]. For networks, the main result of Holzman and Law-Yone [11] can be stated as follows. If the network has independent routes, then, for every cost function, all equilibria are weakly Pareto efficient and, moreover, are strong in the sense that no group of users can make all its members better off by changing their choice of routes. Conversely, if routes are not independent, then there is a cost function for which none of the equilibria is even weakly Pareto efficient. The main difference between [11] and the present paper is that these authors consider games with a *finite* number of players, each of whom has a non-negligible effect on the others. The present paper, by contrast, assumes there is a *continuum* of users. This may be viewed as a mathematical idealization of a very large population of individuals, each with a nearly negligible ability to affect the others. There are several substantial differences between the cases of finite and infinite populations. One difference is that, in the latter but not in the former case, the connection between the independence of the routes and the efficiency of the equilibria also holds for heterogeneous populations, in which not all users have the same cost function.

Users may differ in the innate quality they assign to the various alternatives or in the degree by which they are affected by congestion. For example, some motorists may be concerned primarily with the travel time, and others with the distance traveled. In the case of finite populations, heterogeneity is a potential source of inefficiency. This is demonstrated by the simple two-user, two-facility example in which each user has a different favorite facility, but would rather *not* use it than share it with the other user (see [15]). In this example, there are two (pure-strategy) Nash equilibria, one of which is worse for *both* users. By contrast, if there is a continuum of users, then one Nash equilibrium may strictly Pareto dominate another only if one of the networks in Figures 1 and 2 is embedded in the network. In a network that does not have this property, i. e., one with independent routes, all Nash equilibria are efficient – for heterogeneous as well as homogeneous populations.

An intermediate model between that of a finite population of users and that of a continuum of users is one in which flow is continuous but the population of users is finite (see, for example, [20]). Each user controls a fixed portion of the flow and distributes it so as to minimize the user's total cost. While the results of the present paper are not directly applicable, it is likely that connections similar to those established here between network topology and the efficiency of the equilibria also hold for this model. Yet another line of investigation concerns the *social optimality* of the equilibria, i. e., whether, in equilibrium, the *total* cost is minimized – a condition more demanding than Pareto efficiency. This question may not be answered by only considering the network topology; it is also necessary to know the functional form of the cost functions. For networks that consist of several edges connected in parallel, conditions for social optimality of the equilibria are given in [17].

2 Graph Theoretic Preliminaries

2.1 Basic Terminology

An *undirected multigraph* consists of a finite set V of *vertices* together with a finite set E of *edges*. Each edge e is associated with an unordered pair $\{u, \nu\}$ of distinct vertices, which the edge is said to *join*. These are called the *end vertices* of e. Thus, loops are not allowed, but more than one edge can join two vertices. An edge e and a vertex ν are said to be *incident* with each other if ν is an end vertex of e. A *walk* of length n ($n \geq 0$) is an alternating sequence s of vertices and edges $\nu_0, e_1, \nu_1, \ldots, \nu_{n-1}, e_n, \nu_n$, beginning and ending with vertices, in which each edge is incident with the two vertices immediately preceding and following it. The vertices ν_0 and ν_n are called the *initial* and *terminal* vertices of s, respectively. The walk $\nu_n, e_n, \nu_{n-1}, \ldots, \nu_1, e_1, \nu_0$, which includes the same vertices and edges as s but passes them in reverse order, is denoted $-s$. If t is a walk of the form $\nu_n, e_{n+1}, \nu_{n+1}, \ldots, \nu_{m-1}, e_m, \nu_m$, the initial vertex of which is the same as the terminal vertex of s, then $\nu_0, e_1, \nu_1, \ldots, \nu_{n-1}, e_n, \nu_n, e_{n+1}, \nu_{n+1}, \ldots, \nu_{m-1}$, e_m, ν_m is also a walk, denoted $s + t$. A *section* of s is any walk of the form

$\nu_{n_1}, e_{n_1+1}, \nu_{n_1+1}, \ldots, \nu_{n_2-1}, e_{n_2}, \nu_{n_2}$, with $0 \leq n_1 \leq n_2 \leq n$. If all the vertices (and, hence, all the edges) in a walk s are distinct, then each section of s is uniquely identified by its initial vertex u and terminal vertex ν, and may therefore be denoted by $s_{u\nu}$. If the section is of length zero, i. e., has no edges, then u and ν coincide. If it is of length one, i. e., has a single edge, then u and ν are the two end vertices of this edge. In this case, the section may be viewed as an indication of the direction in which s passes through the edge.

2.2 Two-Terminal Networks

A *two-terminal network* (*network*, for short) is an undirected multigraph together with a distinguished ordered pair of distinct vertices, o (for "origin") and d (for "destination"), such that each vertex and each edge belong to at least one walk in which the initial vertex is o, the terminal vertex is d, and all the vertices are distinct. Any walk with these properties will be called a *route*. The set of all routes in a network is denoted $\mathcal{R}$.

Two networks $G'$and G'' will be said to be *isomorphic* if there is a one-to-one correspondence between the vertices of G' and G'' and between their edges such that (i) the incidence relation is preserved and (ii) the origin and destination in G' are paired with the origin and destination in G'', respectively. A network G' is *embedded* in a network G'' if G'' is isomorphic to G' or to a network obtained from it by a sequence of operations of the following three types (see Figure 4).

1. The subdivision of an edge; i. e., its replacement by two edges with a single common end vertex.
2. The addition of a new edge that joins two existing vertices.
3. The "extension" of the origin or the destination; i. e., the addition of a new edge that joins o or d and another, new vertex, which becomes the new origin or destination, respectively.

Two networks G' and G'' with the same origin-destination pair, but no other common vertices or edges, may be connected *in parallel*. The set of vertices in the resulting network G is the union of the sets of vertices in G' and G'', and similarly for the set of edges. The origin and destination in G are the same as in G' and G''. Two networks G' and G'' with a single common vertex (and, hence, without common edges), which is the destination in G' and the origin in G'', may be connected *in series*. The set of vertices in the resulting network G is the union of the sets of vertices in G' and G'', and similarly for the set of edges. The origin in G coincides with the origin in G', and the destination is the destination in G''.

A network is said to be *series-parallel* if no two routes pass through any edge in opposite directions. The two networks in Figure 2 are series-parallel. The Wheatstone network in Figure 1 is not series-parallel, since there are two routes passing through e_5 in opposite directions. In fact, as the following

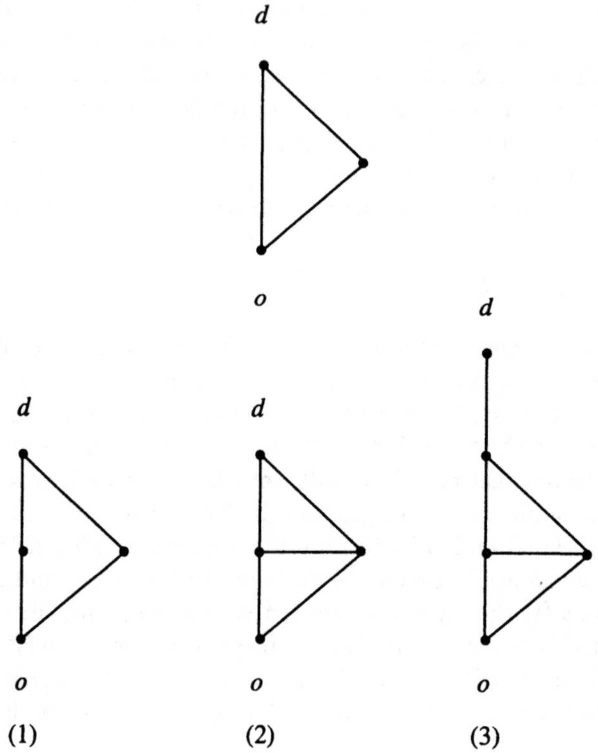

Fig. 4. The upper network is embedded in each of the three lower ones. These are obtained from it by carrying out in sequence the operations of (1) subdividing an existing edge, (2) adding a new edge, and (3) extending the destination

proposition shows, the Wheatstone network is part of *any* non-series-parallel network. This result is very similar to one of Duffin [11, Theorem 1].

Proposition 1. *A network G is series-parallel if and only if the network in Figure 1 is not embedded in it.*

As noted by Riordan and Shannon [21], series-parallel networks can also be defined recursively: a network is series-parallel if and only if it can be constructed from single edges by carrying out any number of times the operations of connecting networks in series or in parallel. Hence the term "series-parallel". The following proposition establishes this.

Proposition 2. *A network G is series-parallel if and only if*
(i) it has a single edge only, or
(ii) it is the result of connecting two series-parallel networks in parallel, or
(iii) it is the result of connecting two series-parallel networks in series.

One corollary of Proposition 2 is that every series-parallel network is planar and, moreover, remains so when a new edge, joining o and d, is added to it.

Equivalently, every series-parallel network can be embedded in the plane in such a way that o and d lie on the exterior face, or boundary. Using Proposition 2, this corollary can easily be proved by induction on the number of edges. A practical way of verifying that a given network is series-parallel is suggested by the third condition in the following proposition.

Proposition 3. *For every network G, the following three conditions are equivalent:*

(*i*) *G is series-parallel;*

(*ii*) *for every pair of distinct vertices u and ν, if u precedes ν in some route r containing both vertices, then u precedes ν in all such routes;*

(*iii*) *the vertices can be indexed in such a way that, along each route, they have increasing indices.*

A network with *independent routes* is one in which every route contains at least one edge that does not belong to any other route. An example of a network with independent routes is shown in Figure 3. As shown below, a network with independent routes is necessarily series-parallel. The converse, however, is false. For example, the two networks in Figure 2 are series-parallel but the routes in them are not independent. In fact, it follows from Proposition 1 and the next proposition that at least one of these networks is part of *any* series-parallel network that does not have independent routes.

Proposition 4. *A network G is a network with independent routes if and only if none of the networks in Figures 1 and 2 is embedded in it.*

The result that every network with independent routes is series-parallel follows as an immediate corollary from Propositions 1 and 4. It can also be deduced from the following recursive characterization of networks with independent routes, due to Law-Yone [14]. This characterization differs from the one for series-parallel networks (Proposition 2) only in condition (*iii*), which is stronger than the corresponding condition there.

Proposition 5. *A network G is a network with independent routes if and only if*

(*i*) *it has a single edge only, or*

(*ii*) *it is the result of connecting in parallel two networks with independent routes, or*

(*iii*) *it is the result of connecting in series a network with independent routes and a network with a single edge.*

Since, in every network, each route r has a unique set of edges, it may be identified with a unique binary vector, coordinate e of which is 1 if edge e belongs to r and 0 otherwise. This vector can be viewed as an element of the vector space $\mathbb{Z}_2^{|E|}$, where $|E|$ is the number of edges in the network and $\mathbb{Z}_2$ is the field of the integers modulo 2. A set of routes will be said to be *linearly independent* if the corresponding set of vectors is linearly independent in $\mathbb{Z}_2^{|E|}$.

Equivalently, a set of routes is linearly independent if it is not possible to write one of the corresponding vectors as the (component-wise) sum modulo 2 of some of the others. As the following proposition shows, networks with independent routes are characterized by the property that the set of *all* routes is linearly independent.[2] Hence the term. Another property characterizing networks with independent routes is that the set of all routes does not contain a "bad configuration" [11]. A *bad configuration* is a triplet of routes such that the first route contains some edge e_1 that does not belong to the second route, the second route contains some edge e_2 that does not belong to the first route, and the third route contains both e_1 and e_2. Still another equivalent property is that pairs of routes never merge only in their middle; in other words, any common section must extend to either o or d.

Proposition 6. *For every network G, the following four conditions are equivalent.*

(*i*) *The set $\mathcal{R}$ of all routes in G is linearly independent.*
(*ii*) *A triplet of routes constituting a bad configuration does not exist.*
(*iii*) *For every pair of distinct routes r and s and every vertex ν common to both routes, either the section $r_{o\nu}$ (which consists of ν and all the vertices and edges preceding it in r) is equal to $s_{o\nu}$, or $r_{\nu d}$ is equal to $s_{\nu d}$.*
(*iv*) *G is a network with independent routes.*

3 Flows and Costs

A flow in a network is a specification of a nonnegative *route flow* f_r for each route r. This can be written in the form of a *flow vector* $\mathbf{f}$ (of dimension equal to the number of routes in the network), coordinate r of which is f_r. Given the flow vector, the flow f_s through any walk s is defined as the total flow in all the routes of which s is a section:

$$f_s = \sum_{\substack{r \in \mathcal{R} \\ s \text{ is a section of } r}} f_r \, . \tag{1}$$

If s is a walk of length zero, consisting of a single vertex, then f_s represents the total flow in all the routes passing through this vertex. In particular, the flow through the origin,

$$f_o = \sum_{r \in \mathcal{R}} f_r \, ,$$

represents the *total origin-destination flow*. It is equal, of course, to f_d, the flow through the destination. If s is a walk of length one, consisting of a single

[2] Note that linear independence is defined with respect to $\mathbb{Z}_2$, not (the real field) $\mathbb{R}$. For example, the network in Figure 1 is not a network with independent routes, despite the fact that vectors representing its four routes are linearly independent in $\mathbb{R}^{|E|}$. However, each of these vectors is equal to the sum modulo 2 of the other three.

edge and its two end vertices, then f_s represents the total flow through the edge in a the direction indicated by s; this will be referred to as an *edge flow*. Note that each edge is associated with a *pair* of edge flows, one in each direction. However, in a series-parallel network, in which all routes pass through an edge in the same direction, only one of these can ever be positive. In a network with independent routes, in which each route contains at least one edge that does not belong to any other route, the edge flows uniquely determine the flow vector.

It should be emphasized that, in this paper, flow is always assumed to originate in a single vertex, o, and terminate in a single vertex, d. Multiple origin-destination pairs are not allowed. This restriction can be partially circumvented by connecting all sources to a single, artificial vertex, from which all flow is assumed to originate, and similarly for the sinks. However, such a construction substantially alters the network topology.

A *cost function* is a vector-valued function $\mathbf{c}$ specifying the cost $c_s(\mathbf{f})$ of each walk s as a function of the (whole) flow vector $\mathbf{f}$.[3] This cost is assumed to satisfy the following monotonicity condition: for every pair of flow vectors $\hat{\mathbf{f}}$ and $\tilde{\mathbf{f}}$, if $\hat{f}_t > \tilde{f}_t$ and $\hat{f}_{-t} \geq \tilde{f}_{-t}$ for all sections t of the walk s, then $c_s(\hat{\mathbf{f}}) \geq c_s(\tilde{\mathbf{f}})$.[4] This implies, in particular, that the cost of a walk only depends on the flow through each of its sections and the flows in the opposite directions. In general, the cost may remain the same even if these flows increase. A cost function $\mathbf{c}$ will be said to be *increasing* if it satisfies the following additional condition: for every route r and every pair of flow vectors $\hat{\mathbf{f}}$ and $\tilde{\mathbf{f}}$, if $\hat{f}_s \geq \tilde{f}_s$ and $\hat{f}_{-s} \geq \tilde{f}_{-s}$ for all sections s of r, and there is at least one section s of length one for which $\hat{f}_s > \tilde{f}_s$, then $c_r(\hat{\mathbf{f}}) > c_r(\tilde{\mathbf{f}})$. A cost function $\mathbf{c}$ will be said to be *additively separable* if, for every route r, every pair of distinct vertices u and v such that u precedes v in r, and every flow vector $\mathbf{f}$, $c_{r_{ov}}(\mathbf{f}) = c_{r_{ou}}(\mathbf{f}) + c_{r_{uv}}(\mathbf{f})$. Additive separability entails that the cost of each route is the sum of the costs of its edges. Note that the cost of an edge generally depends on the *direction* in which it is passed through. However, in each direction, the cost is only a

[3] Note that all walks are assigned costs, not just routes. Thus, the vectors in the domain and in the range of $\mathbf{c}$ have different dimensions. These are equal to the number of routes and to the number of walks in the network, *respectively*. Note also that the costs are not *assumed* to be nonnegative. However, they may be *thought of* as such. Indeed, the assumption that costs cannot be negative is implicit in the definition of equilibrium (in the next section), which only considers routes, i. e., walks from the origin to the destination that do not pass through any vertex more than once.

[4] This condition is rather weak. Since it involves a potentially long list of assumptions, the set of flow vector pairs to which it applies is relatively small. Stronger, and perhaps more intuitive, monotonicity conditions could be used instead. For example, it could be required that the cost of a walk can decrease only if one or more of relevant *edge* flows decrease. However, a weaker definition here makes for stronger results, and is therefore preferable from a methodological point of view. The same remark also applies to the two definitions that follow.

function of the flows through the edge in that direction and in the opposite direction and (possibly) the flows through the end vertices. In a series-parallel network, all routes pass through an edge in the same direction, and therefore the cost of passing through it in the opposite direction, as well as the effect on the cost of the flow in the opposite direction, are immaterial.

In models of transportation networks, the assumption that the cost function is additively separable is customary. Moreover, it is often assumed that there is only one direction in which each edge can be passed through (see, e. g., [23, 6]). Thus, a two-way highway is described by a *pair* of edges. Correspondingly, the description of a transportation network typically involves these two kinds of data: a *directed graph*, which describes both the physical network and the directions in which individual edges may be traveled on; and a system of associated edge cost functions, which gives the travel time on each edge as a function of the edge flow. The model presented in [5] differs in that it assumes all roads to be two-way. Moreover, the cost of traveling on a road in both directions is assumed equal, and is only a function of the sum of the flows in all routes passing through the road in *either* direction. The model in the present paper subsumes both these models. Here, each edge is associated with a *pair* of cost functions – one for each direction. The first transportation model described above corresponds to a case in which the cost of passing through an edge in a particular direction is prohibitively high. The second model corresponds to a case in which the two costs are equal, and only depend on the sum of the edge flows in both directions.

The fact that the present model does not assume additive separability makes it somewhat more general than other models. For example, turning restrictions can be incorporated simply by assigning very high (or infinite) cost to certain routes. The same applies to the possibility that route costs are affected by the flow through their vertices. Such flows may, for example, represent a crude measure of congestion at four-way stop junctions. However, neither the possibility of non-separable costs nor that of junction costs is essential for any of the results below.

4 Equilibrium

A flow vector $\mathbf{f}^*$ is said to be an *equilibrium* if the entire flow from o to d passes through least cost routes only. Mathematically, the equilibrium condition is:

$$\text{for all routes } r \text{ with } f_r^* > 0: \quad c_r(\mathbf{f}^*) = \min_{s \in \mathcal{R}} c_s(\mathbf{f}^*) \,. \tag{2}$$

In this case, the minimum in (2), denoted c^*, is the *equilibrium cost*. In the transportation literature, a flow vector satisfying (2) is known as Wardrop, or user, equilibrium. This condition expresses the principle, formulated by Wardrop and others (see references in [18, p. 151]), that, in equilibrium, the travel time on all used routes is equal, and less than or equal to the travel time that would be experienced by of a single vehicle on any unused route [23].

The equilibrium condition 2 can also be given a variational inequality formulation, as follows (see [18, Theorem 4.5]).

For every flow vector $\mathbf{f}$ with the same total origin-destination flow as $\mathbf{f}^*$:

$$\sum_{r \in \mathcal{R}} c_r(\mathbf{f}^*)(f_r^* - f_r) \leq 0\,.$$

Standard results (e. g., [18, Theorem 1.4]) therefore imply that, if the cost function is continuous, then for every $\delta \geq 0$ there is at least one equilibrium with a total origin-destination flow of δ. In general, there can be more than one such equilibrium. However, as the next proposition shows, if the network is series-parallel and the cost function is additively separable, then the total origin-destination flow δ uniquely determines the equilibrium *cost*. Moreover, in such a setting, increasing the edge costs or the total origin-destination flow weakly increases the equilibrium cost.

Proposition 7. *Let* $\hat{\mathbf{c}}$ *and* $\tilde{\mathbf{c}}$ *be two additively separable cost functions for the same series-parallel network, such that* $\hat{c}_r(\mathbf{f}) \geq \tilde{c}_r(\mathbf{f})$ *for all routes* r *and flow vectors* $\mathbf{f}$*. If* $\hat{\mathbf{f}}$ *and* $\tilde{\mathbf{f}}$ *are equilibria with respect to* $\hat{\mathbf{c}}$ *and* $\tilde{\mathbf{c}}$*, respectively, and the total origin-destination flows satisfy* $\hat{f}_o \geq \tilde{f}_o$*, then the respective equilibrium costs* $\hat{c}$ *and* $\tilde{c}$ *satisfy* $\hat{c} \geq \tilde{c}$*. Consequently, if the cost functions and the total origin-destination flows are the same, then the equilibrium costs are also the same.*

The result that, in a series-parallel network with an additively separable cost function, the equilibrium cost cannot decrease when edge costs increase, holds in the case of elastic as well as inelastic demand. In the latter case, in any equilibrium the total origin-destination flow must equal the fixed demand δ. Therefore, the above result follows directly from Proposition 7. In the former case, the demand is determined as a nonincreasing function of the cost [23, p. 135], [6, p. 102]. That is, the total origin-destination flow may vary, but it is lower in one equilibrium than in another only if the first equilibrium cost is higher than the second. In particular, for $\hat{\mathbf{c}}$ and $\tilde{\mathbf{c}}$ as in Proposition 7 and for any pair of corresponding equilibria $\hat{\mathbf{f}}$ and $\tilde{\mathbf{f}}$: $\hat{f}_o < \tilde{f}_o$ implies $\hat{c} \geq \tilde{c}$. On the other hand, by Proposition 7, $\hat{f}_o \geq \tilde{f}_o$ also implies $\hat{c} \geq \tilde{c}$. Therefore, the last inequality holds unconditionally. This proves that, in the case of elastic as well as inelastic demand, a unique equilibrium cost corresponds to any additively separable cost function, and this cost either increases or remains the same when edge costs are increased.

The case of elastic demand is, in fact, not much different from that of inelastic, fixed demand. This can be demonstrated by making the option of "staying at home" explicit. This option, which carries a *fixed* cost, is optimal if and only its cost is less than or equal to the cost of any route in the network. Therefore, the option of "staying at home" can be represented by a single edge, joining o and d, with a flow-independent cost. By Propositions 2 and 5, adding such an edge to a series-parallel network or one with independent routes does not affect the respective property. Also, if the cost function in the original network is additively separable, then the same is true for the enlarged one. (However, the new cost function is not *increasing*.) These considerations show that, for certain purposes at least, demand may be *assumed* inelastic. In this case, the total origin-destination flow can always be normalized to 1.

5 Efficiency of Equilibrium

Braess's paradox represents an extreme form of inefficiency. It occurs in a network G if there are two additively separable cost functions $\hat{\mathbf{c}}$ and $\tilde{\mathbf{c}}$ such that $\tilde{c}_r(\mathbf{f}) \geq \hat{c}_r(\mathbf{f})$ for all routes r and flow vectors $\mathbf{f}$, but for every equilibrium $\hat{\mathbf{f}}$ with respect to $\hat{\mathbf{c}}$ with a total origin-destination flow of unity and every equilibrium $\tilde{\mathbf{f}}$ with respect to $\tilde{\mathbf{c}}$ with a similar total origin-destination flow, the respective equilibrium costs $\hat{c}$ and $\tilde{c}$ satisfy $\hat{c} < \tilde{c}$. Thus, Braess's paradox occurs when raising the cost curves potentially leads to a *lower* equilibrium cost. As the following theorem shows, this is the case for every network that is not series-parallel. Conversely, for a network that *is* series-parallel, a pair of cost functions as above never exists. From this result, it is easy to deduce (for the case of two-terminal networks) Theorem 11 of [7], which is described in the Introduction.

Theorem 1. *For every network G, the following conditions are equivalent.*
(*i*) *Braess's paradox occurs in G;*
(*ii*) *G is not series-parallel.*

Even though series-parallel networks never exhibit Braess's paradox, they do not always have efficient equilibria. This is demonstrated by the example in Figure 2, in which the equilibrium flow can be rearranged in such a way that the costs of *all* routes actually used become less than the equilibrium cost. As the next theorem shows, the reason this kind of inefficiency occurs in the networks in Figure 2 is that routes in these networks are not independent.

An equilibrium $\mathbf{f}^*$, with equilibrium cost c^*, will be said to be *weakly Pareto efficient* if there is no flow vector $\mathbf{f}$, with the same total origin-destination flow as $\mathbf{f}^*$, such that $c_r(\mathbf{f}) < c^*$ for all routes r with $f_r > 0$. The equilibrium $\mathbf{f}^*$ is *Pareto efficient* if there is no flow vector $\mathbf{f}$, with the same total origin-destination flow as $\mathbf{f}^*$, such that $c_r(\mathbf{f}) < c^*$ for all routes r with $f_r > 0$, with strict inequality for at least one such r.

Theorem 2. *For every network G, the following conditions are equivalent.*[5]
(*i*) *For every cost function, all equilibria are weakly Pareto efficient.*
(*ii*) *For every increasing cost function, all equilibria are Pareto efficient.*
(*iii*) *G is a network with independent routes.*

[5] In (*i*), (*ii*), or both, "cost function" can be replaced by "additively separable cost function". The equivalence of three conditions is not affected by these changes.

The weak Pareto efficiency of the equilibria in a network with independent routes implies that, in such a network, the equilibrium cost is *uniquely* determined by the cost function and the total origin-destination flow. Recall that, by Proposition 7, if the cost function is *additively separable*, then a weaker topological condition than independence of the routes, namely, that the network is series-parallel, is sufficient for uniqueness of the equilibrium cost. (In fact, a somewhat stronger version of additive separability than the one used here implies uniqueness of the equilibrium cost for *any* network, series-parallel or not; see Proposition 8 below, and [2].) However, if the payoff function is not additively separable, then it can be shown that a series-parallel network is not a sufficient condition of uniqueness of the equilibrium cost. See footnote 6 below.

6 Heterogeneity

The model presented in Section 3 can be generalized by dropping the assumption that all users of the same route incur the same cost. This requires modifying the definition of equilibrium as well. When route costs differ across users, the equilibrium condition becomes: each route is only used by those for whom it is a least cost route. A more formal definition follows. Let the population of *users* be the unit interval $[0; 1]$. For each user i, let $\mathbf{c}^i$ be i's *cost function*. The cost function specifies the cost $c_s^i(\mathbf{f})$ of each walk s for user i as a function of the flow vector $\mathbf{f}$. As in the homogeneous case, this is assumed to satisfy the condition that, for every walk s and every pair of flow vectors $\hat{\mathbf{f}}$ and $\tilde{\mathbf{f}}$, if $\hat{f}_t \geq \tilde{f}_t$ and $\hat{f}_{-t} \geq \tilde{f}_{-t}$ for all sections t of s, then $c_s^i(\hat{\mathbf{f}}) \geq c_s^i(\tilde{\mathbf{f}})$. A *strategy profile* σ is a mapping from users to routes such that, for each route r, the set of all users i with $\sigma(i) = r$ is Lebesgue measurable; the measure of this set, which will be denoted $f_r(\sigma)$, is the corresponding route flow. Thus, each strategy profile σ gives rise to a particular flow vector $\mathbf{f}(\sigma)$, with a total origin-destination flow of unity. A strategy profile σ is a *Nash equilibrium* if the following condition holds:

$$\text{for all users } i, \quad c_{\sigma(i)}^i(\mathbf{f}(\sigma)) = \min_{r \in \mathcal{R}} c_r^i(\mathbf{f}(\sigma)) \,. \tag{3}$$

In this case, the minimum in 3 is user i's *equilibrium cost*. In the special case in which all users have the same cost function, this definition essentially reduces to 2.

The case of user-dependent cost functions differs from that of a single, common cost function in a number of ways. For example, the result, that in a series-parallel network Braess's paradox does not occur (Theorem 1), does not extend to the case of heterogeneous populations. This is shown by the following example.[6]

[6] In this example, the cost functions of all the users are additively separable. It is easy to modify the example in such a way that all users have *the same* cost

Example 1. Three types of users travel from o to d in the series-parallel network shown in Figure 5. Type I users are those with $0 \leq i < 2/7$, type II are those with $2/7 \leq i < 4/7$, and type III are those with $4/7 \leq i \leq 1$. The cost functions are additively separable: for each type of users, the cost of each route is the sum of the costs of its edges. The cost of edge e_j for user i depends on i's type and on the fraction x of users using e_j. For type III users, the costs also depend on a positive parameter δ. The costs are given in the Table 1 (with blank cells indicating prohibitively high costs). For $\delta < 6/7$, there is

Table 1.

	e_1	e_2	e_3	e_4	e_5
Type I	x		$8x$		$6x$
Type II		$8x$		x	$6x$
Type III		$\delta + 3x$	$3x$		$6x$

a unique Nash equilibrium, in which type I and type II users go though e_5 and type III users take the route going though e_2 and e_3. For the first two types of users, the equilibrium cost is 24/7. For type III users, it is $18/7 + \delta$ (and thus tends to 24/7 as δ approaches 6/7). For $\delta > 6/7$, there is also a unique Nash equilibrium. However, for all users, the equilibrium cost is *lower* than for $\delta < 6/7$, and equals 18/7. In this Nash equilibrium, type I users take the route going through e_1 and e_3, type II users take the route going though e_2 and e_4, and type III users use e_5. For $\delta = 6/7$, the set of Nash equilibria consists of all convex combinations (in terms of the proportion of users of each type using each route) of the above two equilibria, and similarly for the equilibrium costs. Thus, there is a *continuum* of Nash equilibria, and these equilibria can be Pareto ranked in the sense that, in each equilibrium, everyone's equilibrium cost is higher, or everyone's equilibrium cost is lower, than in each of the other equilibria. Indeed, in each equilibrium, the equilibrium costs are the same for all users.

function, but this cost function is not additively separable. Specifically, in the modified version of Example 1, all but four routes have a very high cost. These are: the route containing e_1 and e_3, the route containing e_2 and e_3, the route containing e_2 and e_4, and the one containing e_5 only. The costs of these routes, for all users, are the same as their costs for type I users, type II users, type III users, and all users, respectively, in the original Example 1. When the parameter δ is equal to 6/7, the two equilibria in the original example correspond to two equilibria in the modified example, with different equilibrium costs. This proves that, for a homogeneous population of users with a non-additively separable cost function, a series-parallel network is not a sufficient condition for uniqueness of the equilibrium cost. Cf. the remarks at the end of Section 5.

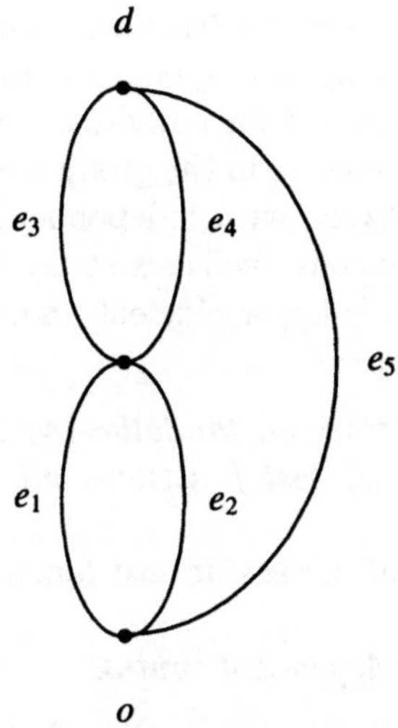

Fig. 5. Population heterogeneity and inefficiency of equilibrium. For details see text

It can be shown that, for $\delta > 6/7$, the unique Nash equilibrium in Example 1 is Pareto efficient, in the obvious sense (see below). For $\delta < 6/7$, the unique equilibrium is clearly not even weakly Pareto efficient. As in the homogeneous case (Theorem 2), this lack of efficiency may be attributed to the fact that, in the network shown in Figure 5, routes are not independent. As Theorem 3 below shows, in a network with independent routes Nash equilibria are always weakly Pareto efficient. And if the cost functions of all users are increasing, then all Nash equilibria satisfy a condition stronger than Pareto efficiency, which will be referred to as hyper-efficiency. (Since homogeneous populations are a special case of heterogeneous ones, the last assertion is true also in the homogeneous case.)

A strategy profile σ is *weakly Pareto efficient* if there is no strategy profile τ such that $c^i_{\tau(i)}(\mathbf{f}(\tau)) < c^i_{\sigma(i)}(\mathbf{f}(\sigma))$ for all users i. It is *Pareto efficient* if, for every strategy profile τ,

$$\begin{aligned} &\text{either } c^i_{\tau(i)}(\mathbf{f}(\tau)) = c^i_{\sigma(i)}(\mathbf{f}(\sigma)) \text{ for all users } i \\ &\text{or } c^i_{\tau(i)}(\mathbf{f}(\tau)) > c^i_{\sigma(i)}(\mathbf{f}(\sigma)) \text{ for some } i\,. \end{aligned}$$

A strategy profile σ will be said to be *hyper-efficient* if, for every strategy profile τ,

$$\begin{aligned} &\text{either } c^i_{\tau(i)}(\mathbf{f}(\tau)) = c^i_{\sigma(i)}(\mathbf{f}(\sigma)) \text{ for all users } i \\ &\text{or } c^i_{\tau(i)}(\mathbf{f}(\tau)) > c^i_{\sigma(i)}(\mathbf{f}(\sigma)) \text{ for some } i \text{ with } \tau(i) \neq \sigma(i)\,. \end{aligned} \tag{H}$$

In words, a strategy profile is hyper-efficient if any effective change of route choices is harmful to some of those who change their routes. Clearly, any hyper-efficient strategy profile σ is a Nash equilibrium. Indeed, it is a strong equilibrium, and even a strictly strong equilibrium.[7] This means that deviations are never profitable, not just for individuals but also for groups of users: Any deviation that makes someone in the group better off must leave someone else in it worse off. In a network with independent routes, and when all cost functions are increasing, the converse is also true. That is, under these conditions, *any* Nash equilibrium is hyper-efficient and, hence, Pareto efficient and a strictly strong equilibrium.

Theorem 3. *For every network G, the following conditions are equivalent.*

(i) For every assignment of cost functions, all Nash equilibria are weakly Pareto efficient.

(ii) For every assignment of increasing cost functions, all Nash equilibria are hyper-efficient.

(iii) G is a network with independent routes.

While independence of the routes implies that all equilibria are efficient, even this condition does not guarantee uniqueness of the equilibrium costs if the population is heterogeneous. The following example shows this.

Example 2. Three types of users travel from o to d in the network shown in Figure 3. Type I users are those with $0 \leq i < 4/13$, type II are those with $4/13 \leq i < 8/13$, and type III are those with $8/13 \leq i \leq 1$. The cost functions are additively separable: for each type of users, the cost of each route is the sum of the costs of its edges. The cost of edge e_j for user i depends on i's type and on the fraction x of users using e_j, as detailed in the Table 2 (with

Table 2.

	e_1	e_2	e_3	e_4	e_5
Type I	$3.1 + x$		$8x$		x
Type II		$8x$		$2.1 + x$	x
Type III		$0.5 + x$	x		x

blank cells indicating prohibitively high costs). There is one Nash equilibrium in which type I users use e_1, type II users use e_2, and type III users take the route going through e_3 and e_5. There is another Nash equilibrium, in which type I users take the route going through e_3 and e_5, type II users take the route going through e_4 and e_5, and type III users use e_2. For type I users, the second equilibrium cost is lower than the first. For type II and type III users, the second equilibrium cost is higher than the first.

[7] A strategy profile σ is a *strictly strong equilibrium* [25] if, for every strategy profile τ, $c^i_{\tau(i)}(\mathbf{f}(\tau)) \geq c^i_{\sigma(i)}(\mathbf{f}(\sigma))$ for all users i or $c^i_{\tau(i)}(\mathbf{f}(\tau)) > c^i_{\sigma(i)}(\mathbf{f}(\sigma))$ for some i with $\tau(i) \neq \sigma(i)$.

The determination of necessary and sufficient *topological* conditions guaranteeing that, for any assignment of increasing and additively separable cost functions, each user's equilibrium cost is unique, is at yet an unresolved problem. A *sufficient* condition for this is that the network consists of several edges connected in parallel [16, 12] or it is the result of connecting several such networks in series. However, this condition is not *necessary*. Note that the above problem is different from that of finding conditions on the *cost functions* that guarantee uniqueness of the equilibrium costs for any *network*. One such (sufficient) condition (see, e. g., [3]) is that the cost of each edge for each user is made up of (1) a nondecreasing function of the flow though the edge in the direction of motion, which is the same for all users, and (2) some user-specific constant, which does not depend on the flow.

Proposition 8. *Suppose there is an additively separable cost function* $\mathbf{c}$ *such that (i) for each user i,* $\mathbf{c}^i$ *is equal to* $\mathbf{c}$ *plus some constant cost function, i. e., one that does not depend on the flow, and (ii) for each walk s of length one,* $c_s(\mathbf{f})$ *is determined solely by* f_s*, as a nondecreasing function. Then, for all users i, user i's equilibrium cost in any two Nash equilibria* σ *and* τ *is the same.*

If the edge costs are *increasing* functions of the respective edge flows, then not only the equilibrium costs but also the equilibrium *edge flows* are unique.

Appendix

The Appendix contains the proofs of the various propositions and theorems in this paper.

Proof of Proposition 1.

The three claims given below together prove the proposition and, in addition, establish the following result.

Lemma 1. *A network is series-parallel if and only if it satisfies the following condition: for every pair of distinct vertices u and* ν*, if u precedes* ν *in some route r containing both vertices, then u precedes* ν *in all such routes. Consequently, for every two routes r and s in a series-parallel network and every vertex* ν *common to both routes, the walk* $r_{o\nu} + s_{\nu d}$ *is a route (i. e., it does not pass through any vertex more than once).*

Claim 1. Every network that satisfies the condition in Lemma 1 is series-parallel.
This is obvious. Indeed, for a network to be series-parallel it only has to satisfy the condition for all pairs of vertices u and ν that are joined by some edge.

Claim 2. The network in Figure 1 is embedded in every network that does not satisfy the above condition.

Suppose there are two routes r and s in a network G, and two vertices u and ν common to both routes, such that u precedes ν in r but follows it in s. Suppose also that u and ν are chosen in such a way that the length of $r_{u\nu}$ is maximal. Then, any vertex u' common to r and s that precedes u in r must precede ν in s, and any vertex ν' common to both routes that follows ν in r must follow u in s (see Figure 6). Let u' be chosen as the *last* vertex before

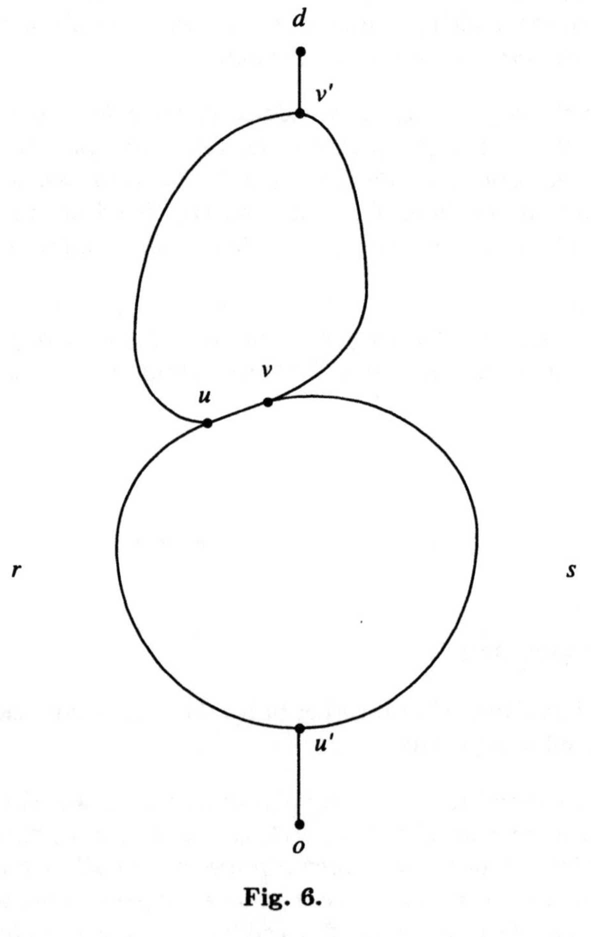

Fig. 6.

u in r which also belongs to s (possibly, $u' = o$), and ν' as the *first* vertex after ν in r which also belongs to s (possibly, $\nu' = d$). All the edges in $r_{u'u}$, and all the vertices in this section of r with the exception of the initial and terminal ones, do *not* belong to s, and the same is true for $r_{\nu\nu'}$. This implies that the network in Figure 1 is embedded in the network that consists of all the vertices and edges in s, $r_{u'u}$, and $r_{\nu\nu'}$. Hence, the same is true for G.

Claim 3. The network in Figure 1 is not embedded in any series-parallel network.

This is clear: any network with an embedded non-series-parallel network is itself not series-parallel. □

Proof of Proposition 2.

Clearly, every network with a single edge, as well as every network that is the result of connecting two series-parallel networks in series or in parallel, is itself series-parallel. Therefore, it only has to be shown that every series-parallel network G with more than one edge satisfies (*ii*) or (*iii*).

Claim 1. In the set of all routes in G, the relation "routes r and s have a vertex in common, other than o and d, or the two routes are identical", is an equivalence relation.

This relation is obviously reflexive and symmetric. It remains to show that it is transitive. That is, if r, s, and t are three routes such that there is some vertex $u \neq o$, d common to r and s and some vertex $\nu \neq o$, d common to s and t, then there is also some vertex, other than o and d, common to r and t. Suppose not. Without loss of generality, it may be assumed that u precedes ν in s and there is no other vertex in the section $s_{u\nu}$ which belongs to either r or t (see Figure 7). This assumption implies that the network in Figure 1 is embedded in the network that consists of all the vertices and edges in r, t, and $s_{u\nu}$. However, by Proposition 1, this contradicts the assumption that G is series-parallel. This contradiction proves Claim 1.

Two cases are possible: either there are two or more equivalence classes with respect to the equivalence relation in Claim 1; or this relation holds between any pair of routes in G. In the former case, pick up one of the equivalence classes, and consider the network G' that consists of all the vertices and edges that belong to at least one route in this equivalence class, as well as the network G'' that consists of all the vertices and edges that belong to at least one route *not* in the class. Each vertex ν, other than o and d, belongs to one, and only one, of these two networks. (Otherwise, ν would belong to two routes in two different equivalence classes, which is impossible by definition of the equivalence relation.) Therefore, each edge also belongs to one, and only one, of them. This implies that G is the result of connecting G' and G'' in parallel. Clearly, since G is series-parallel, the same is true for G' and G''. Hence, G satisfies (*ii*).

In the rest of this proof, it will be assumed that the equivalence relation in Claim 1 holds between any pair of routes in G. It will be shown that there is some vertex, other than o and d, which is common to all routes. Note that the above assumption implies that each route in G has at least two edges. Indeed, since a route with a single edge is not equivalent to any other route, it must be the only route in G. The route's unique edge is then the only edge in G, a contradiction to the assumption that G has more than one edge.

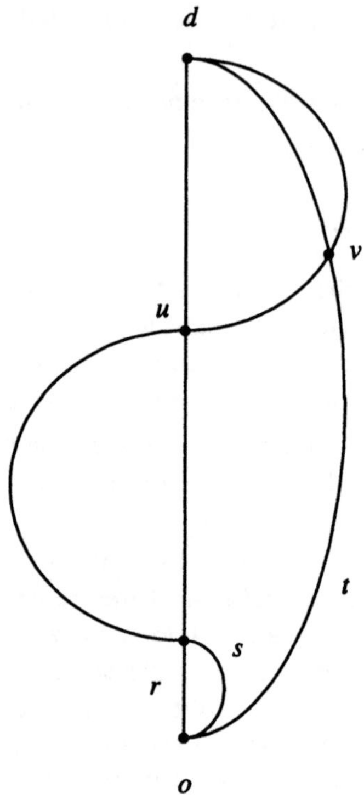

Fig. 7.

Fix a route r and, for every route s, denote by ν_s the first vertex in s, other than o, which also belongs to r. Since, by assumption, the equivalence relation holds between r and s, $\nu_s \neq d$. Let t be a route such that, for all routes s, ν_s is either equal to ν_t or precedes it in r.

Claim 2. The vertex ν_t belongs to all routes.

Suppose the contrary, that ν_t does not belong to some route s. Let u be the last vertex before ν_t in r which also belongs to s. By the way t was chosen, $u \neq o$. By Lemma 1, the first vertex after u in s which also belongs to r follows ν_t in r. This implies that $t_{o\nu_t}$ and s_{ud} do not have a vertex in common. For if a common vertex ν did exist, then the route $t_{o\nu} + s_{\nu d}$ would have the property that the vertex $\nu_{t_{o\nu}+s_{\nu d}}$ (i. e., the first vertex after o in which the route intersects r) follows ν_t in r, which is a contradiction to the way t was chosen (see Figure 8). The fact that $t_{o\nu_t}$ and s_{ud} do not have a common vertex implies that the walk $t_{o\nu_t} + (-r)_{\nu_t u} + s_{ud}$ is a route (i. e., it does not pass through any vertex more than once). However, this contradicts the assumption that G is series-parallel. This contradiction proves Claim 2.

It follows from Claim 2 that G is the result of connecting two series-parallel networks in series: the network G' that consists of ν_t (as destination) and all the vertices and edges that precede it in some route in G; and the network G''

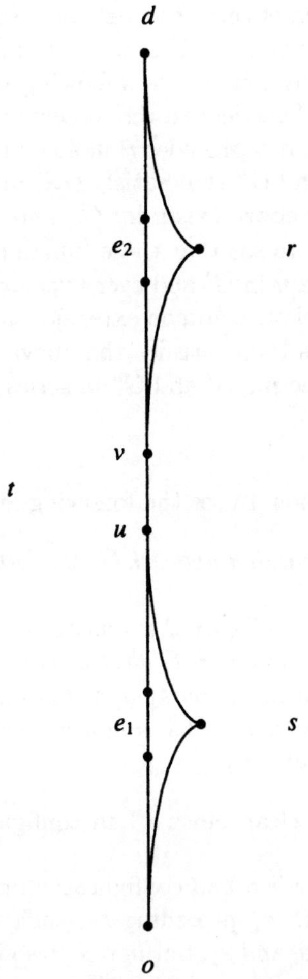

Fig. 8.

that consists of ν_t (as the origin) and all the vertices and edges that follow it in some route in G. By Lemma 1, ν_t is the only vertex common to both networks. Since G' and G'' are clearly series-parallel, G satisfies (*iii*). □

Proof of Proposition 3.

Since, clearly, (*iii*) implies (*ii*) and (*ii*) implies (*i*), it suffices to show that every series-parallel network G satisfies (*iii*). This will be proved by induction on the number of edges in G. If there is only one edge, (*iii*) holds trivially. Suppose, then, that G has more than one edge and that (*iii*) holds for any series-parallel network with a smaller number of edges than G. By Proposition 2, G is the result of connecting two series-parallel networks in series or in parallel. By the induction hypothesis, each of these two networks satisfies (*iii*). It will now be shown that this implies that G itself also satisfies (*iii*).

Condition (*iii*) is equivalent to the following one: There is a one-to-one function φ from the set of vertices to the integers such that, for every pair of distinct vertices u and ν, if u precedes ν in some route r, then $\varphi(u) < \varphi(\nu)$. When two networks G' and G'' that satisfy (*iii*) are connected in series or in parallel, a function φ' as above exists for G', and another one φ'' for G''. It is, moreover, not difficult to see that these functions can be chosen in such a way that, for every vertex u in G' and every vertex ν in G'', $\varphi'(u) = \varphi''(\nu)$ if and only if $u = \nu$. The unique common extension φ of φ' and φ'' to the union of the two sets of vertices then satisfies the above condition for the network that is the result of connecting G' and G'' in series or in parallel. Hence, that network also satisfies (*iii*). □

The proof of Proposition 4 uses the following lemma.

Lemma 2. *For a series-parallel network G, the following three conditions are equivalent:*
- (*i*) *One of the networks in Figure 2 is embedded in G.*
- (*ii*) *There is a triplet of routes in G that constitutes a "bad configuration". (This term is defined in the paragraph that follows Proposition 5).*
- (*iii*) *There is a pair of routes r and s, and a vertex ν common to both routes, such that $r_{o\nu} \neq s_{o\nu}$ and $r_{\nu d} \neq s_{\nu d}$.*

Proof. (*i*) ⇒ (*ii*). This is clear, since a bad configuration exists in both networks in Figure 2.
(*ii*) ⇒ (*iii*). Suppose there is a bad configuration: a route t and two edges e_1 and e_2 in this route, with e_1 preceding e_2, such that e_2, but not e_1, also belongs to another route s, and e_1, but not e_2, also belongs to a third route r. Note that, since the network is series-parallel, r and s must pass through e_1 and e_2, respectively, in the same directions as t. Let u be the first vertex in s which also belongs to t and, in that route, follows e_1 but precedes e_2. Let ν be the last vertex in r which also belongs to t and, in that route, follows e_1 but precedes e_2 (see Figure 9). If s_{ou} and $r_{\nu d}$ have a common vertex ν', then (*iii*) holds: $r_{o\nu'} \neq s_{o\nu'}$ since e_1 belongs to $r_{o\nu'}$ but not to $s_{o\nu'}$; and $r_{\nu' d} \neq s_{\nu' d}$ since e_2 belongs to $s_{\nu' d}$ but not to $r_{\nu' d}$. Suppose, then, that s_{ou} and $r_{\nu d}$ do not have a vertex in common. This implies that u is either equal to ν or precedes it in r. For if u followed ν in r, then the walk $s_{ou} + (-t)_{u\nu} + r_{\nu d}$ would be a route (i. e., it would not pass through any vertex more than once), which is a contradiction to the assumption that the network is series-parallel. Consider the two routes t and $s_{ou} + t_{u\nu} + r_{\nu d}$. The inequality $t_{o\nu} \neq s_{ou} + t_{u\nu}$ holds, since e_1 belongs to the walk on the right but not to the one on the left; and $t_{\nu d} \neq r_{\nu d}$ holds, since e_2 belongs to the walk on the right but not to the one on the left. Therefore, (*iii*) holds, with t and $s_{ou} + t_{u\nu} + r_{\nu d}$ replacing r and s, respectively.

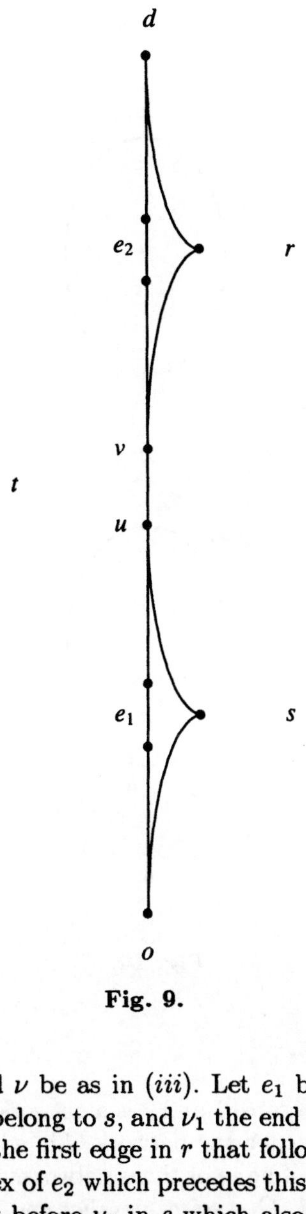

Fig. 9.

$(iii) \Rightarrow (i)$. Let r, s, and ν be as in (iii). Let e_1 be the last edge in r that precedes ν and does not belong to s, and ν_1 the end vertex of e_1 which follows this edge in r. Let e_2 be the first edge in r that follows ν and does not belong to s, and u_2 the end vertex of e_2 which precedes this edge in r (see Figure 10). Let u_1 be the last vertex before ν_1 in s which also belongs to r, and ν_2 the first vertex after u_2 in s which also belongs to r. Consider the network that consists of all the vertices and edges in r, $s_{u_1\nu_1}$, and $s_{u_2\nu_2}$. Since only the initial and terminal vertices of each of the last two sections of s are in r, one of the networks in Figure 2 is embedded in this network. If $\nu_1 = u_2$, then this is the network in Figure 2(a); and if $\nu_1 \neq u2$, then it is the one in Figure 2(b). □

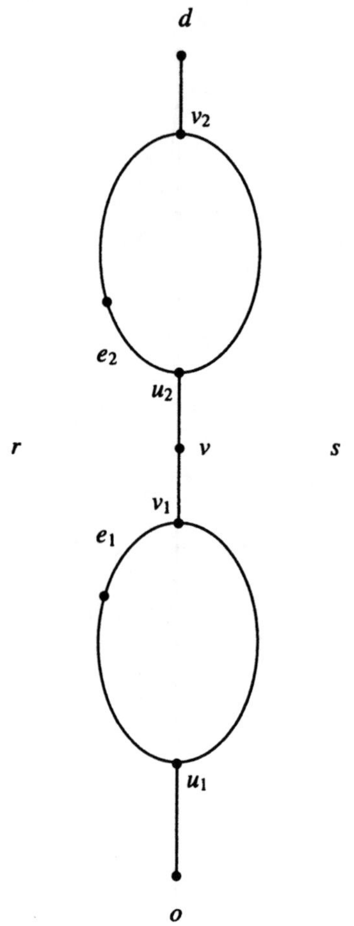

Fig. 10.

Proof of Proposition 4.

None of the networks in Figures 1 and 2 is a network with independent routes. Therefore, the same is true for any network G in which one of these networks is embedded. Conversely, if none of these networks is embedded in a network G, then, by Proposition 1, G is series-parallel and, by Lemma 2, a triplet of routes constituting a bad configuration does not exist. It will now be shown that this implies that G is a network with independent routes. Suppose, by contradiction, that there is some route r, every edge of which also belongs to at least one other route. Let $s \neq r$ be a route having a maximal number of r's edges. Since, clearly, no route other than r can have all of r's edges, there is some edge e_1 in r which is not in s. Let $t \neq r, s$ be a route containing e_1. Since, by construction, t does not contain more of r's edges than does s, there is at least one edge e_2 common to r and s which is not in t. However, this implies that r, s, and t constitute a bad configuration, which is a contradiction. □

Proof of Proposition 5.

One direction is obvious: if a network satisfies (i), (ii), or (iii), then it is a network with independent routes. To prove the converse, assume that G is a network with independent routes. Since this is *not* the case for any of the networks in Figures 1 and 2, none of these networks is embedded in G. By Proposition 1, this implies that G is series-parallel. Also, by Lemma 2: (1) for every pair of routes r and s and every vertex ν common to both routes, $r_{o\nu} = s_{o\nu}$ or $r_{\nu d} = s_{\nu d}$, and (2) a triplet of routes constituting a bad configuration does not exist in G. This implies the following two claims.

Claim 1. If two routes in G have a vertex in common, other than o and d, then their first edge is the same or their last edge is the same.

Claim 2. In the set of all routes in G, the relation "routes r and s have an edge in common" is an equivalence relation.

Claim 2 follows from the fact that, if route r shares an edge e_1 with route s, and s shares an edge e_2 with route t, then at least one of the edges e_1 and e_2 must be common to r and t, otherwise these three routes would constitute a bad configuration.

Two cases are possible: either there are two or more equivalence classes with respect to the equivalence relation in Claim 2; or this relation holds between any pair of routes in G. In the former case, pick up one of the equivalence classes, and consider the network G' that consists of all the vertices and edges that belong to at least one route in this equivalence class, as well as the network G'' that consists of all the vertices and edges that belong to at least one route *not* in the class. It follows from Claim 1 and the definition of the equivalence relation that each vertex, other than o and d, belongs to one, and only one, of these two networks. Therefore, each edge also belongs to one, and only one, of them. This implies that G is the result of connecting G' and G'' in parallel. Since G is a network with independent routes, the same is clearly true for G' and G''. Hence, G satisfies (ii).

In the rest of this proof, it will be assumed that the equivalence relation holds between any pair of routes in G; i.e., any two routes have an edge in common.

Claim 3. All the routes in G have the same first edge, or they all have the same last edge.

To prove this claim, suppose there are two routes r and s which do *not* have the same last edge. Since, by assumption, the two routes have *some* edge in common, it follows from Claim 1 that the first edge in r coincides with the first edge in s. Call this edge e. By a similar argument, any route t which does not have e as its first edge must have the same last edge as r. Similarly, it must have the same last edge as s. However, r and s do *not* have the same last edge. This contradiction proves that *all* routes must have e as their first edge, thus completing the proof of Claim 3.

It follows from Claim 3 that there is some edge e, with o or d as one of its end vertices, which belongs to all routes. This implies that e is the only edge incident with o or d, respectively. Therefore, either G has a single edge, or it is the result of connecting in series the network that consists of e and its two end vertices and the network that consists of all the other vertices and edges in G plus the end vertex of e which is *not* o or d. Clearly, the latter is a network with independent routes. Hence, G satisfies (i) or (iii). □

Proof of Proposition 6.

For each of the networks in Figures 1 and 2, none of the four conditions holds. Therefore, it follows from Proposition 4 that (i), (ii), and (iii) do not hold for *any* network G which is not a network with independent routes. Conversely, if G *is* a network with independent, then (i) clearly holds. And since G is series-parallel and none of the networks in Figure 2 is embedded in it, (ii) and (iii) hold by Lemma 2. □

The following two lemmas are used in the proof of Proposition 7.

Lemma 3. *Let G be a series-parallel network, and $\hat{\mathbf{f}}$ and $\tilde{\mathbf{f}}$, two flow vectors. If the total origin-destination flows satisfy $\hat{f}_o > 0$ and $\hat{f}_o \geq \tilde{f}_o$, then there is some route r such that, for all sections s of r of length zero or one (i. e., those containing only one vertex or only one edge), $\hat{f}_s > 0$ and $\hat{f}_s \geq \tilde{f}_s$. If $\hat{f}_o > \tilde{f}_o$, then a similar result holds with the last pair of inequalities replaced by $\hat{f}_s > \tilde{f}_s$.*

Proof. The proof of the lemma proceeds by induction on the number of edges. For a network with a single edge, the result is trivial. Consider, then, a series-parallel network G with two or more edges, such that the result holds for any two flow vectors in any series-parallel network with a smaller number of edges than G. (This is the induction hypothesis.) By Proposition 3, G is the result of connecting two series-parallel networks, G' and G'', in series or in parallel. Consider, first, the case in which G' and G'' are connected in series, so that the destination in G', ν, coincides with the origin in G''. The set $\mathcal{R}'$ of all routes in G' is then equal to $\{r_{o\nu} \mid r \in \mathcal{R}\}$. Every flow vector $\mathbf{f}$ in G induces a flow vector $\mathbf{f'}$ in G', which is defined by the equations $f'_{r'} = f_{r'}$ $(r' \in \mathcal{R}')$. The flow f'_s through any walk s in G' is given by an equation similar to (1), namely,

$$f'_s = \sum_{\substack{r' \in \mathcal{R}' \\ s \text{ is a section of } r'}} f'_{r'} \,. \tag{4}$$

It is, however, not difficult to see that, for every such s, $f'_s = f_s$. In particular, if the flow vectors $\hat{\mathbf{f}}$ and $\tilde{\mathbf{f}}$ satisfy the pair of inequalities $\hat{f}_o > 0$ and $\hat{f}_o \geq \tilde{f}_o$, or the inequality $\hat{f}_o > \tilde{f}_o$, then similar inequalities or inequality hold for the flow vectors $\hat{\mathbf{f}}'$ and $\tilde{\mathbf{f}}'$ they induce in G'. It then follows from the induction hypothesis that there is a route r' in G' such that, for all sections s of r' of length zero or one, $\hat{f}_s > 0$ and $\hat{f}_s \geq \tilde{f}_s$, or $\hat{f}_s > \tilde{f}_s$, respectively. By similar considerations, there is a route r'' in G'' such that similar inequalities or inequality hold for all sections s of r'' of length zero or one. Therefore, the same is true for the route $r = s' + s''$ in G. This completes the proof for the case in which G is the result of connecting the two series-parallel networks in series.

Suppose, now, that G is the result of connecting G' and G'' in parallel, so that o and d are also the origin and destination, respectively, in G' and G''. The set $\mathcal{R}$ of all routes in G is then the disjoint union of the set $\mathcal{R}'$ of all routes in G' and the set $\mathcal{R}''$ of all routes in G''. Every flow vector $\mathbf{f}$ in G induces a flow vector $\mathbf{f}'$ in G', which is defined as in the previous paragraph, and a flow vector $\mathbf{f}''$ in G'', which is defined in a similar manner. For every walk s in G', the flow f'_s through s is given by (4). If s is *not* one of the zero-length walks o and d, then $f'_s = f_s$. However, unlike in the case previously considered, f'_o and f'_d (which are both equal to $\sum_{r' \in \mathcal{R}'} f_{r'}$) are generally less than f_o and f_d, and the same is true for f''_o and f''_d. In fact, $f_o = f'_o + f''_o$. It follows from this equality that if $\hat{\mathbf{f}}$ and $\tilde{\mathbf{f}}$ are two flow vectors satisfying $\hat{f}_o > 0$ and $\hat{f}_o \geq \tilde{f}_o$ then $\hat{f}'_o > 0$ and $\hat{f}'_o \geq \tilde{f}'_o$, or $\hat{f}''_o > 0$ and $\hat{f}''_o \geq \tilde{f}''_o$. Then, by the induction hypothesis, there is some route r, either in G' or in G'' (and, hence, in G), such that, for all sections s of r of length zero or one, $\hat{f}_s > 0$ and $\hat{f}_s \geq \tilde{f}_s$. Similarly, if $\hat{f}_o > \tilde{f}_o$, then there is a route r in G' or in G'' such that, for all sections s of r of length zero or one, $\hat{f}_s > \tilde{f}_s$. □

Lemma 4. *Let G be a series-parallel network, $\mathbf{c}$ an additively separable cost function, and $\mathbf{f}^*$ a corresponding equilibrium. For every vertex ν, there is a number $F(\nu)$ such that, for all routes r passing through ν, if every edge in r belongs to at least one least cost route s, then $c_{r_{o\nu}}(\mathbf{f}^*) = F(\nu)$. In particular, for every route r that is itself a least cost route, $c_{r_{o\nu}}(\mathbf{f}^*) = F(\nu)$ for every vertex ν in r. Consequently, $F(d)$ is the equilibrium cost.*

Proof. Let r be a route in which every edge e belongs to at least one least cost route s. For every vertex ν in r, the following two claims hold.

Claim 1. For every least cost route s passing through ν, $c_{s_{o\nu}}(\mathbf{f}^*) \leq c_{r_{o\nu}}(\mathbf{f}^*)$. Otherwise, $c_s(\mathbf{f}^*)$ would be greater than $c_{r_{o\nu}}(\mathbf{f}^*) + c_{s_{\nu d}}(\mathbf{f}^*)$, which is the cost of the route $r_{o\nu} + s_{\nu d}$, a contradiction to the assumption that s is a least cost route.

Claim 2. For every least cost route s passing through ν, $c_{s_{o\nu}}(\mathbf{f}^*) \geq c_{r_{o\nu}}(\mathbf{f}^*)$. This will be proved by induction on the length of $r_{o\nu}$. The induction hypothesis is that, for every vertex u that precedes ν in r, Claim 2 holds with u

replacing ν. Suppose, by contradiction, that there is some least cost route t passing through ν such that $c_{t_{o\nu}}(\mathbf{f}^*) < c_{r_{o\nu}}(\mathbf{f}^*)$. Clearly, this is possible only if $\nu \neq o$. Let e be the edge immediately preceding ν in r, and u its other end vertex. Let s be a least cost route passing through e. By the induction hypothesis, $c_{s_{ou}}(\mathbf{f}^*) \geq c_{r_{ou}}(\mathbf{f}^*)$. Since $s_{u\nu} = r_{u\nu}$, and hence $c_{s_{u\nu}}(\mathbf{f}^*) = c_{r_{u\nu}}(\mathbf{f}^*)$, additive separability implies $c_{s_{o\nu}}(\mathbf{f}^*) \geq c_{r_{o\nu}}(\mathbf{f}^*)$, and hence $c_{s_{o\nu}}(\mathbf{f}^*) > c_{t_{o\nu}}(\mathbf{f}^*)$. However, Claim 1, when applied to t instead of r, gives $c_{s_{o\nu}}(\mathbf{f}^*) \leq c_{t_{o\nu}}(\mathbf{f}^*)$, which is a contradiction. This contradiction proves Claim 2.

For every vertex ν that belongs to some least cost route s, set $F(\nu) = c_{s_{o\nu}}(\mathbf{f}^*)$. The value of $F(\nu)$ does not depend on the particular choice of s. Indeed, by Claims 1 and 2, if r is any route passing through ν in which every edge e belongs to at least one least cost route, $c_{r_{o\nu}}(\mathbf{f}^*) = c_{s_{o\nu}}(\mathbf{f}^*)$. For all vertices ν that do not belong to any least cost route, set $F(\nu) = 0$. □

Proof of Proposition 7

Let $\hat{\mathbf{f}}$ and $\tilde{\mathbf{f}}$ be equilibria with respect to $\hat{\mathbf{c}}$ and $\tilde{\mathbf{c}}$ respectively, such that the total origin-destination flows satisfy $\hat{f}_o \geq \tilde{f}_o$. Since, by assumption, $\hat{c}_r(\mathbf{f}) \geq \tilde{c}_r(\mathbf{f})$ for all routes r and flow vectors $\mathbf{f}$, if $\hat{\mathbf{f}}$ is equal to $\tilde{\mathbf{f}}$ then the respective equilibrium costs $\hat{c}$ and $\tilde{c}$ satisfy $\hat{c} \geq \tilde{c}$. Suppose, then, that the two flow vectors are distinct (and, hence, $\hat{f}_o > 0$). By Lemma 3, there is some route r such that, for all sections s of r of length zero or one, $\hat{f}_s > 0$ and $\hat{f}_s \geq \tilde{f}_s$. The first inequality implies that every edge e in r belongs to some route t with $\hat{f}_t > 0$. Therefore, by the equilibrium condition (2) and Lemma 4, $\hat{c}_r(\hat{\mathbf{f}}) = \hat{c}$. The second inequality, together with the assumption that the cost functions are additively separable, implies $\hat{c}_r(\hat{\mathbf{f}}) \geq \hat{c}_r(\tilde{\mathbf{f}})$. Since $\hat{c}_r(\tilde{\mathbf{f}}) \geq \tilde{c}_r(\tilde{\mathbf{f}})$ and, by definition, $\tilde{c}_r(\tilde{\mathbf{f}}) \geq \tilde{c}$, this proves that $\hat{c} \geq \tilde{c}$. The second assertion of the proposition immediately follows from the first. □

Proof of Theorem 1.

The implication $(i) \Rightarrow (ii)$ is given by Proposition 7. To prove the reverse implication, recall that, by Proposition 1, if a network is not series-parallel, then the network in Figure 1 is embedded in it. As shown, in that particular network Braess's paradox does occur. That the same is true for any network in which that network is embedded can easily be seen by considering the following rules for assigning costs to the new edges created by the three operations defining embedding: for each of the two edges created when an existing edge is subdivided, the cost is one half that of the original edge; for an edge joining two existing vertices, the cost is infinite (or, at least, very high); and for the edge created when the origin or the destination are "extended", the cost is an arbitrary increasing function of the edge flow. □

The following lemma is used in the proof of Theorem 2.

Lemma 5. *A series-parallel network is a network with independent routes if and only if, for every pair of distinct flow vectors* $\hat{\mathbf{f}}$ *and* $\tilde{\mathbf{f}}$ *with equal total origin-destination flows, there is some route* r *such that* $\hat{f}_s \geq \tilde{f}_s$ *for all sections* s *of* r*, and* $\hat{f}_r \geq \tilde{f}_r$.

Proof. For either of the series-parallel networks in Figure 2, consider the following two flow vectors $\hat{\mathbf{f}}$ and $\tilde{\mathbf{f}}$. For the route r_1 passing through the edges e_1 and e_3, $\hat{f}_{r_1} = 0$ and $\tilde{f}_{r_1} = 2$; for the route r_2 passing through e_2 and e_4, $\hat{f}_{r_2} = 0$ and $\tilde{f}_{r_2} = 0$; for the route r_3 passing through e_1 and e_4, $\hat{f}_{r_3} = 1$ and $\tilde{f}_{r_3} = 0$; and for the route r_4 passing through e_2 and e_3, $\hat{f}_{r_4} = 1$ and $\tilde{f}_{r_4} = 0$. The only routes r such that $\hat{f}_r > \tilde{f}_r$ are r_3 and r_4. However, r_3 includes e_1 and r_4 includes e_3, and for these two edges the edge flow in $\hat{\mathbf{f}}$ is *less* than in $\tilde{\mathbf{f}}$. In view of Propositions 1 and 4, this example proves that for *any* series-parallel network in which routes are not independent there is a pair of distinct flow vectors $\hat{\mathbf{f}}$ and $\tilde{\mathbf{f}}$ for which a route r as above does not exist.

The converse, that, for any network G with independent routes, a route r as above does exist for every pair of distinct flow vectors with equal total origin-destination flows, will be proved by induction on the number of edges. The induction hypothesis is that this is true for any network with independent routes and a smaller number of edges than G. Let $\hat{\mathbf{f}}$ and $\tilde{\mathbf{f}}$ be two distinct flow vectors in G with equal total origin-destination flows. Since the flow vectors are distinct, G must have at least two edges. Therefore, it follows from Proposition 5 that G is the result of connecting two series-parallel networks, G' and G'', in series or in parallel. In the former case, one of these, say G'', has only one edge, e. Then, by similar arguments to those used in the proof of Lemma 3, the induction hypothesis implies that there is some route r' in G' such that $\hat{f}_s \geq \tilde{f}_s$ for all sections s of r', and $\hat{f}_{r'} \geq \tilde{f}_{r'}$. The route r in G obtained by appending e and d to r' has the same property. This is because every section s of r is (1) a section of r', (2) the zero-length walk consisting of d alone, or (3) the result of appending e and d to some section t of r' - in which case the flows through t and s are always equal. This completes the proof for the case in which G is the result of connecting G' and G'' in series. If G is the result of connecting the two networks in parallel, then, again by similar arguments to those used in the proof of Lemma 3, it follows from the induction hypothesis that there is some route r in G' or in G'' such that $\hat{f}_s \geq \tilde{f}_s$ for all sections s of r, and $\hat{f}_r \geq \tilde{f}_r$. Since r is also a route in G, this completes the proof. □

Proof of Theorem 2.

Suppose that G is a network with independent routes. Let $\mathbf{c}$ be a cost function, $\mathbf{f}^*$ a corresponding equilibrium with nonzero total origin-destination flow, and c^* the equilibrium cost. If $\mathbf{f}$ is another flow vector, with the same total origin-destination flow as $\mathbf{f}^*$, then, by Lemma 5, there is some route r such that $f_s \geq f_s^*$ for all sections s of r, and $f_r > f_r^*$. Since routes in G are independent, there is some section s of r of length one such that $f_s = f_r > f_r^* = f_s^*$. It follows that $c_r(\mathbf{f}) \geq c_r(\mathbf{f}^*)$, and if $\mathbf{c}$ is increasing, then $c_r(\mathbf{f}) > c_r(\mathbf{f}^*)$. Since $f_r > 0$, and, by definition, $c_r(\mathbf{f}^*) \geq c^*$, this proves that $\mathbf{f}^*$ is weakly Pareto efficient, and, moreover, is Pareto efficient if $\mathbf{c}$ is increasing.

Suppose now that routes in G are not independent. By Proposition 4, one of the networks in Figures 1 and 2 is embedded in G. As shown, for each of these three networks there is an increasing, additively separable cost function and a corresponding equilibrium that is not even weakly Pareto efficient. The same is true for every network in which one of these networks is embedded; the proof of this is based on the same arguments as in the proof of Theorem 1. □

Proof of Theorem 3.

Suppose that G is a network with independent routes. For a given assignment of cost functions, let σ be an equilibrium, and τ another strategy profile. If $\mathbf{f}(\tau) = \mathbf{f}(\sigma)$, then it follows from the equilibrium condition (3) that $c^i_{\tau(i)}(\mathbf{f}(\tau)) \geq c^i_{\sigma(i)}(\mathbf{f}(\sigma))$ for all users i, and equality holds if $\tau(i) = \sigma(i)$. Hence, condition (H) holds. Suppose, then, that $\mathbf{f}(\tau) \neq \mathbf{f}(\sigma)$. By Lemma 5, there is some route r such that $f_s(\tau) \geq f_s(\sigma)$ for all sections s of r, with strict inequality for r itself and, because of the independence of the routes, for some section s of r of length one. For all users i, $c^i_r(\mathbf{f}(\tau)) \geq c^i_r(\mathbf{f}(\sigma)) \geq \min_{s\in\mathcal{R}} c^i_s(\mathbf{f}(\sigma)) = c^i_{\sigma(i)}(\mathbf{f}(\sigma))$, and if i's cost function is increasing, then the first inequality is strict. Since $f_r(\tau) > 0$, the set of all users i with $\tau(i) = r$ has positive measure. This proves that the equilibrium σ is weakly Pareto efficient. Moreover, since $f_r(\tau) > f_r(\sigma)$, the set of all users i such that $\tau(i) = r$ but $\sigma(i) \neq r$ also has positive measure. As shown above, for each such i, if i's cost function is increasing, then $c^i_{\tau(i)}(\mathbf{f}(\tau)) > c^i_{\sigma(i)}(\mathbf{f}(\sigma))$. This proves that if *all* users have increasing cost functions, then (H) holds, and hence the equilibrium σ is hyper-efficient.

Suppose now that routes in G are not independent. There are increasing, additively separable cost functions for G and a corresponding Nash equilibrium that is not even weakly Pareto efficient. The proof of this is the same as the one given for Theorem 2. □

Proof of Proposition 8.

Suppose that $\mathbf{c}$ is an additively separable cost function satisfying assumptions (*i*) and (*ii*), and σ and τ two Nash equilibria. For each user i and route r, let $\sigma_r(i)$ be defined as 1 or 0 according to whether r is equal to or different from $\sigma(i)$, respectively. Then, (3) implies $0 \geq \sum_{r\in\mathcal{R}}(\sigma_r(i) - \tau_r(i))c^i_r(\mathbf{f}(\sigma))$, and a similar inequality holds with σ and τ interchanged. Hence, $0 \geq \sum_{r\in\mathcal{R}}(\sigma_r(i) - \tau_r(i))(c^i_r(\mathbf{f}(\sigma) - c^i_r(\mathbf{f}(\tau))) = \sum_{r\in\mathcal{R}}(\sigma_r(i) - \tau_r(i))(c_r(\mathbf{f}(\sigma) - c_r(\mathbf{f}(\tau)))$ by assumption (i). Integration over i gives $0 \geq \sum_{r\in\mathcal{R}}(f_r(\sigma) - f_r(\tau))(c_r(\mathbf{f}(\sigma) - c_r(\mathbf{f}(\tau)))$. Since $\mathbf{c}$ is additively separable, this implies

$$0 \geq \sum_{r \in \mathcal{R}} \Big(f_r(\sigma) - f_r(\tau) \Big) \sum_{\substack{s \\ s \text{ is a length one} \\ \text{section of r}}} \Big(c_s(\mathbf{f}(\sigma)) - c_s(\mathbf{f}(\tau)) \Big) =$$

$$= \sum_{\substack{s \\ s \text{ is a walk} \\ \text{of length one}}} \Big(c_s(\mathbf{f}(\sigma)) - c_s(\mathbf{f}(\tau)) \Big) \sum_{\substack{r \in \mathcal{R} \\ s \text{ is a section of r}}} \Big(f_r(\sigma) - f_r(\tau) \Big) =$$

$$= \sum_{\substack{s \\ s \text{ is a walk} \\ \text{of length one}}} \Big(c_s(\mathbf{f}(\sigma)) - c_s(\mathbf{f}(\tau)) \Big) \Big(f_s(\sigma) - f_s(\tau) \Big) .$$

On the other hand, it follows from assumption (*ii*) that each term in the last sum is nonnegative. Therefore, all terms must be zero. This implies that, for all walks s of length one, $c_s(\mathbf{f}(\sigma)) = c_s(\mathbf{f}(\tau))$. For if, for example, s is such that $c_s(\mathbf{f}(\sigma)) > c_s(\mathbf{f}(\tau))$, then $f_s(\sigma) > f_s(\tau)$ by assumption (*ii*), and hence the corresponding term in the above sum is positive. (Note that, if the cost of s is in fact an *increasing* function of f_s, then similar considerations give the stronger conclusion $f_s(\sigma) = f_s(\tau)$.) Since $\mathbf{c}$ is additively separable, it now follows from the assumptions that $c_r^i(\mathbf{f}(\sigma)) = c_r^i(\mathbf{f}(\tau))$ for all routes r and users i. In particular, the equilibrium costs in σ and τ are the same. □

References

1. Aashtiani, H. Z., Magnanti, T. L. (1981): Equilibria on a congested transportation network. SIAM Journal on Algebraic and Discrete Methods **2**, 213–226
2. Altman, E., Kameda, H. (2001): Equilibria for multiclass routing problems in multi-agent networks (mimeo)
3. Arnott, R., Small, K. (1994): The economics of traffic congestion. American Scientist, bf 82, 446–455
4. Beckmann, M., McGuire, C. B., Winston, C. B. (1956): Studies in the Economics of Transportation. Yale Univ. Press, New Haven, CT
5. Bell, M. G. H., Iida, Y. (1997): Transportation Network Analysis. Wiley, New Haven, CT
6. Calvert, B., Keady, G. (1993): Braess's paradox and power-law nonlinearities in networks. Journal of the Australian Mathematical Society (Series B) **35**, 1–22
7. Cohen, J. E., Horowitz, P. (1991): Paradoxical behaviour of mechanical and electrical networks, Nature **352**, 699–701
8. Cohen, J. E., Jeffries, C. (1997): Congestion resulting from increased capacity in single-server queuing networks. IEEE/ACM Transactions on Networking **5**, 305–310
9. Dafermos, S., Nagurney, A. (1984): On some traffic equilibrium theory paradoxes. Transportation Research **18B**, 101–110
10. Duffin, R. J. (1965): Topology of series-parallel networks. Journal of Mathematical Analysis and Applications **10**, 303–318
11. Frank, M. (1981). The Braess paradox. Mathematical Programming **20**, 283–302
12. Holzman, R., Law-Yone, N. (1997): Strong equilibrium in congestion games. Games and Economic Behavior **21**, 85–101

13. Konishi, H. (2001): Uniqueness of user equilibrium in transportation networks with heterogenous commuters (mimeo)
14. Law-Yone, N. (1995): Strong equilibrium in congestion games. M.Sc. Thesis, the Technion – Israel Institute of Technology (in Hebrew)
15. Milchtaich, I. (1996): Congestion games with individual-specific payoff functions. Games and Economic Behavior **13**, 111–124
16. Milchtaich, I. (2000): Generic uniqueness of equilibrium in large crowding games. Mathematics of Operations Research **25**, 349–364
17. Milchtaich, I. (2001): Social optimality and cooperation in large congestion games (mimeo)
18. Nagurney, A. (1999): Network Economics: A Variational Inequality Approach, 2nd ed. Kluwer, Boston
19. Newell, G. F. (1980): Traffic Flow on Transportation Networks. MIT Press, Cambridge, MA
20. Orda, A., Rom, R., Shimkin, N. (1993): Competitive routing in multiuser communication networks. IEEE/ACM Transactions on Networking **5**, 510–521
21. Riordan, J., Shannon, C. E. (1942): The number of two-terminal series-parallel networks. Journal of Mathematics and Physics **21**, 83–93
22. Roughgarden, T., Tardos, E. (2000): How bad is selfish routing? (mimeo)
23. Sheffi, Y. (1985): Urban Transportation Networks. Prentice-Hall, Englewood Cliffs, NJ
24. Steinberg, R., Zangwill, W. I. (1983): The prevalence of the Braess' paradox. Transportation Science **17**, 301–318
25. Voorneveld, M., Borm, P., van Megen, F., Tijs, S., Facchini, G. (1999): Congestion games and potentials reconsidered. International Game Theory Review **1**, 283–299

Essential Components of the Set of Weakly Pareto-Nash Equilibria for Multiobjective Generalized Games*

Hui Yang

Department of Mathematics, Guizhou University,
Guiyang, Guizhou, 550025, P.R. China
E-mail: hui-yang@163.com

Summary. In this paper, we introduce the notion of essential component of the set of weakly Pareto-Nash equilibria for multiobjective generalized games. We prove that, for every multiobjective generalized game (satisfying some continuity and convexity conditions), there exists at least one essential component of its weakly Pareto-Nash equilibria.

Key words: Multiobjective generalized game, weakly Pareto-Nash equilibrium, essential component

1 Introduction and Preliminaries

Recently, in game theory, much attention has been attracted to multicriteria models, i.e., games with multiple noncommensurable criteria, since they can be better applied to real-world situations. Since Shapley (Shapley, 1959) proposed the concept of equilibrium for multiobjective games, many concepts of solution have been proposed and many results on existence of such solutions have been obtained, see (Chose and Prasad, 1989), (Wang, 1993) and (Yu and Yuan, 1998) and references therein.

Also in game theory, the stability and perfection of Nash equilibria have become important topics, see (Vandame, 1991) and references therein. Many concepts of refinement for Nash equilibria has been proposed. Unfortunately, all of such refinements do not satisfy most so-called plausible requirements. Kohlberg and Mertens (Kohlberg and Mertens, 1986) introduced the notion of of stable set (what we call essential set here) of Nash equilibria, i.e., the subset is robust to perturbations of the game's payoffs or strategy sets, precisely speaking, every game nearby has a Nash equilibrium nearby such subset. They

* This project is supported by Natural Science Foundation of Guizhou Province.

proved that every finite game has finite components of its Nash equilibria and at least one component is essential. In recent years, such problems have been studied intensively and some existence results for essential components of the set of Nash equilibria and weakly Pareto-Nash equilibria have been obtained, see (Yu and Xiang, 1999), (Yu and Luo, 1999) and (Yang and Yu, 2002).

Aim of this paper is to establish the existence of essential components of the set of weakly Pareto-Nash equilibria for multiobjective generalized games. First we introduce the notion of essential components of the solution set of vector quasi-equilibrium problems and we prove the existence of such essential components. As applications, we show that every multiobjective generalized game (satisfying some continuity and convexity conditions) possesses at least one essential components of its weakly Pareto-Nash equilibria. Our results include corresponding results in the literature as special cases.

Let H be a real Banach space and C be a cone of H. A cone C is *convex* if and only if $C + C = C$, and *pointed* if and only if $C \cap (-C) = \{\theta\}$, where θ denotes the zero element of H. Let A be a subset of H, we denote $intA$ the topological interior of A in H. In the following context, E denotes a real Banach space and H denotes a real Banach space with a closed, convex and pointed cone C with $intC \neq \emptyset$.

Let X be a nonempty subset of E and $\varphi : X \times X \to H$ be a vector-valued function satisfying $\varphi(x,x) \notin intC$. Then a vector equilibrium problem is to find $x^* \in X$ such that

$$\varphi(x^*, y) \notin intC, \forall y \in X.$$

If, furthermore, we consider a set-valued map $G : X \to 2^X$, then a vector quasi-equilibrium problem is to find $x^* \in X$ such that $x^* \in G(x^*)$ and

$$\varphi(x^*, y) \notin intC, \forall y \in G(x^*).$$

If $H = R$ and $C = [0, +\infty)$, then a vector equilibrium (resp., quasi-equilibrium) problem reduces to an equilibrium (resp., quasi-equilibrium) problem.

In order to establish the existence theorem of the solutions for a vector quasi-equilibrium problem, we first recall some notions. The following two definitions can be found in (Luc, 1989).

Definition 1. *Let X be a nonempty subset of E and $f : X \to H$ be a vector-valued function. f is said to be C-continuous at $x_0 \in X$ if, for any open neighborhood V of θ in H, there exists an open neighborhood U of x_0 in X such that, for all $x \in U$,*

$$f(x) \in f(x_0) + V + C.$$

f is said to be C-continuous on X if it is C-continuous at any point of X.

Definition 2. *Let X be a nonempty convex subset of E and $f : X \to H$ be a vector-valued function. f is said to be C-concave if, for any $x_1, x_2 \in X$ and any $\lambda \in [0,1]$,*

$$f(\lambda x_1 + (1-\lambda)x_2) - (\lambda f(x_1) + (1-\lambda) f(x_2)) \in C.$$

f is said to be C-convex if $-f$ is C-concave.

Throughout this paper, 2^X denotes the collection of all nonempty subsets of X.

Definition 3. *Let X and Y be two Hausdorff topological spaces and $F : Y \to 2^X$ be a set-valued map, then*

(i) F is upper semicontinuous at $y_0 \in Y$ if for each open set U in X with $U \supset F(y_0)$, there exists an open neighborhood $O(y_0)$ of y_0 such that $U \supset F(y)$ for any $y \in O(y_0)$;

(ii) F is upper semicontinuous on Y if F is upper semicontinuous at every point in Y; and

(iii) F is an usco map if F is upper semicontinuous on Y and $F(y)$ is compact for every $y \in Y$;

(iv) F is lower semicontinuous at $y_0 \in Y$ if for each open set U in X with $U \cap F(y_0) \neq \emptyset$, there exists an open neighborhood $O(y_0)$ of y_0 such that $U \cap F(y) \neq \emptyset$ for each $y \in O(y_0)$; and

(v) F is continuous if F is both upper and lower semicontinuous.

The following lemma is Lemma 1.1 of (Yang and Yu, 2002).

Lemma 1. *Let H be a Banach space with a closed, convex and pointed cone C with $intC \neq \emptyset$. Then we have $intC + C \subset intC$.*

Lemma 2. *If f is C-concave, then the set $D := \{x \in X : f(x) \in intC\}$ is convex.*

Proof. Let $x_1, x_2 \in D$, then $f(x_1) \in intC$ and $f(x_2) \in intC$. Since $intC$ is convex, we have $\lambda f(x_1) + (1-\lambda) f(x_2) \in intC$ for any $\lambda \in [0,1]$. It follows from the C-concavity of f that $f(\lambda x_1 + (1-\lambda)x_2) \in \lambda f(x_1) + (1-\lambda) f(x_2) + C$. By Lemma 1.1, $f(\lambda x_1 + (1-\lambda)x_2) \in intC$, i.e., $\lambda x_1 + (1-\lambda)x_2 \in D$. Hence D is convex and our proof is complete. □

Lemma 3. *Let X be a nonempty compact subset of Banach space E , H be a Banach space and C be a closed convex pointed cone with $intC \neq \emptyset$. Suppose that $G : X \to 2^X$ is a lower semicontinuous set-valued map and $\varphi : X \times X \to H$ is C-continuous on $X \times X$. Then $D := \{x \in X | \forall y \in G(x), \varphi(x,y) \notin intC\}$ is a closed subset.*

Proof. Let $\{x_n\}$ be an arbitrary sequence in D with $x_n \to x_0$. If $x_0 \notin D$, then there exists $y_0 \in G(x_0)$ such that $\varphi(x_0, y_0) \in intC$. By the lower semicontinuity of G, there exists $y_n \in G(x_n)$ such that $y_n \to y_0$. By the C-continuity of φ, we have $\varphi(x_n, y_n) \in intC$ for sufficiently large n, a contradiction. Therefore $x_0 \in D$. Hence D is closed and the proof is complete. □

2 Essential Components of the Solution Set for VQEP

In this section, we first give an existence result of solutions for vector quasi-equilibrium problems (briefly, VQEP). Then we introduce the notion of essential components of the solution set of VQEPs and prove that every VQEP (satisfying some continuity and convexity conditions) possesses at least one essential component of it solution set.

The following lemma is Theorem 1.1 of (Yang and Yu, 2002).

Lemma 4. *Let X be a nonempty convex compact subset of E. Suppose that $\varphi : X \times X \to H$ satisfies the following conditions:*
(i) for each fixed $y \in X, x \to \varphi(x,y)$ is C-continuous;
(ii) for each fixed $x \in X, y \to \varphi(x,y)$ is C-concave; and
(iii) for each $x \in X, \varphi(x,x) \notin intC$.
Then there exists $x^* \in X$ such that $\varphi(x^*,y) \notin intC$ for all $y \in X$.

Theorem 1. *Let X be a nonempty convex compact subset of E. Suppose that $G : X \to 2^X$ is continuous with nonempty closed convex values, and $\varphi : X \times X \to H$ satisfies the following conditions:*
(i) φ is C-continuous on $X \times X$;
(ii) for each fixed $x \in X, y \to \varphi(x,y)$ is C-concave; and
(iii) for each $x \in X, \varphi(x,x) \notin intC$.
Then there exists $x^* \in X$ such that $x^* \in G(x^*)$ and $\varphi(x^*,y) \notin intC$ for all $y \in G(x^*)$.

Proof. By way of contradiction, suppose that for any $x \in X$, either $x \notin G(x)$ or there is $y \in G(x)$ such that $\varphi(x,y) \in intC$. Saying that $x \notin G(x)$ implies that there exists $p \in E^*$ such that $\langle p,x\rangle > \sup_{y\in G(x)}\langle p,y\rangle$. Set

$$\begin{cases} \Delta_0 := \{x \in X \mid \exists y \in G(x), \varphi(x,y) \in intC\} \\ \Delta_p := \{x \in X \mid \langle p,x\rangle > \sup_{y\in G(x)}\langle p,y\rangle\}, p \in E^*, \end{cases}$$

Then, obviously, we have

$$X \subset \Delta_0 \cup [\cup_{p\in E^*}\Delta_p].$$

By Lemma 1.3, Δ_0 is open. By Proposition 2 of (Aubin and Ekeland, 1984)(p.122), Δ_p is open for each $p \in E^*$. Since X is compact, there exist $p_1, p_2, ..., p_n \in E^*$ such that

$$X \subset \Delta_0 \cup [\bigcup_{i=1}^{n} \Delta_i], \Delta_i = \Delta_{p_i}.$$

Now we introduce functions $a_i : X \to [0,1], i = 1,2,...,n$, defined by

$$a_i(x) = \frac{d(x, X \setminus \Delta_i)}{\sum_{i=0}^{n} d(x, X \setminus \Delta_i)}, \forall x \in X,$$

where d is the metric induced by the norm of E. Then every $a_i(x)$ is continuous, $a_i(x) > 0$ when $x \in \Delta_i$, $a_i(x) = 0$ when $x \notin \Delta_i$, and $\sum_{i=0}^{n} a_i(x) = 1$.

We then introduce the function

$$\psi(x,y) = a_0(x)\varphi(x,y) + [\sum_{i=1}^{n} a_i(x)\langle p_i, x-y\rangle]z,$$

where z is a given point in $intC$. By (i), (ii) and (iii), ψ is C-continuous on $X \times X$, C-concave with respect to y, and satisfies $\psi(x,x) \notin intC$ for all $x \in X$. By Lemma 2.1, there exists $x^* \in X$ such that

$$\psi(x^*,y) \notin intC, \forall y \in X. \tag{2.1}$$

We shall complete our proof by contradicting (2.1). Indeed, let $I = \{i : i \neq 0, x^* \in \Delta_i\}$. If $x^* \in \Delta_0$, then we have $a_0(x^*) > 0$ and there exists $y^* \in G(x^*)$ such that $\varphi(x^*,y^*) \in intC$. Note that $\langle p_i, x^* - y\rangle > 0$ for each $y \in G(x^*)$ and

$$\psi(x^*,y^*) = a_0(x^*)\varphi(x^*,y^*) + [\sum_{i\in I} a_i(x^*)\langle p_i, x^* - y^*\rangle]z.$$

Then $\sum_{i\in I} a_i(x^*)\langle p_i, x^* - y^*\rangle = 0$ if $I = \emptyset$ and $\sum_{i\in I} a_i(x^*)\langle p_i, x^* - y^*\rangle > 0$ if $I \neq \emptyset$. Therefore we have that

$$\psi(x^*,y^*) = \varphi(x^*,y^*) \in intC, \text{ if } I = \emptyset$$

and by Lemma 1.1,

$$\psi(x^*,y^*) \in intC + intC \subset intC, \text{ if } I \neq \emptyset,$$

both of which contradict (2.1)

If $x^* \notin \Delta_0$, then $a_0(x^*) = 0, I \neq \emptyset$ and $\sum_{i\in I} a_i(x^*)\langle p_i, x^* - y\rangle > 0$ for each $y \in G(x^*)$. Thus we have

$$\psi(x^*,y) = [\sum_{i\in I} a_i(x)\langle p_i, x^* - y\rangle]z \in intC, \forall y \in G(x^*),$$

a contradiction to (2.1). The proof is complete. □

Now let X be a nonempty convex compact subset of a Banach space E and H be a Banach space with a closed, convex and pointed cone C with $intC \neq \emptyset$. Let Y be the collection of all vector-valued functions $\varphi : X \times X \to H$ such that (i) $\varphi(x,y)$ is C-continuous on $X \times X$; (ii) for each $x \in X$, $y \to \varphi(x,y)$ is C-concave; (iii) for each $x \in X$, $\varphi(x,x) = \theta$; and (iv) $\sup_{(x,y)\in X\times X} \|\varphi(x,y)\| < +\infty$.

Let K be the collection of all set-valued maps $G : X \to 2^X$ such that (i) for each $x \in X, G(x)$ is convex and closed and (ii) G is continuous on X.

Let $M = Y \times K$. For any $u^1 = (\varphi^1, G^1), u^2 = (\varphi^2, G^2) \in Y$, define

$$\rho(u^1, u^2) = \sup_{(x,y)\in X\times X} \|\varphi^1(x,y) - \varphi^2(x,y)\| + \sup_{x\in X} h(G^1(x), G^2(x)),$$

where h is the Hausdorff metric defined on X. Clearly, (M, ρ) is a metric space.

For each $u \in M$, denote by $F(u)$ the set of all solutions of vector quasi-equilibrium problem u. Then by Theorem 2.1, $F(u) \neq \emptyset$ and thus F defines a set-valued map from M into X

Theorem 2. *$F : M \to 2^X$ is an usco map.*

Proof. Since X is compact, by Theorem 7.1.16 of (Klein and Thompson, 1984), it suffices to show that F is a closed map, i.e., the graph $Graph(F)$ of F is closed in $M \times X$, where

$$Graph(F) = \{(u, x) \in M \times X : x \in F(u)\}.$$

Let $\{(u^n, x^n)\}$ be any sequence in $Graph(F)$ with $(u^n, x^n) \to (u^*, x^*) \in M \times X$. Denote $u^n = (\varphi^n, G^n), u^* = (\varphi^*, G^*)$, then $\varphi^n \to \varphi^*, G^n \to G^*, x^n \in G^n(x^n), x^n \to x^*$ and $\varphi^n(x^n, y) \notin intC$ for any $y \in G^n(x^n)$. Suppose that $(u^*, x^*) \notin Graph(F)$, then there exists some $y^* \in G^*(x^*)$ such that $\varphi(x^*, y^*) \in intC$. Since $intC$ is an open set, there exists an open ball $B(\theta, t)$ of θ such that $\varphi(x^*, y^*) + B(\theta, t) \subset intC$. Because

$$\begin{aligned} h(G^n(x^n), G^*(x^*)) &\leq h(G^n(x^n), G^*(x^n)) + h(G^*(x^n), G(x^*)) \\ &\leq \rho(u^n, u^*) + h(G(x^n), G^*(x^*)) \to 0, \end{aligned}$$

we have $h(G^n(x^n), G^*(x^*)) \to 0$. Also, since

$$\begin{aligned} d(x^*, G^*(x^*)) &\leq d(x^*, x^n) + d(x^n, G^n(x^n)) + h(G^n(x^n), G^*(x^*)) \\ &= d(x^*, x^n) + h(G^n(x^n), G^*(x^*)) \to 0, \end{aligned}$$

we obtain that $x^* \in G^*(x^*)$. Since φ^* is C-continuous on $X\times X$ and $(x^n, y^*) \to (x^*, y^*)$, there exists N_0 such that, for any $n \geq N_0$,

$$\varphi^*(x^n, y^*) \in \varphi^*(x^*, y^*) + B(\theta, \frac{1}{2}t) + C,$$

where $B(\theta, \frac{1}{2}t)$ is the ball with center at θ and radius $\frac{1}{2}t$. Moreover, since $\varphi^n \to \varphi^*$, there exists N_1 with $N_1 \geq N_0$ such that, for any $n \geq N_1$, $\varphi^n(x, y) \in \varphi^*(x, y) + B(\theta, \frac{1}{2}t)$ for all $(x, y) \in X \times X$. Hence we have, for any $n \geq N_1$,

$$\varphi^n(x^n, y^*) \in \varphi^*(x^n, y^*) + B(\theta, \frac{1}{2}t) \subset \varphi^*(x^*, y^*) + B(\theta, t) + C \subset intC$$
$$+C \subset intC,$$

a contradiction. Therefore $(u^*, x^*) \in Graph(F)$. Now we can conclude that $Graph(F)$ is closed and the proof is complete. □

For each $u \in M$, the *component of a point* $x \in F(u)$ is the union of all the connected subsets of $F(u)$ containing x. Note that the components are connected closed subsets of $F(u)$ and thus are connected compact, see (Engelking, 1989), p.356. It is easy to see that the components of two distinct points of $F(u)$ either coincide or are disjoint, so that all components constitute a decomposition of $F(u)$ into connected pairwise disjoint compact subsets, i.e.,

$$F(u) = \bigcup_{\alpha \in \Lambda} S_\alpha$$

where, Λ being an index set, for each $\alpha \in \Lambda, S_\alpha$ is a nonempty connected compact subset of $F(u)$ and, for any $\alpha, \beta \in \Lambda(\alpha \neq \beta)$, $S_\alpha \cap S_\beta = \emptyset$.

Definition 4. *Let $u \in M$ and S_α be a component of $F(u)$. For some $\alpha \in \Lambda, S_\alpha$ is said to be an essential component of $F(u)$ if, for each open set $O \supset S_\alpha$, there exists $\delta > 0$ such that for any $u' \in M$ with $\rho(u,u') < \delta, F(u') \cap O \neq \emptyset$.*

The following lemma is Lemma 3.1 of (Yu and Luo, 1999).

Lemma 5. *Let B, D be two nonempty convex compact subsets of a metric space E. Then*

$$h(B, \lambda B + \mu D) \leq h(B, D),$$

where h is the Hausdorff metric defined on $E, \lambda \geq 0, \mu \geq 0$ and $\lambda + \mu = 1$.

Theorem 3. *For each $u \in M$, there exists at least one essential component of $F(u)$.*

Proof. By way of contradiction, suppose that there is an element $u = (\varphi, G) \in M$ such that no one of components of its weakly Pareto-Nash equilibria is essential, i.e., for each $\alpha \in \Lambda$, there is an open set $O_\alpha \supset F_\alpha(u)$ such that for any $\delta > 0$, there is $u_\alpha = (\varphi_\alpha, G_\alpha)$ with $\rho(u,u') < \delta$, but $F(u_\alpha) \cap O_\alpha = \emptyset$.

The same arguments as in the proof of Theorem 3 of (Jiang, 1963) state that, by the compactness of $F(u)$, there are two finite open coverings $\{V_i\}_{i=1}^n$ and $\{W_i\}_{i=1}^n$ of $F(u)$ satisfying

(i) $\overline{W}_i \subset V_i$, where $\overline{W}_i$ denotes the closure of W_i;

(ii) $V_i \cap V_j = \emptyset, \forall i \neq j$;

(iii) V_i containes at least one of F_{α_i} and $O_{\alpha_i} \supset V_i$.

By the upper semicontinuity of F (by Theorem 2.2), for open set $\cup_{i=1}^n W_i$ containing $F(u)$, there is a $\delta_0 > 0$ such that $\cup_{i=1}^n W_i \supset F(v)$ for each $v \in M$ with $\rho(u,v) < \delta_0$. Therefore there exists $u_{\alpha_i} \in M$ with $\rho(u, u_{\alpha_i}) < \delta$ such that $S(u_{\alpha_i}) \cap O_{\alpha_i} = \emptyset$ and thus $S(u_{\alpha_i}) \cap V_i = \emptyset$.

Denote $u_{\alpha_i} = (\varphi_{\alpha_i}, G_{\alpha_i}), i = 1, 2, ..., n$ and define a VQEP $v = (\varphi', G')$ as follows:

$$\varphi'(x,y) = \begin{cases} \varphi(x,y), & \text{if } x \in X \setminus \cup_{i=1}^n V_i, y \in X, \\ \varphi_{\alpha_i}(x,y), & \text{if } x \in \overline{W}_i, y \in X, \\ \lambda_i(x)\varphi(x,y) + \mu_i(x)\varphi_{\alpha_i}(x,y), & \text{if } x \in V_i \setminus \overline{W}_i, y \in X, \end{cases}$$

$$G'(x) = \begin{cases} G(x), & \text{if } x \in X \backslash \cup_{i=1}^{n} V_i, \\ G_{\alpha_i}(x), & \text{if } x \in \overline{W}_i, \\ \lambda_i(x)G(x) + \mu_i(x)G_{\alpha_i}(x), & \text{if } x \in V_i \setminus \overline{W}_i, \end{cases}$$

where

$$\lambda_i(x) = \frac{d(x, \overline{W}_i)}{d(x, \overline{W}_i) + d(x, X \backslash \cup_{i=1}^{n} V_i)},$$

$$\mu_i(x) = \frac{d(x, X \backslash \cup_{i=1}^{n} V_i)}{d(x, \overline{W}_i) + d(x, X \backslash \cup_{i=1}^{n} V_i)}.$$

Note that $\lambda_i(x), \mu_i(x)$ is continuous, $\lambda_i(x) \geq 0, \mu_i(x) \geq 0$, and $\lambda_i(x) + \mu_i(x) = 1, \forall x \in X$. Thus, by the definition of φ', F', we have (1) $\varphi'(x,y)$ is C-continuous on $X \times X$; (2) $\forall x \in X, y \to \varphi'(x,y)$ is C-concave; (3) $\varphi'(x,x) = \theta, \forall x \in X$; (4) $\sup_{(x,y)\in X \times X} \|\varphi'(x,y)\| < +\infty$; (5) $G'(x)$ is nonempty closed convex subset of X; (6) G' is continuous on X. Hence $v \in M$.

By Lemma 2.3,

$$h(G(x), \lambda_i(x)G(x) + \mu_i(x)G_{\alpha_i}(x)) \leq h(G(x), G_{\alpha_i}(x)).$$

Also, since

$$\|\varphi(x,y) - (\lambda_i(x)\varphi(x,y) + \mu_{\alpha_i}(x,y))\| = \mu_i(x)\|\varphi(x,y) - \varphi_{\alpha_i}(x,y)\|,$$

we have

$$\rho(u,v) = \sup_{(x,y)\in X \times X} \|\varphi(x,y) - \varphi'(x,y)\| + \sup_{x \in X} h(G(x), G'(x)) < \delta,$$

and thus $S(v) \subset \cup_{i=1}^{n} W_i$.

Take any $x_0 \in S(v)$, i.e., $x_0 \in G'(x_0), \varphi'(x_0,y) \notin intC, \forall y \in G'(x_0)$. Then there exists i_0 such that $x_0 \in W_{i_0} \subset \overline{W}_{i_0}$, $G'(x_0) = G_{\alpha_{i_0}}(x_0), \varphi'(x_0,y) = \varphi_{\alpha_{i_0}}(x_0,y) \notin intC, \forall y \in G'(x_0) = G_{\alpha_{i_0}}(x_0)$, i.e., $x_0 \in S_{\alpha_{i_0}}(u_{\alpha_{i_0}})$, $x_0 \in S_{\alpha_{i_0}}(u_{\alpha_{i_0}}) \cap W_{i_0} \subset S_{\alpha_{i_0}}(u_{\alpha_{i_0}}) \cap V_{i_0}$, a contradiction.

Hence there is at least one essential component of $S(u)$. □

3 Essential Components of the Set of Weakly Pareto-Nash equilibria

Let $N := \{1, 2, ..., n\}$ be the set of players. For each $i \in N$, let X_i be a nonempty compact convex subset of a normed space E_i. Denote $X := \prod_{i=1}^{n} X_i, X_{\hat{i}} := \prod_{j \in N \backslash i} X_j$. A finite-player multiobjective generalized game in its strategic form (also called a normal form) $\Gamma := (X_i, G_i, F^i)_{i \in N}$, is defined as follows: for each $i \in N$, X_i is the set of player i's strategies; $G_i : X_{\hat{i}} \to 2^{X_i}$ is player i's feasible strategy correspondence; each $F^i : X \to R^{k_i}$ is the vector-valued payoff of player i, where k_i is a positive integer.

For each positive integer m, denote

$$R^m_+ := \{(u^1, ..., u^m) \in R^m : u^j \geq 0 \text{ for all } j = 1, ..., m\}.$$

and

$$intR^m_+ := \{(u^1, ..., u^m) \in R^m : u^j > 0 \text{ for all } j = 1, ..., m\}.$$

For R^m, we take the norm $\|r\| = \sum_{i=1}^m |r_i|$ where $r = (r_1, ..., r_m) \in R^m$. Without loss of generality, we assume that $k_1 \leq k_2 \leq \cdots \leq k_n$.

Definition 5. *A strategy profile $x^* \in X$ is called a weakly Pareto-Nash equilibrium point of a multiobjective generalized game $\Gamma := (X_i, G_i, F^i)_{i\in N}$ if, for each $i \in N, x_i^* \in G_{\hat{i}}(x_{\hat{i}}^*)$ and*

$$F^i(y_i, x_{\hat{i}}^*) - F^i(x_i^*, x_{\hat{i}}^*) \notin intR^{k_i}_+, \text{ for all } y_i \in G_{\hat{i}}(x_{\hat{i}}^*).$$

If $k_i = 1$ for all $i = 1, ..., n$, then the multiobjective generalized game is just a usual generalized game in the usual sense and the weakly Pareto-Nash equilibria reduce to Nash equilibria.

In order to prove our existence theorem of weakly Pareto-Nash equilibria for multiobjective generalized games, we give the following lemmas which can be proved easily and hence the proof are omitted.

Lemma 6. *Let $f : X \to R^m$ be a vector-valued function, where $f = (f_1, \cdots, f_m)$. Then f is R^m_+-continuous if and only if f_i is lower semicontinuous for every $i = 1, ..., m$.*

Lemma 7. Lemma 3.2 *Let X be a convex subset of a normed space and $f : X \to R^m$ be a vector function, where $f = (f_1, \cdots, f_m)$. Then f is R^m_+-concave if and only if f_i is concave for every $i = 1, ..., m$.*

By applying Theorem 2.1, we obtain the following existence theorem.

Theorem 4. *Let $\Gamma := (X_i, G_i, F^i)_{i\in N}$ be a multiobjective generalized game, where $F^i = (f_1^i, ..., f_{k_i}^i)$. Suppose that Γ satisfies the following conditions:*

(i) for each $i \in N, X_i$ is a nonempty convex compact subset of a normed space E_i;

(ii)for each $i \in N, G_i$ is continuous on $X_{\hat{i}}$ with nonempty closed convex values;

(iii)for each $i \in N$ and each $j = 1, ..., k_i$, f_j^i is continuous on X; and

(iv) for fixed $i \in N$ and each fixed $x_{\hat{i}} \in X_{\hat{i}}, u_i \to f_j^i(u_i, x_{\hat{i}})$ is concave for every $j = 1, ..., k_i$.

Then there exists a weakly Pareto-Nash equilibrium point x^* of Γ.

Proof. Consider the following vector quasi-equilibrium problem $u = (\varphi, G)$, where $G : X \to 2^X, \varphi : X \times X \to R^{k_n}$ are defined by: $\forall x, y \in X$,

$$G(x) = \prod_{i=1}^n G_i(x_{\hat{i}}), \varphi(x, y) = \sum_{i=1}^n \varphi_i(x, y) \tag{3.1}$$

where

$$\varphi_i(x,y) = (\underbrace{F^i(y_i,x_{\hat{i}}) - F^i(x_i,x_{\hat{i}})}_{k_i \text{ components}};$$

$$\underbrace{f_1^i(y_i,x_{\hat{i}}) - f_1^i(x_i,x_{\hat{i}}),\cdots,f_1^i(y_i,x_{\hat{i}}) - f_1^i(x_i,x_{\hat{i}})}_{k_n - k_i \text{ components}}) \in R^{k_n}. \quad (3.2)$$

It can be routinely checked that
(i) G is continuous on X with nonempty closed convex values;
(ii)φ is $R_+^{k_n}$-continuous on $X \times X$ (by Lemma 3.1);
(iii) for each fixed $x \in X, y \to \varphi(x,y)$ is $R_+^{k_n}$-concave (by Lemma 3.2); and
(iv) for each $x \in X, \varphi(x,x) = 0 \notin intR_+^{k_n}$.
Therefore by Theorem 2.1, there exists $x^* \in X$ such that $x^* \in G(x^*)$ and $\varphi(x^*,y) \notin intR_+^{k_n}$ for all $y \in G(x^*)$.

For each $i \in N$ and each $y_i \in G_{\hat{i}}(x_{\hat{i}}^*)$, set $y = (y_i, x_{\hat{i}}^*) \in G(x^*)$, then $\varphi_i(x^*,y) = \varphi(x^*,y) \notin intR_+^{k_n}$. If $F^i(y_i,x_{\hat{i}}^*) - F^i(x_i^*,x_{\hat{i}}^*) \in intR_+^{k_i}$, then $f_j^i(y_i,x_{\hat{i}}^*) - f_j^i(x_i^*,x_{\hat{i}}^*) \in intR_+$ for each $j = 1,...,k_i$ and $\varphi_i(x^*,y) \in intR_+^{k_n}$ which contradicts that $\varphi_i(x^*,y) \notin intR_+^{k_n}$. Hence $F^i(y_i,x_{\hat{i}}^*) - F^i(x_i^*,x_{\hat{i}}^*) \notin intR_+^{k_i}, \forall i \in N, \forall y_i \in G_i(x_{\hat{i}}^*)$, i.e., x^* is a weakly Pareto-Nash equilibrium point of the game Γ. The proof is complete. □

Now for each $i \in N$, let X_i be a nonempty convex compact subset of a normed space E_i and Z be the collection of vector functions $z = (F^1,...,F^n)$ such that the conditions (ii), (iii), (iv) in Theorem 3.1 hold and (iv) $\sup_{x\in X}\sum_{i=1}^n \|F^i(x)\| < +\infty$.. Let L be the collection of all correspondences $(G_1,G_2,...,G_n)$ such that (i) for each $i \in N$ and for each $x_{\hat{i}} \in X_{\hat{i}}, G_i(x_{\hat{i}})$ is a nonempty compact convex subset of X_i; and (ii) for each $i \in N, G_i$ is continuous on $X_{\hat{i}}$.

Let $P = Z \times L$. For any $p_1 = (F_1^1,...,F_1^n; G_1^1,...,G_n^1)$, $p_2 = (F_2^1,...,F_2^n; G_1^2,...,G_n^2) \in P$, define

$$\rho_1(p_1,p_2) = \sup_{x\in X}\sum_{i=1}^n \|F_1^i(x) - F_2^i(x)\| + \sup_{x\in X}\sum_{i=1}^n h_i(G_i^1(x_{\hat{i}}),G_i^2(x_{\hat{i}})),$$

where for each $i \in N, h_i$ is the Hausdorff metric defined on X_i. Clearly, (P,ρ_1) is a metric space. Each $p \in P$ determines a multiobjective generalized game, denoted also by p. Denote by $E(p)$ the set of all weakly Pareto-Nash equilibria of the game p. By Theorem 3.1, for each $p \in P$, $E(p) \neq \emptyset$. Similar to Definition 2.1, we can define essential component of $E(p)$. In order to establish the existence of essential components of $E(p)$, we need the following lemma, see (Yang and Yu, 2002):

Lemma 8. *For metric spaces $(M,\rho), X$ and (P,ρ_1), let $F : M \to 2^X$ and $B : P \to 2^X$ be two set-valued maps. Suppose that there exists a continuous map*

$S : P \to M$ such that $B(p) \supset F(S(p))$ for each $p \in P$. Suppose furthermore that there exists at least one essential component of $F(u)$ for each $u \in M$. Then there exists at least one essential component of $B(p)$ for each $p \in P$.

Theorem 5. *For each $p \in P$, there exists at least one essential component of $E(p)$.*

Proof. For any $p \in P$, define $S : P \to M$ by $S(p) = u = (\varphi, G)$, where (G, φ) is defined by (3.1) and (3.2) as in the proof of Theorem 3.1. For any $p_1 = (F_1^1, ..., F_1^n; G_1^1, ..., G_1^n), p_2 = (F_2^1, ..., F_2^n; G_2^1, ..., G_2^n) \in P$, if $\rho_1(p_1, p_2) < \varepsilon$, then

$$\begin{aligned}\rho(T(p_1), T(p_2)) = & \sup_{(x,y)\in X\times X} \| \sum_{i=1}^{n} [\varphi_i^1(x,y) - \varphi_i^2(x,y)] \| \\ & + \sup_{x\in X} h(\prod_{i=1}^{n} F_i^1(x_{\hat{i}}), \prod_{i=1}^{n} F_i^2(x_{\hat{i}})) \\ & \leq 2k_n \sup_{x\in X} \sum_{i=1}^{n} ||F_i^1(x) - F_i^2(x)|| + \sup_{x\in X} \sum_{i=1}^{n} h_i(G_i^1(x_{\hat{i}}), G_i^2(x_{\hat{i}})) \\ & \leq 2k_n \rho_1(p_1, p_2) < 2k_n \varepsilon,\end{aligned}$$

Hence, S is continuous. The same argument as in the proof of Theorem 3.1 states that $E(p) \supset F(S(p))$. Thus by Theorem 2.3 and Lemma 3.3, there exists at least one essential component of $E(p)$. □

References

1. Shapley, L.S., (1959): Equilibria in games with vector payoffs. Naval Research Logistics Quarterly, **6**, 57-61.
2. Chose, D., Prasad, U.R. (1989): Concepts in two-person multicriteria games. J. Opti. Theory Appl. **63**, 167-189.
3. Wang, S.Y. (1993): Existence of a Pareto equilibrium. J. Opti. Theory Appl. **79**, 373-384.
4. Yu, J., Yuan, G. X.-Z. (1998): The study of Pareto equilibria for multiobjective games by fixed point and Ky Fan minimax inequality methods. Computers Math. Appl., **35**, 17-24.
5. Van Damme, E. (1991): Stability and Perfection of Nash Equilibria. Springer-Verlag, Berlin.
6. Kohlberg, E., Mertens, J.F. (1986): On the strategic stability of equilibria. Econometrica, **54**, 1003-1037.
7. Yu, J., Xiang, S.W. (1999): On essential components of the Nash equilibria. Nonlinear Analysis, TMA, **38**, 259-264.
8. Yu, J. and Q. Luo (1999): On essential components of the solution set of generalized games. J. Math. Anal. Appl., **230**, 303-310.
9. Yang, H. and J. Yu (2002a): On Essential Components of the Set of Weakly Pareto-Nash Euilibrium Points. Applied Math. Letters. **15**, 553-560.
10. Luc, D.T., (1989): Theory of Vector Optimization. Springer-Verlag, Berlin.

11. Aubin, J.P. and Ekeland, I. (1984): Applied Nonlinear Analysis. John Wiley and Sons, New York.
12. E. Klein, E. and A.C. Thompson, E. (1984): Theory of correspondences. John Wiley & Sons.
13. Engelking, E. (1989): General Topology. Heldermann Verlag Berlin.
14. J.H. Jiang, J.H. (1963): Essential component of the set of fixed points of the multivalued maps and its applications the theory of games. Scientia Sinica, **12**, 951-964.

Part III

Applications to the Natural and Social Sciences

Discretization of Information Collecting Situations and Continuity of Compensation Rules

R. Brânzei[1], F. Scotti[2], S. Tijs[3], and A. Torre[2]

[1] "Al. I. Cuza" University, Faculty of Computer Science,
11, Carol I Bd., 6600 Iacomputablesi, Romania
E-mail: branzeir@infoiasi.ro
[2] Università di Pavia, Dipartimento di Matematica,
via Ferrata 1, 27100 Pavia, Italy
E-mail: scotti@dimat.unipv.it
E-mail: atorre@dimat.unipv.it
[3] Tilburg University,
CentER and Department of Econometrics and Operations Research,
11, Carol I Bd., 6600 Iacomputablesi, Romania
E-mail: S.H.Tijs@kub.nl

Summary. A canonical procedure is described which associates to each infinite information collecting situation a related information collecting situation with finite state and action spaces, in such a way that the two corresponding IC-games are near to each other. Compensations for informants are then shown to be also near to each other in the two IC-situations, provided that they are based on continuous compensation rules.

Key words: Bi-monotonic allocation scheme, Cooperative game, Core, Discretization, Information collecting situation

1 Introduction

In this paper information collecting (IC) situations and games (cf. [2, 4, 12]) are central. They model decision-making situations where the outcome of any decision depends on the state of the nature and the decision-maker is imperfectly informed. Collecting information from available agents who are more informed about the situation creates the potential for better decisions. More refined information to support the decision-making process yields, in principle, additional reward which is the source to compensate the informants.

Different procedures for collecting information have given rise to natural compensation rules in the context of cooperative game theory. Relevant for the class of information collecting games are marginal based compensation rules and bi-monotonic allocation schemes.

The state space and the action space in an information collecting situation can be infinite. We consider possibilities of approximating such an IC-situation with a finite IC-situation where the state space and the action space are finite and we relate the original IC-game with the IC-games of finite approximations. The considered approximations turn out to be good in the sense that the corresponding games are nearby. So continuity properties of relevant solutions for the class of IC-games are interesting. We compare relevant solutions such as the core [5], the set of bi-monotonic allocation schemes [4] and marginal-based allocation rules [12] of the approximate game with those of the original game, extending results from [6].

The outline of the paper is as follows. We introduce IC-situations and the corresponding games, look at properties and also at the subclass of IC-games, where a certain concavity condition holds. Furthermore we study compensation rules and continuity properties of solution concepts which are relevant for the class of IC-games and we present procedures for discretizing the state space (going from an infinite state space to a finite one) and for reducing the dimension of the action space, respectively.

2 Information Collecting Situations and Games

Information collecting situations model cooperative behavior of agents when one of them (the decision-maker) is facing uncertainty due to informational deficiencies when making decisions and the others (the informants) can provide additional information about the state of the nature.

In an information collecting (IC) situation the decision-maker has to decide which action to take in order to maximize his expected reward which depends upon both his choice and the true state of the world not precisely known by the agent. An IC situation C is described by the tuple

$$C = \langle N, n, (\Omega, \mathcal{F}, \mu), \{\mathcal{I}_i\}_{i \subset N}, A, r \colon \Omega \times A \to \Re \rangle \,,$$

where agent $n \in N$ is the decision-maker (action taker) who has to choose an action a from infinite action set A and can consult the several informants in $N \setminus \{n\}$. $(\Omega, \mathcal{F}, \mu)$ is a measure space, where Ω is the set of possible states which are relevant to the decision situation, $\mathcal{F}$ is a σ-algebra of subsets of Ω and μ is a probability measure, which describes the prior belief of the decision maker over all the states ω in Ω. The information of each (partially) informed agent in N about the state $\omega \in \Omega$ at hand is represented by the information partition $\mathcal{I}_i$, a finite partition of Ω into $\mathcal{F}$-measurable sets $I_i(\omega)$ with positive measure. If $\omega \in \Omega$ is the true state, then agent i knows that the event $I_i(\omega)$ happens, where $I_i(\omega)$ is that element (atom) of the partition $\mathcal{I}_i$ of Ω that contains ω. Decision-maker n receives the reward $r(\omega, a)$ if ω turns out to be the true state and he chooses action a. We assume that the reward function is a bounded $\mathcal{F}$-measurable function and that agent n is risk-neutral.

Related to an IC-situation with decision-maker n, we define a cooperative game, the IC-game. The IC-game (N, v) is defined by the player set $N = \{1, 2, \ldots, n\}$ and the characteristic function $v\colon 2^N \to \Re$, with domain the family 2^N of subsets of N, where $v(S) = 0$ if $n \notin S$ and

$$v(S) = \sum_{I \in \mathcal{I}_S} \sup_{a \in A} \int_I r(\omega, a)\, d\mu(\omega)$$

for all $S \in 2^N$ with $n \in S$. Here $\mathcal{I}_S = \wedge_{i \in S} \mathcal{I}_i$ is the coarsest partition of Ω that is a refinement of $\mathcal{I}_i$ for each $i \in S$:

$$\mathcal{I}_S = \left\{ \bigcap_{i \in S} I_i \mid I_i \in \mathcal{I}_i,\ \bigcap_{i \in S} I_i \neq \emptyset \right\}.$$

Notice that alone agent n can attain the expected payoff

$$v(\{n\}) = \sum_{I \in \mathcal{I}_n} \sup_{a \in A} \int_I r(\omega, a)\, d\mu(\omega).$$

A characterization of IC-games is given in [2, 4], where it is shown that a game is an IC-game with decision-maker n if and only if it is a n-monotonic game in which the decision-maker has veto power, that is $v(S) \leq v(T)$, for all $n \in S \subset T \subset N$, and $v(S) = 0$ for each $S \subset N$ with $n \notin S$. We denote by MV_n the set of such games.

For each coalition $S \subset N$ and $i \in S$, the marginal contribution of player i in the game $\langle (N, v) \rangle$ is defined by

$$M_i(S, v) = v(S) - v(S \setminus \{i\}).$$

A game $\langle N, v \rangle$ in MV_n is a so-called *big-boss* game if it satisfies

$$v(N) - v(S) \geq \sum_{i \in N \setminus S} M_i(N, v)$$

for all $S \subset N$ with $n \in S$.

Interesting are those IC-games with the n-concavity property. An IC-game (N, v) is called n-concave if

$$M_i(S, v) = v(S) - v(S \setminus \{i\}) \geq v(T) - v(T \setminus \{i\}) = M_i(T, v)$$

for all $i \in N \setminus \{n\}$, and for all $S \subset T \subset N$ with $\{i, n\} \subset S$.

We denote this subclass of IC-games by MV_nC_n. It is shown in [4] that a non-negative IC-game with decision maker n is n-concave if and only if it is a total big-boss game (cf. [7]).

A game $\langle N, v \rangle$ in MV_n is called a *total big-boss* game if it satisfies

$$v(T) - v(S) \geq \sum_{i \in T \setminus S} M_i(T, v)$$

for all $S \subset T \subset N$ with $n \in S$.

2.1 Compensation Rules and Continuity Considerations

If the decision-maker n works together with the agents in $N \setminus \{n\}$ to improve his reward, then the question "How to compensate the informants?" arises. The introduced IC-game opens the possibility to consider compensation schemes which correspond to solution concepts for cooperative games.

In particular, we focus on marginal based allocation rules, where the compensation of each informant is based on the marginal contribution induced by his information to the decision-maker's reward and on bi-monotonic allocation schemes which take the decision-maker's veto power into account. Both of these compensation solutions are related with the core of an IC-game, which is always non-empty.

Denote by P_n the set $\{S \subset N \mid n \in S\}$ of coalitions containing the decision-maker n.

An allocation scheme $[b_{S,i}]_{S \in P_n, i \in S}$ is a *bi-mas* of v if

$$(b_{S,i})_{i \in S} \in C(S,v), \quad b_{S,n} \leq b_{T,n} \text{ and } b_{S,i} \geq b_{T,i}$$

for all $S, T \in P_n$ with $S \subset T$ and all $i \in S \setminus \{n\}$. Here $C(S,v)$ denotes the core of the subgame (S,v).

An imputation x is *bi-mas extendable* if there is a bi-mas $[b_{S,i}]_{S \in P_n, i \in S}$, such that $x_j = b_{N,j}$ for all $j \in N$.

We denote by BM_n^N the set of all games (N,v) which have at least one bi-mas, and where n is a veto-player.

Let

$$E_n = \{(S,i) \in (2^N \setminus \{\emptyset\}) \times N \mid n, i \in S\}.$$

Then consider the multifunction

$$BIMAS\colon BM_n^N \to \Re^{E_n}.$$

$BIMAS$ is the non-empty-valued multifunction which assigns to each $v \in BM_n^N$ the set of all bi-monotonic allocation schemes $[b_{S,i}]_{S \in P_n, i \in S}$ of v, where for each $S \in P_n$ the "row" $(b_{S,i})_{i \in S}$ is a core element of the subgame (S,v) with extra conditions of the form $b_{S,i} \geq b_{T,i}$ for each $i \in N \setminus \{n\}$ and $b_{S,n} \leq b_{T,n}$, which are linear inequalities.

By the related stability theorems in this field [12, 6] we obtain

Theorem 1. *The multifunction $BIMAS$ is upper and lower semicontinuous.*

Now we concentrate on marginal based allocation rules, where the reward compensation of each informant is a fraction of his marginal contribution.

In a marginal based allocation rule informant i receives $\alpha_i M_i(N,v)$ with $\alpha_i \in [0,1]$ and the decision-maker is given the remainder

$$v(N) - \sum_{i \in N \setminus \{n\}} \alpha_i M_i(N,v).$$

For every fixed $\alpha = (\alpha_1, \alpha_2, \ldots, \alpha_{n-1}) \in [0,1]^{N\setminus\{n\}}$, we define the marginal allocation rule

$$\Psi^\alpha \colon MV_nC_n \to \Re^n ,$$

$$\Psi^\alpha(v) = \Big(\alpha_1 M_1(N,v), \ldots, \alpha_{n-1}M_{n-1}(N,v), v(N) - \sum_{i\in N\setminus\{n\}} \alpha_i M_i(N,v)\Big) .$$

Theorem 2. *For every fixed α, the function Ψ^α is $(2n-1)$-Lipschitz.*

3 The Canonical Discretization of an IC-Situation

In this section we describe a natural way to discretize the infinite state space Ω of an IC-situation

$$\mathcal{C} = \langle N, n, (\Omega, \mathcal{F}, \mu), \{\mathcal{I}_i\}_{i\in N}, A, r\colon \Omega \times A \to \Re\rangle$$

and to arrive at a strategic equivalent IC-situation

$$\mathcal{C}' = \langle N, n, (\Omega', \mu'), \{\mathcal{I}'_i\}_{i\in N}, A, r'\colon \Omega' \times A \to \Re\rangle ,$$

where the state space Ω' is finite. The idea behind the discretization is simple. Each of the players in the IC-situation $\mathcal{C}$ is provided with a finite information partition $\mathcal{I}_i$ of Ω. The common refinement $\mathcal{I}_N = \wedge_{i\in N}\mathcal{I}_i$ is also a finite partition of Ω. Each atom of this common refinement consists of points which are neither distinguishable by agents nor by coalitions who pool their information. So it is natural to replace such an atom by one point and then these points generate Ω'. The probability measure, the information partitions and the reward function can then be adjusted, leading to the canonical discretization.

We can find for each IC-situation $\mathcal{C}$ with infinite state and action spaces a related finite approximation $\mathcal{C}''$ such that the two involved IC-games are as near to each other as one wants. The continuity properties of the different compensation schemes guarantee then that the corresponding compensations in the two games are also near to each other. A similar discretization procedure can be developed for information sharing games, introduced in [9]. In this context it is interesting to note that the multifunction $PMAS$, which assigns to a game the set of its population monotonic allocation schemes (cf. [10]) is upper and lower continuous by the same argument used for the multifunction $BIMAS$.

References

1. Brânzei, R., Tijs, S., Timmer, J. (2001a): Collecting Information to Improve Decision Making. International Game Theory Review **3**, 1–12

2. Brânzei, R., Tijs, S., Timmer, J. (2001b): Information Collecting Situations and Bi-monotonic Allocation Schemes. Mathematical Methods of Operations Research **54**, 303–313
3. Gillies, D. B. (1953): Some Theorems on n-Person Games. PhD Thesis, Princeton University Press, Princeton, New Jersey
4. Lucchetti, R., Patrone, F., Tijs, S. H., Torre, A. (1987): Continuity Properties of Solution Concepts for Cooperative Games. OR Spektrum **9**, 101–107
5. Muto, S., Nakayama, M., Potters, J., Tijs, S. (1988): On Big Boss Games. The Economic Studies Quarterly **39**, 303–321
6. Schmeidler, D. (1969): The Nucleolus of a Characteristic Function Game. SIAM Journal on Applied Mathematics **17**, 1163–1170
7. Slikker, M., Norde, H., Tijs, S. (2000): Information Sharing Games. CentER Discussion Paper No. 2000-100, Tilburg University, The Netherlands
8. Sprumont, Y. (1990): Population Monotonic Allocation Schemes for Cooperative Games with Transferable Utility. Games and Economic Behavior **2**, 378–394
9. Tijs, S. H. (1981): Bounds for the Core and the τ-Value. In: Moeschlin, O., Pallaschke, D. (Eds.): Game Theory and Mathematical Economics. North-Holland, Amsterdam, 123–132
10. Tijs, S., Timmer, J., Brânzei, R. (2001): Compensations in Information Collecting Situations. CentER Discussion Paper No. 2000-02, Tilburg University, The Netherlands
11. Voorneveld, M., Tijs, S., Grahn, S. (2000): Monotonic Allocation Schemes in Clan Games. CentER Discussion Paper No. 2000-80, Tilburg University, The Netherlands
12. Walkup, D. W., Wets, R. J.-B. (1969): Lifting Projections of Convex Polyedra. Pacific J. Math. **28**, 465–475

Some Variation Calculus Problems in Dynamic Games on 2D Surfaces*

Arik Melikyan[1] and Naira Hovakimyan[2]

[1] Institute for Problems in Mechanics,
Russian Academy of Sciences,
Vernadsky Ave., 101-1, Moscow 117526, Russia
E-mail:melik@ipmnet.ru

[2] School of Aerospace Engineering,
Georgia Institute of Technology,
Atlanta, GA 30332-0150
E-mail:naira.hovakimyan@ae.gatech.edu

Summary. When the game space is a manifold rather than a Euclidean space, and there exist two or more minimal geodesic lines, connecting the players for certain positions, with the same lengths, the construction of optimal phase portraits cannot be done based on Euclidean solution. This paper demonstrates how variational calculus, including the focal point technique, can be used in differential games on surfaces. In particular, it was shown that in games on the two-dimensional surfaces, the existence of two equal geodesics in for some positions of the players gives rise to so-called secondary domain where the optimal motion of the players is a motion along a geodesic. Each player exploits his own geodesic line, which is different from the one connecting them. Optimal phase portraits in differential games on manifolds are constructed and the structure of the game values are derived.

1 Introduction

Geodesic lines of a manifold are crucial for construction of optimal phase portraits in differential games on that manifold. If the manifold, i. e. the game space, is the Euclidean space itself, then the optimal pursuit-evasion time is the one along the *unique* minimal geodesic connecting the players [4]. When the game space is a manifold rather than a Euclidean space, and there exist two or more minimal geodesic lines, connecting the players for certain positions, with the same lengths, the construction of optimal phase portraits *cannot be done based on Euclidean solution. An interesting and important* problems of this type are the games on two-dimensional surfaces. The existence of two equal geodesics for some positions of the players gives rise to

* This work was partially supported by the grant No. 01-01-00376 of the Russian Foundation for Basic Research.

so-called secondary domain where the optimal motion of the players is still a motion along a geodesic, while each player exploits his own geodesic line, which is different from the connecting them one. The singular (equivocal) surface, separating the primary and secondary domains, represents optimal trajectories which are envelopes of a family of geodesics; such surfaces also are called switching envelopes. As shown in [5, 7], in the games on 2D cones in addition to these two types of domains there might exist a domain $D = D_3$ of a third type, where the game value depends only on the coordinates of the pursuer. In [6] it is shown that such phenomenon takes place in pursuit-evasion games on a family of 2D second order surfaces of revolution, including cones and hyperboloids of two sheets. Earlier games on hyperboloids were investigated in [2, 3]. This paper demonstrates how the methods of variational calculus, including the focal point technique, can be used in differential games on surfaces.

2 Dynamic Games with Simple Motion on Manifolds

Let two points-players P (pursuer) and E (evader) perform so-called simple motion [4] on a manifold M having non-degenerate Riemannian metric defined on it. We will consider two-dimensional surfaces in three-dimensional Euclidean space as a game space. Let $x, y \in R^2$ be local coordinates, describing the positions of the players. Then the dynamic equations for the players, performing simple motions, can be written as:

$$\dot{x} = u\,, \quad \dot{y} = v\,, \quad x, y, u, v \in R^2\,, \tag{1}$$

subject to the constraints:

$$\sqrt{\langle G(x)u, u\rangle} \leq 1\,, \quad \sqrt{\langle G(y)v, v\rangle} \leq \nu\,.$$

Here $G(x)$ is a 2×2-matrix, metric tensor of the manifold (surface) specifying a Riemannian metric. The equation (1) presumes that the maximal speed of the pursuer is 1 and that of the evader is $\nu \geq 0$. The constant ν may also be regarded as the speed ratio $\nu = v_E/v_P$. We assume that $\nu \leq 1$.

A pursuit-evasion game is considered for the system (1) with zero capture radius, i. e. the game starting at time $t = 0$ ends for some $t = T$ if the capture condition takes place: $x(T) = y(T)$ and the cost function is T.

3 Primary Solution, Geodesic Variational Problem

The basic variation calculus problem arising in dynamic games on manifolds is the problem on minimal geodesic line(s). Generally, in the main part $D_1 \subset M \times M$ of the phase space $M \times M$ of the game, called *primary domain*, the game value $V(x, y)$ equals to so-called primary time:

$$V_1(x,y) = \frac{L(x,y)}{1-\nu} \qquad (x,y) \in D_1 \,, \tag{2}$$

where $L(x,y)$ is the minimal Riemannian distance between P and E defined as a solution of the following variation calculus problem:

$$L(x,y) = \min_{\xi(\sigma)} \int_{\sigma_0}^{\sigma_1} \sqrt{\langle G(\xi)\dot{\xi},\dot{\xi}\rangle} d\sigma \,, \qquad \xi(\sigma_0) = y \,, \quad \xi(\sigma_1) = x \,. \tag{3}$$

Here $\xi(\sigma)$ is a piece-wise smooth curve $\xi(\sigma) \in M$, $\sigma_0 \le \sigma \le \sigma_1$. The solutiuons (generally non-unique) of the Euler equation

$$K_\xi - \frac{d}{d\sigma} K_{\dot{\xi}} = 0 \,, \quad \xi(\sigma_0) = y \,, \quad \xi(\sigma_1) = x \,, \quad \left(K(\xi,\dot{\xi}) \equiv \sqrt{\langle G(\xi)\dot{\xi},\dot{\xi}\rangle} \right)$$

for the problem (3) are called geodesic lines, while $L(x,y)$ is the length of the shortest geodesic line(s).

In the primary domain D_1 the optimal motion of the players is along the geodesic line connecting them. Such strategy is not well defined for positions on the surface $\Gamma \subset M \times M$, for which there exist more than one shortest geodesic line connecting the players and having equal lengths.

Manifold Γ. Typically, the manifold Γ has a boundary (an edge) $\partial\Gamma$, which divides the phase space into two parts. One part consists of the positions with the unique shortest geodesic line, the second part consists of the positions with non-unique shortest geodesic line. For the surfaces of revolution and many other $2D$ game spaces for certain positions there exist two geodesic lines L^+, L^- with equal lengths: $L^+(x,y) = L^-(x,y)$. We will call this surface Γ_1. Notice that it is a *dispersal* surface.

Manifold B. An optimality requirement for the primary solution reveals an edge B of the dispersal manifold Γ_1, which separates within Γ the dispersal part Γ_1. The points of the manifolds Γ_1 (of co-dimension 3) and $B \subset \Gamma$ (of co-dimension 2) satisfy the conditions:

$$\begin{aligned} \Gamma_1 : \quad & L^+ = L^- \,, \quad |a^+ - a^-|_x - \nu|b^+ - b^-|_y \le 2(1-\nu) \,, \\ B : \quad & L^+ = L^- \,, \quad |a^+ - a^-|_x - \nu|b^+ - b^-|_y = 2(1-\nu) \,, \end{aligned} \tag{4}$$

where $L^\pm$ are the lengths of the two geodesics, $a^\pm$ and $b^\pm$ are outward unit tangent vectors to these geodesics at the points P and E; the norms are understood as:

$$|a|_x = \sqrt{\langle G(x)a,a\rangle} \,, \qquad |b|_y = \sqrt{\langle G(y)b,b\rangle} \,.$$

On the manifolds M for which all the positions satisfy the conditions (4), i. e. the set B is empty, the game value is represented by the formula (2) in the whole phase space. This is, for instance, the case for a cylinder. For an initial position on the surface Γ_1 (which coincides with Γ) the players can start the optimal motion either along L^+ or L^-, by the decision of P.

4 Secondary Solution

In the case when the manifold B is non-empty and there exist positions violating conditions (4):

$$L^+ = L^-, \quad |a^+ - a^-|_x - \nu|b^+ - b^-|_y > 2(1-\nu) ,$$

two branches of singular equivocal surface (switching envelopes) arise Γ_2^+ and Γ_2^-. These surfaces can be constructed using the following Cauchy problem in terms of so-called singular characteristics in inverse time [5]:

$$\begin{aligned} &\dot{z} = -H_r , \qquad && \dot{r} = H_z + \frac{\{\{HF_1\}H\}}{\{\{F_1H\}F_1\}}\Big(r - \frac{\partial V_1}{\partial z}(z)\Big) , \\ &z(0) = z_0 , && r(0) = \frac{1}{2}\Big(\frac{\partial V_1^+}{\partial z}(z_0) + \frac{\partial V_1^-}{\partial z}(z_0)\Big) , \quad z_0 \in B . \end{aligned} \tag{5}$$

Here $H = H(z,r)$ is the left-hand side of the HJBI equation of the pursuit-evasion game (1) (the Hamiltonian):

$$\begin{aligned} H(z,r) = -\sqrt{\langle G^{-1}(x)p,p\rangle} + \nu\sqrt{\langle G^{-1}(y)q,q\rangle} + 1 = 0 , \\ p = \partial V/\partial x, \quad q = \partial V/\partial y, \quad z = (x,y) \in R^4, \quad r = (p,q) \in R^4 . \end{aligned} \tag{6}$$

The function $F_1 = F_1(z,V)$ has the form:

$$F_1(z,V) = V - V_1(z) .$$

Here and in (5) one has to use the function V_1^+ (V_1^-) for the branch Γ_2^+ (Γ_2^-), where $V_1^\pm$ are given by (2) with $L^\pm$. In (5) one has the Jacobi (Poisson) bracket:

$$\{FG\} = \langle F_z + rF_V, G_r\rangle - \langle G_z + rG_V, F_r\rangle$$

for two functions $F(z,V,r)$, $G(z,V,r)$.

The secondary solution $V = V_2(z)$, $z \in D_2$ is defined as the solution of the HJBI equation (6) with the boundary condition specified on the equivocal surface:

$$V_2(z) = V_1(z) , \quad z \in \Gamma_2 .$$

The corresponding Cauchy problem is an irregular problem [5] because the characteristics quit the boundary surface Γ_2 tangentially.

One can show that the optimal trajectories in D_2 for both players are geodesic lines different from the one which connects the points P and E.

For more detailed construction algorithm see [5, 7].

5 The Region of the Third Type

In [5, 7] it is shown that in a game on conical surface in addition to the above primary and secondary domains there exists a domain of the third type where the game value does not depend on the coordinates of the evader. The consideration in [5, 7] is based on certain geometry of the problem. However, the possibility of the existence of such a domain one can observe from the analysis of the HJBI equation.

In pursuit-evasion differential games with simple motion the HJBI equation [4] often has the form similar to (6):

$$-F(x,p) + \nu F(y,q) + 1 = 0\,, \tag{7}$$

where x, y are the coordinate vectors of the pursuer P and the evader E, p, q are gradients of the game value $V(x,y)$ with respect to x, y,

$$p = \partial V/\partial x\,, \quad q = \partial V/\partial y\,.$$

The function $F(x,p)$ in (7) has a sense of a norm, see (4), (6), and thus is non-negative and vanishes as soon as p and q equal to zero:

$$F(x,p) \geq 0\,,\; F(y,q) \geq 0\,, \quad F(x,0) = 0\,,\; F(y,0) = 0\,.$$

Due to these properties in some region D of the x,y-space the following condition holds:

$$\frac{\partial V}{\partial y} = 0\,, \quad -F(x,p) + 1 = 0\,, \quad (x,y) \in D_3\,, \tag{8}$$

while the relations

$$\frac{\partial V}{\partial x} = 0\,, \quad \nu F(x,q) + 1 = 0$$

are not possible. The equation (8) means that the game value in the region D_3 depends only on x and does not depend on y, and consequently, the pursuer has a definite optimal behavior, while the evader may exploit an arbitrary control till reaching the boundary of D_3. In other words, in D one has an optimal control problem rather than a dynamic game.

As one can see in the sequel, the relation (8) may have different analytical form, while the essence is still independence of the game value on the evader's position.

6 The Value in D_3 for the Game on a Cone

Putting $b = 0$ in (9) one gets a conical surface with the apex at the origin $O(0,0,0)$ of the coordinate system. One of the convenient ways of description of the dynamics is to use so-called relative coordinates R, r, φ, where R and r are the distances between the point O and the points P and E, respectively, and φ is the angle between the segments PO and EO on the plane development of the cone.

The game dynamics in terms of the relative coordinates has the form:

$$\dot{R} = u_1, \quad \dot{r} = v_1, \quad \dot{\varphi} = \frac{v_2}{r} - \frac{u_2}{R}, \qquad (9)$$
$$u_1^2 + u_2^2 \le 1, \quad v_1^2 + v_2^2 \le \nu^2.$$

Here u_i and v_i are the components of the players' velocities in certain orthogonal coordinate system, generally, different from the one used in the form (1). Based on (9) the HJBI equation takes the form:

$$-\sqrt{V_R^2 + V_\varphi^2/R^2} + \nu\sqrt{V_r^2 + V_\varphi^2/r^2} + 1 = 0.$$

In a region D of the relative coordinates the independence of the value on the evader's position means that:

$$V_\varphi = 0, \quad V_r = 0$$

which, under the assumption $V_R > 0$, simplifies the HJBI equation to the differential equation

$$-V_R + 1 = 0$$

with the general solution

$$V = R + C$$

for some constant C. One can show that for $R = 0$ one has $V = 0$, so that $C = 0$.

7 Game on a Surface of Revolution

Consider a surface of revolution M given in 3-dimensional space with the coordinates (x_1, x_2, x_3) by the equation:

$$x_3 = f\left(\sqrt{x_1^2 + x_2^2}\right), \qquad (10)$$

where $f(z)$ is a twice differentiable function for $0 \le z < \infty$, such that

$$f'(0) = 0, \quad f'(z) > 0, \quad 0 < z < \infty. \qquad (11)$$

Under these conditions the surface M has twice differentiable generatrix. As a particular case of the surface (10) we will consider

$$f(z) = \sqrt{(z^2 + b)/k},$$

which gives the upper nappe of the hyperboloid (the cone) of revolution

$$x_1^2 + x_2^2 + b = kx_3^2, \quad b, k > 0, \qquad (12)$$

where b, k are parameters, positive real numbers. For $b = 0$ the equation (12) defines a cone of revolution. For $b < 0$ one has a hyperboloid of one sheet, and the solution of a pursuit-evasion game on it is given in [3].

Consider as a game space the surface M. One can choose the first two components $x = (x_1, x_2)$, $y = (y_1, y_2)$ as local coordinates, describing the positions of the players. Then the dynamic equations for the players, performing simple motions, can be written in the form (1), were $G(x)$ is a 2×2-matrix, metric tensor of the surface. One can compute the entries $g_{ij}(x)$ of this matrix using the quadratic form:

$$\langle G(x)dx, dx\rangle = dx_1^2 + dx_2^2 + dx_3^2 \,,$$

where dx_3 must be excluded using (10) and $z = \sqrt{x_1^2 + x_2^2}$:

$$g_{11} = 1 + \frac{f'^2(z)x_1^2}{x_1^2 + x_2^2}\,, \quad g_{22} = 1 + \frac{f'^2(z)x_2^2}{x_1^2 + x_2^2}\,, \quad g_{12} = g_{21} = \frac{f'^2(z)x_1x_2}{x_1^2 + x_2^2}\,. \tag{13}$$

The HJBI equation for the pursuit-evasion game on the surface M with simple dynamics (1) has the form (6).

Introduce relative coordinates R, r, φ, where R (respectively r) is the length of the shortest geodesic line connecting the apex $A(0, 0, f(0))$ of the surface M with the point P (point E); φ is the angle between the tangent lines to these two geodesics at the point A in the horizontal plane $x_3 = f(0)$. Note that the variables R, r have the same meaning as in (10), whereas φ is different from that in (10).

Dynamic equations in relative coordinates have the form:

$$\dot{R} = u_1\,, \quad \dot{r} = v_1\,, \quad \dot{\varphi} = \frac{v_2}{h(r)} - \frac{u_2}{h(R)}\,,$$
$$u_1^2 + u_2^2 \le 1, \quad v_1^2 + v_2^2 \le \nu^2\,,$$

where the function h can be defined using metric tensor $G(x)$. An exact specification of this function is not important for the sequel.

The HJBI equation in relative coordinates could be written as:

$$\min_u \max_v \left[V_R w(R, u_1) + V_r w(r, v_1) + V_\varphi (h(r, v_2) - h(R, u_2))\right] + 1 = 0\,.$$

For the region D, where, by definition, the equalities $V_\varphi = 0$, $V_r = 0$ are fulfilled one has:

$$\min_u V_R w(R, u_1) + 1 = 0\,.$$

The above minimum, under the assumption $V_R > 0$, is attained at $w(R, u_1) = -1$, which corresponds to the motion of the player P along the geodesic AP towards the apex A with maximal speed 1. Thus one has the simplified equation and its general solution in the form:

$$-V_R + 1 = 0\,, \quad V(R, r, \varphi) = R + C\,. \tag{14}$$

To compute the constant C one needs to find points where the domains D_1, D_2 and D_3 intersect. This requires the following geometric analysis.

8 The Edges of the Manifolds Γ and B

Consider first the boundary $\partial\Gamma$. One can suppose (and the considerations in the sequel confirm this) that the two geodesics, L^+ and L^-, coincide at the positions from $\partial\Gamma$. That means that the points P and E are, by one of the definitions, mutually conjugate points with respect to the geodesic functional (3). This phenomenon was investigated numerically in [2]. Thus the determination of $\partial\Gamma$ means that one has to find all mutually conjugate points P and E. Due to rotational symmetry and the equality $L^+ = L^-$ in (4) these points can be regarded as points of a generatrix of the surface M that lies, say, in the coordinate plane $x_2 = 0$, and according to (10), is defined as:

$$x_3 = f(x_1), \quad x_1 \geq 0, \qquad x_3 = f(-x_1), \quad x_1 \leq 0. \tag{15}$$

The two coinciding geodesics PE represent a curve segment of the curve (15), which by virtue of (11) is a twice differentiable curve.

The projection of the curve PE onto the plane $x_3 = const$ is an extremal of the functional (3). The Lagrangian (integrand) in (3) is a homogeneous function in $\dot{\xi}$, which causes certain degeneracy. The latter could be removed if one introduces x_1 as an independent variable and transforms the functional (3) in terms of one scalar function $x_2 = x_2(x_1)$, or using the classical notation $x_2 = y$, $x_1 = x$, $y = y(x)$, to the following form:

$$L = \int_{x_0}^{x_1} F(x,y,y')dx\,, \quad F = \sqrt{g_{11}(x,y) + 2g_{12}(x,y)y' + g_{22}(x,y)y'^2}\,, \tag{16}$$

where g_{ij} are the entries of the symmetric (2×2)-matrix $G(x,y)$.

One can verify that the curve under the investigation, $y = y(x) \equiv 0$, $x_0 \leq x \leq x_1$, is an extremal for (16), i.e. satisfies the Euler equation:

$$F_y - \frac{d}{dx}F_{y'} = 0\,.$$

The focal point $P(x_1, 0)$ conjugate to $E(x_0, 0)$ with respect to the functional (16) is defined using the solution $h(x)$ of the following initial value problem:

$$\frac{d}{dx}(Ph') + (Q' - R)h = 0\,, \quad h(x_0) = 0\,, \quad h'(x_0) = 1\,, \tag{17}$$

$$P(x) = F_{y'y'} = \frac{1}{\sqrt{g_{11}}}\,, \quad Q(x) = F_{y'y} = \frac{f'^2(x)}{x\sqrt{g_{11}}}\,,$$

$$R(x) = F_{yy} = \frac{f'(x)f''(x)}{x\sqrt{g_{11}}} - \frac{f'^2(x)}{x^2\sqrt{g_{11}}}\,, \quad g_{11}(x,0) = 1 + f'^2(x)\,.$$

The value x_1 is the next root of $h(x)$:

$$h(x_1) = 0\,, \quad h(x) > 0\,, \quad x_0 < x < x_1\,.$$

The above linear ODE is known as the Jacobi equation for the functional (16); it is an equation in variations for the Euler equation. For the construction of $\partial\Gamma$ it is sufficient to find corresponding x_1 for all $x_0 > 0$.

The edge ∂B of the manifold B is a submanifold of $\partial\Gamma$ for which the last equality in (4) is fulfilled. Since the tangent vectors a^+ and a^-, b^+ and b^- coincide in (4) and the equality is fulfilled trivially, one needs a limit form of it as a point under consideration tends to $\partial\Gamma$ from the interior of Γ.

Using in (4) second order expansions in terms of small $|a^+ - a^-|$ and $|b^+ - b^-|$, as in [6], one can obtain the following condition for the points of ∂B:

$$h(x_1) = 0\,, \quad h'(x_1) = -\sqrt{\nu}\frac{P(x_0)}{P(x_1)} = -\sqrt{\nu}\sqrt{\frac{g_{11}(x_1,0)}{g_{11}(x_0,0)}}\,. \tag{18}$$

Here the connection (17) between $P(x)$ and $g_{11}(x,0)$ is used. The relations (18) represent a system of two equations with respect of two unknowns x_0, x_1. The solution of this system defines the manifold ∂B up to revolution about the vertical x_3-axis.

In the definition of the conjugate point (focus) x_1 the value $h'(x_1)$ is free. In (18) it takes a prescribed value, which, generally, defines a unique pair x_0, x_1.

One can verify that the equation (17) has two following linearly independent solutions:

$$a(x) = x\,, \quad b(x) = x\varphi(x) \qquad \left(\varphi(x) = \int \sqrt{g_{11}(x,0)}\,\frac{dx}{x^2}\right).$$

Using these expressions one can simplify the system (18) to the form:

$$x_0 = -\sqrt{\nu}\,x_1\,, \quad \varphi(x_1) + \varphi(\sqrt{\nu}x_1) = 0\,. \tag{19}$$

Note, that for a paraboloid of revolution, when $f(z) = kz^2/2$ in (10), one has:

$$\varphi(x) = \sqrt{1 + k^2x^2}/x - k\ln(kx + \sqrt{1 + k^2x^2})\,.$$

For the hyperboloid with $f(z) = \sqrt{(z^2 + b)/k}$ in (10) the antiderivative $\varphi(x)$ is expressed in elliptic functions.

As one can see for the solution x_0, x_1 of the system (19) the numbers x_0 and x_1 have different signs, i.e. the corresponding points E and P are on different sides with respect to the apex A. Thus in (R, r, φ) variables such a solution defines a point N:

$$R_1 = \int_0^{|x_1|} \sqrt{g_{11}(x,0)}\,dx\,, \quad r_1 = \int_0^{|x_0|} \sqrt{g_{11}(x,0)}\,dx\,, \quad \varphi = \pi\,, \tag{20}$$

where R_1, r_1 are the lengths of a subarcs of the curve (15).

9 Computation of C, Structure of the Game Value

The edge $N = \partial B$ of the manifold B defined in (4) lies on the boundary between the domains D and D_1. The distance between the players at the position N is $L = R_1 + r_1$. Thus equating the expressions (2) and (14) for the game value in the domains D and D_1 at the point N one gets:

$$V = R_1 + C = \frac{R_1 + r_1}{1 - \nu} = V_1 , \qquad C = \frac{\nu R_1 + r_1}{1 - \nu} ,$$

which gives the following expression for the game value $V(R)$:

$$V_3 = V(R) = R + \frac{\nu R_1 + r_1}{1 - \nu} . \tag{21}$$

Generally, the game space consists of three domains: D_1, D_2 and $D = D_3$. All these regions intersect at the manifold ∂B, which is a point in relative variables. The regions D_1 and D_3 are specified in the above considerations. The secondary domain D_2 and corresponding secondary solution $V = V_2$ are constructed using the relations (5) and an irregular Cauchy problem, as described in section 2.

The boundary between the regions D_1 and D_2 is the surface Γ_2 constructed by means of (5). The boundary between the regions D_1 and D_3 one can construct equating the expressions (2) and (21). The boundary between D_2 and D_3 requires more intensive numerical efforts. For the case of cone ($b = 0$) these surfaces are found in [5, 7].

References

1. Giaquinta, M., Hildebrandt, S. (1996): Calculus of Variations I, II. Springer, Berlin
2. Hovakimyan, N. V., Harutunian, L. L. (1998): Game Problems on Rotation Surfaces. In: Petrosjan, L. A. et al. (Eds.): Game Theory and Applications, Vol. IV, 62–74
3. Hovakimyan, N. V., Melikyan, A. A. (2000): Geometry of Pursuit-Evasion on Second Order Rotation Surfaces. Dynamics and Control **10**, 297–312
4. Isaacs, R. (1965): Differential Games. John Wiley, New York
5. Melikyan, A. A. (1998): Generalized Characteristics of First Order PDEs: Applications in Optimal Control and Differential Games. Birkhauser, Boston
6. Melikyan, A. A. (2001): Some Properties of the Bellman–Isaacs Equation for the Games on the Surfaces of Revolution. International Game Theory Review (to appear)
7. Melikyan, A. A., Hovakimyan, N. V., Harutunian, L. L. (1998): Games of Simple Pursuit and Approach on Two Dimensional Cone. JOTA **98(3)**, 515–543

On The Chinese Postman Game

Daniel Granot[1], Herbert Hamers[2], Jeroen Kuipers[3], and Michael Maschler[4]

[1] Faculty of Commerce, University of British Columbia,
Vancouver, BC, Canada
E-mail: daniel.granotcommerce.ubc.ca

[2] Center and Department of Econometrics, Tilburg University,
Tilburg, The Netherlands
E-mail: H.J.M.Hamerskub.nl

[3] Dept. of Mathematics, University of Maastricht,
Maastricht, The Netherlands
E-mail: kuipersMATH.MAASTRICHT.NL

[4] Department of Mathematics and Center for Rationality and
Interactive Decision Theory, The Hebrew University of Jerusalem,
Jerusalem, Israel
E-mail: maschlervms.huji.ac.il

Summary. The paper examines an extended model of the Chinese Postman (CP) game which allows for prizes and public edges, that is, edges not belonging to any of the players. The prizes, which are associated with the edges, are collected only upon the first traversal of the edges, and their inclusion in the model is analogous to the introduction of prizes in the traveling salesmen problem. A class of CP-games induced by Eulerian graphs, without public edges, having the 4-cut property is analyzed. In particular, it is shown that both the core and nucleolus are the Cartesian products of the cores and nucleoli of CP-games defined on cycle graphs, which are induced by elementary subpaths in the original graph. Moreover, for CP-games induced by Eulerian graphs having the 4-cut property, but without public edges, one can test core membership in linear time and the nucleolus can be computed in quadratic time.

1 Introduction

A class of delivery games was introduced by Hamers et al. [9] to analyze a cost allocation problem which arises in some delivery problems on graphs. These delivery problems are associated with the Chinese postman problem [5, 10] and can be described as follows. A server is located at some fixed vertex of a graph, G, to be referred to as the post office, and each edge of G belongs to a different player. The players need some service, e. g. mail delivery, and the nature of this service requires the server to travel from the post office and visits all edges (players) before returning to the post office. The cost allocation problem associated with this delivery problem is concerned with a fair allocation of the cost of a cheapest Chinese postman tour in the graph. That is, the cost of a cheapest tour, which starts at the post office, visits each edge of G at least once and returns to the post office. Following what is by now an established line of research, Hamers et al. [9] formulated this cost allocation problem as a cooperative game $(N; c)$, referred to as the Chinese postman (CP) game, where N is the set of players (edges) in the graph, and $c\colon 2^N \to \mathcal{R}$ is the characteristic function. For each $S \subseteq N$ $c(S)$ is the cost of a cheapest tour, which starts at the post office, visits each edge in S at least once and returns to the post office. Solution concepts in cooperative game theory were then evaluated as possible cost allocation schemes for the above delivery problem.

One of the most prominent solution concepts in cooperative game theory is the core of a game. It consists of all vectors which distribute the cost incurred to N, $c(N)$, among the players in such a way that no subset of players can be better off by seceding from the rest of the players and act on their own behalf. A cooperative game whose core is not empty is said to be balanced, and if the core of any subgame of it is nonempty, it is said to be totally balanced.

In general, a CP game associated with an undirected graph could have an empty core. However, Hamers et al. [9] have shown that a CP game induced by a connected weakly Eulerian graph is balanced. Here, a graph G is called weakly Eulerian if each biconnected component in G is Eulerian. Further, Hamers [8] has shown that if a connected undirected graph is weakly cyclic, that is, every edge therein is contained in at most one cycle, then the associated CP game is submodular.

Define a graph to be Chinese Postman-submodular, Chinese Postman-totally balanced or Chinese Postman-balanced (or, for short, CP-submodular, CP-totally balanced and CP-balanced), if the corresponding CP game is submodular, totally balanced, or balanced, respectively, for all edge costs and all locations of the post office. Granot et al. [6] proved that an undirected graph is CP-submodular if and only if it is CP-totally balanced, which holds if and only if it is weakly cyclic. An undirected graph was shown by them to be CP-balanced if and only if it is weakly Eulerian. In contrast with the undirected case, they proved that any connected directed graph is CP-balanced. Further, it was proven by Granot et al. [6] that a CP game induced by a directed graph is submodular if and only if the directed graph is weakly cyclic. In a directed weakly cyclic graph each arc is contained in exactly one circuit.

In this paper we study an extended model of the CP-game which allows for public edges (i. e., edges not belonging to any of the players) and prizes. The prizes, which are associated with the edges, are collected only upon the first traversal of the edges, and their inclusion in the model is analogous to the introduction of prizes in the traveling salesmen problem [2, 3]. Moreover, the introduction of prizes and public edges in our extended model is natural in our context, as we show that the Davis–Maschler reduced game of an extended

CP-game with respect to a core vector is a CP-game having prizes and public edges.

We study in this paper the class of CP-games induced by Eulerian graphs, without public edges, having the 4-cut property[5]. For this class of games, both the core and nucleolus are the Cartesian products of the cores and nucleoli of CP-games defined on cycle graphs, which are induced by elementary subpaths in the original graph. It is shown in [7] that a CP-game defined on a cycle graph has only a linear (in the number of edges/players) number of non redundant core constraints, and it is further developed therein a quadratic algorithm to calculate the nucleolus of such a game. Coupled with our results on Eulerian graphs having the 4-cut property, we can conclude that for CP-games induced by Eulerian graphs having the 4-cut property, but without public edges, one can test core membership in linear time and the nucleolus can be computed in quadratic time.

2 Preliminary and Notation

We present in this section some elementary definitions in game theory and graph theory which are needed for subsequent analysis. A (cost) cooperative game is a pair $\Gamma = (N; c)$, where N is a finite set of players, c is a mapping, $c\colon 2^N \to \mathcal{R}$, with $c(\emptyset) = 0$. A subset of N will be sometimes referred to as a coalition. A function $h\colon 2^N \to \mathcal{R}$ is said to be *submodular* or *convex* if

$$h(T \cup \{j\}) - h(T) \leq h(S \cup \{j\}) - h(S) \tag{1}$$

for all $j \in N$ with $S \subset T \subseteq N \setminus \{j\}$. A game $(N; c)$ is submodular (convex) if and only if the map $c\colon 2^N \to \mathcal{R}$ is submodular.

An allocation $x = (x_i)_{i \in N} \in \mathcal{R}^N$ is a *core-element* of $(N; c)$ if $\sum_{i \in N} x_i = c(N)$ and $\sum_{i \in S} x_i \leq c(S)$ for all $S \in 2^N$. The *core* of a game, Core $(N; c)$, consists of all core-elements. A game is called *balanced* if its core is non-empty and it is *totally balanced* if for each $S \subseteq N, (S; c_S)$ is balanced, where c_S is the restriction of c to the family of subsets of S.

The set of all imputations of Γ is denoted by $X(\Gamma)$. Thus, $X(\Gamma) = \{x\colon \sum_{j=1}^{n} x_i = c(N), x_i \leq c(\{i\}), i = 1, \ldots, n\}$. For a game $\Gamma = (N; c)$ and $x \in \mathcal{R}^N$, let $\theta(x; \Gamma)$ be the $|2^N|$-dimensional vector whose components are the excesses, $c(S) - x(S)$, for $S \in 2^N$, arranged in a nondecreasing order, where $x(S) \equiv \sum_{k \in S} x_k$. Let $\succeq$ denote the *lexicographically greater than* relationship between vectors of the same dimension. The nucleolus of Γ is given by,

$$\nu(\Gamma) = \{x \in X(\Gamma) : \ \theta(x; \Gamma) \succeq \theta(y; \Gamma), \ \text{ for all } \ y \in X(\Gamma)\} . \tag{2}$$

Let $G = (V(G), E(G))$ be an undirected graph, where $V(G)$ and $E(G)$ denote the set of vertices and the set of edges of G, respectively. A *walk* in G is a finite sequence of vertices and edges of the form $v_1, e_1, v_2, \ldots, e_k, v_{k+1}$

[5] Definitions are provided in subsequent sections

with $k \geq 0, v_1, \ldots, v_{k+1} \in V(G)$, $e_1, \ldots, e_k \in E(G)$, such that $e_j = (v_j, v_{j+1})$ for all $j \in \{1, \ldots, k\}$. Such a walk is said to be closed if $v_1 = v_{k+1}$. A *path* in G is a walk in which all vertices (except, possibly v_1 and v_{k+1}) and edges are distinct. A closed path, i.e., a path in which $v_1 = v_{k+1}$, containing at least one edge is called a cycle.

3 The Model

An extended Chinese postman enterprise is given by $\Gamma = (G, v_p, a, p, N)$, where G is a connected undirected graph, containing a special vertex v_o called "post-office", a nonnegative cost function $a(\cdot)$ defined on the edges of G, a nonnegative prize function $p(\cdot)$, defined on the edges of G, a set of players N, each residing and occupying an edge of G, so that each edge contains at most one player. Edges without players residing in them are called "public edges". The postman makes closed walks, starting at the post office and ending there. Such closed walks will also be referred to as tours. While making the tours the postman pays the cost of the walk which is the sum of the costs of the edges he traverses minus the sum of the prizes encountered. However, there is only a one-time prize per edge, so if during the walk he traverses an edge, say twice, he pays twice the cost of that edge and collects the price just once. The players are expected to reimburse the postman for his expenses.

For each Chinese postman enterprise $\Gamma = (G, v_o, a, p, N)$ we associate a (cost) game $(N; c_\Gamma)$, where N is the set of players and $c_\Gamma = c : 2^N \to \mathcal{R}$, the cost function, is defined by $c(S)$ = *the cost of a least expensive tour that serves all members of* S.

We will refer to $(N; c)$ as the Chinese postman (CP) game. The following lemma follows directly from the definition.

Lemma 1. *The Chinese postman game is monotonic; namely, if* $S \subseteq T$ *then* $c(S) \leq c(T)$.

Corollary 1. *The condition* $c(S) \geq 0$ *is equivalent to the condition* $c(\emptyset) \geq 0$.

In this paper we assume that the graph G of each extended Chinese postman enterprise is Eulerian; namely, the degree (valency) of each vertex is even. One reason for this restriction is the following theorem due to Hamers et al. [9].

Theorem 1. *If the graph of a Chinese postman enterprise is Eulerian and if there are no public edges and no prizes then the core is not empty.*

This theorem ceases to be true if public edges and/or prizes are introduced.

An important tool for the study of the core and the nucleolus of a cost game $(N; c)$ is the reduced game $(S; \hat{c}_S)$ on S at x, which was first studied by Davis and Maschler [4]. Here S is a non-empty subset of N and $\hat{c}_S(R)$, for x in the core, is defined to be:

$$\hat{c}_S(R) = \min_{Q \subseteq S^c} [c(R \cup Q) - x(Q)], \quad S^c := N \setminus S, \; R \subseteq S. \tag{3}$$

Theorem 2. *If $(N;c)$ corresponds to an extended Chinese Postman enterprise $\Gamma = [G, v_o, a, p, N]$, and $x \in Core(N; c_\Gamma)$, then the reduced game $(S; \hat{c}_S)$ corresponds to the enterprise $\hat{\Gamma} = [G, v_o, a, p + \hat{x}, S)$, where $\hat{x}_i = 0$ if $i \in S$, and $\hat{x}_i = x_i$ if $i \notin S$. Thus, $\hat{\Gamma}$ is obtained from G by making the edges, which were previously occupied by members of $N \setminus S$, public, and the prize of each such edge i increases by x_i.*

4 Chinese Postman on Eulerian Graphs Having the 4-Cut Property

We study in this section CP-games induced by Eulerian graphs having the 4-cut property. We show that the core and nucleolus of such CP games is the Cartesian product of cores and nucleoli defined on cycle graphs. In [7] we show that the core of a CP-game defined on a cycle or a chain graph has only a linear number of non redundant core constraints, and we develop therein an algorithm which computes the nucleolus of such games in quadratic time. Thus, results we derive in this section, coupled with those obtained in [7], imply that the nucleolus of CP-games induced by Eulerian graphs having the 4-cut property can be computed in quadratic time, and that verifying whether a vector is in the core of this class of games can be done in linear time.

Theorem 3. *Let $(N;c)$ be a Chinese postman game corresponding to an extended Chinese postman enterprise $\Gamma = (G, v_0, a, p, N)$ in which G is a simple cycle. Then, the following statements are equivalent.*

1. *$c(S) \geq 0$ for all $S \subseteq N$.*
2. *$Core[(N;c)] \neq \emptyset$.*
3. *$(N;c)$ is a submodular game.*

Definition 1. *An elementary subpath in a graph G is a subpath in G wherein all interior vertices have degree 2, and which is maximal under inclusion.*

If P is an elementary subpath of G between v_1 and v_2, we will refer to v_1 and v_2 as the *extreme vertices* of P.

Definition 2. *A proper edge cut-set of G is any edge cut-set of G, in which at most one edge belongs to the same elementary subpath of G.*

Definition 3. *A graph G is said to have the 4-cut property if the cardinality of every proper edge cut-set therein is at least 4.*

Theorem 4. *Let G be a Eulerian graph with costs and prizes and no public edges having the 4-cut property. If the core of the CP-game associated with G is not empty then for every elementary subpath, P, of G and every core allocation vector x, $x(P) = k(P)$, where $k(P)$ is the cost of traversing once the subpath P.*

We can now characterize the core and nucleolus of an extended Chinese postman enterprise defined on a Eulerian graph G having the 4-cut property which does not have public edges.

Theorem 5. *Let $(N;c)$ be the CP game associated with the extended CP enterprise $\Gamma = (G, v_0, a, p, N)$ in which G is a Eulerian graph having no public edges. Let $\mathcal{P}$ be the set of simple elementary subpaths in G. For each $P \in \mathcal{P}$ with extreme vertices $\hat{v}_P$, $\hat{w}_P$, denote by Γ_P an enterprise $\Gamma_P = (G_P, v_P, a\backslash P, p\backslash P, N_P)$, where G_P is a simple cycle whose edges and their costs and prizes are in 1-1 correspondence with those in G and are, respectively, occupied by the players that occupied the edges of P. Here, $\hat{v}_P$ and $\hat{w}_P$ correspond to the single post office v_P. Under these conditions, Core $(\Gamma) = \prod_{P \in \mathcal{P}}$ Core (Γ_P), where $\prod$ is the Cartesian product.*

Theorem 6. *Under the notation and conditions of the previous theorem, if the core of Γ is not empty then the nucleolus of Γ is the Cartesian product of the nucleoli of Γ_P:*

$$\nu(\Gamma) = \prod_{P \in \mathcal{P}} \nu(\Gamma_P)\,. \tag{4}$$

Corollary 2. *Let $\Gamma = (G, v_0, a, p, N)$ be an extended Chinese postman enterprise in which $G = (V, E)$ is the complete graph G_{2n+1}, $n \geq 2$, and there are no public edges. Then, Core (Γ) is not empty if and only if $c(e) - p(e) \geq 0$ for all $e \in E$. Further, if the latter condition is satisfied, then $\nu(\Gamma) = \{(c(e) - p(e)) : e \in E\}$.*

Finally, consider extended Chinese postman enterprises with corresponding CP games $(N;c)$ induced by cycle graphs with prizes and public edges. It can be shown that the core of such games can be described in terms of only $3n-1$ non redundant core constraints, and an $\mathcal{O}(n^2)$ algorithm for calculating the nucleolus for such games can be constructed [7], where n is the number of players. Therefore,

(i) the core of a CP game induced by a Eulerian graph having the 4-cut property can be described in terms of a linear (in the number of players/edges) number of non-redundant core constraints,

(ii) one can check in linear time whether a given vector is in the core of such a game, and

(iii) if the core is not empty, then the nucleolus of such a game can be calculated in quadratic time.

References

1. Balas, E. (1989): The Prize Collecting Traveling Salesman Problem. Networks **19**, 621–636

2. Bienstock, D., Goemans, M., Simchi-Levi, D., Williamson, D. (1993): A Note on the Prize Collecting Traveling Salesman Problem. Mathematical Programming **59**, 413–420
3. Davis, M., Maschler, M. (1965): The Kernel of a Cooperative Game. Naval Research Logistics Quarterly **12**, 223–259
4. Edmonds, J., Johnson, E. (1973): Matching, Euler Tours and the Chinese Postman. Mathematical Programming **5**, 88–124
5. Granot, D., Hamers, H., Kuipers, J., Maschler, M. (2002): Chinese Postman Games on Some Special Graphs. Working paper, Faculty of Commerce and Business administration, University of British Columbia
6. Granot, D., Hamers, H., Tijs, S. (1999): On Some Balanced, Totally Balanced and Submodular Games. Mathematical Programming **86**, 355–366
7. Hamers, H. (1997): On the Concavity of Delivery Games. European Journal of Operations Research **99**, 445–458
8. Hamers, H., Borm, P., van de Leensel, R., Tijs, S. (1994): The Chinese Postman and Delivery Games. European Journal of Operations Research
9. Kwan, Mei-Ko (1962): Graphic Programming Using Odd or Even Points. Chin. Math. **1**, 273–277

Farsighted Stability in Duopoly Markets with Product Differentiation

Takeshi Masuda[1] and Shigeo Muto[2]

[1] Tokyo Metropolitan University, Faculty of Economics
1-1-1 Minami-Ohsawa, Hachiohji-shi, Tokyo, 192-0397, Japan
E-mail: masuda@bcomp.metro-u.ac.jp

[2] Tokyo Institute of Technology,
Graduate School of Decision Making and Technology,
Department of Value and Decision Science
2-12-1 Oh-okayama, Meguro-ku, Tokyo, 152-8552, Japan
E-mail: muto@valdes.titech.ac.jp

Summary. Farsighted stability as proposed by Harsanyi and Chwe is examined in duopoly markets with product differentiation. Irrespective of competition style(Cournot or Bertrand) or irrespective of properties of goods(substitutes or complements) the von Neumann–Morgenstern farsighted stable set gives a Pareto efficient and individually rational outcome. Moreover, the largest consistent set due to Chwe consists of all individually rational outcomes.

Key words: game theory, farsighted stability, von Neumann–Morgenstern stable set, largest consistent set, duopoly, product differentiation

1 Introduction

This paper studies what is brought in duopoly markets with product differentiation if firms behave with farsightedness.

Greenberg [2] introduced the von Neumann–Morgenstern stability to the analysis of non-cooperative games. Muto and Okada (1996, 1998) applied Greenberg's idea to Cournot and Bertrand duopoly markets.

Later Chwe [1] criticized myopic behavior of players assumed in [4] and also in [2]. He introduced players' farsightedness into the definition of the stable set. He defined two solutions; the stable set with a sequence of deviations and its modification, which he called a consistent set. Suzuki and Muto [6] applied these two solutions to Prisoner's Dilemma as well as duopoly markets, and they found that the stable set with a sequence of deviations gives Pareto efficient and individually rational outcomes while the (largest) consistent set consists of all individually rational outcomes.

In this paper, we apply these two solution concepts to duopoly markets with product differentiation. We essentially find that every Pareto efficient outcome is itself a stable set with a sequence of deviations while the largest consistent set consists of all strategy combinations that give each firm a profit greater than or equal to the minimax payoff.

2 Farsighted Stability

Let $\Gamma = \{N, (A_i)_{i\in N}, (u_i)_{i\in N}\}$ be a two-person strategic game, where $N = \{1,2\}$ is the set of players, A_i is the set of strategies of player i, and $u_i : A_1 \times A_2 \to \Re$ is the payoff of player i.

For any two strategy profile $a = (a_1, a_2) \in A = A_1 \times A_2$ and $b = (b_1, b_2) \in A$, we say a is *induced from* b, *via player* i, denoted $b \xrightarrow[i]{} a$, if $a_j = b_j$ for $j \neq i$; i. e., a is reached from b by player i's unilateral move. We say a *indirectly dominates* b, denoted $a\, indom\, b$, if there exist a sequence of strategy profiles $b = a^0, a^1, \ldots, a^M = a$ and a sequence of players $i^1, i^2, \ldots, i^M$ such that for all $m = 1, \ldots, M$: $a^{m-1} \xrightarrow[i^m]{} a^m$ and $u_{i^m}(a^{m-1}) < u_{i^m}(a)$. Hence a indirectly dominates b when there are a sequence of strategy profiles starting from b and ending at a and a corresponding sequence of deviating players such that in each deviation a deviating player is better off at the end point a. In this paper we assume all moves are individual; no coalitional move is considered. The same results hold even if coalitional moves are assumed together with individual moves.

When $m = 1$, we say a *directly dominates* b, denoted $a\, dom\, b$. The direct domination is considered in [4] and [2]. The indirect domination above is borrowed from [1] which is slightly different from the definition of indirect domination by Harsanyi [3].

Chwe [1] incorporated farsighted behavior of players into stability by von Neumann and Morgenstern [4]. By replacing direct domination in the von Neumann–Morgenstern stable set by indirect domination, we obtain a farsighted stable set. A set $K \subseteq A$ is called a *farsighted stable set* if (1) for any $a, b \in K$, neither $a\, indom\, b$ nor $b\, indom\, a$, and (2) for any $b \in A - K$, there exists $a \in K$ such that $a\, indom\, b$. Properties (1) and (2) are called *internal* and *external stability*, respectively.

Chwe [1] introduced a consistent set by altering the definition of internal and external stability. A *consistent set* is a set $L \subseteq A$ satisfying the following two properties. (3) For any $a \in L$, $b \in A$ and $i \in N$ such that $a \xrightarrow[i]{} b$, there exists $c \in L$, where $c = b$ or $c\, indom\, b$, such that $u_i(a) \geq u_i(c)$. (4) For any $a \in A - L$, there exist $b \in A$ and $i \in N$ such that $a \xrightarrow[i]{} b$ and for all $c \in L$ where $c = b$ or $C\, indom\, b$, $u_i(a) < u_i(c)$. We call properties (3) and (4) *internal* and *external consistency*, respectively. The *largest consistent set* is a consistent set that contains all other consistent sets.

It should be noted that in the definition of the consistent set if we replace "there exists" in (3) by "for all" and "for all" in (4) by "there exists" then we obtain the definition of the farsighted stable set. Thus it can be shown that any farsighted stable set is contained in the largest consistent set. See proposition 3 in [1], p. 309. Moreover the largest consistent set is always unique, but farsighted stable sets are generally not unique. See [1].

3 Duopoly Markets with Differentiated Products

In this paper we consider the following duopoly model with differentiated products examined by Singh and Vives [5]. There are two firms. Let $N = \{1, 2\}$ be the set of firms. Each firm $i = 1, 2$ produces a differentiated good with the same marginal cost $c > 0$. Denote a production level of each firm $i = 1, 2$ by $x_i \geq 0$ and its unit price by $p_i \geq 0$. The representative consumer maximizes his or her utility separable and linear in money M

$$U(x_1, x_2) - M = a(x_1 + x_2) - \frac{b}{2}((x_1)^2 + 2\theta x_1 x_2 + (x_2)^2) - M ,$$

subject to the budget constraint $p_1 x_1 + p_2 x_2 \leq M$. We assume that $a > c$, $b > 0$, and $-1 < \theta < 1$. Hence firms face the demand structure described by

$$p_i = \max(a - bx_i - b\,\theta x_j, 0) \text{ for } i, j = 1, 2, \ i \neq j .$$

Direct demands of firms are given by

$$x_i = \max\left(\frac{(1-\theta)a - p_i + \theta p_j}{b(1-\theta^2)}, 0\right) \text{ for } i, j = 1, 2, \ i \neq j .$$

Therefore differentiated products are substitutes if $0 < \theta < 1$, independent if $\theta = 0$, and complements if $-1 < \theta < 0$. This paper considers the cases where differentiated products are substitutes and compliments.

In Cournot quantity competition, firms' strategies are their production levels. Firms maximize their profits

$$\pi_i(x_1, x_2) = (\max(a - bx_i - b\,\theta x_j, 0) - c)\, x_i \text{ for } i, j = 1, 2, \ i \neq j .$$

A Cournot duopoly model is formulated as the strategic game

$$\Gamma^C = \{N, (X_i)_{i \in N}, (\pi_i)_{i \in N}\} ,$$

where $N = 1, 2$ is the set of players, $X_i = [0, \infty)$ is the set of production levels of firm i, i. e., the set of strategies of player i, and $\pi_i(x_1, x_2)$ is the payoff of player i.

In Bertrand price competition, firms' strategies are their price levels. Firms maximize their profits

$$\rho_i(p_1, p_2) = (p_i - c) \max\left(\frac{(1-\theta)a - p_i + \theta p_j}{b(1-\theta^2)}, 0\right) \quad \text{for } i, j = 1, 2, \; i \neq j .$$

A Bertrand duopoly model is formulated as the strategic game

$$\Gamma^B = \{N, (Y_i)_{i \in N}, (\rho_i)_{i \in N}\} ,$$

where $N = 1, 2$ is the set of players, $Y_i = [0, \infty)$ is the set of price levels of firm i, i. e., the set of strategies of player i, and $\rho_i(p_1, p_2)$ is the payoff of player i.

For a two-person strategic game $\Gamma = \{N, (A_i)_{i \in N}, (u_i)_{i \in N}\}$, a strategy profile $a \in A$ is *Pareto efficient* if there is no $a' \in A$ such that $u_i(a') \geq u_i(a)$ for all $i = 1, 2$ and $u_i(a') > u_i(a)$ for some $i = 1, 2$. A strategy profile $a \in A$ is *individually rational* if for all $i, j = 1, 2, i \neq j$:

$$u_i(a) \geq \max_{a'_i \in A_i} \min_{a'_j \in A_j} u_i(a') .$$

If strict inequalities hold, a strategy profile $a \in A$ is *strictly individually rational.* A strategy profile $a \in A$ is *enforceable* if for all $i, j = 1, 2, i \neq j$:

$$u_i(a) \geq \min_{a'_j \in A_j} \max_{a'_i \in A_i} u_i(a') .$$

If strict inequalities hold, a strategy profile $a \in A$ is *strictly enforceable.* Let P be the set of Pareto efficient strategy profiles, I be the set of individually rational strategy profiles, S be the set of strictly individually rational strategy profiles, T be the set of enforceable strategy profiles, and W be the set of strictly enforceable strategy profiles. It is easy to see that the set of strictly enforceable strategy profiles W is contained in the set of strictly individually rational strategy profiles S since for all $i, j = 1, 2, i \neq j$:

$$\max_{a'_i \in A_i} \min_{a'_j \in A_j} u_i(a') \leq \min_{a'_j \in A_j} \max_{a'_i \in A_i} u_i(a') .$$

We consider the Cournot duopoly game Γ^C. Firms' best responses are given by

$$x_i = \max\left(\frac{a - b\theta x_j - c}{2b}, 0\right) \quad \text{for } i, j = 1, 2, \; i \neq j .$$

Hence the Nash equilibrium $(x_1^*, x_2^*,)$ is uniquely given:

$$(x_1^*, x_2^*,) = \left(\frac{a-c}{(\theta+2)b}, \frac{a-c}{(\theta+2)b}\right) .$$

Firms' best responses and the unique Nash equilibrium are depicted in figure 1, which corresponds to the case where differentiated products are substitutes. Figure 2 is for the case where differentiated products are compliments.

For the Cournot duopoly game, the following lemmas 1, 2, 3, 4 and 5 hold. Lemma 1 is for firms' profits. Lemmas 2 and 3 are for Pareto efficiency and enforceability in the case where differentiated products are substitutes, respectively. Lemmas 4 and 5 are for Pareto efficiency and strict enforceability in the case where differentiated products are compliments, respectively. Denote $x \in X = X_1 \times X_2$.

Lemma 1. *We consider the Cournot duopoly game Γ^C. For $i, j = 1, 2$, $i \neq j$:*

$$\begin{aligned} \pi_i(x) > 0 \quad &\text{if and only if} \quad x_i > 0 \ \text{and} \ a - bx_i - b\theta x_j - c > 0\,, \\ \pi_i(x) = 0 \quad &\text{if and only if} \quad x_i = 0 \ \text{and} \ a - bx_i - b\theta x_j - c = 0\,, \\ \pi_i(x) < 0 \quad &\text{if and only if} \quad x_i < 0 \ \text{and} \ a - bx_i - b\theta x_j - c < 0\,. \end{aligned}$$

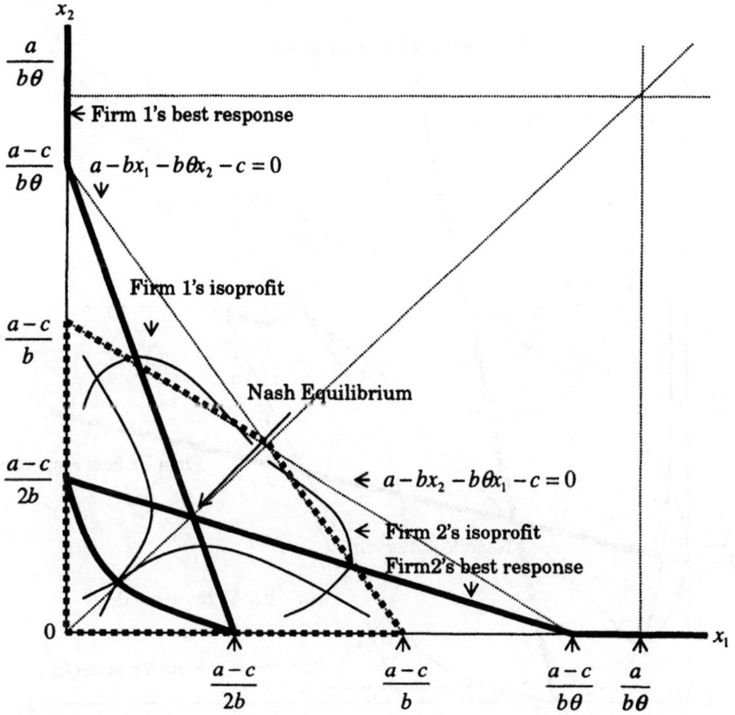

Fig. 1. For the Cournot duopoly game where differentiated products are substitutes, (a) firms' best responses, (b) the unique Nash equilibrium, (c) the set of Pareto efficient strategy profiles, (d) the set of enforceable strategy profiles, and (e) firms' isoprofits are illustrated. (a), (b), (c), (d), and (e) are described by thick lines, an intersection of thick lines, a thick curve, a region enclosed by dotted lines, and thin curves, respectively

Lemma 2. *We consider the Cournot duopoly game Γ^C where differentiated products are substitutes, $0 < \theta < 1$. Let*

$$\phi(x) = 2\theta b^2(x_1)^2 + 2\theta b^2(x_2)^2 + 4b^2 x_1 x_2 \\ -(\theta+2)(a-c)bx_1 - (\theta+2)(a-c)bx_2 + (a-c)^2 .$$

Then a strategy profile $x \in X$ is Pareto efficient only if:

$$\phi(x) = 0 \text{ and } a - bx_i - b\theta x_j - c > 0 \text{ for all } i,j = 1,2,\ i \neq j .$$

Remark A Pareto efficient strategy profile $x \in P$ implies:

$$a - bx_i - b\theta x_j - c > 0 \text{ for all } i,j = 1,2,\ i \neq j .$$

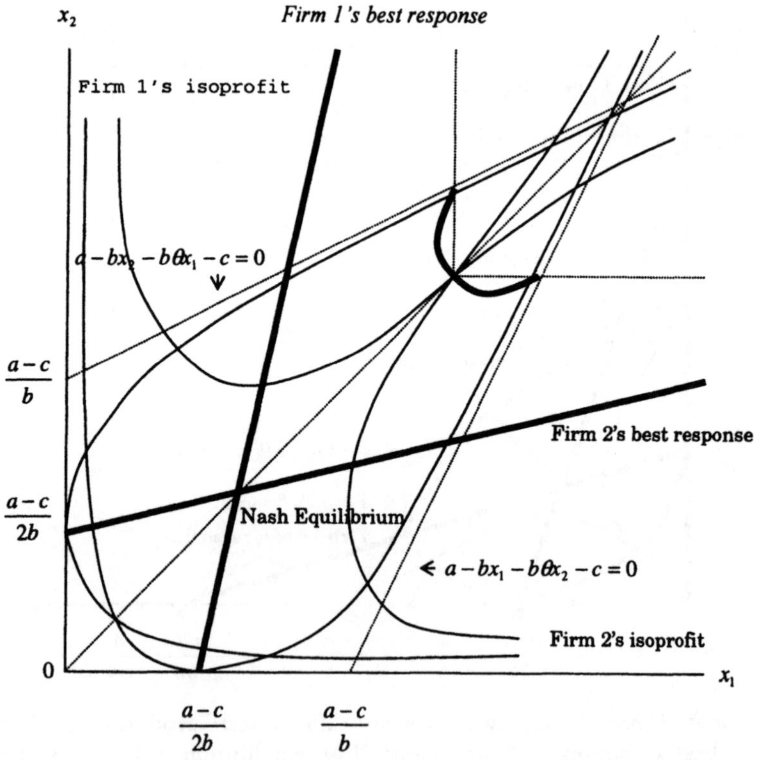

Fig. 2. For the Cournot duopoly game where firms' products are compliments, (a) firms' best responses, (b) the unique Nash equilibrium, (c) the subset of the set of Pareto efficient strategy profiles, (d) the set of strictly enforceable strategy profiles, and (e) firms' isoprofits are illustrated. (a), (b), (c), (d), and (e) are described by thick lines, an intersection of thick lines, a thick curve, a region enclosed by dotted curves, and thin curves, respectively

Lemma 3. *We consider the Cournot duopoly game Γ^C where differentiated products are substitutes, $0 < \theta < 1$. A strategy profile $x \in X$ is enforceable if and only if*

$$a - bx_i - b\,\theta x_j - c \geq 0 \;\; and \;\; x_i \geq 0 \;\; for\ all \;\; i, j = 1, 2\,,\; i \neq j\,.$$

Remark For all $i, j = 1, 2$, $i \neq j$: $\min_{x'_j \in X_j} \max_{x'_i \in X_i} \pi_i(x') = \max_{x'_i \in X_i} \min_{x'_j \in X_j} \pi_i(x') = 0$.

Remark The set of Pareto efficient strategy profiles P is contained in the set of enforceable strategy profiles T.

Lemma 4. *We consider the Cournot duopoly game Γ^C where differentiated products are complements, $-1 < \theta < 0$. Let*

$$\begin{aligned}\phi(x) = {} & 2\theta b^2 (x_1)^2 + 2\theta b^2 (x_2)^2 + 4b^2 x_1 x_2 \\ & -(\theta + 2)(a - c)bx_1 - (\theta + 2)(a - c)bx_2 + (a - c)^2\,.\end{aligned}$$

Then a strategy profile $x \in X$ is Pareto efficient and

$$x_i > 0 \;\; and \;\; a - bx_i - b\,\theta x_j - c > 0, \;\; for\ all \;\; i, j = 1, 2\,,\; i \neq j$$

only if $\phi(x) = 0$.

Lemma 5. *We consider the Cournot duopoly game Γ^C where differentiated products are complements, $-1 < \theta < 0$. Then a strategy profile $x \in X$ is strictly enforceable if and only if*

$$(a - bx_i - b\,\theta x_j - c)x_i > \frac{(a - c)^2}{4b} \;\; for\ all \;\; i, j = 1, 2\,,\; i \neq j\,.$$

Remark Denote $\nu_i = \min_{x'_j \in X_j} \max_{x'_i \in X_i} \pi_i(x')$. For all $i, j = 1, 2$, $i \neq j$:

$$\nu_i = \max_{x'_i \in X_i} \min_{x'_j \in X_j} \pi_i(x') = \frac{(a - c)^2}{4b} > 0\,.$$

In the Cournot duopoly game, whether firms' products are substitutes or compliments, the unique Nash equilibrium fails to attain Pareto efficiency. See figures 1 and 2.

We consider the Bertrand duopoly game Γ^B. Firms' best responses are given as follows. For $i, j = 1, 2$, $i \neq j$,

$$\begin{aligned} & p_i = \frac{(1 - \theta)a + c + \theta p_j}{2}\,, && \text{if } (1 - \theta)a + \theta p_j > c\,, \\ & p_i \geq (1 - \theta)a + \theta p_j\,, && \text{if } (1 - \theta)a + \theta\, p_j \leq c\,. \end{aligned}$$

Hence if differentiated products are substitutes, then the Nash equilibrium (p_1^*, p_2^*) is uniquely given,

$$(p_1^*, p_2^*) = \left(\frac{(1-\theta)(a-c)}{2-\theta} + c, \ \frac{(1-\theta)(a-c)}{2-\theta} + c \right) .$$

Firms' best responses and the unique Nash equilibrium is described in figure 3, which corresponds to the case where the substitutionability of firms' products is sufficiently high, $(1-\theta)a \leq c$. Figure 4 is for the case where the substitutionability of firms' products is sufficiently low, $(1-\theta)a > c$. If firms' products are compliments, then multiple Nash equilibria exist, i.e., besides the above one all strategy profiles (p_1^*, p_2^*) such that $p_i^* \geq (c-(1-\theta)a)/\theta$ for all $i = 1, 2$ are Nash equilibria. In figure 5, firms' best responses and multiple Nash equilibria are illustrated.

For the Bertrand duopoly game, the following lemmas 6, 7, 8, 9 and 10 hold. Lemma 6 is for firms' profits. Lemmas 7 and 9 are for Pareto efficiency in the cases where firms' products are substitutes and compliments, respectively. Lemmas 8 and 10 are for strict enforceability in the cases where firms' products are substitutes and compliments, respectively. Denote $p \in Y = Y_1 \times Y_2$.

Lemma 6. *We consider the Bertrand duopoly game Γ^B. For $i, j = 1, 2, i \neq j$,*

$$\begin{aligned}
\rho_i(p) > 0 &\quad \text{if and only if} \quad p_i > c \text{ and } p_i < (1-\theta)a + \theta p_j ; \\
\rho_i(p) = 0 &\quad \text{if and only if} \quad p_i = c \ \text{ or } \ p_i \geq (1-\theta)a + \theta p_j ; \\
\rho_i(p) < 0 &\quad \text{if and only if} \quad p_i < c \text{ and } p_i < (1-\theta)a + \theta p_j .
\end{aligned}$$

Lemma 7. *We consider the Bertrand duopoly game Γ^B where differentiated products are substitutes, $0 < \theta < 1$. Let*

$$\begin{aligned}
\phi'(p) = {} & 2\theta(p_1 - c)^2 + 2\theta(p_2 - c)^2 - 4(p_1 - c)(p_2 - c) + \\
& + (2-\theta)(1-\theta)(a-c)(p_1 - c) + (2-\theta)(1-\theta)(a-c)(p_2 - c) \\
& - (1-\theta)^2(a-c)^2 .
\end{aligned}$$

Then a strategy profile $p \in Y$ is Pareto efficient only if

$$\phi'(p) = 0 \text{ and for all } i, j = 1, 2, \ i \neq j: \ p_i > c \ \text{ and } \ p_i < (1-\theta)a + \theta p_j .$$

Lemma 8. *We consider the Bertrand duopoly game Γ^B where differentiated products are substitutes, $0 < \theta < 1$. The following three properties hold.*
(a) A strategy profile $p \in Y$ is individually rational if and only if

$$p_i \geq c \text{ for all } \ i = 1, 2 .$$

(b) A strategy profile $p \in Y$ is strictly individually rational if and only if

$$p_i > c \text{ and } \ p_i < (1-\theta)a + \theta p_j \text{ for all } \ i, j = 1, 2, \ i \neq j .$$

(c) Suppose the substotutionability of firms' products is sufficiently high, $(1-\theta)a \leq c$. Then a strategy profile $p \in Y$ is strictly enforceable if and only if

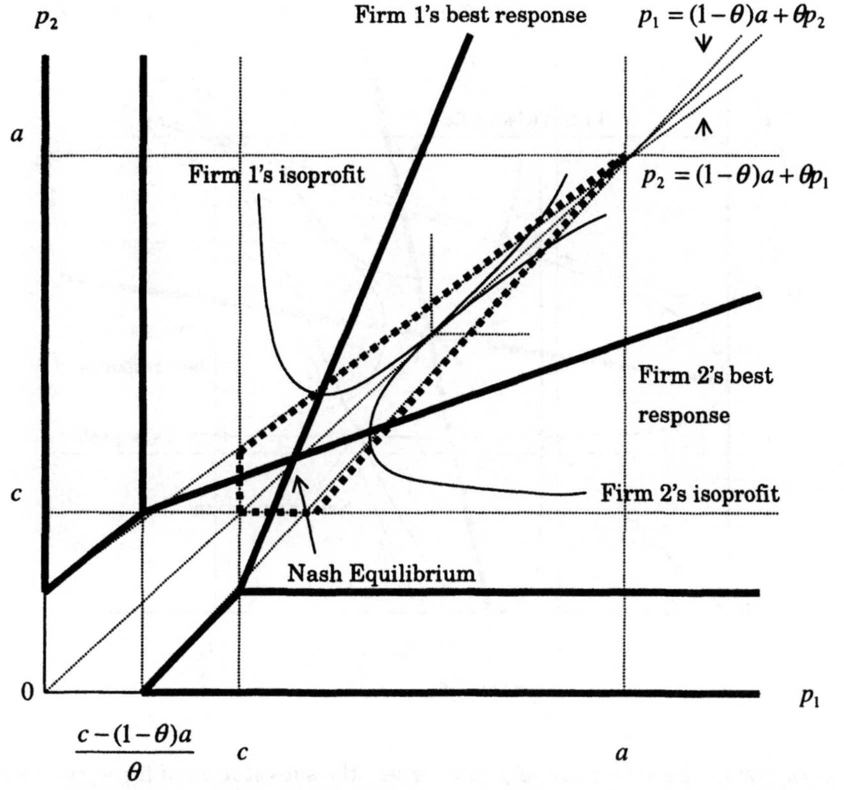

Fig. 3. For the Bertrand duopoly game where differentiated products are relatively close substitutes, (a) firms' best responses, (b) the unique Nash equilibrium, (c) the set of Pareto efficient strategy profiles, (d) the set of strictly enforceable strategy profiles, and (e) firms' isoprofits are illustrated. (a), (b), (c), (d), and (e) are described by thick lines, an intersection of thick lines, a thick curve, a region enclosed by dotted lines, and thin curves, respectively

$$p_i > c \text{ and } p_i < (1-\theta)a + \theta p_j \text{ for all } i,j = 1,2,\ i \neq j\,.$$

Suppose the substitutionability of firms' products is sufficiently low, $(1-\theta)a > c$. Then a strategy profile $p \in Y$ is strictly enforceable if and only if:

$$(p_i - c)\left(\frac{(1-\theta)a - p_i + \theta p_j}{b(1-\theta^2)}\right) > \frac{((1-\theta)a-c)^2}{4b(1-\theta^2)} \text{ for all } i,j = 1,2,\ i \neq j\,.$$

Remark For all $i,j = 1,2,\ i \neq j$: $\max_{p'_i \in Y_i} \min_{p'_j \in Y_j} \rho_i(p') = 0$.

Remark Denote $\nu_i = \min_{p'_j \in Y_j} \max_{p'_i \in Y_i} \rho_i(p')$.

(a) If $(1-\theta)a \leq c$, then for all $i = 1,2$: $\nu_i = 0$.

(b) If $(1-\theta)a > c$, then for all $i = 1,2$: $\nu_i = \frac{((1-\theta)a-c)^2}{4b(1-\theta^2)} > 0$.

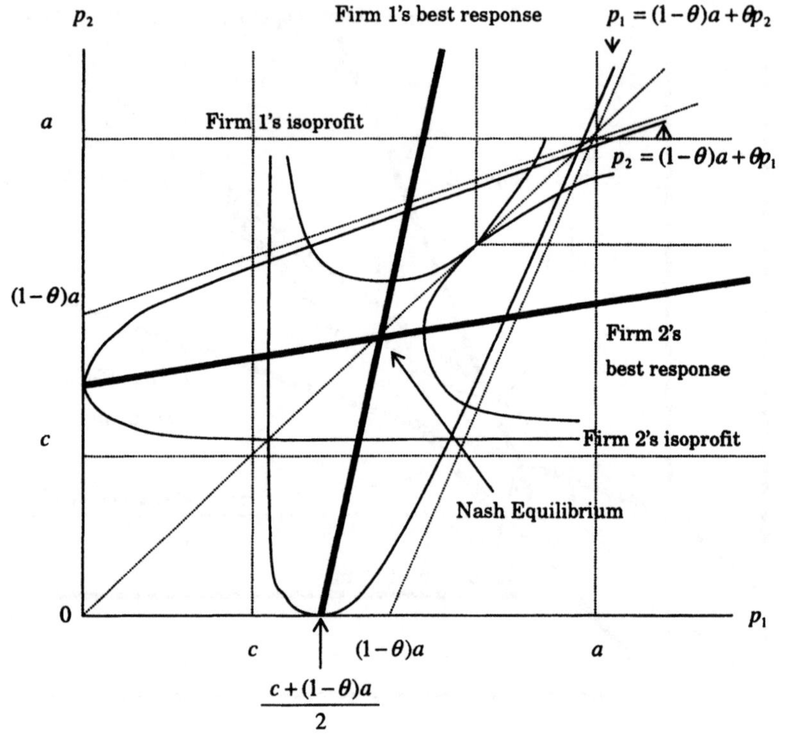

Fig. 4. For the Bertrand duopoly game where the substitution of firms' products is relatively low, (a) firms' best responses, (b) the unique Nash equilibrium, (c) the set of Pareto efficient strategy profiles, (d) the set of strictly enforceable strategy profiles, and (e) firms' isoprofits are illustrated. (a), (b), (c), (d), and (e) are described by thick lines, an intersection of thick lines, a thick curve, a region enclosed by dotted curves, and thin curves, respectively

Lemma 9. *We consider the Bertrand duopoly game Γ^B where differentiated products are compliments, $-1 < \theta < 0$. Let*

$$\begin{aligned}\phi'(p) = {} & 2\theta(p_1 - c)^2 + 2\theta(p_2 - c)^2 - 4(p_1 - c)(p_2 - c) \\ & +(2-\theta)(1-\theta)(a-c)(p_1 - c) + (2-\theta)(1-\theta)(a-c)(p_2 - c) - \\ & -(1-\theta)^2(a-c)^2 .\end{aligned}$$

Then a strategy profile $p \in Y$ is Pareto efficient and

$$p_i > c \quad and \quad p_i < (1-\theta)a + \theta p_j \quad for\ all \quad i, j = 1, 2,\ i \neq j\ ,$$

only if $\phi'(p) = 0$.

Lemma 10. *We consider the Bertrand duopoly game Γ^B where differentiated products are compliments, $-1 < \theta < 0$. The following two property hold.*
(a) A strategy profile $p \in Y$ is enforceable if and only if

$$p_i \geq c \;\; \textit{for all} \;\; i = 1, 2 \,, \;\; \textit{or} \;\; p_i \geq (1-\theta)a + \theta p_j \;\; \textit{for all} \;\; i, j = 1, 2 \,, \; i \neq j \,.$$

(b) A strategy profile $p \in Y$ is strictly enforceable if and only if

$$p_i > c \;\; \textit{and} \;\; p_i < (1-\theta)a + \theta p_j \;\; \textit{for all} \;\; i, j = 1, 2 \,, \; i \neq j \,.$$

Remark For all $i, j = 1, 2$, $i \neq j$: $\min_{p'_j \in Y_j} \max_{p'_i \in Y_i} \rho_i(p') = \max_{p'_i \in Y_i} \min_{p'_j \in Y_j} \rho_i(p') = 0$.

Also in the Bertrand duopoly game, whether differentiated products are substitutes or compliments, Pareto efficiency is never achieved in Nash equilibria. In figures 3 and 4, the unique Nash equilibrium does not belong in the set of Pareto efficient strategy profiles. In figure 5 there is no Nash equilibrium that belong in the set of Pareto efficient strategy profiles.

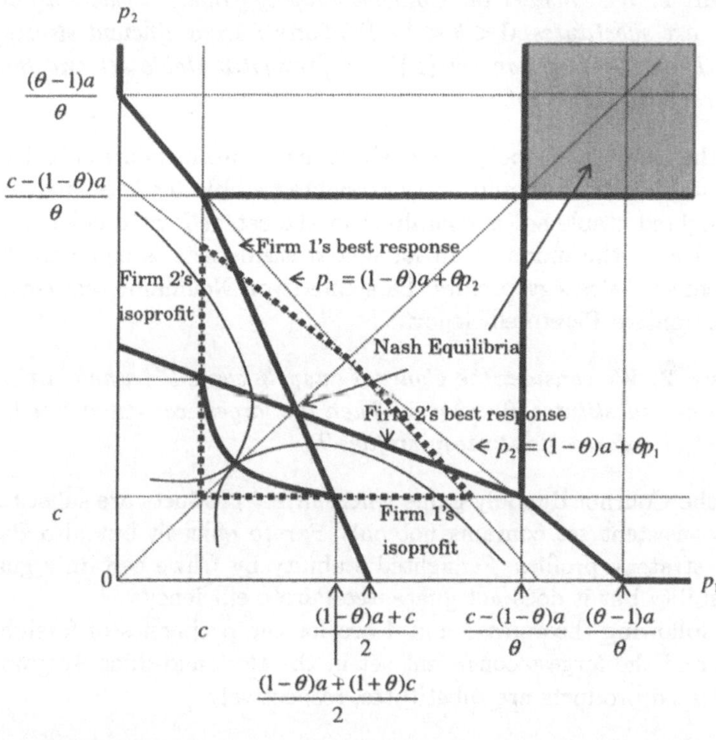

Fig. 5. For the Bertrand duopoly game where firms' products are compliments, (a) firms' best responses, (b) Nash equilibria, (c) the set of Pareto efficient and strictly enforceable strategy profiles, (d) the set of strictly enforceable strategy profiles, and (e) firms' isoprofits are illustrated. (a), (b), (c), (d), and (e) are described by thick lines, an intersection of thick lines and a gray area, a thick curve, a region enclosed by dotted lines, and thin curves, respectively

4 Farsighted Behavior in Duopoly Markets with Substitutes

This section considers what will result from firms' farsighted behavior in both Cournot and Bertrand duopoly markets, where firms' products are substitutes. Our principal findings are as follows. Whether firms engage in Cournot quantity competition or Bertrand price competition, every farsighted stable set is a singleton and it produces only a Pareto efficient outcome. By contrast, the largest consistent set contains not only Pareto efficient but also Pareto inefficient outcomes.

For the Cournot duopoly game where differentiated products are substitutes, the following theorems 1 and 2 hold. The former is for farsighted stable sets and the latter is for the largest consistent set.

Theorem 1. *We consider the Cournot duopoly game Γ^C where differentiated products are substitutes, $0 < \theta < 1$. Take any Pareto efficient strategy profile $x \in P$. Then the singleton set $\{x\}$ is a farsighted stable set and there is no other farsighted stable set.*

For the Cournot duopoly game where firms' products are substitutes, farsighted stable sets exist and every farsighted stable set is a singleton. Also, any farsighted stable set is contained in the set of Pareto efficient strategy profiles. Hence the union of all farsighted stable sets is equal to the set of Pareto efficient strategy profiles. Farsighted von Neumann–Morgenstern stability guarantees Pareto efficiency.

Theorem 2. *We consider the Cournot duopoly game Γ^C where differentiated products are substitutes, $0 < \theta < 1$. Then the largest consistent set L is equal to the set of enforceable strategy profiles T.*

For the Cournot duopoly game where firms' products are substitutes, the largest consistent set contains not only Pareto efficient but also Pareto inefficient strategy profiles. Farsighted stability by Chwe certainly guarantees enforceability but it does not guarantee Pareto efficiency.

The following theorems 3 and 4 are for the properties of farsighted stable sets and the largest consistent set in the Bertrand duopoly game where differentiated products are substitutes, respectively.

Theorem 3. *We consider the Bertrand duopoly game Γ^B where firms' products are substitutes, $0 < \theta < 1$. Take any Pareto efficient and strictly enforceable strategy profile $p \in P \cap W$. Then the singleton set $\{p\}$ is a farsighted stable set and there is no other farsighted stable set.*

Theorem 4. *We consider the Bertrand duopoly game Γ^B where firms' products are substitutes, $0 < \theta < 1$. The largest consistent set L is equal to the set of strictly enforceable strategy profiles W.*

Also for the Bertrand duopoly game where differentiated products are substitutes, every farsighted stable set is a singleton and it is contained in the set of Pareto efficient and strictly enforceable outcomes. By contrast, the largest consistent set is equal to the set of strictly enforceable outcomes, and hence it contains also Pareto inefficient outcomes. Farsighted von Neumann–Morgenstern stability guarantees Pareto efficiency, although farsighted stability by Chwe does not.

5 Farsighted Behavior in Duopoly Markets with Compliments

This section examines farsightedly stable outcomes in both Cournot and Bertrand duopoly markets where differentiated products are compliments. Our main results are as follows. Whether in the Cournot game or the Bertrand game, every farsighted stable set is a singleton and it is contained in the set of Pareto efficient and strictly enforceable strategy profiles, while the largest consistent set is equivalent to the set of strictly enforceable strategy profiles.

The following theorems 5, 6, 7 and 8 hold. Theorems 5 and 7 are for farsighted stable sets in Cournot and Bertrand duopoly markets, respectively. Theorems 6 and 8 are for the largest consistent set in Cournot and Bertrand duopoly markets, respectively.

Theorem 5. *We consider the Cournot duopoly game Γ^C where differentiated products are complements, $-1 < \theta < 0$. Take any Pareto efficient and strictly enforceable strategy profile $x \in P \cap W$. Then the singleton set $\{x\}$ is a farsighted stable set and there is no other farsighted stable set.*

Theorem 6. *We consider the Cournot duopoly game Γ^C where differentiated products are complements, $-1 < \theta < 0$. The largest consistent set L is equal to the set of strictly enforceable strategy profiles W.*

Theorem 7. *We consider the Bertrand duopoly game Γ^B where firms' products are complements, $-1 < \theta < 0$. Take any Pareto efficient and strictly enforceable strategy profile $p \in P \cap W$. Then the singleton set $\{p\}$ is a farsighted stable set and there is no other farsighted stable set.*

Theorem 8. *We consider the Bertrand duopoly game Γ^B where firms' products are complements, $-1 < \theta < 0$. The largest consistent set L is equal to the set of strictly enforceable strategy profiles W.*

Also in duopoly markets where firms' products are compliments, whether firms engage in quantity competition or price competition, every farsighted stable set is a singleton and it produces only a Pareto efficient and strictly enforceable outcome, while the largest consistent set is equal to the set of strictly enforceable outcomes, i.e., it contains not only Pareto efficient also

Pareto inefficient outcomes. Farsighted von Neumann–Morgenstern stability guarantees Pareto efficiency. By contrast, farsighted stability by Chwe may guarantee strict enforceability but does not guarantee Pareto efficiency.

References

1. Chwe, M. S.-Y. (1994): Farsighted Coalitional Stability. Journal of Economic Theory **63**, 299–325
2. Greenberg, J. (1990): The Theory of Social Situations: an Alternative Game Theoretic Approach. Cambridge University Press
3. Harsanyi, J. (1974): An Equilibrium-Point Interpretation of Stable Sets and a Proposed Alternative Definition. Management Science **20**, 1472–1495
4. Neumann, J. von, Morgenstern, O. (1953): The Theory of Games and Economic Behavior, the 3rd ed. Princeton University Press
5. Singh, N., Vives, X. (1984): Price and Quantity Competition in a Differentiated Duopoly. Rand Journal of Economics **15**, 546–554
6. Suzuki, A., Muto, S. (1998): Farsightedness Leads to Efficiency in Duopoly Markets. In: Proceedings of the 8-th International Symposium on Dynamic Games and Applications, July 5–8, 1998, 421–427

International Emissions Trading with Agent-Based Simulation and Web-Based Gaming

Hideyuki Mizuta[1] and Yoshiki Yamagata[2]

[1] Tokyo Research Laboratory, IBM Research,
Shimotsuruma 1623-14, Yamato, Kanagawa 242-8502, Japan
E-mail: e28193@jp.ibm.com

[2] Climate Change Research Project,
National Institute for Environmental Studies,
Onogawa 16-2, Tsukuba, Ibaraki 305-0053, Japan
E-mail: yamagata@nies.go.jp

Summary. The cooperative utilization of agent-based computational simulation and gaming simulation reproduces complex phenomena in the economic and social environment. It supports experiments with various controlling methods, to evaluate the design, and to extract the fundamental results. We develop a simple agent-based simulation framework "ASIA" and apply it to economic and environmental studies including the international greenhouse gas (GHG) emissions trading. It is found the worldwide cost reduction follow through the individual adjustment interpreted as movement in agents' cost landscape. We also introduce a Web application for the gaming simulation connected with the agent system, which can be used for the international emissions trading experiments with human players.

Key words: kyoto protocol, emissions trading, agent-based approach, gaming simulation, cost landscape

1 Introduction

In real society, the dynamic behavior and interactions between people are very complicated and may often seem irrational. Further complicating the situation, the recent progress and popularity of network communication technologies greatly widens the diversity of participants and affects the market mechanism itself, and increases the dynamic fluctuations of economic systems. In the past, traditional economic and social theories have only considered idealized representative participants in equilibrium states. It is very difficult to analyze dynamically changing situations involving heterogeneous subjects using such static and homogeneous methods. In the last decade, many researchers, including physicists and computer scientists, are starting to apply new approaches to investigate such complex dynamics in their studies of social science [5].

One of these approaches is the agent-based approach which reconstructs complex macro societies aggregating micro agents through the micro-macro loop. In many cases, this approach utilizes computer simulation technology to aggregate the micro behavior and equate the micro economic agent with the software agent designed with object-oriented technology.

On the other hand, gaming simulations with human players are also attracting researchers. By using the methodology of gaming, it becomes possible to examine the behavior of real human players under controlled circumstances. This is also expected to provide common testbeds for a variety of researchers.

In a series of reports [2, 3, 4], we introduced a Java framework, ASIA (Artificial Society with Interacting Agents), and several agent-based simulations including a simple commodities market, a dynamic online auction, and the international greenhouse gas (GHG) emission trading discussed in the Kyoto Protocol.

Based on the agreement at COP7, the Kyoto Protocol is expected to be ratified soon. For efficient achievement of the Protocol, the Kyoto mechanisms such as IET (International Emissions Trading), JI (Joint Implementation), and CDM (Clean Development Mechanism) are important. However there are still many concerns about how the mechanisms will function in the real international situation.

In this paper, we will investigate the cost landscape and the agents' behavior for several simple cases of the international GHG emissions trading. We will also consider a Web application for the gaming simulation of the emissions trading with human players to investigate and improve the behavior model for agents.

2 Agent-Based Simulation Framework: ASIA

For effective implementations of the agent-based economic and social simulations, we developed a simple framework, Artificial Society with Interacting Agents (ASIA), using Java. This framework provides only very simple and fundamental functionality for social simulations.

We constructed our framework with a layered structure. The Agent Layer contains a basic agent class and the fundamental environment for the agents. The environment provides the fundamental facilities for agents and users to create agents, to dispose of agents, and to send messages through a MessageManager class. The MessageManager collects and distributes messages sequentially with its own thread according to the predetermined schedule. Agents also have their own threads to process the distributed messages. The Social Layer describes the basic role of agents in the society and gives the example of message exchanges for trade. We implemented Central, Participant, and Watcher agents and a simple market process using RFB and BID messages. The Central agent creates, registers and initiates Participant agents and Watcher agents.

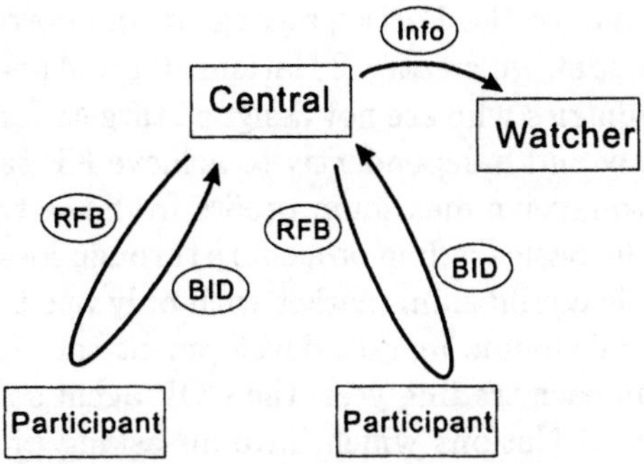

Fig. 1. Message Transactions in the Social Layer

One sample trade procedure can be executed as follows (see Fig. 1).To begin a trade, the Central agent sends a Request For Bid (RFB) message to each Participant. Upon receiving a RFB message, a Participant agent replies with a BID message. The Central agent collects all of the BID messages and proceeds to the trade transaction if the users have customized the descendant appropriately. Finally, each Watcher agent receives information about the trade and report it to the users in the desired format.

3 Greenhouse Gas Emissions Trading

In this section, we consider the application of the agent-based simulation for the international greenhouse gas (GHG) emissions trading under the Kyoto Protocol (KP).

To prevent the global warming, 160 countries agreed to the KP on limiting GHG emissions at COP3 in 1997. KP sets targets for Annex I countries at assigned reductions below the 1990 levels, with the targets to be met during the commitment period 2008–2012. For example, Japan and the US should reduce 6% and 8% of their emissions, respectively. The KP allows international GHG emission trading, where countries who cannot reach the reduction targets can buy the emissions rights from other countries who can easily satisfy the target. Such a market mechanism is expected to reduce the worldwide cost for GHG reduction because of the large range in the marginal abatement cost curves (MACs) for reducing GHG emissions.

J. Grütter developed the CERT model [1] which calculate the equilibrium price with various options and parameters for MACs. The CERT model treats only one trade in 2010 and each country must achieve the targets in that year. Because this model is implemented with a spreadsheet and macros, it is difficult to expand the model to treat successive trades and to assign different strategies to different countries.

The structure of the simulation system is as follows. COP agent is a descendant of the Central agent and manages the international trading. Nation agents are descendants of the Participant agent and correspond to countries or groups. In this model, we create 12 Nations; 6 are Annex I countries and 6 are Non Annex I countries who are not assigned targets for reduction. Nations behave autonomously and independently to achieve KP targets assigned with minimum costs or to receive maximum profits from the trades.

Figure 2 shows the basic trading procedure through message exchanges. We consider both a static equilibrium market with only one trade in 2010, as was discussed in [1], and dynamic market development through the commitment period 2008–2012. In each trading year, the COP agent sends Request for Bid (RFB) messages to all Nations which have an asking price. Upon receiving the RFB message, a Nation agent examine the asking price and his MAC to decide the amount of the domestic reduction. Then he sends back a Bid message to the COP agent which says how much he wants to buy or to sell at the asked price. After repeating this RFB-BID process, the COP agent will find the equilibrium price where the demand and the supply balance, and send the Trade message to approve the trades for the year. Thus, the equilibrium price at one year is determined when the MAC functions and the assigned reduction of all participants are given.

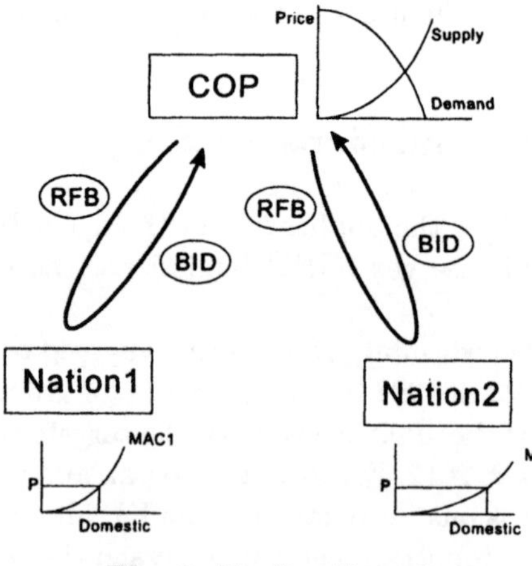

Fig. 2. Trading Procedure

Then, we consider the multiple trading periods. Nation i divide the assigned reduction R_i into each year $n = 0, 1, 2, \ldots,$

$$\sum_n R_{in} = R_i \, .$$

As described previously, we can find the equilibrium price P_n^* at each year using a partition of the assigned reduction R_{in} and a MAC function at the moment. To consider the dynamics of MAC, we introduce a technology function $t_{in}(p)$ which gives the amount of reduction using the available technology at a given cost p for the Nation i at the year n. MAC is given as the inverse function of the integral of the technology function.

$$\text{MAC (amount)} \equiv D_{in}^{-1}(P) ,$$

$$D_{in}(P) \equiv \int_0^P t_{in}(p)dp ,$$

where $D_{in}(P)$ is the cumulated amount of the domestic reduction using technologies which is cheaper than the given price P. The difference between the assigned reduction R_{in} and the domestic reduction $D_{in}(P)$ will be traded among counties,

$$T_{in}(P) \equiv R_{in} - D_{in}(P) .$$

At the equilibrium price P_n^*, there is no excess demand,

$$\sum_j T_{jn}(P_n^*) = 0 .$$

This condition gives the equilibrium price using the assigned reductions at each year. Hence we can consider that the equilibrium price P_n^*, the domestic reduction at equilibrium D_{in}^*, and the trading amount at equilibrium T_{in}^* as functions of the assigned reduction R_{in}. The reduction cost at one year is given by

$$\tilde{C}_n(R_{in}) \equiv C_{in}(D_{in}^*(R_{in})) + T_{in}^*(R_{in})P_{in}^*(R_{in}) ,$$

where $C_{in}(x)$ is a domestic cost function defined by the integral of MAC.

At each year, all countries determine the amount of the domestic reduction with which the values of MAC for all countries agree with one international value, that is, the equilibrium price, to minimize the worldwide reduction cost. Similarly, the total cost over the commitment period will be minimized by choosing the partition R_{in} $(n = 0, 1, 2, \dots)$ of the assigned reduction so that the differential coefficient of the total cost $\tilde{C}'_{in}(R_{in}) = P_n^* + T_{in}^*/\tau_n^*$ becomes constant over $n = 0, 1, 2, \dots$, with $\tau_n^* \equiv \sum_j t_{jn}(P_n^*)$.

As a simple dynamic process for the reduction technology $t_{in}(p)$, we adopt reusability $0 \le \alpha \le 1$ and deflation $0 \le \gamma \equiv 1/\beta \le 1$. Once the technology whose cost is lower than the price P^* is used, the reusability of the technology will be restricted with the coefficient α. On the other hand, the technical innovations and deflation decreases the cost of the technology. With $\bar{P}_{in} \equiv \max\{\gamma_i^n P_0^*, \gamma_i^{n-1} P_1^*, \dots, \gamma_i P_{n-1}^*\}$, we can obtain the technology function as

$$t_{in}(p) \equiv \begin{cases} \alpha_i \beta_i^n t_{i0}(\beta_i^n p) , & p < \bar{P}_{in} , \\ \beta_i^n t_{i0}(\beta_i^n p) , & \text{otherwise} . \end{cases}$$

We assume the initial technology function $t_{i0}(p)$ with two coefficients a_i and b_i to reproduce the quadratic MAC function in the CERT model,

$$t_{i0}(p) \equiv \frac{1}{\sqrt{b_i^2 + 4a_i p}} .$$

In the simulation, we fix the parameters $\{a_i\}$, $\{b_i\}$ and $\{R_i\}$ for 12 countries as given in the CERT model and use randomly distributed $\{\alpha_i\}$ and $\{\beta_i\}$. Each Nation agent i determine the initial partition of the reduction $\{R_{in}\}$ and update them after the commitment period so that the variance of the marginal reduction cost $\{\tilde{C}'_{in}(R_{in})\}$ decreases.

Figure 3 shows an example of the simulation views. We can see brief reports on 12 agents and price changes from 2008 to 2012.

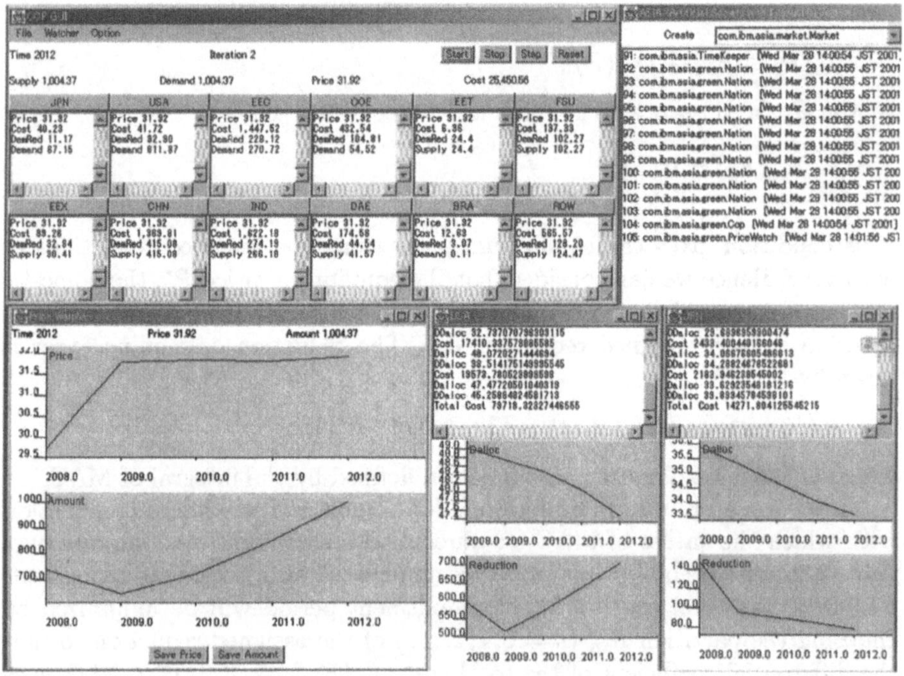

Fig. 3. Dynamic GHG Emissions Trading over the Commitment Period, 2008–2012

Figure 4 also shows changes of the total reduction cost for the entire world and for each country. In the beginning of a simulation, all countries fix their partition as the average value through the trading period. By adjusting the partition after all of the trades so that the marginal reduction costs becomes a constant over the trading periods, each country expects that the total cost will be optimized. Though each country tries selfishly to decrease only its own cost, the total cost for the world can be reduced via this process.

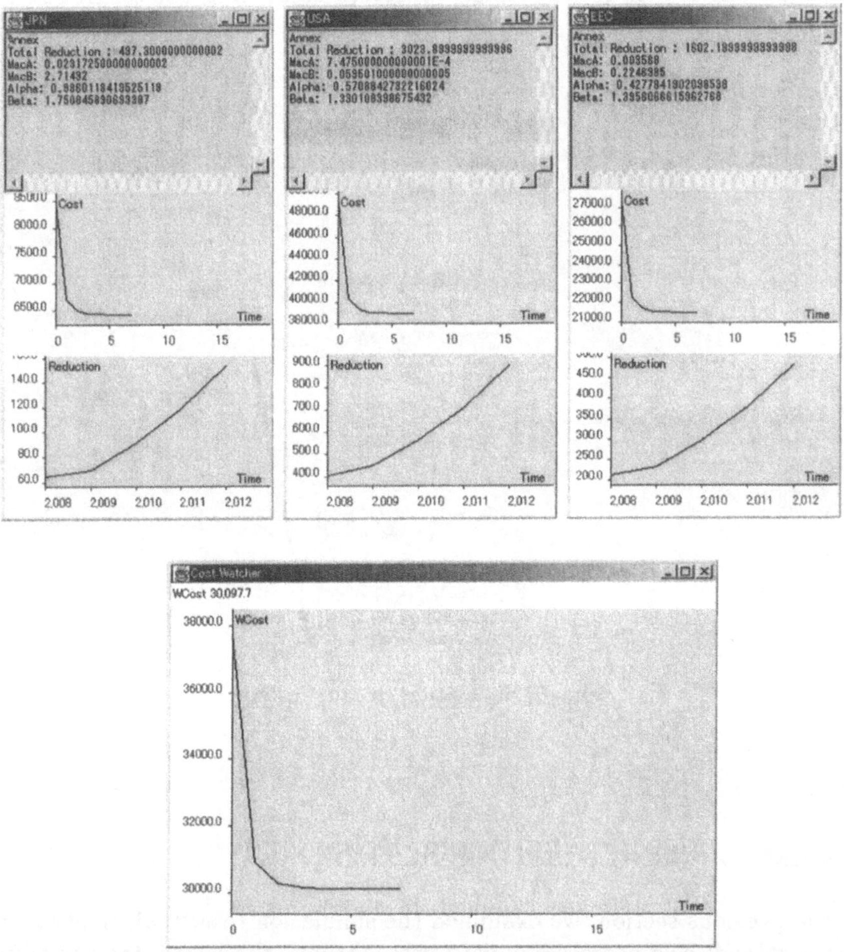

Fig. 4. Changes of Costs via Partition Adjustments

This simulation contains a hierarchical structure of iterations with different parameters and strategies. We have organized the simulation as described below and stored results into a database representing the relations.

First, we define one experiment using identical experimental parameters such as the number of nations, cost functions, and trading periods. An experiment consists of games, which show the results with different strategies of the nations, but with the same parameters. One game includes several trades according to the predetermined trading periods. Typically we use 5 years of trades. Finally, each trade involves an iteration of RFB-BID transactions to determine the equilibrium price. Figure 5 shows a screenshot of four views on one experiment with a hierarchical structure containing games, trades, rfbs and bids.

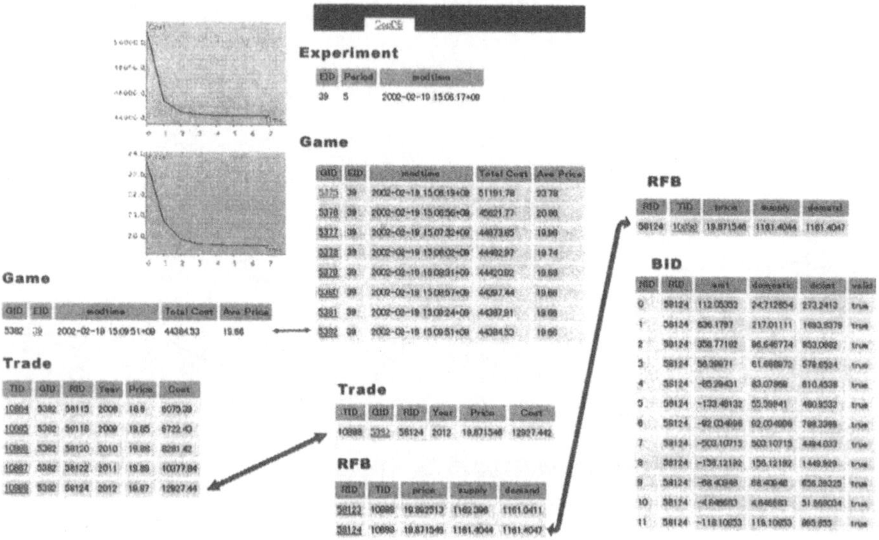

Fig. 5. Simulation Results in DB

4 Cost Landscape and Agent Behavior

In the previous section, we examined the simulation results which show that the worldwide cost decreases in most cases through the selfish strategy adjustments of the agents. To more deeply investigate the nature of international emissions trading, we consider the cost landscapes for pairs of two Nation agents with two trading periods as a basis. The extension to more general trading periods and numbers of agents remains as future work, but will be done on this foundation. In particular, we can expect to extend the trading periods backward by gradually adding an external trading year before the optimally settled trades. Also, we can investigate the market impact of the strategies of one particular agent out of many Nation agents similarly with the following analysis for the impact matrices in the simplest situation.

As described in the previous section, a total reduction target R_i and a dynamic MAC function are assigned to each Nation agent. We considered the partition $\sum_n R_{in} = R_i$ of the reduction target as agent i's strategy.

Now we consider only two trading periods ($n = 0, 1$). Hence, the reduction amount for the first trading year R_{i0} determines the strategy i, that is, $R_{i1} \equiv R_i - R_{i0}$.

Once the strategies of the two agents ($i = 0, 1$) are settled, the equilibrium trading price P_n^* and the corresponding amount of domestic reduction D_{in}^* and trading T_{in}^* through the trading periods ($n = 0, 1$) are determined. The total costs of the Nations $\tilde{C}_i$ ($i = 0, 1$) are given by the sum of the domestic reduction cost and the trading expenditure,

$$\tilde{C}_i(R_{i0}) \equiv \sum_{n=0}^{1} C_{i0}(D_{in}^*(R_{i0})) + T_{in}^*(R_{i0})P_{in}^*(R_{i0}) ,$$

where $C_{in}(x)$ is a domestic cost function defined by the integral of the MAC. This cost may have a negative value when a Nation obtains sufficient profit through the trade to cancel the domestic cost. By visualizing this total cost $\tilde{C}_i$ against strategies R_{00} and R_{10}, we obtain the cost landscapes.

Finally, we consider the variance matrix V_j^i formed by the average of the variance of the costs of agent i with changes in the strategy of agent j. We obtained the cost landscape with discrete $m = 51$ values for each strategy. We denote the cost for agent i with strategies $R_{00} \equiv r_k^0$ ($k = 0 \ldots 50$) and $R_{10} \equiv r_l^1$ ($l = 0 \ldots 50$) by c_{kl}^i. The value of V_j^i is given by

$$V_0^i \equiv \frac{1}{m} \sum_l \sum_k \left(c_{kl}^i - \frac{1}{m} \sum_k c_{kl}^i \right)^2 ,$$

$$V_1^i \equiv \frac{1}{m} \sum_k \sum_l \left(c_{kl}^i - \frac{1}{m} \sum_l c_{kl}^i \right)^2 .$$

By normalizing V_j^i with the ranges of r_k^i and r_l^i, and taking the ratio of the square root of each element to make the diagonal element to be 1, we obtain the normalized impact matrix p_j^i,

$$p_j^i \equiv \begin{pmatrix} 1 & \tilde{v}_1^0/\tilde{v}_0^0 \\ \tilde{v}_0^1/\tilde{v}_1^1 & 1 \end{pmatrix} ,$$

where $\tilde{v}_j^i$ is given by

$$\tilde{v}_j^i \equiv \sqrt{\frac{V_j^i}{|r_{m-1}^j - r_0^j|}} .$$

In Fig. 6, the cost landscape for the pair of JPN and EEC is shown. As the colors change from black and red to yellow and white, the cost increases. Points connected with lines indicate the learning paths of agents, and the corresponding changes of the total cost and average price are shown in Fig. 7. For this pair, the agent learning works quite effectively and the strategies converges near the optimal values of total cost. This is because EEC has the dominant impact on its cost landscape as shown in the second graph of Fig. 6, and can determine his strategy independently to optimize his cost. As a result, JPN has to choose a relatively efficient strategy which is located on the saddle point in the first graph of Fig. 6. The impact matrix p_j^i for this cost landscape,

$$p_j^i = \begin{pmatrix} 1 & 0.583 \\ 0.264 & 1 \end{pmatrix},$$

also indicates the strong market influence of EEC.

Experiments

EID	Period	modtime	xopt
61	2	2002-05-20 02:14:48+09	sweep 50 patheid 75

Nations

EID	NID	name	target	reduction	a	b	alpha	beta
61	0	JPN	273.54	99.46	0.0231725	2.71492	1.0	1.2
61	1	EEC	821.56	320.44	0.003588	0.2246985	1.0	1.2

Sweep Games

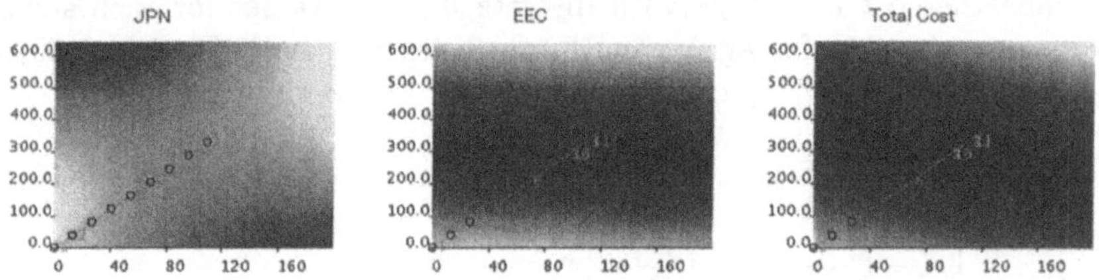

Fig. 6. Cost Landscape for JPN and EEC

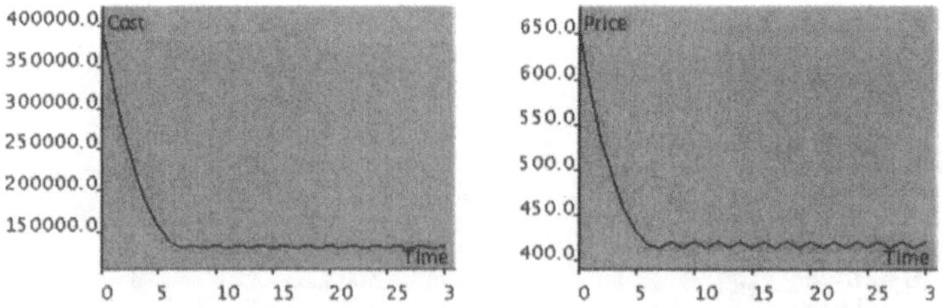

Fig. 7. Cost and Price Changes for JPN and EEC

On the other hand, Fig. 8 shows the cost landscape and learning path for the pair of JPN and FSU. The situation is quite different for this pair. As shown in the cyclic learning path and Fig. 9, this market oscillates strongly and does not converge to the optimal strategies. The reason seems to be that both agents have the same influences on the cost landscapes, as indicated by the diagonal patterns of their graphs. Furthermore, since FSU has a negative reduction target which almost cancels JPN's reduction target, in most cases the market remains with excess supply at price 0. Hence, we will next consider the case when FSU is assigned a positive reduction target. In this situation, the impact matrix p^i_j

$$p^i_j = \begin{pmatrix} 1 & 0.779 \\ 1.432 & 1 \end{pmatrix},$$

also shows small diversity.

Experiments

EID	Period	modtime	xopt
77	2	2002-05-24 23:17:49+09	sweep 50 patheid 76

Nations

EID	NID	name	target	reduction	a	b	alpha	beta
77	0	JPN	273.54	99.46	0.0231725	2.71492	1.0	1.2
77	1	FSU	816.364	-91.364	0.0034385	0.006279	1.0	1.2

Sweep Games

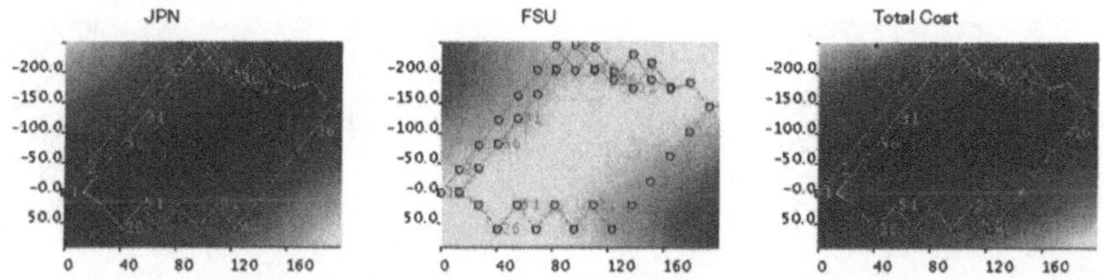

Fig. 8. Cost Landscape for JPN and FSU.

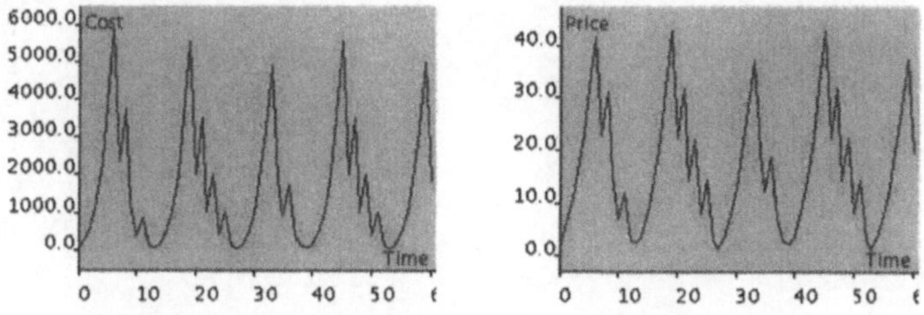

Fig. 9. Cost and Price Changes for JPN and FSU

Figure 10 shows the cost landscape for the pair of JPN and FSU which has a positive reduction target based on changing the BAU emission estimate. In this case, the diagonal pattern seems to disappear, replaced by a vertical pattern which means JPN has a strong influence in the cost landscape. We also confirm that the learning path and changes in cost and price become more stable (see Fig. 11 and the divergence of the impact matrix p^i_j,

$$p^i_j = \begin{pmatrix} 1 & 0.764 \\ 2.034 & 1 \end{pmatrix},$$

becomes larger than in the previous case.

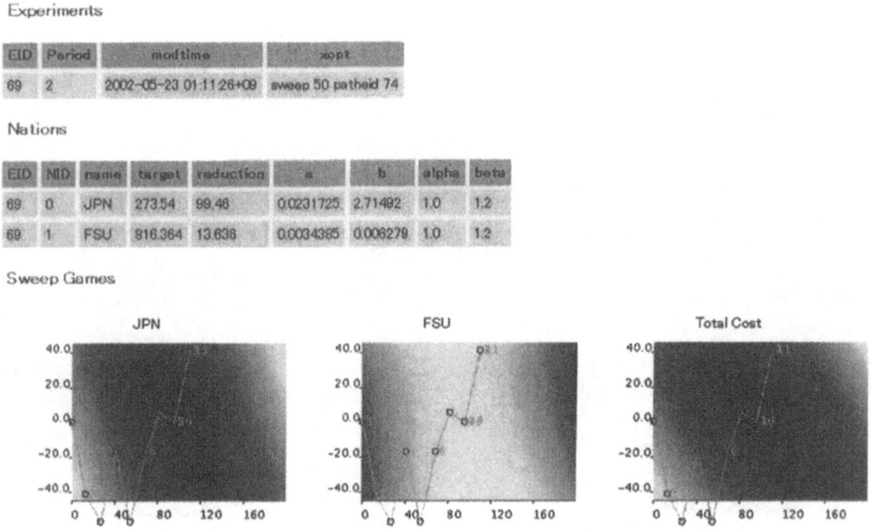

Experiments

EID	Period	modtime	xopt
69	2	2002-05-23 01:11:26+09	sweep 50 patheid 74

Nations

EID	NID	name	target	reduction	a	b	alpha	beta
69	0	JPN	273.54	99.46	0.0231725	2.71492	1.0	1.2
69	1	FSU	816.364	13.636	0.0034385	0.006279	1.0	1.2

Fig. 10. Cost Landscape for JPN and FSU (with positive target)

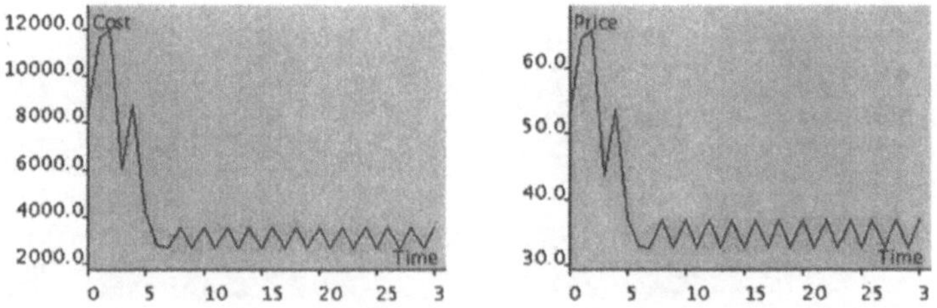

Fig. 11. Cost and Price Changes for JPN and FSU (with positive target)

5 Web Applications for Gaming

So far, we have only considered the behavior of autonomous software agents. The behavior of these agents is quite simple, though there is some heterogeneity which makes analysis difficult. We need to refine the model to be more realistic while retaining essential dynamics.

Gaming simulations with human players in an environment similar to the agents' environment are expected to help us in constructing plausible behavior models and extracting the essential dynamics. We implemented a remote accessing interface in the ASIA framework which enables human players to collaborate with software agents, and constructed a Web application for a gaming simulation modeling international emissions trading. In this section, we introduce this gaming system (see Fig. 12).

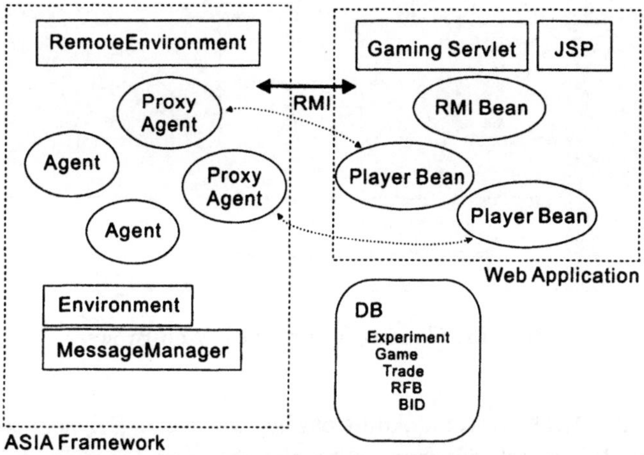

Fig. 12. Web Application and the Agent System

For remote access, we use RMI services and proxy agents at the Agent Layer. Through the RMI server, remote clients can create and dispose of agents. The RMI server also provides the naming service for agents and proxy messaging. A remote client who wants to talk with these agents needs to create a proxy agent with a message queue. The proxy agent acts in the same way as a normal agent in the ASIA environment, sends messages on behalf of the remote client, and passes the received messages back to the remote client. We also provide proxy versions of Participant and Watcher agents in the Social Layer.

This means, remote clients are able to connect to the agent system and join the social simulation. We have developed a Web application using servlets for gaming so that most client PCs with standard Web browsers can easily access it. By using the standard HTTP and SHTTP protocol, we can easily control access and use secure transactions.

In this system, as with typical Web applications, the servlets work as controllers and JSPs provide views. To access the RMI server, servlets and JSPs use a Java bean which acts as an RMI client. To retain information about the players, this application also requires Java beans corresponding to proxy

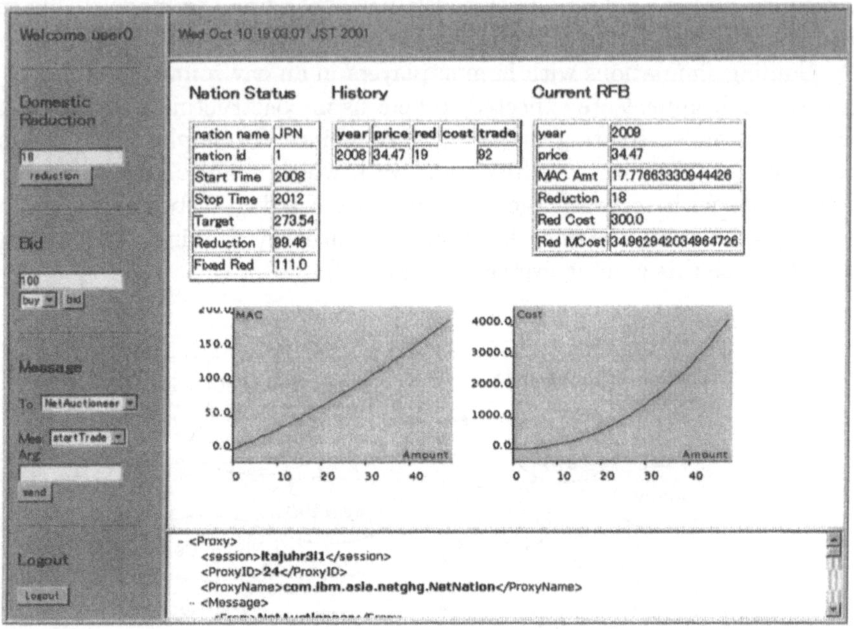

Fig. 13. Gaming System with Web Browser

agents in the ASIA system. Additionally, we use a DB access bean which connects to the agent database with the structure described in section 3. Messages like RFBs received by the agents are passed to the beans for the players in XML format, are interpreted, and displayed by servlets and JSPs. On the other side, player's inputs such as BIDs are packed into messages and sent through the proxy agent. In this simulation, players should determine the amounts of the domestic reduction of the greenhouse gases and the amounts of the excess demand for international emissions trading according to the presented price in the RFB at each iteration. Information such as the abatement cost graph, the MAC, the total reduction target and the trading history are also given.

6 Concluding Remarks

We developed a dynamic simulation for international GHG emissions trading with our agent-based simulation framework, ASIA. In a simulation study of international emissions trading, we observed the price formulation for each trading year, and the dynamic improvement of strategies which reduce the total cost. The implementation of various types of the agent-based simulation can be easily performed with this framework, since it offers simple and fundamental facilities for agents including messaging, multi-threading, and an example of social negotiation transactions in separate layers.

In this paper, we have investigated the cost landscape in the simplified situation with two agents and two trading periods. We have observed the connection between the stability of the agents' behavior and the cost landscapes. In particular, we have considered the cost impact matrices.

We have also developed a Web application for the gaming simulation of the emissions trading which works along with the agent system through a network. We plan to execute this gaming simulations at several universities in 2002. We are expecting that such a gaming simulation with this system will help us to construct improved heterogeneous behavior models and to propose more stable and less expensive trading processes for the Kyoto Protocol.

References

1. Grütter, J. M. (2000): World Market for GHG Emission Reductions. Prepared for the World Bank's National AIJ/JI/CDM Strategy Studies Program
2. Mizuta, H., Steiglitz, K., Lirov, E. (1999): Effects of Price Signal Choices on Market Stability. In: The 4th Workshop on Economics with Heterogeneous Interacting Agents (WEHIA'99). Journal of Economic Behavior and Organization (to appear)
3. Mizuta, H., Yamagata, Y. (2001): Agent-Based Simulation and Greenhouse Gas Emissions Trading. In: Peters, B. A., Smith, J. S., Medeiros, D. J., Rohrer, M. W. (Eds.): Proceedings of the 2001 Winter Simulation Conference, 535–540
4. Mizuta, H., Yamagata, Y. (2002): Agent-Based Simulation and Gaming System for International Emissions Trading. In: Namatame, A., et al. (Eds.): Agent-Based Approaches in Economic and Social Complex Systems. IOS Press, Amsterdam, 69–78
5. Namatame, A., Terano, T., Kurumatani, K. (Eds.) (2002): Agent-Based Approaches in Economic and Social Complex Systems. IOS Press, Amsterdam

Comparison of two Economic Models for a Business-to-Business Exchange*

Nikolay Zenkevich[1] and Suzhou Huang[2]

[1] St. Petersburg State University, Center of Game Theory, Faculty of Applied Mathematics and Control Processes, Bibliotechnaya pl. 2, St. Petersburg, 198504, Russia
E-mail: zenkevich@isdgrus.ru

[2] e-Technology Dept., Ford Research Laboratory, Dearborn, MI 48121, USA
E-mail: shuang10@ford.com

Summary. We model business-to-business (B2B) exchange as intermediary within the principal-agent framework. As the principal, the exchange chooses the trading mechanism. As agent, sellers and buyers voluntarily play deal-formation games according to the rule specified by the trading mechanism. Participations in the exchange is endogenized by allowing sellers and buyers to possess outside options. We examine two trading mechanisms in detail, Perry's bargaining model, and Chatterjee and Samuelson's bilateral double auction with sealed bids. A two-part-tariff charge system, entry fee plus commission percentage, is imposed on sellers and buyers. We find that individual behaviours in the deal-formation game vary drastically with the exact form of trading mechanism, relative bargaining power between sellers and buyers, and the existence of outside options. On the other hand, the optimal commission percentages are found to be remarkably insensitive to these micro-level details. Entry fees, though as the effective as commission percentages in raising revenue, strongly discourage participations.

Key words: Business-to-Business Exchange, seller, buyer, trading mechanism, deal-formation mechanism, Bayesian Nash equilibrium, optimal mechanism design, outside option.

1 Introduction

The economic model follows the paradigm of the principal-agent framework. In this framework the B2B exchange is the principal and sellers/buyers are agents. The principal specifies a deal-formation game for all the agents to play,

* Opinions expressed herein are entirely those of the authors, rather than those of Ford Motor Company.

with the purpose of maximizing its own revenue subject to agents' constraints in incentive compatibility and individual rationality.

In general, the choice of the deal-formation mechanism should be a consequence of the revenue maximization process. However, this work will not explicitly pursue this optimal mechanism design. Instead, we only examine several well known deal-formation processes and explore how these deal-formation mechanisms impact on the well-beings of all participants at the exchange.

We mostly concentrate on situations where sources of revenue for the exchange are simple commission fees proportional to the realized market price in each deal, with a percentage α charged to sellers and a percentage β charged to buyers. Then, the optimal commission percentages are determined by maximizing the expected revenue at a given deal formation process. For simplicity, the matching between sellers and buyers is treated exogenously as random.

We also separately examine the effect due to entry fees. However, several factors one needs to keep in mind when applying entry fees beyond the level whose purpose is to discourage potential moral hazard.

1. Entry fees imply a complex price system.
2. Entry fees strongly discourage participation, because sellers and buyers who come to the exchange have positive probability to lose money, which is particularly acute when these markets participants are risk averse.
3. In a typical B2B environment the market is very thin (not too many sellers and buyers that can seriously satisfy each other), any additional discouragement other than the outside options could unravel the market at the exchange.

2 The Model and its Solution Concept

There are three kinds of players in the model who are all risk neutral, the monopolistic exchange as the principal, sellers and buyers as agents. After sinking a large fixed cost, the exchange incurs a negligible marginal cost for sponsoring on-line procurement transactions. The charge system for the participation at the exchange has a form of two-part-tariff. Every seller (buyer) who decides to enter the deal-formation process pays an entry fee $e_S(e_B)$ to the exchange. In addition, if a deal is formed the seller (buyer) also pays a commission equal to the transaction price times a fixed percentage $\alpha(\beta)$.

We assume that every potential player at the exchange can choose to either participate at the exchange or pursue outside option, but not both. This is to reflect the fact that firms need to incur large costs to switch modes of transaction in procurement. Even though there is no direct competition, the exchange does face indirect competitions, because sellers and buyers have outside options and hence may not participate at the exchange. We further make following simplifying assumptions: each seller/buyer can participate at the exchange only once, and matching participants at the exchange is a non-deliberate process. The former essentially gives up any opportunity to study dynamic aspects of the problem. The latter almost diminishes the role of the exchange as the match-maker. Unfortunately, these assumptions are necessary in order to make the problem sufficiently containable.

Each seller has one unit of undifferentiated good to sell. The cost incurred to produce the good, c, is a private information. Sellers' costs are uniformly distributed on $[\omega, 1]$, with $\omega \in (0, 1)$. Seller c has an expected utility $\xi(1-c)$ if she decides to pursue the outside option, with $\xi \geq 0$. The utility for seller c is given by

$$u_S = \begin{cases} (1-\alpha)p_S - c - e_S & \text{deal with price } p_S\,, \\ -e_S & \text{no deal}\,. \end{cases}$$

Each buyer wants to buy one unit of the good. The valuation of the good, v, is a private information. Buyers' valuations are uniformly distributed on $[\omega, 1]$. Buyer v has an expected utility $\eta(v - \omega)$ if he decides to pursue his outside option, with $\eta \geq 0$. The utility for buyer v is given by

$$u_B = \begin{cases} v - (1+\beta)p_B - e_B & \text{deal with price } p_B\,, \\ -e_B & \text{no deal}\,. \end{cases}$$

Instead of trying to characterize the optimal trading mechanism the exchange would like to deploy, which is way beyond the scope of this paper, we only examine two specific deal-formation processes: Perry's bargaining model with infinite rounds of alternative offers and counter offers [2]; and Chatterjee and Samuelson's bilateral double auction with sealed bids [1]. These deal-formation processes are chosen to be representative, but otherwise mostly for convenience.

There are three steps we need to follow to find the solution. First, the seller and buyer behaviour in each trading mechanism are solved again incorporating the effects of charges and outside options. Second, charges to participants at the exchange are aggregated to form the revenue of the exchange. Third, the revenue is maximized with respect to the variables that the exchange controls. Specific issues that will be examined are the following: how the deal formation process impacts every player's behaviour; how the bargaining powers of each side of the market affects the outcome; how the existence of outside options influences the participation; what sets the scale for the revenue of the exchange; whether all the conclusions are qualitatively robust with respect to various micro-level details.

Since the deal-formation games involve incomplete information, the equilibrium concept is the standard Bayesian Nash. However, because the participation at the exchange is endogenized in our model, sellers and buyers need to assess who are the likely players of the deal-formation games. This implies additional consistency conditions for the pertinent Bayesian equilibrium.

There is a duality that implies a symmetry between sellers and buyers strategies in the market [3]. The duality imposes a strong check on the final results.

3 Perry's Bargaining Mechanism

In Perry's deal formation mechanism, the first offerer is one who incurs less transaction cost among the two bargaining parties. We assume that the probability for sellers to be in the advantageous position is $\kappa \in [0,1]$ and the probability for buyers $1-\kappa$.

With outside options, high-cost sellers in $c \in [c_M, 1]$ and low-variation buyers in $v \in [\omega, v_m]$ will not participate in the exchange. Due to individual rationality c_M and v_m obey $c_M \in [\omega, c_M^0]$ and $v_m \in [v_m^0, 1]$, where $c_M^0 = (1-\alpha)/(1+\beta)$ and $v_m^0 = \omega(1+\beta)/(1-\alpha)$. See Figure 1 for an illustration of notations.

When seller $c \in [\omega, c_M]$ is making the first offer, her price and associated expected utility are given by, respectively,

$$p_S(c) = \frac{1}{2}\left(\frac{1}{1+\beta} + \frac{c}{1-\alpha}\right) \quad \text{and} \quad \pi_S^{(1)}(c) = \frac{(1+\beta)(c_M^0 - c)^2}{4(1-\alpha)(1-v_m)}\, \theta(c_M^0 - c)\,.$$

When buyer $v \in [v_m, 1]$ is making the first offer, her price and associated expected utility are given by, respectively,

$$p_B(v) = \frac{1}{2}\left(\frac{\omega}{1-\alpha} + \frac{v}{1+\beta}\right) \quad \text{and} \quad \pi_B^{(1)}(v) = \frac{(1-\alpha)(v - v_m^0)^2}{4(1+\beta)(c_M - \omega)}\, \theta(v - v_m^0)\,.$$

Fig. 1. Illustrations of (a) the expected utility for seller c and (b) the expected utility for buyer v in Perry's bargaining model with commission fees and outside options. Only the upper envelops (thick lines) are individually rational

When seller $c \in [\omega, c_M]$ is receiving a buyer's offer, her acceptance criterion is to accept when $(1-\alpha)$ times the buyer's offer price is more than her cost, with an expected utility given by,

$$\pi_S^{(2)}(c) = \frac{(1+\beta)(c_M^0 + \omega - 2c)^2}{4(1-\alpha)(1-v_m)}\, \theta(c_M^0 + \omega - 2c)\,.$$

When buyer $v \in [v_m, 1]$ is receiving a seller's offer, his acceptance criterion is to accept when $(1+\beta)$ times the buyer's offer price is less than his valuation, with an expected utility given by,

$$\pi_B^{(2)}(v) = \frac{(1-\alpha)(2v-1-v_m^0)^2}{4(1+\beta)(c_M-\omega)}\,\theta(2v-1-v_m^0)\,.$$

The total expected surpluses of seller c and buyer v are given by

$$\pi_S(c) = \kappa\pi_S^{(1)}(c) + (1-\kappa)\pi_S^{(2)}(c)\,,$$

$$\pi_B(c) = (1-\kappa)\pi_B^{(1)}(c) + \kappa\pi_B^{(2)}(v)\,.$$

The indifference conditions for the seller $c = c_M$ and for the buyer $v = v_m$ are given by $\pi_S(c_M) = \xi(1-c_M)$ and $\pi_B(v_m) = \eta(v_m - \omega)$, respectively. Let $c_M \equiv c_M^0 - \Delta c_M$ and $v_m \equiv v_m^0 + \Delta v_m$, where the leading corrections for small ξ and η are given by, as long as the relative bargaining power is not too lopsided (or κ is not too close to 0 or 1).

$$\Delta c_M = \frac{2}{1+\beta}\sqrt{(\alpha+\beta)\left((1-\alpha)-(1+\beta)\omega\right)\frac{\xi}{\kappa}} + \cdots\,;$$

$$\Delta v_m = \frac{2}{1-\alpha}\sqrt{(\alpha+\beta)\left((1-\alpha)-(1+\beta)\omega\right)\frac{\omega\eta}{1-\kappa}} + \cdots\,.$$

The expected revenue for the exchange can be decomposed as

$$\Pi_E(\alpha,\beta) \equiv \theta\left(\frac{1-\alpha}{1+\beta}-\omega\right)\left[\Pi_E^0(\alpha,\beta) + \Delta\Pi_E(\alpha,\beta)\right],$$

where Π_E^0 is the counterpart when there are no outside options.

$$\Pi_E^0 = \frac{(\alpha+\beta)\left((1-\alpha)(1+\kappa)+(1+\beta)(2-\kappa)\omega\right)\left((1-\alpha)-(1+\beta)\omega\right)^2}{12(1-\alpha)^2(1+\beta)^2(1-\omega)^2}$$

$$\Delta\Pi_E = -\frac{\alpha+\beta}{(1-\omega)^2}\left\{\frac{\kappa}{1-\alpha}\frac{(\Delta c_M)^2}{4} + \frac{(1-\kappa)\omega}{1+\beta}\frac{(\Delta v_m)^2}{4}\right\} + \cdots\,.$$

It is insightful to examine the situation when the margin $1-\omega \equiv \Delta\omega$ is small. Let $\alpha \equiv \alpha^1\Delta\omega$, $\beta \equiv \beta^1\Delta\omega$. We then have

$$\Pi_E^0 = \Delta\omega\frac{(\alpha^1+\beta^1)(1-\alpha^1-\beta^1)^2}{4} + \mathcal{O}(\Delta^2\omega)\,,$$

$$\Delta\Pi_E = \Delta\omega(\alpha^1+\beta^1)^2(1-\alpha^1-\beta^1)(\xi+\eta) + \mathcal{O}(\Delta^2\omega)\,,$$

which immediately lead to $\alpha^* + \beta^* = \Delta\omega/3 + \mathcal{O}(\Delta^2\omega)$. The aggregate seller and buyer surplus are defined as:

$$\Pi_S \equiv \frac{1-v_m}{1-\omega}\int_\omega^{c_M}\frac{dc}{1-\omega}\,\pi_S(c) \quad \text{and} \quad \Pi_B \equiv \frac{c_M-\omega}{1-\omega}\int_{v_m}^1\frac{dv}{1-\omega}\,\pi_B(v)\,.$$

Evaluated at the optimal commission percentages (with $\xi \to 0$, $\eta \to 0$ and $\kappa = 1/2$) we have, to the leading order in $\Delta\omega$, $\Pi_S^* = \Pi_B^* = \Delta\omega/54$.

4 Chatterjee and Samuelson's Double Auction Mechanism

The realized price is related to seller c's offer $S(c)$ and buyer v's bid $B(v)$ as follows

$$p = [\kappa S(c) + (1-\kappa)B(v)]\,\theta\,(B(v) - S(c))\,.$$

With outside options, high-cost sellers in $c \in [c_M, 1]$ and low-valuation buyers in $v \in [\omega, v_m]$ will not participate in the exchange. Due to individual rationality c_M and v_m obey $c_M \in [\omega, c_M^0]$ and $v_m \in [v_m^0, 1]$, where c_M^0 and v_m^0 are given by

$$c_M^0 = \frac{(1-\kappa)\omega}{2} + \frac{(1+\kappa)(1-\alpha)}{2(1+\beta)} \quad \text{and} \quad v_m^0 = \frac{\kappa}{2} + \frac{(2-\kappa)(1+\beta)\omega}{2(1-\alpha)}\,.$$

We will only consider the case when $c_M > v_m$. This is the situation when uncertainty is always present for each possible deal. See Figure 2 for an illustration of bidding strategies and notations.

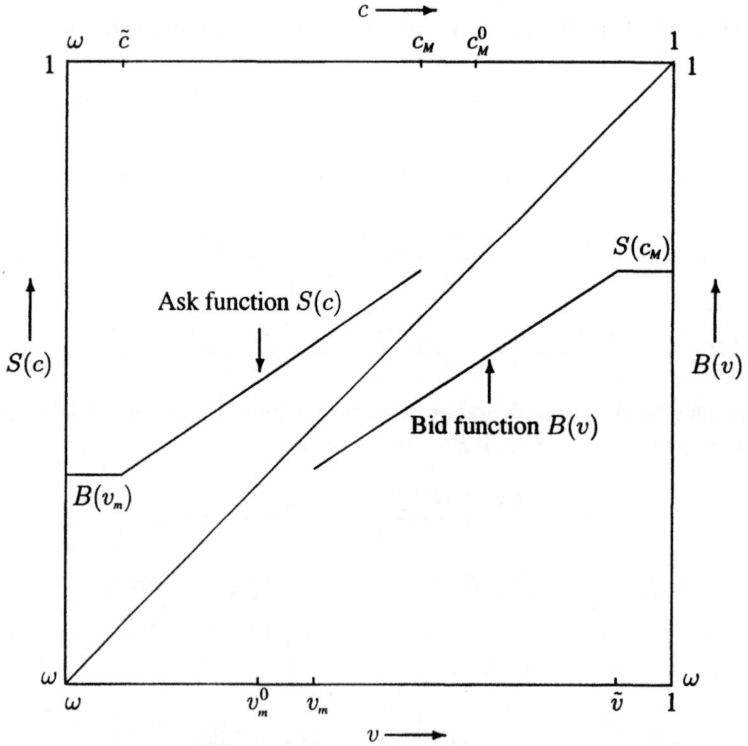

Fig. 2. Illustrations of the bidding strategies S(c) and B(v) in Chatterjee and Samuelsons bilateral double auction with commissions and outside options. The thin diagonal line indicates the truthful behavior

For seller $c \in [\omega, c_M]$, her ask function is

$$S(c) = \begin{cases} B(v_m) & \text{for } c \in [\omega, \tilde{c}) \,, \\ \dfrac{\kappa c_M^0 + c}{(1+\kappa)(1-\alpha)} & \text{for } c \in [\tilde{c}, c_M) \,, \end{cases}$$

$$\text{where} \quad \tilde{c} = \omega + \frac{(1+\kappa)(1-\alpha)}{(2-\kappa)(1+\beta)}(v_m - v_m^0) \quad \text{and} \quad \tilde{c} \in [\omega, c_M] \,.$$

For buyer $v \in [v_m, \omega]$, her bid function is

$$B(v) = \begin{cases} \dfrac{(1-\kappa)v_m^0 + v}{(2-\kappa)(1+\beta)} & \text{for } v \in [v_m, \tilde{v}] \,, \\ S(c_M) & \text{for } v \in (\tilde{v}, 1] \,, \end{cases}$$

$$\text{where} \quad \tilde{v} = 1 - \frac{(2-\kappa)(1+\beta)}{(1+\kappa)(1-\alpha)}(c_M^0 - c_M) \quad \text{and} \quad \tilde{v} \in [v_m, 1] \,.$$

The expected surplus for seller $c \in [\tilde{c}, c_M]$ is given by

$$\pi_S(c) = \frac{(2-\kappa)(1+\beta)}{(1+\kappa)(1-\alpha)} \left\{ \frac{c_M^0 - c}{1 - v_m} [(1-\alpha)(\kappa S(c) \right.$$

$$\left. +(1-\kappa)S(c_M)) - c] - \frac{(1-\kappa)(c_M - c)^2}{2(1+\kappa)(1-v_m)} \right\} .$$

The expected surplus for buyer $v \in [v_m, \tilde{v}]$ is given by

$$\pi_B(v) = \frac{(1+\kappa)(1-\alpha)}{(2-\kappa)(1+\beta)} \left\{ \frac{v - v_m^0}{c_M - \omega} [v - (1+\beta)(\kappa B(v_m) \right.$$

$$\left. +(1-\kappa)B(v))] - \frac{\kappa(v - v_m)^2}{2(2-\kappa)(c_M - \omega)} \right\} .$$

The indifference conditions for the seller $c = c_M$ and for the buyer $v = v_m$ are given by $\pi_S(c_M) = \xi(1 - c_M)$ and $\pi_B(v_m) = \eta(v_m - \omega)$, respectively.

Let $c_M \equiv c_M^0 - \Delta c_M$ and $v_m \equiv v_m^0 + \Delta v_m$, where the leading corrections for small ξ and η are given by

$$\Delta c_M = \sqrt{\frac{(1+\kappa)^2(1-\alpha)}{(2-\kappa)(1+\beta)}(1-c_M^0)(1-v_m^0)\frac{\xi}{\kappa}} + \cdots ;$$

$$\Delta v_m = \sqrt{\frac{(2-\kappa)^2(1+\beta)}{(1+\kappa)(1-\alpha)}(c_M^0 - \omega)(v_m^0 - \omega)\frac{\eta}{1-\kappa}} + \cdots .$$

These expressions are valid when κ is not close to either 0 or 1, so that the overlap always exists (or $c_M > v_m$).

The expected revenue for the exchange can be decomposed as

$$\Pi_E(\alpha,\beta) \equiv \theta\left(\frac{1-\alpha}{1+\beta}-\omega\right)\left[\Pi_E^0(\alpha,\beta)+\Delta\Pi_E(\alpha,\beta)\right],$$

where Π_E^0 is the counterpart when there are no outside options.

$$\Pi_E^0 = \frac{(2-\kappa)(1+\kappa)}{2}\left\{\frac{(\alpha+\beta)((1-\alpha)(1+\kappa)+(1+\beta)(2-\kappa)\omega)\,((1-\alpha)-(1+\beta)\omega)^2}{12(1-\alpha)^2(1+\beta)^2(1-\omega)^2}\right\}$$

$$\Delta\Pi_E = -\frac{\alpha+\beta}{(1-\omega)^2}\left\{\frac{2-\kappa}{(1+\kappa)(1-\alpha)}\frac{(\Delta c_M)^2}{2}+\frac{(1+\kappa)\omega}{(2+\kappa)(1+\beta)}\frac{(\Delta v_m)^2}{2}\right\}+\cdots$$

The problem of maximizing $\Pi_E^0(\alpha,\beta)$ is identical to that of Perry's model. Here again $(c_M^0, v_m^0)^*$ is only going to depend on α^* and β^* through the ratio $r^* = (1-\alpha^*)/(1+\beta^*)$. Perturbatively,

$$r^* = 1-\frac{1}{3}\Delta\omega-\frac{2(5-\kappa)}{81}(\Delta\omega)^2-\frac{2(58-25\kappa+7\kappa^2)}{2187}(\Delta\omega)^3+\mathcal{O}(\Delta^4\omega),$$

$$(c_M^0)^* = 1-\frac{2-\kappa}{3}\Delta\omega-\frac{(5-\kappa)(1+\kappa)}{81}(\Delta\omega)^2+\mathcal{O}(\Delta^4\omega),$$

$$(v_m^0)^* = 1-\frac{2-\kappa}{3}\Delta\omega-\frac{(2-\kappa)(4+\kappa)}{81}-\frac{(14-\kappa)(2-\kappa)}{243}(\Delta\omega)^3+\mathcal{O}(\Delta^4\omega),$$

which leads to

$$(c_M^0 - v_m^0)^* = \frac{1-2\kappa}{27}(\Delta\omega)^2+\frac{(14-\kappa)(2-\kappa)}{243}(\Delta\omega)^3+\mathcal{O}(\Delta^4\omega).$$

So, the overlap ($c_M > v_m$) is guaranteed when $\Delta\omega$, ξ and η are small and $\kappa < 1/2$. It is interesting to note that Perry's mechanism and Chatterjee and Samu-
elson's mechanism are equivalent when $\kappa = 0$ or $\kappa = 1$. This equivalence can be checked explicitly by verifying that optimal strategies, participation bounds and surplus coincide.

An analogous analysis of the case with pure entry fees can be made as in Perry's deal formation process. Furthermore, the optimal entry fees also turn out to be incompatible with the assumption $c_M > v_m$.

5 Summary and Conclusion

This model directly addresses the economics of a B2B exchange as an intermediary. It provides explicit mechanisms to model revenues and optimal charges. In each of these mechanisms all sellers and buyers face incomplete information and have to act strategically at the interim stage.

There is a duality that implies a peculiar symmetry between of sellers' behaviour and buyers' behaviour, which entails a symmetric contribution to the exchange's expected revenue and optimal commission fees.

More efficient agents, i.e., sellers with lower costs and buyers with higher valuations, are more likely to participate at the exchange. Less efficient players are more likely to pursue their outside options.

For fixed commission percentages, outside options (for both sellers and buyers) tend to decrease the revenue. In addition, the optimal commission percentages decrease with the presence of outside options. This is intuitively clear because outside options exert a competitive pressure on the exchange.

Although the deal formation process has a strong impact on the magnitude of the revenue and participations, the optimal commission percentages appear to be rather robust with respect to various deal formation processes: $(\alpha + \beta)^* \sim \Delta\omega/3$. Even though entry fees are as effective for raising revenue for the exchange as commission fees, great cares need to be exercised in their applications. This is because that entry fees strongly discourage participations.

The collective bargaining power of each side of the market has a strong impact on profits of all parties. However, the impact is small to the optimal commission percentages.

Due to incentive compatibility and individual rationality (with outside options), all sellers and buyers can only benefit from the existence of the exchange. If the exchange's revenue can cover the fixed cost (assuming zero marginal cost), establishing the exchange is a Pareto improvement over an environment where only outside options exist.

Table 1 (pure commission fee case) and Table 2 (pure entry fee case) contain information for efficiency analysis. Overall, Chatterjee and Samuelson's bilateral double auction with pure commission fees is the most efficient deal-formation mechanism among the two mechanisms examined.

Table 1. Comparison of deal-formation mechanisms in the pure commission fee case with $\xi \to 0$, $\eta \to 0$ and even bargaining powers. All results are to the leading order in $\Delta\omega$

Mechanism	α^*	β^*	c_M^*	v_m^*	Π_E^*	Π_S^*	Π_B^*
Perry	$\Delta\omega/6$	$\Delta\omega/6$	$1-\Delta\omega/3$	$1-2\Delta\omega/3$	$\Delta\omega/27$	$\Delta\omega/54$	$\Delta\omega/54$
C&S	$\Delta\omega/6$	$\Delta\omega/6$	$1-\Delta\omega/2$	$1-2\Delta\omega/2$	$\Delta\omega/24$	$\Delta\omega/48$	$\Delta\omega/48$

There are many aspects of the business at a B2B exchange that are not explicitly modeled. These include multi-dimensional aspects (differentiated products); combinatorial aspects (complementarity and bundling of products); dynamic aspects; information revelation processes; second chance after the first round fails to form a deal, matching mechanisms between sellers and buyers; market participants' direct influence (other than the indirect influence through outside options) to the deal formation rule at the exchange; and other services the exchange can provide.

Table 2. Comparison of deal-formation mechanisms in the pure entry fee case with $\xi \to 0$, $\eta \to 0$ and even bargaining powers. All results are to the leading order in $\Delta\omega$

Mechanism	e_S^*	e_B^*	c_M^*	v_m^*	Π_E^*	Π_S^*	Π_B^*
Perry	$\Delta\omega/6$	$\Delta\omega/6$	$1-2\Delta\omega/3$	$1-\Delta\omega/3$	$\Delta\omega/27$	$\Delta\omega/54$	$\Delta\omega/54$
C&S	$\Delta\omega/6$	$\Delta\omega/6$	$1-2\Delta\omega/3$	$1-\Delta\omega/3$	$\Delta\omega/27$	$\Delta\omega/54$	$\Delta\omega/54$

It may be also worthwhile to cast the problem in the framework of optimal mechanism design.

Appendix

In this appendix we present results for cases in which only entry fees are charged.

A.1. Perry's Bargaining Mechanism

When the exchange also charges entry fees, e_S to sellers and e_B to buyers, participation constraints become $\pi_S(c) \geq e_S+\xi(1-c)$ for seller c and $\pi_B(v) \geq e_B+\eta(v-\omega)$ for buyer v, respectively. The implied indifference conditions are $\pi_S(c_M) = e_S+\xi(1-c_M)$ and $\pi_B(v_m) = e_B+\eta(v_m-\omega)$. The case of pure entry fees can be worked out explicitly in the limit of $\xi \to 0$ and $\eta \to 0$. In can be shown that the overlap $c_M > v_m$ is incompatible with pure entry fee. This is because the optimal offer prices will result to negative cumulative probability.

Assuming no overlap $c_M < v_m$: the optimal offer prices are boundary solutions $p_S(c) = v_m$ and $p_B(v) = c_M$, with the corresponding surpluses

$$\pi_S(c) = \kappa(v_m-c)+(1-\kappa)(c_M-c) \quad \text{and} \quad \pi_B(v) = (1-\kappa)(v-c_M)+\kappa(v-v_m)$$

Indifference conditions become $\pi_S(c_M) = \kappa(v_m-c_M) = e_S$ and $\pi_B(c) = (1-\kappa)(v_m-c_M) = e_B$, which in turn yields

$$\Pi_E(e_S, e_B) = \frac{(e_S+e_B)(c_M-\omega)(1-v_m)}{(1-\omega)^2} = \frac{(v_m-c_M)(c_M-\omega)(1-v_m)}{(1-\omega)^2}.$$

A straightforward maximization gives $v_m^* = (2+\omega)/3$, $c_M^* = (1+2\omega)/3$, $e_S^* = \kappa\Delta\omega/3$, $e_B^* = (1-\kappa)\Delta\omega/3$ and $\Pi_E^* = \Delta\omega/27$. The aggregate seller and buyer surpluses (with ξ, $\eta \to 0$ and $\kappa = 1/2$):

$$\Pi_S^* \equiv \frac{1-v_m^*}{1-\omega}\int_{\omega}^{c_{M^*}} \frac{dc}{1-\omega}\left(\pi_S^*(c)-e_S^*\right) = \frac{\Delta\omega}{54},$$

$$\Pi_B^* \equiv \frac{c_M^* - \omega}{1-\omega} \int_{v_m^*}^{1} \frac{dv}{1-\omega} \left(\pi_B^*(v) - e_B^*\right) = \frac{\Delta\omega}{54}.$$

It is interesting to verify that the total possible surplus

$$\int_{v_m^*}^{1} \frac{dv}{1-\omega} \int_{\omega}^{c_{M^*}} \frac{dc}{1-\omega}(v-c) = 2\frac{\Delta\omega}{27}$$

which implies that all potential gains in trading are realized. This is because that no offers are rejected when there is no overlap.

A.2. Chatterjee and Samuelson's Double Auction Mechanism

An analogous analysis of the case with pure entry fees can be made as in Perry's deal formation process. However, the optimal entry fees turn out to be such that the assumption $c_M > v_m$ is violated. This means that the optimal entry fees derived using the explicit expressions in this section are not valid.

References

1. Chatterjee, K., Samuelson, W. (1983): Bargaining Under Incomplete Information. Operations Research **31**, 835–851
2. Perry, M. (1986): An Example of Price Formation in Bilateral Situations: A Bargaining Model with Incomplete Information. Econometrica **54**, 313–321
3. Rustichini, A., Satterthwaite, M., Williams, S. (1994): Convergence to Efficiency in a Simple Market with Incomplete Information. Econometrica **62**, 1041–1063

Part IV

Computational Aspects of Games

Computability of Nash Equilibrium

Hidetoshi Tashiro*

Graduate School of Economics, Hitotsubashi University,
Kunitachi, Tōkyō, 186-0004, Japan
E-mail: `gem5104@srv.cc.hit-u.ac.jp; tashiroh@gol.com`

Summary. This paper considers the computability of the Nash equilibria of a game, i.e. the possibility of an algorithm to play the game with respect to Nash equilibria. We consider a two-person game, in which both players have countably many feasible actions and their particular payoff functions are computable in the sense that there are algorithms to compute the values of ones for any given action profile. It is proved that there exists no algorithm to decide whether or not a given action profile is a Nash equilibrium. Moreover, we show that the set of the Nash equilibria is not empty, but there exists no algorithm to enumerate it allowing repetitions. These results mean that no Nash equilibrium of the game cannot be computed by any human being in practice, although the existence of the Nash equilibria can be proved in theory. Indeed, it is impossible to supply the players with algorithms regarding how they should decide whether or not a given action profile is a Nash equilibrium of the game or how they should enumerate all Nash equilibria. On the other hand, I show that if players's payoff functions are any rational number valued polynomials bounded from the above with rational number coefficients, then the Nash equilibria of the game are computable.

Key words: Nash equilibrium, computability, algorithm
JEL Classification Numbers: C72

1 Introduction

The purpose of this paper is to show a limitation of the existence theorem of Nash equilibrium from the viewpoint of computability theory. I study com-

* The author is grateful to Mamoru Kaneko, Kotaro Suzumura, and especially to Takashi Nagashima. Furthermore, he would like to acknowledge Thomas Brenner, Christian Ewerhart, Peter Hammond, Kaori Hasegawa, Midori Hirokawa, Wiebe van der Hoek, Ryo Kashima, Luchuan A. Liu, Kin Chung Lo, Akihiko Matsui, Shigeo Muto, Mikio Nakayama, John Nash, Jr., Yongsheng Xu, and Itzhak Zilcha for their advice, comments, encouragement, and discussions. Of course, any possible errors and misunderstandings are due to the author.

putability of Nash equilibrium, which is important from at least two viewpoints. First, the introduction of computability-bounded rationality requires us to take into account the computability limitations of human beings.[2] Second, in the application of game theory to economic analysis it is useful to know which computations are possible, and which are not.

The widely accepted Church(–Turing) Thesis[3] asserts that every function or relation that can be computed by means of a Turing machine is one computable by human beings. Capturing the Church(–Turing) Thesis in the context of game theory, it is natural to ask whether or not there exists a Nash equilibrium that is computable, i. e. one that is realistic for players, like human beings, with perceptual and communication limitations of a computational nature.[4] Remark that the existence of an algorithm to compute a function or a relation implies that computers can be programmed to compute it.

Section 2 of this paper intends to extend the discussion of Rabin classical article [9] on the computability of subgame perfect equilibrium to the computability of Nash equilibrium. On the other hand, Tashiro [12] considers the computability of subgame perfect equilibrium.

2 Non-Computability of Nash Equilibria of a Game with Computable Payoff Functions

Both players, 1 and 2, have countably many feasible actions, i. e. each player's action space is $\mathbb{N} = \{0, 1, 2, \dots\}$. Players 1 and 2's actions are denoted by x and y, respectively. Furthermore, players 1 and 2's payoff functions are denoted by $f(x,y)$ and $g(x,y)$, respectively.

Both players 1 and 2 simultaneously choose their actions x and y, respectively. The pair (x,y) determines players' payoffs $f(x,y)$ and $g(x,y)$.

Definition 1. A pair (x^*, y^*) is said to be a *Nash equilibrium* of the game defined above if and only if $\forall x \left[f(x^*,y^*) \geq f(x,y^*)\right]$ and $\forall y \left[g(x^*,y^*) \geq g(x^*,y)\right]$. (x^*,y^*) is called a *Nash equilibrium action profile* of the game.

A set is said to be *computable* if there is an algorithm for deciding whether or not any given object belongs to the set. [5] The Nash equilibria of a game are said to be *computable* if and only if the set of the Nash equilibria is computable. Remark that the non-computability of the Nash equilibria does not exclude the possibility that there is an algorithm to decide whether or not *some particular* action profile is a Nash equilibrium one. To make this

[2] See Binmore [1].

[3] See [7, pp. 317–323].

[4] For the epistemological importance of the concept of computability, see [7, § 62]. Furthermore, see [3, 4] and [8, Section I.8].

[5] Clearly, the universal set $\mathbb{N}$ and the empty set $\emptyset$ are both computable. There are countably many computable sets. See [11, p. 57].

point clear, I consider Abel's theorem in algebra. The theorem states that the general quintic equation $a_0x^5 + a_1x^4 + a_2x^3 + a_3x^2 + a_4x + a_5 = 0$ is unsolvable algebraically, i. e. there exists no algebraic algorithm to solve any given equation of the fifth degree. However, some particular quintic equation, e. g. $x^5 - 1 = 0$, is solvable algebraically. However, this paper studies the general possibility of an algorithm to compute the Nash equilibria of any given game with computable payoff functions.

In order to construct a game such that while every payoff function is computable in the sense that there is an algorithm to compute the value of it for any action profile, the Nash equilibria of the game are not computationally playable, I introduce a particular function such that the function itself is computable but the domain is not a computable set, i. e. there exists no algorithm to decide whether or not a given object belongs to it. Computability theory, which is traditionally called "theory of recursive functions" or "recursion theory," provides us with the following fact, which is not only fundamental in computability theory but also important to our analysis.

Fact 1 *There can be effectively constructed a particular function τ from a subset of $\mathbb{N}$ to $\mathbb{N}$ such that the following properties hold:*

1. *the function τ iteself is computable,*[6]
2. *the domain $Dom(\tau) = \{x \mid \exists y\,[y = \tau(x)\,]\}$ is not computable,*[7] *and,*
3. *the set $Dom(\tau) \cap \{x \mid \tau(x) \leq x\}$ is computable.*[8]

Theorem 1. *Let both players 1 and 2's payoff functions f and g be given as follows:*

$$f(x,y) = \begin{cases} 1 & \text{if } y \in \mathrm{Dom}(\tau),\ x = \tau(y) \text{ and } y < x, \\ 0 & \text{otherwise,} \end{cases} \tag{1}$$

and

[6] See [7, p. 281]. Kleene defines a particular relation $T(z,x,y)$ such that $T(z,x,y)$ itself is computable, i. e. there is an algorithm to decide for given natural numbers z, x and y whether or not $T(z,x,y)$ holds. $T(z,x,y)$ is called *Kleene's T-predicate*, which is a logical model of the working of computer of von Neumannian type. Set $\tau(x) = \min\{y \mid T(x,x,y)\}$. Then, computers can be programmed to compute τ.

[7] Kleene [7, p. 301], shows that the relation $\exists y T(x,x,y)$ is not computable, i. e. there exists no algorithm to decide for given x whether or not $\exists y T(x,x,y)$ holds. Thus, by the definition of τ stated in Footnote 6, $\mathrm{Dom}(\tau) = \{x \mid \exists y T(x,x,y)\}$ is not computable.

[8] Kleene [7, p. 228], shows that given x, if a relation $R(y)$ is computable, then the relation $\exists y(R(y)$ and $y \leq x)$, or $\exists y_{y \leq x} R(y)$, is computable. Thus, since T is computable, the relation $\exists y(T(x,x,y)$ and $y \leq x)$, or $\exists y_{y \leq x} T(x,x,y)$, is computable. Hence, so is $\mathrm{Dom}(\tau) \cap \{x \mid \tau(x) \leq x\}$.

$$g(x,y) = \begin{cases} 1 & \text{if } \mathrm{x} \in \mathrm{Dom}(\tau),\ \mathrm{y} = \tau(\mathrm{x}) \text{ and } \mathrm{x} < \mathrm{y}, \\ 0 & \text{otherwise.} \end{cases} \tag{2}$$

Then, both f and g are computable but the Nash equilibria are not computable. In other words, while there are algorithms to compute both players' payoff functions, there exists no algorithm to decide whether or not a given action profile is a Nash equilibrium of the game.

Proof. By Fact 1.1, the function τ is computable. Then, by (1) and (2), both f and g are computable functions.

Let E denote the set of Nash equilibria of the game. Now I prove that $(x^*,y^*) \in E$ is equivalent to the following:

$$y^* \in \mathrm{Dom}(\tau) \cap \{y \mid y < \tau(x)\} \Longrightarrow x^* = \tau(y^*) \text{ and } y^* < x^*\ , \tag{3}$$

and

$$x^* \in \mathrm{Dom}(\tau) \cap \{x \mid x < \tau(y)\} \Longrightarrow y^* = \tau(x^*) \text{ and } x^* < y^*\ , \tag{4}$$

where x and y are any natural numbers. By (1), f values 0 or 1. Thus, the condition $\forall x\,[f\,(x^*,y^*) \geq f\,(x,y^*)]$ holds if and only if it follows that

$$\exists x\,[f\,(x,y^*) = 1] \implies f\,(x^*,y^*) = 1\,. \tag{5}$$

Suppose that (5) holds. Let $y^* \in \mathrm{Dom}(\tau) \cap \{y \mid y < \tau(x)\}$. Then $y^* \in \mathrm{Dom}(\tau)$ and $y^* < \tau(x)$. Since x is arbitrary, I have $y^* \in \mathrm{Dom}(\tau)$ and $y^* < \tau(y^*)$. Hence, by (1), I have $f(\tau(y^*),y^*) = 1$. Therefore by supposition $f\,(x^*,y^*) = 1$ holds. Consequently (5) implies (3). Conversely, suppose that (3) holds. Take any x such that $f\,(x,y^*) = 1$ holds. Then, by (1), I have $y^* \in \mathrm{Dom}(\tau)$, $x = \tau(y^*)$ and $y^* < x$. Hence, by (3), $x^* = \tau(y^*)$ and $y^* < x^*$. Therefore, again by (1), I have $f\,(x^*,y^*) = 1$. Consequently, (3) implies (5). Similarly, the condition $\forall y\,[g\,(x^*,y^*) \leq g(x^*,y)]$ is equivalent to (4). Consequently, $(x^*,y^*) \in E$ if and only if both (3) and (4) hold.

By the above equivalence, $(x^*,x^*) \in E$ if and only if it follows that

$$x^* \in \mathrm{Dom}(\tau) \cap \{x \mid x < \tau(x)\} \Longrightarrow x^* = \tau(x^*) \text{ and } x^* < x^*.$$

Thus I have the following equivalence:

$$(x^*,x^*) \in N \iff x^* \notin \mathrm{Dom}(\tau) \cap \{x \mid x < \tau(x)\}\,.$$

Now suppose that E were computable. Then so is the relation $(x^*,x^*) \in E$. Thus, by the above equivalence, the set $\mathrm{Dom}(\tau) \cap \{x \mid x < \tau(x)\}$ is computable. Hence, by Fact 1.3, the most right hand side of the following equation is computable:

$$\begin{aligned}\mathrm{Dom}(\tau) &= \mathrm{Dom}(\tau) \cap (\{x \mid \tau(x) \leq x\} \cup \{x \mid \tau(x) > x\}) \\ &= (\mathrm{Dom}(\tau) \cap \{x \mid \tau(x) \leq x\}) \cup (\mathrm{Dom}(\tau) \cap \{x \mid \tau(x) > x\})\,.\end{aligned}$$

This contradicts to Fact 1.2, which states the non-computability of $\mathrm{Dom}(\tau)$. Hence, E is not computable. Therefore, the Nash equilibria of the game are not computable. □

The non-computability of the Nash equilibria shown in Theorem 1 implies the existence.

Theorem 2. *Let both players 1 and 2's payoff functions f and g be given as* (1) *and* (2). *Then there exists a Nash equilibrium of the game.*

Proof. Obviously, an algorithm that always answers 'no' for any given input can decide whether or not a given object belongs to the empty set $\emptyset$. Hence, $\emptyset$ is computable.[9] Consequently, if a set is not computable, then it is not empty. Therefore, since by Theorem 1 the set E of Nash equilibria of the game is not computable, it should be nonempty. □

Theorems 1 and 2 imply that there is a game such that while every player's payoff function is computable and there exists a Nash equilibrium of the game, the Nash equilibria are non-computable in the sense that it is impossible to provide the players with any finite and effective instructions regarding how to decide whether or not a given action profile is a Nash equilibrium of the game.

3 Computable Enumerability of Nash Equilibrium

A set is said to be *computably* or *recursively enumerable* if it is empty or it is the range of a computable function. Thus, if the set E of the Nash equilibria is computably enumerable, then there is a computable function Φ which enumerates it (allowing repetitions), i. e. such that $E = \{\Phi(0), \Phi(1), \Phi(2), \ldots\}$.

Computability theory provides us with the following fact, which is not only fundamental in computability theory but also important to our analysis:

Fact 2

1. *A set is computable if and only if both the set and its complement are computably enumerable.* [10]
2. *A nonempty set S is computably enumerable if and only if the relation $x \in S$ is expressible in the form $\exists y R(x, y)$ with a computable relation R.*[11]

Theorem 3. *Let both players 1 and 2's payoff functions f and g be the same as given in Theorem 1. Then, while f and g are computable, the set of the Nash equilibria are not computably enumerable. In other words, while there are algorithms to compute both players' payoff functions, there exists no algorithm to enumerate the Nash equilibria of the game.*

---

9 See [8, p. 139], or [11, p. 57].
10 See [8, p. 140].
11 See [7, p. 306].

Proof. Let E denote the set of Nash equilibria of the game. By Fact 2.1, I have only to prove that the complement E^c of E is computably enumerable. By the definition of Nash equilibrium, $(x,y) \in E^c$ if and only if $\exists u\,[f(x,y) < f(u,y)]$ or $\exists v\,[g(x,y) < g(x,v)]$, i.e.

$$\exists (u,v)\ [f(x,y) < f(u,y) \text{ or } g(x,y) < g(x,v)]\ .$$

Since both f and g are computable, the relation $f(x,y) < f(u,y)$ or $g(x,y) < g(x,v)$ is computable. Hence, by Fact 2.2, E^c is computably enumerable. □

Theorem 3 implies that there is a game such that while every player's payoff function is computable and there exists a Nash equilibrium of the game, it is impossible to provide the players with any finite and effective instructions regarding how to enumerate any finite number of the Nash equilibria of the game.

4 Polynomial Payoff Functions Yielding Computable Nash Equilibria

Despite of Theorems 1 and 3, if the player's payoff functions f and g belong to a certain class, then the Nash equilibria are computable.

Denote by $\mathbb{Z}$ the set of all integers $\ldots, -2, -1, 0, 1, 2, \ldots$.

Theorem 4. *Let both players 1 and 2's payoff functions $f\colon \mathbb{N} \times \mathbb{N} \to \mathbb{Z}$ and $g\colon \mathbb{N} \times \mathbb{N} \to \mathbb{Z}$ be any integral number valued polynomials with integral number coefficients. Moreover, let f and g have their upper bounds $\bar{f} \in \mathbb{Z}$ and $\bar{g} \in \mathbb{Z}$, respectively. In other words, assume that $f(x,y) \le \bar{f}$ and $g(x,y) \le \bar{g}$ for any $(x,y) \in \mathbb{N} \times \mathbb{N}$. Then, the Nash equilibria of the game are computable.*

Proof. Set $F(x,y) = \bar{f} - f(x,y)$. Then $F(x,y)$ is minimized if and only if $f(x,y)$ is maximized.

Since $f(x,y)$ is a integral number valued polynomials with integral number coefficients, $F(x,y)$ is represented as follows:

$$F(x,y) = \sum_{i=0}^{n} F_i(y)x^i = F_0(y)x^0 + F_1(y)x^1 + \cdots + F_n(y)x^n\ , \tag{6}$$

where, for every $i = 0, \ldots, n$, $F_i(y)$ is a polynomial with integral number coefficients, and n is a natural number. Notice that $F_i(y)$ can be negative for some $i = 0, \ldots, n$.

Take a $y \in \mathbb{N}$ and fix it. Consider the following two mutually exclusive cases.

Case (a): $F_1(x) = \cdots = F_n(x) = 0$;

Case (b): $F_1(x) \neq 0$, ..., *or* $F_n(x) \neq 0$.

Note that it is computable which case holds. By (6), in Case (a) I have $F(x,y) = F_0(y) = F(0,y)$ for any y. Then F is minimized for any y. Thus

I assume that Case (b) holds. Let $m(y)$ be the least number of i such that $F_i(y) \neq 0$ for given y. Note that the function $m(y)$ is computable.

Now I prove the following implication:

$$\text{if } F(x,y) = \min_w F(w,y)\,, \text{ then } x \leq M\,, \tag{7}$$

where M is defined as follows:

$$M = \frac{2\,|F_0(y)| + |F_1(y)| + \cdots + |F_n(y)|}{|F_{m(y)}(y)|}\,. \tag{8}$$

Suppose that $x > M$. Then by (8) I have $x > 1$. Thus it follows that

$$\begin{aligned} F(x,y) &\geq |F_{m(y)}(y)| \left(x - \frac{|F_0(y)| + \ldots + |F_{m(y)-1}(y)|}{|F_{m(y)}(y)|} \right) x^{m(y)-1} \\ &\geq |F_{m(y)}(y)| \left(x - \frac{|F_0(y)| + \ldots + |F_{m(y)-1}(y)|}{|F_{m(y)}(y)|} \right)\,. \end{aligned}$$

Now I have

$$\begin{aligned} & x - \frac{|F_0(y)| + \ldots + |F_{m(y)-1}(y)|}{|F_{m(y)}(y)|} \\ &> \frac{2|F_0(y)| + |F_1(y)| + \ldots + |F_{m(y)}(y)|}{|F_{m(y)}(y)|} - \frac{|F_0(y)| + \ldots + |F_{m(y)-1}(y)|}{|F_{m(y)}(y)|} \\ &= \frac{|F_0(y)| + |F_{m(y)}(y)|}{|F_{m(y)}(y)|}\,. \end{aligned}$$

Hence, it follows that

$$\begin{aligned} F(x,y) &> |F_{m(y)}(y)| \frac{|F_0(y)| + |F_{m(y)}(y)|}{|F_{m(y)}(y)|} = |F_0(y)| + |F_{m(y)}(y)| \\ &> |F_0(y)| \geq F(0,y) \geq \min_w F(w,y)\,. \end{aligned}$$

Consequently (7) holds.

By (7), for given y, in the finite interval $0 \leq x \leq M$, I can find an x to minimize $F(x,y)$. Hence, for given y, in the finite interval $0 \leq x \leq M$, I can find an x to maximize $f(x,y)$.

Let $\phi(y)$ be the least x such that $f(x,y)$ is maximized for given y. Then $\phi(y)$ is formally given as follows:

$$\phi(y) = \min\{\, x \mid f(x,y) = \max_w f(x,w) \,\}\,.$$

In other words, ϕ is a best reply of player 1 to player 2's given action.

Hence,

$$\begin{aligned} \phi(y) &= \min\{\, x \mid f(x,y) = \max_w f(w,y) \,\} \\ &= \min\{\, x \mid \forall w\, [\, F(x,y) \geq F(w,y)\,]\,\} \\ &= \min\{\, x \mid \forall w\, [\, 0 \leq w \leq M \textit{ implies } F(x,y) \geq F(w,y)\,]\,\}\,. \end{aligned}$$

Therefore, the function ϕ is computable. Then it is computable whether or not a given (x, y) satisfies $x = \phi(y)$.

Similarly, let $\psi(x)$ be the least y such that $g(x, y)$ is maximized for given x. By the similar calculation, the function ψ is computable. Then it is computable whether or not a given (x, y) satisfies $y = \psi(y)$.

Therefore, it is computable whether or not a given (x, y) satisfies $x = \phi(y)$ and $y = \psi(y)$. Consequetly, it is computable whether or not a given (x, y) is a Nash equilibrium of the game. □

Theorem 4, which is a positive result on the computability of Nash equilibrium, can be extended to the case that both players' payoff functions are any integral number valued polynomials with *rational* coefficients.[12]

Denote by $\mathbb{Q}$ the set of all rational numbers.

Corollary 1. *Let both players 1 and 2's payoff functions* $f: \mathbb{N} \times \mathbb{N} \to \mathbb{Q}$ *and* $g: \mathbb{N} \times \mathbb{N} \to \mathbb{Q}$ *be any rational number valued polynomials with rational number coefficients. Moreover, let* f *and* g *have their upper bounds* $\bar{f} \in \mathbb{Q}$ *and* $\bar{g} \in \mathbb{Q}$, *respectively. In other words, assume that* $f(x, y) \leq \bar{f}$ *and* $g(x, y) \leq \bar{g}$ *for any* $(x, y) \in \mathbb{N} \times \mathbb{N}$. *Then, the Nash equilibria of the game are computable.*

References

1. Binmore, K. (1987): Modeling Rational Players, Part I. Economics and Philosophy **2**, 179–214
2. Davis, M. (1958): Computability and Unsolvability. McGraw-Hill, New York
3. Gödel, K. (1936): Über die Länge von Beweisen (On the Length of Proofs). In: Ergebnisse eines Mathematischen Kolloquiums, Heft 7, 23–24; reprinted with English translation in (1986): Feferman, S., Dawson, J. W. Jr., Kleene, S. C., Moore, G. H., Solovay, R. M., van Heijenoort, J. (Eds.): Kurt Gödel Collected Works, Volume I. Oxford University Press, New York, 396–398
4. Gödel, K. (1946): Remarks before the Princeton Bicentennial Conference on Problems in Mathematics (unpublished); reprinted in (1990):Feferman, S., Dawson, J. W. Jr., Kleene, S. C., Moore, G. H., Solovay, R. M., van Heijenoort, J. (Eds.): Kurt Gödel Collected Works, Volume II. Oxford University Press, New York, 150–153
5. Hennie, F. (1977): Introduction to Computability, Addison-Wesley, Reading
6. Kaneko, M., Nagashima, T. (1996): Game Logic and its Applications I. Studia Logica **57**, 325–354
7. Kaneko, M., Nagashima, T. (1997): Game Logic and its Applications II. Studia Logica **58**, 273–303
8. Kleene, S. C. (1952): Introduction to Metamathematics. North-Holland, Amsterdam
9. Odifreddi, P. (1989): Classical Recursion Theory. North-Holland, Amsterdam
10. Rabin, M. O. (1957): Effective Computability of Winning Strategies. In: Dresher, M., Tucker, A. W., Wolfe, P. (Eds.): Contributions to the Theory of Games, Volume III. Princeton University Press, Princeton, 147–157

[12] This fact was suggested by Professor Itzhak Zilcha.

11. Rogers, H. (1967): Theory of Recursive Functions and Effective Computability. McGraw-Hill, New York
12. Tashiro, H. (1998): Computational Playability of Backward Induction Solutions II. In: The North American Summer Meeting of the Econometric Society (June, 1998). Montreal, Canada

Numerical Algorithm for Solving Cross-Coupled Multiparameter Algebraic Riccati Equations of Multimodeling Systems Related to Nash Games

Hiroaki Mukaidani[1], Tetsu Shimomura[2], and Hua Xu[3]

[1] Hiroshima City University, Faculty of Information Sciences,
3–4–1, Ozukahigashi, Asaminami-ku, Hiroshima, 731–3194, Japan
E-mail: mukaida@im.hiroshima-cu.ac.jp

[2] Hiroshima University, Graduate School of Education,
1–1–1, Kagamiyama, Higashi-Hiroshima, 739–8524, Japan
E-mail: tshimo@hiroshima-u.ac.jp

[3] The University of Tsukuba, Graduate School of Business Sciences,
3–29–1 Otsuka, Bunkyou-ku, Tokyo, 112–0012, Japan
E-mail: xuhua@gssm.otsuka.tsukuba.ac.jp

Summary. We study linear quadratic Nash games for infinite horizon multiparameter singularly perturbed systems (MSPS). We propose a new algorithm which is based on the Kleinman algorithm for solving the generalized cross-coupled multiparameter algebraic Riccati equations (GCMARE). It is shown that the resulting algorithm guarantees quadratic convergence. Simulation results show that the proposed algorithm succeeds in dramatically improving the convergence rate compared with previous results.

Key words: Multiparameter singularly perturbed systems (MSPS), Linear quadratic Nash games, Generalized cross-coupled multiparameter algebraic Riccati equations (GCMARE), Kleinman algorithm

1 Introduction

The linear quadratic Nash games have been studied intensively in many papers (see e. g., [1, 18]). Starr and Ho [18] derived the closed-loop perfect-state linear Nash equilibrium strategies for a class of analytic differential games. It is well-known that in order to obtain the Nash equilibrium strategies, we must solve the cross-coupled algebraic Riccati equations (CARE). Li and Gajić [9] proposed an algorithm called the Lyapunov iterations for solving the CARE.

However, there are no results for the convergence rate of the Lyapunov iterations. It is easy to verify that the convergence speed is very slow when the simulation is carried out. In order to improve the convergence rate of the Lyapunov iterations, Mukaidani et al. [15] proposed the Riccati iterations which is based on the algebraic Riccati equation (ARE) for solving the CARE. On the other hand, Freiling et al. [1] proposed the algorithm which is different from [15] for solving the CARE. However, the convergence of these Riccati iterations have not been proved exactly.

Multimodeling control problems have been investigated extensively (see e. g., [4, 5, 6, 7]). The multimodeling problems arise in large scale dynamic systems. For example, these multimodel situations in practice are illustrated by the multi-area power system [5]. Linear quadratic Nash games for such models, that is, multiparameter singularly perturbed systems (MSPS) and the singularly perturbed systems (SPS), have been studied by using composite controller design [4, 6, 7, 19]. When the parameters represent small unknown perturbations whose values are not known exactly, the composite design is very useful. However, the composite Nash equilibrium solution achieves only a performance which is $O(\|\mu\|)$ (where $\|\mu\|$ denotes the norm of the vector $\|[\varepsilon_1 \ \varepsilon_2]\|$) close to the full-order performance. Moreover, since the closed-loop solution of the reduced Nash problem depends on the path along $\varepsilon_1/\varepsilon_2$ as $\|\mu\| \to 0$, we cannot conclude that the closed-loop solution of the full problem converges to the closed-loop solution of the reduced problem [7]. Therefore, as long as the small perturbation parameters ε_j are known, much effort should be made towards finding the exact strategies which guarantees the Nash equilibrium without the ill-conditioning.

In recent years, the recursive algorithm for solving the CARE for the SPS has been investigated [13, 14]. It has been shown that the recursive algorithm is very effective to solve the CARE when the system matrices are functions of a small perturbation parameter ε. However, the recursive algorithm converges only to the approximation solution because the convergence solutions depend on the 0-order solutions. In addition, the recursive algorithm has the property of the linear convergence. Thus, the convergence speed is very slow. Moreover, it should be noted that so far the numerical algorithm for solving the cross-coupled multiparameter algebraic Riccati equations (CMARE) has not been proposed.

In this paper, we study the linear quadratic Nash games for infinite horizon MSPS from a viewpoint of solving the CMARE. After defining the generalized cross-coupled multiparameter algebraic Riccati equations (GCMARE), we first investigate the uniqueness and boundedness of the solution to the GCMARE and establish its asymptotic structure. The proof of the existence of the solution to the GCMARE with asymptotic expansion is obtained by an implicit function theorem [2]. The main contribution of this paper is to propose a new iterative algorithm for solving the GCMARE. Since the new algorithm is based on the Kleinman algorithm [16, 17], it is shown that the new algorithm has a quadratic convergence property. The quadratic convergence of the resulting algorithm is proved by using the Newton–Kantorovich theorem [20]. Using the new algorithm, we will improve the convergence speed compared with the previous results [1, 13, 15]. Finally, the simulation results show that the proposed algorithm succeed in improving the convergence rate dramatically.

Notation. The notations used in this paper are fairly standard. The superscript T denotes matrix transpose. I_n denotes the $n \times n$ identity matrix. $\|\cdot\|$ denotes its Euclidean norm for a matrix. $\det M$ denotes the determinant of M. $\operatorname{vec} M$ denotes an ordered stack of the columns of M [11]. $\otimes$ denotes Kronecker product. U_{lm} denotes a permutation matrix in Kronecker matrix sense [11] such that $U_{lm}\operatorname{vec} M = \operatorname{vec} M^T \ (M \in \mathbf{R}^{l\times m})$.

2 Problem Formulation

Consider a linear time-invariant MSPS

$$\dot{x}_0(t) = \sum_{i=0}^{2} A_{0i}x_i(t) + \sum_{i=1}^{2} B_{0i}u_i(t),\ x_0(0) = x_0^0\,, \tag{1a}$$

$$\varepsilon_1\dot{x}_1(t) = A_{10}x_0(t) + A_{11}x_1(t) + B_{11}u_1(t),\ x_1(0) = x_1^0\,, \tag{1b}$$

$$\varepsilon_2\dot{x}_2(t) = A_{20}x_0(t) + A_{22}x_2(t) + B_{22}u_2(t),\ x_2(0) = x_2^0\,, \tag{1c}$$

with quadratic cost functions

$$J_i(u_i,\ u_j) = \frac{1}{2}\int_0^\infty [y_i^T(t)y_i(t) + u_i^T(t)R_{ii}u_i(t) + u_j^T(t)R_{ij}u_j(t)]dt\,, \tag{2a}$$

$$R_{ii} > 0,\quad R_{ij} \geq 0,\quad i,j = 1,2,\quad i \neq j\,,$$

$$y_i(t) = C_{i0}x_0(t) + C_{ii}x_i(t) = C_ix(t),\quad x(t) = \begin{bmatrix} x_0(t) \\ x_1(t) \\ x_2(t) \end{bmatrix}, \tag{2b}$$

where $x_j \in \mathbf{R}^{n_j}$, $j = 0,1,2$ are the state vector, $u_j \in \mathbf{R}^{m_j}$, $j = 1,2$ are the control input. All the matrices are constant matrices of appropriate dimensions.

ε_1 and ε_2 are two small positive singular parameters of the same order of magnitude such that

$$0 < k_1 \leq \alpha \equiv \frac{\varepsilon_1}{\varepsilon_2} \leq k_2 < \infty\,. \tag{3}$$

Let us introduce the partitioned matrices

$$A_e = \Phi_e^{-1} A, \quad B_{1e} = \Phi_e^{-1} B_1, \quad B_{2e} = \Phi_e^{-1} B_2,$$
$$S_{ie} = B_{ie} R_{ii}^{-1} B_{ie}^T = \Phi_e^{-1} S_i \Phi_e^{-1}, \quad G_{je} = B_{je} R_{jj}^{-1} R_{ij} R_{jj}^{-1} B_{je}^T = \Phi_e^{-1} G_j \Phi_e^{-1},$$
$$i, j = 1, 2, \quad i \neq j,$$
$$\Phi_e = \begin{bmatrix} I_{n_0} & 0 & 0 \\ 0 & \varepsilon_1 I_{n_1} & 0 \\ 0 & 0 & \varepsilon_2 I_{n_2} \end{bmatrix}, \quad A = \begin{bmatrix} A_{00} & A_{01} & A_{02} \\ A_{10} & A_{11} & 0 \\ A_{20} & 0 & A_{22} \end{bmatrix},$$
$$B_1 = \begin{bmatrix} B_{01} \\ B_{11} \\ 0 \end{bmatrix}, \quad B_2 = \begin{bmatrix} B_{02} \\ 0 \\ B_{22} \end{bmatrix},$$
$$S_1 = B_1 R_{11}^{-1} B_1^T = \begin{bmatrix} S_{001} & S_{011} & 0 \\ S_{011}^T & S_{111} & 0 \\ 0 & 0 & 0 \end{bmatrix}, \quad S_2 = B_1 R_{22}^{-1} B_2^T = \begin{bmatrix} S_{002} & 0 & S_{022} \\ 0 & 0 & 0 \\ S_{022}^T & 0 & S_{222} \end{bmatrix},$$
$$G_1 = B_1 R_{11}^{-1} R_{21} R_{11}^{-1} B_1^T = \begin{bmatrix} G_{001} & G_{011} & 0 \\ G_{011}^T & G_{111} & 0 \\ 0 & 0 & 0 \end{bmatrix},$$
$$G_2 = B_2 R_{22}^{-1} R_{12} R_{22}^{-1} B_2^T = \begin{bmatrix} G_{002} & 0 & G_{022} \\ 0 & 0 & 0 \\ G_{022}^T & 0 & G_{222} \end{bmatrix},$$
$$Q_1 = \begin{bmatrix} Q_{001} & Q_{011} & 0 \\ Q_{011}^T & Q_{111} & 0 \\ 0 & 0 & 0 \end{bmatrix}, \quad Q_2 = \begin{bmatrix} Q_{002} & 0 & Q_{022} \\ 0 & 0 & 0 \\ Q_{022}^T & 0 & Q_{222} \end{bmatrix}.$$

We now consider the linear quadratic Nash games for infinite horizon MSPS (1) under the following basic assumptions (see e. g., [2, 4]).

Assumption 1 *There exists an* $\mu^* > 0$ *such that the triplet* (A_e, B_{ie}, C_i), $i = 1, 2$ *are stabilizable and detectable for all* $\mu \in (0, \mu^*]$, *where* $\mu := \sqrt{\varepsilon_1 \varepsilon_2}$.

Assumption 2 *The triplet* (A_{ii}, B_{ii}, C_{ii}), $i = 1, 2$ *are stabilizable and detectable.*

Assumption 3 *The matrices* A_{ii}, $i = 1, 2$ *are nonsingular.*

These conditions are quite natural since at least one control agent has to be able to control and observe unstable modes. The purpose is to find a linear feedback controller (u_1^*, u_2^*) such that

$$J_i(u_i^*, u_j^*) \leq J_i(u_i, u_j^*), \quad i, j = 1, 2, \quad i \neq j. \tag{4}$$

The Nash inequality shows that u_i^* regulates the state to zero with minimum output energy. The following lemma is already known [18].

Lemma 1. *Under the assumption 1, there exists an admissible controller such that* (4) *hold iff the following full-order CMARE*

$$A_e^T X_e + X_e A_e + Q_1 - X_e S_{1e} X_e - X_e S_{2e} Y_e - Y_e S_{2e} X_e + Y_e G_{2e} Y_e = 0, \tag{5a}$$
$$A_e^T Y_e + Y_e A_e + Q_2 - Y_e S_{2e} Y_e - Y_e S_{1e} X_e - X_e S_{1e} Y_e + X_e G_{1e} X_e = 0, \tag{5b}$$

have stabilizing solutions $X_e \geq 0$ and $Y_e \geq 0$, where

$$X_e = \begin{bmatrix} X_{00} & \varepsilon_1 X_{10}^T & \varepsilon_2 X_{20}^T \\ \varepsilon_1 X_{10} & \varepsilon_1 X_{11} & \sqrt{\varepsilon_1\varepsilon_2} X_{21}^T \\ \varepsilon_2 X_{20} & \sqrt{\varepsilon_1\varepsilon_2} X_{21} & \varepsilon_2 X_{22} \end{bmatrix}, \; Y_e = \begin{bmatrix} Y_{00} & \varepsilon_1 Y_{10}^T & \varepsilon_2 Y_{20}^T \\ \varepsilon_1 Y_{10} & \varepsilon_1 Y_{11} & \sqrt{\varepsilon_1\varepsilon_2}\, Y_{21}^T \\ \varepsilon_2 Y_{20} & \sqrt{\varepsilon_1\varepsilon_2}\, Y_{21} & \varepsilon_2 Y_{22} \end{bmatrix}.$$

Then, the closed-loop linear Nash equilibrium solutions to the full-order problem are given by

$$u_1^*(t) = -R_{11}^{-1} B_{1e}^T X_e x(t) \,, \tag{6a}$$
$$u_2^*(t) = -R_{22}^{-1} B_{2e}^T Y_e x(t) \,. \tag{6b}$$

Note that it is difficult to solve the CMARE (5) because of the different magnitude of their coefficient caused by the small perturbed parameter ε_j and high dimensions.

3 Asymptotic Structure

To obtain the solutions of the CMARE (5), we introduce the following useful lemma.

Lemma 2. *The CMARE* (5) *is equivalent to the following GCMARE* (7), *respectively.*

$$A^T X + X^T A + Q_1 - X^T S_1 X - X^T S_2 Y - Y^T S_2 X + Y^T G_2 Y = 0 \,, \tag{7a}$$
$$A^T Y + Y^T A + Q_2 - Y^T S_2 Y - Y^T S_1 X - X^T S_1 Y + X^T G_1 X = 0 \,, \tag{7b}$$

where

$$X_e = \Phi_e X = X^T \Phi_e \,, \; X = \begin{bmatrix} X_{00} & \varepsilon_1 X_{10}^T & \varepsilon_2 X_{20}^T \\ X_{10} & X_{11} & \sqrt{\alpha}^{-1} X_{21}^T \\ X_{20} & \sqrt{\alpha} X_{21} & X_{22} \end{bmatrix}, \; X_{ii} = X_{ii}^T \,, \; i = 0, 1, 2 \,,$$

$$Y_e = \Phi_e Y = Y^T \Phi_e \,, \; Y = \begin{bmatrix} Y_{00} & \varepsilon_1 Y_{10}^T & \varepsilon_2 Y_{20}^T \\ Y_{10} & Y_{11} & \sqrt{\alpha}^{-1} Y_{21}^T \\ Y_{20} & \sqrt{\alpha} Y_{21} & Y_{22} \end{bmatrix}, \; Y_{ii} = Y_{ii}^T \,, \; i = 0, 1, 2 \,.$$

Proof. Since the proof is similar to the proof of Lemma 3 in [12], it is omitted. □

After substituting X and Y into the GCMARE (7), we obtain the following equations as $\varepsilon_j \to +0$, $j = 1, 2$, where $\bar{X}_{lm}$, $\bar{Y}_{lm}$, $lm = 00, 10, 20, 11, 21, 22$ are the 0-order solutions.

$$A^T \bar{X} + \bar{X}^T A + Q_1 - \bar{X}^T S_1 \bar{X} - \bar{X}^T S_2 \bar{Y} - \bar{Y}^T S_2 \bar{X} + \bar{Y}^T G_2 \bar{Y} = 0 \,, \tag{8a}$$
$$A_{11}^T \bar{Y}_{11} + \bar{Y}_{11} A_{11} - \bar{Y}_{11} S_{111} \bar{X}_{11} - \bar{X}_{11} S_{111} \bar{Y}_{11} + \bar{X}_{11} G_{111} \bar{X}_{11} = 0 \,, \tag{8b}$$
$$A_{11}^T \bar{X}_{11} + \bar{X}_{11} A_{11} - \bar{X}_{11} S_{111} \bar{X}_{11} + Q_{111} = 0 \,, \tag{8c}$$
$$A^T \bar{Y} + \bar{Y}^T A + Q_2 - \bar{Y}^T S_2 \bar{Y} - \bar{Y}^T S_1 \bar{X} - \bar{X}^T S_1 \bar{Y} + \bar{X}^T G_1 \bar{X} = 0 \,, \tag{9a}$$
$$A_{22}^T \bar{X}_{22} + \bar{X}_{22} A_{22} - \bar{X}_{22} S_{222} \bar{Y}_{22} - \bar{Y}_{22} S_{222} \bar{X}_{22} + \bar{Y}_{22} G_{222} \bar{Y}_{22} = 0 \,, \tag{9b}$$
$$A_{22}^T \bar{Y}_{22} + \bar{Y}_{22} A_{22} - \bar{Y}_{22} S_{222} \bar{Y}_{22} + Q_{222} = 0 \,, \tag{9c}$$

where

$$\bar{X} = \begin{bmatrix} \bar{X}_{00} & 0 & 0 \\ \bar{X}_{10} & \bar{X}_{11} & 0 \\ \bar{X}_{20} & 0 & \bar{X}_{22} \end{bmatrix}, \quad \bar{Y} = \begin{bmatrix} \bar{Y}_{00} & 0 & 0 \\ \bar{Y}_{10} & \bar{Y}_{11} & 0 \\ \bar{Y}_{20} & 0 & \bar{Y}_{22} \end{bmatrix}. \tag{10}$$

Note that there exist the positive semidefinite solution $\bar{X}_{11}$, $\bar{X}_{22}$, $\bar{Y}_{11}$ and $\bar{Y}_{22}$ under the assumption 2. The following theorem will establish the relation between the solutions X and Y and the solutions $\bar{X}_{lm}$ and $\bar{Y}_{lm}$ for the reduced-order equations (8) and (9).

Theorem 1. *Suppose that*

$$\det \nabla\mathcal{F}(\bar{\mathcal{P}}) \neq 0\,, \tag{11}$$

where $\bar{X}_{21} = 0$, $\bar{Y}_{21} = 0$ *and*

$$\begin{aligned}
\nabla\mathcal{F}(\bar{\mathcal{P}}) &= \frac{\partial \mathrm{vec}\,\mathcal{F}(\mathcal{P})}{\partial (\mathrm{vec}\,\mathcal{P})^T}\bigg|_{\mathcal{P}=\bar{\mathcal{P}}} \\
&= (\tilde{A} - \tilde{S}\bar{\mathcal{P}} - \mathcal{J}\bar{\mathcal{P}}^T\tilde{S}\mathcal{J})^T \otimes I_N + \left[I_N \otimes (\tilde{A} - \tilde{S}\bar{\mathcal{P}} - \mathcal{J}\bar{\mathcal{P}}^T\tilde{S}\mathcal{J})^T\right]U_{2N2N} \\
&\quad - (\mathcal{J}\tilde{S}\mathcal{J}\bar{\mathcal{P}} - \tilde{G}\bar{\mathcal{P}}\mathcal{J})^T \otimes \mathcal{J} - \left[\mathcal{J} \otimes (\mathcal{J}\tilde{S}\mathcal{J}\bar{\mathcal{P}} - \tilde{G}\bar{\mathcal{P}}\mathcal{J})\right]^T U_{2N2N}\,,
\end{aligned}$$

$$\mathcal{F}(\mathcal{P}) := \tilde{A}^T\mathcal{P} + \mathcal{P}^T\tilde{A} + \tilde{Q} - \mathcal{P}^T\tilde{S}\mathcal{P} - \mathcal{J}\mathcal{P}^T\tilde{S}\mathcal{J}\mathcal{P} - \mathcal{P}^T\mathcal{J}\tilde{S}\mathcal{P}\mathcal{J} + \mathcal{J}\mathcal{P}^T\tilde{G}\mathcal{P}\mathcal{J}\,,$$

$$\bar{\mathcal{P}} := \begin{bmatrix} \bar{X} & 0 \\ 0 & \bar{Y} \end{bmatrix}, \quad \mathcal{P} = \begin{bmatrix} X & 0 \\ 0 & Y \end{bmatrix}, \quad \tilde{A} = \begin{bmatrix} A & 0 \\ 0 & A \end{bmatrix}, \quad \tilde{Q} = \begin{bmatrix} Q_1 & 0 \\ 0 & Q_2 \end{bmatrix},$$

$$\tilde{S} = \begin{bmatrix} S_1 & 0 \\ 0 & S_2 \end{bmatrix}, \quad \tilde{G} = \begin{bmatrix} G_1 & 0 \\ 0 & G_2 \end{bmatrix}, \quad \mathcal{J} = \begin{bmatrix} 0 & I_N \\ I_N & 0 \end{bmatrix}, \quad N = n_0 + n_1 + n_2\,.$$

Under the assumptions 1–3, the GCMARE (7) admits the stabilizing solutions X and Y such that these matrices possess a power series expansion at $\|\mu\| = 0$. That is,

$$X = \begin{bmatrix} \bar{X}_{00} & 0 & 0 \\ \bar{X}_{10} & \bar{X}_{11} & 0 \\ \bar{X}_{20} & 0 & \bar{X}_{22} \end{bmatrix} + O(\|\mu\|) = \bar{X} + O(\|\mu\|)\,, \tag{12a}$$

$$Y = \begin{bmatrix} \bar{Y}_{00} & 0 & 0 \\ \bar{Y}_{10} & \bar{Y}_{11} & 0 \\ \bar{Y}_{20} & 0 & \bar{Y}_{22} \end{bmatrix} + O(\|\mu\|) = \bar{Y} + O(\|\mu\|)\,. \tag{12b}$$

Proof. First, note that $\mathcal{F}(\mathcal{P}) \equiv 0$ is equivalent to the GCMARE (7). We apply the implicit function theorem [2] to the GCMARE (7). To do so, it is enough to show that the corresponding Jacobian is nonsingular at $\|\mu\| = 0$. It can be shown, after some algebra, that the Jacobian of (7) in the limit as $\|\mu\| \to +0$ is given by

$$\begin{aligned}
J_{\bar{\mathcal{P}}} &= \lim_{\|\mu\| \to +0} \frac{\partial \mathrm{vec}\, \mathcal{F}(\mathcal{P})}{\partial (\mathrm{vec}\, \mathcal{P})^T} \\
&= (\tilde{A} - \tilde{S}\bar{\mathcal{P}} - \mathcal{J}\bar{\mathcal{P}}^T \tilde{S} \mathcal{J})^T \otimes I_N + [I_N \otimes (\tilde{A} - \tilde{S}\bar{\mathcal{P}} - \mathcal{J}\bar{\mathcal{P}}^T \tilde{S}\mathcal{J})^T] U_{2N2N} \\
&\quad - (\mathcal{J}\tilde{S}\mathcal{J}\bar{\mathcal{P}} - \tilde{G}\bar{\mathcal{P}}\mathcal{J})^T \otimes \mathcal{J} - [\mathcal{J} \otimes (\mathcal{J}\tilde{S}\mathcal{J}\bar{\mathcal{P}} - \tilde{G}\bar{\mathcal{P}}\mathcal{J})]^T U_{2N2N} \\
&= \nabla \mathcal{F}(\bar{\mathcal{P}}) \, .
\end{aligned}$$

Therefore, $\det J_{\bar{\mathcal{P}}} \neq 0$, i.e., $J_{\bar{\mathcal{P}}}$ is nonsingular at $\|\mu\| = 0$. The conclusion of Theorem 1 is obtained directly by using the implicit function theorem. □

4 Kleinman Algorithm

In order to improve the convergence rate of the Lyapunov iterations, we propose the following new algorithm which is based on the Kleinman algorithm [8].

$$\Phi^{(n)T}\mathcal{P}^{(n+1)} + \mathcal{P}^{(n+1)T}\Phi^{(n)} - \Theta^{(n)T}\mathcal{P}^{(n+1)}\mathcal{J} - \mathcal{J}\mathcal{P}^{(n+1)T}\Theta^{(n)} + \Xi^{(n)} = 0 \, ,$$

$$n = 0, 1, 2, \cdots \, , \tag{13}$$

$$\Leftrightarrow \begin{cases} \Phi_1^{(n)T} X^{(n+1)} + X^{(n+1)T}\Phi_1^{(n)} - \Theta_2^{(n)T} Y^{(n+1)} - Y^{(n+1)T}\Theta_2^{(n)} + \Xi_1^{(n)} = 0 \, , \\ \Phi_2^{(n)T} Y^{(n+1)} + Y^{(n+1)T}\Phi_2^{(n)} - \Theta_1^{(n)T} X^{(n+1)} - X^{(n+1)T}\Theta_1^{(n)} + \Xi_2^{(n)} = 0 \, , \end{cases}$$

where

$$\begin{aligned}
\Phi^{(n)} &:= \tilde{A} - \tilde{S}\mathcal{P}^{(n)} - \mathcal{J}\tilde{S}\mathcal{P}^{(n)}\mathcal{J} = \begin{bmatrix} \Phi_1^{(n)} & 0 \\ 0 & \Phi_2^{(n)} \end{bmatrix} , \\
\Theta^{(n)} &:= \tilde{S}\mathcal{J}\mathcal{P}^{(n)} - \tilde{G}\mathcal{P}^{(n)}\mathcal{J} = \begin{bmatrix} 0 & \Theta_1^{(n)} \\ \Theta_2^{(n)} & 0 \end{bmatrix} , \\
\Xi^{(n)} &:= \tilde{Q} + \mathcal{P}^{(n)T}\tilde{S}\mathcal{P}^{(n)} + \mathcal{J}\mathcal{P}^{(n)T}\tilde{S}\mathcal{J}\mathcal{P}^{(n)} \\
&\quad + \mathcal{P}^{(n)T}\mathcal{J}\tilde{S}\mathcal{P}^{(n)}\mathcal{J} - \mathcal{J}\mathcal{P}^{(n)T}\tilde{G}\mathcal{P}^{(n)}\mathcal{J} = \begin{bmatrix} \Xi_1^{(n)} & 0 \\ 0 & \Xi_2^{(n)} \end{bmatrix} ,
\end{aligned}$$

$$\begin{aligned}
\mathcal{P}^{(n)} &= \begin{bmatrix} X^{(n)} & 0 \\ 0 & Y^{(n)} \end{bmatrix} , \\
X^{(n)} &= \begin{bmatrix} X_{00}^{(n)} & \varepsilon_1 X_{10}^{(n)T} & \varepsilon_2 X_{20}^{(n)T} \\ X_{10}^{(n)} & X_{11}^{(n)} & \sqrt{\alpha}^{-1} X_{21}^{(n)T} \\ X_{20}^{(n)} & \sqrt{\alpha} X_{21}^{(n)T} & X_{22}^{(n)} \end{bmatrix} , \\
Y^{(n)} &= \begin{bmatrix} Y_{00}^{(n)} & \varepsilon_1 Y_{10}^{(n)T} & \varepsilon_2 Y_{20}^{(n)T} \\ Y_{10}^{(n)} & Y_{11}^{(n)} & \sqrt{\alpha}^{-1} Y_{21}^{(n)T} \\ Y_{20}^{(n)} & \sqrt{\alpha} Y_{21}^{(n)T} & Y_{22}^{(n)} \end{bmatrix} ,
\end{aligned}$$

and the initial condition $\mathcal{P}^{(0)}$ has the following form

$$\mathcal{P}^{(0)} = \begin{bmatrix} X^{(0)} & 0 \\ 0 & Y^{(0)} \end{bmatrix} = \begin{bmatrix} \bar{X}_{00} & \varepsilon_1 \bar{X}_{10}^T & \varepsilon_2 \bar{X}_{20}^T & 0 & 0 & 0 \\ \bar{X}_{10} & \bar{X}_{11} & 0 & 0 & 0 & 0 \\ \bar{X}_{20} & 0 & \bar{X}_{22} & 0 & 0 & 0 \\ 0 & 0 & 0 & \bar{Y}_{00} & \varepsilon_1 \bar{Y}_{10}^T & \varepsilon_2 \bar{Y}_{20}^T \\ 0 & 0 & 0 & \bar{Y}_{10} & \bar{Y}_{11} & 0 \\ 0 & 0 & 0 & \bar{Y}_{20} & 0 & \bar{Y}_{22} \end{bmatrix} . \quad (14)$$

Note that the considered algorithm (13) is original. The new algorithm (13) can be constructed setting

$$\mathcal{P}^{(n+1)} = \mathcal{P}^{(n)} + \Delta\mathcal{P}^{(n)}$$

and neglecting $O(\Delta\mathcal{P}^{(n)T}\Delta\mathcal{P}^{(n)})$ term. Kleinman algorithm is well-known and is widely used to find a solution of the algebraic equations, and its local convergence properties are well understood. We are concerned with good choices of the starting points which guarantee to find a required solution of a given GCMARE (7). Our new idea is to set the initial conditions to the matrix (14). The fundamental idea is based on $\|\mathcal{P} - \mathcal{P}^{(0)}\| = O(\|\mu\|)$ which is derived from (12). Consequently, we will get the required solution with rate of the quadratic convergence via the Kleinman algorithm. Moreover, using the Newton–Kantorovich theorem [20], we will also prove the existence of the unique solution for the GCMARE (7). The main result of this section is as follows.

Theorem 2. *Under the assumptions 1–3, the new iterative algorithm* (13) *converges to the exact solution $\mathcal{P}^*$ of the GCMARE* (7) *with the rate of quadratic convergence. Furthermore, the unique bounded solution $\mathcal{P}^{(n)}$ of the GCMARE* (7) *is in the neighborhood of the exact solution $\mathcal{P}^*$. That is, the following conditions are satisfied.*

$$\|\mathcal{P}^{(n)} - \mathcal{P}^*\| \le O(\|\mu\|^{2^n}) , \quad n = 0, 1, 2, \cdots , \quad (15a)$$

$$\|\mathcal{P}^{(n)} - \mathcal{P}^*\| \le \frac{1}{\beta\mathcal{L}}\left[1 - \sqrt{1 - 2\theta}\right] , \quad n = 0, 1, 2, \cdots , \quad (15b)$$

where

$$\mathcal{P} = \mathcal{P}^* = \begin{bmatrix} X^* & 0 \\ 0 & Y^* \end{bmatrix}, \quad \mathcal{L} := 6\|\tilde{S}\| + 2\|\tilde{G}\|, \quad \beta := \|[\nabla\mathcal{F}(\mathcal{P}^{(0)})]^{-1}\| ,$$

$$\eta := \|[\nabla\mathcal{F}(\mathcal{P}^{(0)})]^{-1}\| \cdot \|\mathcal{F}(\mathcal{P}^{(0)})\|, \quad \theta := \beta\eta\mathcal{L} .$$

Proof. The proof is given directly by applying the Newton–Kantorovich theorem [20] for the GCMARE (7). Taking the partial derivative of the function $\mathcal{F}(\mathcal{P})$ with respect to $\mathcal{P}$ yields

$$\begin{aligned} \nabla\mathcal{F}(\mathcal{P}) &= \frac{\partial \mathrm{vec}\, \mathcal{F}(\mathcal{P})}{\partial(\mathrm{vec}\, \mathcal{P})^T} \\ &= (\tilde{A} - \tilde{S}\mathcal{P} - \mathcal{J}\mathcal{P}^T\tilde{S}\mathcal{J})^T \otimes I_N + [I_N \otimes (\tilde{A} - \tilde{S}\mathcal{P} - \mathcal{J}\mathcal{P}^T\tilde{S}\mathcal{J})^T]U_{2N2N} \\ &\quad - (\mathcal{J}\tilde{S}\mathcal{J}\mathcal{P} - \tilde{G}\mathcal{P}\mathcal{J})^T \otimes \mathcal{J} - [\mathcal{J} \otimes (\mathcal{J}\tilde{S}\mathcal{J}\mathcal{P} - \tilde{G}\mathcal{P}\mathcal{J})]^T U_{2N2N} . \end{aligned} \quad (16)$$

It is obvious that $\nabla\mathcal{F}(\mathcal{P})$ is continuous at for all $\mathcal{P}$. Thus, it is immediately obtained from the above equation (16) that

$$\|\nabla\mathcal{F}(\mathcal{P}_1) - \nabla\mathcal{F}(\mathcal{P}_2)\| \le \mathcal{L}\|\mathcal{P}_1 - \mathcal{P}_2\| \,. \tag{17}$$

Moreover, using the fact that

$$\begin{aligned}
&\nabla\mathcal{F}(\mathcal{P}^{(0)}) \\
&= (\tilde{A} - \tilde{S}\mathcal{P}^{(0)} - \mathcal{J}\mathcal{P}^{(0)T}\tilde{S}\mathcal{J})^T \otimes I_N \\
&\quad +[I_N \otimes (\tilde{A} - \tilde{S}\mathcal{P}^{(0)} - \mathcal{J}\mathcal{P}^{(0)T}\tilde{S}\mathcal{J})^T]U_{2N2N} \\
&\quad -(\mathcal{J}\tilde{S}\mathcal{J}\mathcal{P}^{(0)} - \tilde{G}\mathcal{P}^{(0)}\mathcal{J})^T \otimes \mathcal{J} - [\mathcal{J} \otimes (\mathcal{J}\tilde{S}\mathcal{J}\mathcal{P}^{(0)} - \tilde{G}\mathcal{P}^{(0)}\mathcal{J})]^T U_{2N2N} \\
&= \nabla\mathcal{F}(\bar{\mathcal{P}}) + O(\|\mu\|) \,,
\end{aligned} \tag{18}$$

it follows that $\nabla\mathcal{F}(\mathcal{P}^{(0)})$ is nonsingular under the condition (11) for sufficiently small $\|\mu\|$. Therefore, there exists β such that $\beta = \|[\nabla\mathcal{F}(\mathcal{P}^{(0)})]^{-1}\|$. On the other hand, since $\mathcal{F}(\mathcal{P}^{(0)}) = O(\|\mu\|)$, there exists η such that $\eta = \|[\nabla\mathcal{F}(\mathcal{P}^{(0)})]^{-1}\| \cdot \|\mathcal{F}(\mathcal{P}^{(0)})\| = O(\|\mu\|)$. Thus, there exists θ such that $\theta = \beta\eta\mathcal{L} < 2^{-1}$ because of $\eta = O(\|\mu\|)$. Now, let us define

$$t^* \equiv \frac{1}{\beta\mathcal{L}}[1 - \sqrt{1-2\theta}] = \frac{1}{(6\|\tilde{S}\| + 2\|\tilde{G}\|) \cdot \|[\nabla\mathcal{F}(\mathcal{P}^{(0)})]^{-1}\|}[1 - \sqrt{1-2\theta}] \,. \tag{19}$$

Using the Newton–Kantorovich theorem, we can show that $\mathcal{P}^*$ is the unique solution in the subset $\mathcal{S} \equiv \{\mathcal{P}\colon \|\mathcal{P}^{(0)} - \mathcal{P}\| \le t^*\}$. Moreover, using the Newton–Kantorovich theorem, the error estimate is given by

$$\|\mathcal{P}^{(n)} - \mathcal{P}^*\| \le \frac{(2\theta)^{2^n}}{2^n\beta\mathcal{L}}, \quad n = 1, 2, \cdots \,. \tag{20}$$

Substituting $2\theta = O(\|\mu\|)$ into (20), we have (15a). Furthermore, substituting $\mathcal{P}^*$ into $\mathcal{P}$ of the subset $\mathcal{S}$, we can also get (15b). Therefore, (15) holds for the small $\|\mu\|$. □

Remark According to the reference [3], it is well-known that the solution of the GCMARE (7) is not unique and several non-negative solutions exist. In this paper, it is very important to note that if the initial conditions $\Phi_e X^{(0)}$ and $\Phi_e Y^{(0)}$ are the positive semidefinite solutions, the new algorithm (13) converge to the positive semidefinite solution in the same way as the Lyapunov iterations [9].

Remark In order to obtain the initial condition (14), we have to solve the algebraic equations (8a) and (9a) which are independent of the perturbation parameters ε_j. In this situation, we can also apply the Kleinman algorithm to these equations.

5 Numerical Example

In order to demonstrate the efficiency of our proposed algorithm, we have run a simple numerical example. Let us consider the following MSPS

$$\begin{bmatrix} \dot{x}_0 \\ \hline \varepsilon_1 \dot{x}_1 \\ \varepsilon_2 \dot{x}_2 \end{bmatrix} = \left[\begin{array}{cc|cc} 0 & 1 & 1 & 0 \\ -1 & -2 & 0 & 1 \\ \hline 2 & 1 & 1 & 0 \\ 4 & 1 & 0 & 1 \end{array}\right] \begin{bmatrix} x_0 \\ \hline x_1 \\ x_2 \end{bmatrix} + \begin{bmatrix} 0 & 0 \\ 0 & 0 \\ \hline 1 & 2 \\ 0 & 0 \end{bmatrix} u_1 + \begin{bmatrix} 0 & 0 \\ 0 & 0 \\ \hline 0 & 0 \\ 2 & 1 \end{bmatrix} u_2 \, , \quad (21)$$

with the performance index

$$J_1 = \frac{1}{2} \int_0^\infty (x_0^2 + x_1^2 + u_1^2 + u_2^2) dt \, , \quad (22a)$$

$$J_2 = \frac{1}{2} \int_0^\infty (x_0^2 + 10^{-1} x_2^2 + 2u_1^2 + 4u_2^2) dt \, . \quad (22b)$$

The numerical results are obtained for small parameter $\varepsilon_1 = \varepsilon_2 = 10^{-3}$. We give the solutions of the initial condition (14) and the GCMARE (7), respectively.

$$X^{(0)} = \begin{bmatrix} 9.1617\times10^{-1} & 4.1314\times10^{-2} & 9.3732\times10^{-4} & 1.6360\times10^{-3} \\ 4.1314\times10^{-2} & 1.9623\times10^{-1} & 2.9852\times10^{-4} & 5.8721\times10^{-4} \\ 9.3732\times10^{-1} & 2.9852\times10^{-1} & 6.8990\times10^{-1} & 0 \\ 1.6360 & 5.8721\times10^{-1} & 0 & 4.0035\times10^{-1} \end{bmatrix},$$

$$Y^{(0)} = \begin{bmatrix} 9.5207\times10^{-1} & 7.1816\times10^{-2} & 1.9630\times10^{-3} & 6.2847\times10^{-3} \\ 7.1816\times10^{-2} & 2.4131\times10^{-1} & 6.7471\times10^{-4} & 1.7818\times10^{-3} \\ 1.9630 & 6.7471\times10^{-1} & 9.7155\times10^{-1} & 0 \\ 6.2847 & 1.7818 & 0 & 1.6485 \end{bmatrix},$$

$$X = X^{(3)} = \begin{bmatrix} 9.2334\times10^{-1} & 4.4163\times10^{-2} & 9.3875\times10^{-4} & 1.6344\times10^{-3} \\ 4.4163\times10^{-2} & 1.9723\times10^{-1} & 2.9929\times10^{-4} & 5.8739\times10^{-4} \\ 9.3875\times10^{-1} & 2.9929\times10^{-1} & 6.9028\times10^{-1} & 5.5980\times10^{-4} \\ 1.6344 & 5.8739\times10^{-1} & 5.5980\times10^{-4} & 4.0092\times10^{-1} \end{bmatrix},$$

$$Y = Y^{(3)} = \begin{bmatrix} 9.7769\times10^{-1} & 7.9449\times10^{-2} & 1.9695\times10^{-3} & 6.2780\times10^{-3} \\ 7.9449\times10^{-2} & 2.4349\times10^{-1} & 6.7728\times10^{-4} & 1.7831\times10^{-3} \\ 1.9695 & 6.7728\times10^{-1} & 9.7267\times10^{-1} & 2.3041\times10^{-3} \\ 6.2780 & 1.7831 & 2.3041\times10^{-3} & 1.6502 \end{bmatrix}.$$

Table 1 shows the results of the error $\|\mathcal{F}(\mathcal{P}^{(n)})\|$ per iterations. We find that the solutions of the GCMARE (7) converge to the exact solution with accuracy of $\|\mathcal{F}(\mathcal{P}^{(n)})\| < 10^{-12}$ after 3 iterative iterations. Moreover, it is interested in pointing out that the result of Table 1 shows that the algorithms (13) are quadratic convergence. Table 2 shows the necessary number of iterations for the convergence of the Lyapunov iterations [9] versus the new algorithm under the same accuracy of $\|\mathcal{F}(\mathcal{P}^{(n)})\| < 10^{-12}$. It can be seen that the convergence rate of the resulting algorithm is stable for all ε_j since the initial conditions $\mathcal{P}^{(0)}$ is quite good. On the other hand, the Lyapunov iterations converge to the exact solutions very slowly.

Table 1. $\|\mathcal{F}(\mathcal{P}^{(n)})\|$

n	$\varepsilon_1 = \varepsilon_2 = 10^{-1}$	$\varepsilon_1 = \varepsilon_2 = 10^{-2}$	$\varepsilon_1 = \varepsilon_2 = 10^{-3}$	$\varepsilon_1 = \varepsilon_2 = 10^{-4}$
0	1.1	1.1×10^{-1}	1.1×10^{-2}	1.1×10^{-3}
1	1.2	1.3×10^{-2}	1.3×10^{-4}	1.3×10^{-6}
2	6.4×10^{-2}	1.4×10^{-5}	1.5×10^{-9}	2.3×10^{-13}
3	1.1×10^{-4}	6.0×10^{-12}	5.0×10^{-14}	–
4	1.6×10^{-10}	6.1×10^{-14}	–	–
5	4.6×10^{-14}	–	–	–

Table 2. Number of iterations such that $\|\mathcal{F}(\mathcal{P}^{(n)})\| < 10^{-12}$

$\varepsilon_1 = \varepsilon_2$	Lyapunov iterations	Kleinman algorithm
10^{-1}	19	5
10^{-2}	12	4
10^{-3}	12	3
10^{-4}	11	2
10^{-5}	11	2
10^{-6}	11	2

6 Conclusions

The linear quadratic Nash games for infinite horizon MSPS have been studied. In order to solve the GCMARE, we have proposed the new iterations method which is based on the Kleinman algorithm. The proposed algorithm has the property of the quadratic convergence. It has been shown that the Kleinman algorithm can be used well to solve the GCMARE under the appropriate initial condition. When the dimension of the MSPS is quite large, the algorithm appearing in Theorem 2 seems to be formidable. However, this is, in fact, quite numerically tractable for small dimension of the MSPS. Comparing with Lyapunov iterations [9], even if the singular perturbation parameter is extremely small, we have succeeded in improving the convergence rate dramatically.

It is expected that the Kleinman algorithm for solving the GCMARE is applied to the wider class of the control law synthesis involving the solution of the CMARE with indefinite sign quadratic term such as the mixed H_2/H_∞ control problem [10]. This problem will be addressed in future investigations.

References

1. Freiling, G., Jank, G., Abou-Kandil, H. (1996): On Global Existence of Solutions to Coupled Matrix Riccati Equations in Closed-Loop Nash Games. IEEE Trans.

Automatic Control **AC–41(2)**, 264–269
2. Gajić, Z., Petkovski, D., Shen, X. (1990): Singularly Perturbed and Weakly Coupled Linear System – a Recursive Approach. In: Lecture Notes in Control and Information Sciences. Springer–Verlag, Berlin, Vol. 140
3. Jank, G., Kun, G. (1998): Solutions of Generalized Riccati Differential Equations and Their Approximation, Computational Methods and Function. In: Papamichael, N. et al. (Eds.): Theory (CMFT'97). World Scientific Publishing Corporation, 1–18
4. Khalil, H. K. (1980): Multimodel Design of a Nash Strategy. J. Opt. Theory and Appl. **31(4)**, 553–564
5. Khalil, H. K., Kokotovic, P. V. (1978): Control Strategies for Decision Makers Using Different Models of the Same System. IEEE Trans. Automatic Control **AC–23(2)**, 289–298
6. Khalil, H. K., Kokotovic, P. V. (1979a): Feedback and Well-Posedness of Singularly Perturbed Nash Games. IEEE Trans. Automatic Control **AC–24(5)**, 699–708
7. Khalil, H. K., Kokotovic, P. V. (1979b): Control of Linear Systems with Multiparameter Singular Perturbations. Automatica **15(2)**, 197–207
8. Kleinman, D. L. (1968): On the Iterative Technique for Riccati Equation Computations. IEEE Trans. Automatic Control **AC–13(2)**, 114–115
9. Li, T.-Y., Gajić , Z. (1994): Lyapunov Iterations for Solving Coupled Algebraic Lyapunov Equations of Nash Differential Games and Algebraic Riccati Equations of Zero-Sum Games. In: New Trends in Dynamic Games and Applications. Birkhauser, Boston, 333–351
10. Limebeer, D. J. N., Anderson, B. D. O., Hendel, B. (1994): A Nash Game Approach to Mixed H_2/H_∞ Control. IEEE Trans. Automatic Control **AC–39(1)**, 69–82
11. Magnus, J. R., Neudecker, H. (1999): Matrix Differential Calculus with Applications in Statistics and Econometrics. John Wiley and Sons, New York
12. Mukaidani, H., Xu, H., Mizukami, K. (1999): Recursive Approach of H_∞ Control Problems for Singularly Perturbed Systems under Perfect and Imperfect State Measurements. Int. J. Systems Sciences **30(5)**, 467–477
13. Mukaidani, H., Xu, H., Mizukami, K. (2000a): A New Algorithm for Solving Cross-Coupled Algebraic Riccati Equations of Singularly Perturbed Nash Games. In: Proc. 39th IEEE Conf. Decision and Control. Sydney, 3648–3653
14. Mukaidani, H., Xu, H., Mizukami, K. (2000b): Recursive Algorithm for Mixed H_2/H_∞ Control Problem of Singularly Perturbed Systems. Int. J. Systems Sciences **31(10)**, 1299–1312
15. Mukaidani, H., Xu, H., Mizukami, K. (2000c): A New Algorithm for Solving Cross-Coupled Algebraic Riccati Equations of Singularly Perturbed Systems for Mixed H_2/H_∞ Control Problem. In: Proc. 9th Int. Symp. Dynamic Games and Applications. Adelaide, 365–374
16. Mukaidani, H., Xu, H., Mizukami, K. (2001): New Iterative Algorithm for Algebraic Riccati Equation Related to H_∞ Control Problem of Singularly Perturbed Systems. IEEE Trans. Automatic Control **46(10)**, 1659–1666
17. Mukaidani, H., Xu, H., Mizukami, K. (2002): A Revised Kleinman Algorithm to Solve Algebraic Riccati Equation of Singularly Perturbed Systems. Automatica **38(3)**, 553–558
18. Starr, A. W., Ho, Y. C. (1969): Nonzero-Sum Differential Games. J. Optimization Theory and Application **3(3)**, 184–206

19. Xu, H., Mukaidani, H., Mizukami, K. (1999): An Order Reduction Procedure to Composite Nash Solution of Singularly Perturbed Systems. In: IFAC World Congress, Vol. F, 355–360
20. Yamamoto, T. (1986): A Method for Finding Sharp Error Bounds for Newton's Method under the Kantorvich Assumptions. Numerische Mathematik **49**, 203–220

Effects of Symmetry on Paradoxes in Non-Cooperative Distributed Computing

Hisao Kameda[1] and Yoshihisa Hosokawa[2]

[1] University of Tsukuba,
Institute of Information Sciences and Electronics,
Tsukuba Science City, Ibaraki 305-8573, Japan
E-mail: kameda@is.tsukuba.ac.jp

[2] University of Tsukuba,
Graduate School of Systems and Information Engineering,
Tsukuba Science City, Ibaraki 305-8573, Japan
E-mail: hosokawa@osdp.is.tsukuba.ac.jp

Summary. Distributed computer systems consist of nodes (hosts, computers), under which a communication follows from connected nodes. Jobs (or transactions) arrive at each node. Jobs can be forwarded through the communication means to other nodes for remote processing. We assume that there is a decision maker at each node who strives to optimize non-cooperatively the mean response time (the cost) for a job arriving at the node, by choosing the job-processing rate at the node and job forwarding rates. This is regards in a non-cooperative game. Numerical examples of a Braess-like paradox in which the performance for all decision makers in the system in a Nash equilibrium is worse with a communication means than without it, have been reported. Unlike the original Braess paradox, this behavior occurs in the models examined only in the case of finitely many users (decision makers) and not in the case of infinitesimal users, and the degree of performance degradation can increase without bound. Examples are presented which show that against our previous intuition the Braess-like paradox is strongest, i. e. the degree of performance degradation of all users in the Brass-like paradox is highest, in the completely symmetrical system wherein the parameter values describing each user are identical.

Key words: Braess paradox, cost optimization, distributed computer system, distributed decision, load balancing, Nash equilibrium, Pareto inefficiency, prisoners' dilemma

1 Introduction

We can choose one among several distinct objectives for performance optimization in many systems including communication networks, distributed computer systems, transportation flow networks, etc. Among them, we have three typical objectives or optima:

(1) the overall optimum, system-optimum, cooperative optimum or social optimum, where a certain overall and single measure, like the total cost or the overall average response time, is to be optimized. We call it the *overall optimum* here.

(2) the individual optimum, Wardrop equilibrium, or user optimum, where each of infinitesimal individuals, users, or jobs cannot receive any benefit by changing its own decision. Each of infinitesimal users seeks individual optimization. It is further assumed that the decision of a single user has a negligible impact on the performance of other users. We call it the *individual optimum* or *Wardrop equilibrium* here.

(3) the class optimum, Nash non-cooperative equilibrium, where each of a finite number of classes, users, or players cannot receive any benefit by changing its decision. Each of a finite number ($N(> 1)$) of players seeks its own optimization non-cooperatively. The decision of a single player may have a non-negligible impact on the performance of other players. We call it the *class optimum* or *Nash equilibrium* here.

Actually, (3) is reduced to (1) when the number of players reduces to 1 ($N = 1$) and approaches (2) when the number of players becomes infinitely many ($N \to \infty$) [7].

We can think that the total processing capacity of a system will increase when the capacity of a part of the system increases, and so we expect improvements in performance objectives accordingly in that case. The famous Braess paradox tells us that this is not always the case; i. e. increased capacity of a part of the system may sometimes lead to the degradation in the benefits of all users in the Wardrop equilibrium [3, 4, 5, 6, 8, 16]. We can expect that, in the Nash equilibrium where players seek their own optimization non-cooperatively, the similar type of paradox occurs (with large N), whenever it occurs for the Wardrop equilibrium ($N \to \infty$). Indeed, Korilis et al. [14, 15] found examples wherein the Braess-like paradox appears in a Nash equilibrium where all user classes are identical in the same topology for which the original Braess paradox (for the Wardrop equilibrium) was in fact obtained.

Kameda et al. [9] presented the existence of a paradox similar to Braess's that appears in the class optimum (Nash equilibrium) but does not occur in the Wardrop equilibrium, in the same environment. This paper pursues the same line of investigation and show some analytical results and numerical examples where such a paradox appears in class optima but not in Wardrop. It presents some tendencies or properties that can be seen from those analytical and numerical results. It is surprising that in these results, for the class optimum, the Braess-like paradox appears most strongly in symmetrical cases where no forwarding of jobs occurs in the overall and individual (Wardrop) optimum. Note, in passing, that there are cases of symmetrical systems where the degree of paradoxical cost degradation can increase without bound [13]. It is difficult now to interpret intuitively this counter-intuitive phenomenon.

One interpretation might be that general Pareto-inefficiency of Nash equilibria might be particularly apparent in symmetry.

The minimum among the ratios of the cost degradation for all the players (users) by increasing the connection, is considered the measure showing the degree of paradoxical cost degradation. The reason for considering this measure is elaborated in Subsection 1.1.

The rest of this paper is organized as follows. The description of systems considered is given in Section 2. The properties obtained analytically on a two-node system with linear node delays are given in Section 3. For multinode systems with nonlinear delays, some analytical properites and numerical results for systems of 2 and 4 nodes are presented in Section 4. Both results show that our measure of the paradoxical cost degradation is worst when completely symmetry holds among the decision makers. Concluding remarks are given in Section 5.

1.1 Motivation and Implications of the Measure of Paradixical Cost Degradation

The ratio of the cost for a user after adding capacity to that before doing it, is considered the ratio of cost degradation for the user due to adding capacity. The minimum of the ratios of cost degradation among users, is used as our measure of paradoxical cost degradation for the following reason. Even in overall optimization where cost degradation for all users never occurs, there is some possibility that a part of users suffer cost degradation while other users can definitely enjoy cost improvement. Therefore, situations where only a part of users suffer cost degradation may not be considered paradoxical as far as other users enjoy cost improvement. There may exist various measures for overall optimization. Consider overall measures that are weighted means of costs for users. Most commonly used is the one with identical weight factors for all user costs, e. g. the overall mean response time for a job. Consider, furthermore, the situation where one user has cost degradation and another has not. Then, one overall measure that has a large weight factor for the cost of the former user, degrades, and another measure that has a large weight factor for the latter user, improves. Thus even though an overall measure degrades, the entire situation does not look paradoxical if a part of users enjoy cost improvement. Thus, the minimum of the ratios of cost degradation among users is considered the measure of paradoxical cost degradation, since all users suffer the ratio of cost degradation not less than the minimum, in that case. Then, all of overall weighted mean measures degrade if and only if this measure degrades. These are illustrated as follows. Consider that there exist user $1, 2, \ldots, n$. Denote by C_i the cost of user i. Denote by C_i^b and by C_i^a, respectively, the costs of user i before and after the connection is added. Denote $k_i = C_i^a / C_i^b$, $k_{\min} = \min_i k_i$. Consider an overall measure $S = \sum_i \alpha_i C_i$ where α_i is the weight factor for user i and $\sum_i \alpha_i = 1$. Denote by S^b and S^a, respectively, the values of S before and after the connection is added. Then,

$$S^a = \sum_i \alpha_i C_i^a = \sum_i \alpha_i k_i C_i^b \geq \sum_i \alpha_i k_{min} C_i^b = k_{min} \sum_i \alpha_i C_i^b = k_{min} S^b .$$

Thus, $S^a \geq k_{min} S^b$. That is, all the overall measures suffer the ratio not less than k_{min}. But, if $k_i = k_{min} < 1$, there is an overall measure S' for which $S'^a / S'^b < 1$ with sufficiently large $\alpha_i (\simeq 1)$.

2 The Description of Systems Considered

Consider a system consisting of m nodes (hosts) and a communication means that connects nodes. Nodes are numbered $1, 2, \ldots, m$. Classify jobs arriving at node i into class i, $1, 2, \ldots, m$. Node i has an external Poisson arrival with rate ϕ_i, out of which the rate x_{ii} of jobs are processed at node i. The rate x_{ij} $(i \neq j)$ of jobs are forwarded through the communication means to another node j to be processed there, and the results of processing those jobs are returned back through the communication means to node i. Then it follows that $\sum_p x_{ip} = \phi_i$, $x_{ij} \geq 0$, $i, j = 1, 2, \ldots, m$. Denote by $\boldsymbol{x}_i$, the vector $(x_{i1}, x_{i2}, \cdots, x_{im})$ and by $\boldsymbol{x}$, $(\boldsymbol{x}_1, \boldsymbol{x}_2, \cdots, \boldsymbol{x}_m)$. Thus, $\boldsymbol{x} = (x_{11}, x_{12}, \cdots, x_{1m}, x_{21}, x_{22}, \cdots, x_{2m}, \cdots, x_{mm})$. Denote the set of $\boldsymbol{x}$'s that satisfy the constraints by $\boldsymbol{C}$ and let $\Phi = \sum_p \phi_p$. Each node has one decision maker, also numbered i $(i = 1, 2, \ldots, m)$. Within these constraints, a set of values of x_{ij} $(i, j = 1, 2, \ldots, m)$ are chosen to achieve optimization. The resulting load on node i is $\sum_q x_{qi}$ and is denoted by β_i. The system is equivalent to a network that has m origin-destination pairs with one common detination for all pairs. See Figure 1 for the case of $m = 3$.

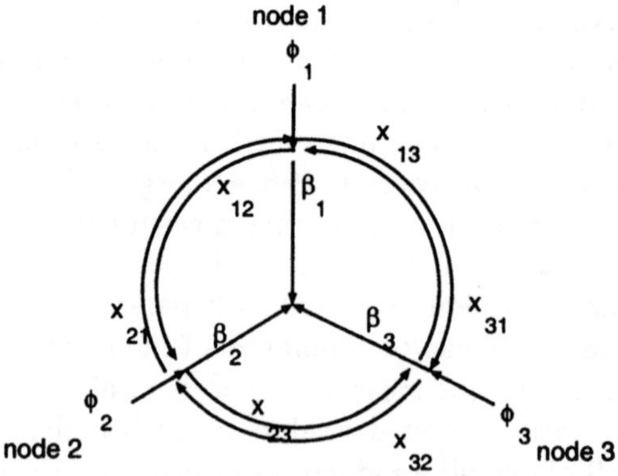

Fig. 1. The system model for $m = 3$

The expected processing (including queueing) time of a job that is processed at node i, is denoted by $D_i(\beta_i)$. The expected communication (including queueing) time of forwarding a job arriving at node i to node$j$ $(j \neq i)$, is denoted by $G_{ij}(\boldsymbol{x})$. Thus the expected response time of a job that arrives at node i is

$$T_i(\boldsymbol{x}) = \frac{1}{\phi_i} \sum_k x_{ik} T_{ik}(\boldsymbol{x}) \,,$$

$$\text{where} \quad T_{ii}(\boldsymbol{x}) = D_i(\beta_i) \,,$$

$$\text{and for } j \neq i \,, \quad T_{ij}(\boldsymbol{x}) = D_j(\beta_j) + G_{ij}(\boldsymbol{x}) \,.$$

Then, the overall expected response time of a job that arrives at the system is

$$T(\boldsymbol{x}) = \frac{1}{\Phi} \sum_i \phi_i T_i(\boldsymbol{x}) \,.$$

As to the communication means, consider the following two types (A) and (B).

(A) It consists of a single or multiple communication line that has no queueing delay. Thus the expected communication time of a job arriving at node i and being processed at node j ($\neq i$) is expressed as

$$G_{ij}(\boldsymbol{x}) = t \,.$$

(B) It consists of a single-channel communication line that is used commonly in forwarding and sending back of jobs that arrive at all nodes. Assume that the expected time length of forwarding and sending back a job is to be

$$G_{ij}(\boldsymbol{x}) = \frac{t}{1 - \lambda t} \,,$$

if $\lambda t < 1$, and is otherwise infinite, where $\lambda = \sum_p \sum_{q,(q \neq p)} x_{pq}$ is the communication traffic through the line. That is, it is assumed that the communication channel is modeled by a processor sharing server with service rate $1/t$, i. e. the mean communication (without queueing) time is t, and thus, the capacity of each communication line is $1/t$.

We refer to the length of time between the instant when a job arrives at a node and the instant when it leaves the node, where it has arrived, after all processing and communication (if any) are over, as *the response time for the job* arriving at the node.

We have three optima, the overall, the individual, and the class.

(1) The overall optimum is given by such $\bar{\boldsymbol{x}}$ as satisfies the following,

$$T(\bar{\boldsymbol{x}}) = \min T(\boldsymbol{x}) \quad \text{with respect to} \quad \boldsymbol{x} \in \boldsymbol{C} \,. \tag{1}$$

(2) The individual optimum is given by such $\hat{\boldsymbol{x}}$ as satisfies the following for all i,

$$T_i(\hat{\boldsymbol{x}}) = \min_p \{T_{ip}(\hat{\boldsymbol{x}})\} \,, \quad \text{such that } \hat{\boldsymbol{x}} \in \boldsymbol{C} \,. \tag{2}$$

(3) The class optimum is given by such $\tilde{\boldsymbol{x}}$ as satisfies the following for all i,

$$\tilde{T}_i = T_i(\tilde{\boldsymbol{x}}) = \min_{\boldsymbol{x}_i} T_i(\tilde{\boldsymbol{x}}_{-(i)}; \boldsymbol{x}_i), \text{ such that } (\tilde{\boldsymbol{x}}_{-(i)}; \boldsymbol{x}_i) \in \boldsymbol{C}. \quad (3)$$

where $(\tilde{\boldsymbol{x}}_{-(i)}; \boldsymbol{x}_i)$ denotes the mm-dimensional vector in which the elements corresponding to $\tilde{\boldsymbol{x}}_i$ have been replaced, respectively, by $\boldsymbol{x}_i$. In [1], it is shown that the three problems (1), (2) and (3) are equivalent to some variational inequalities. For the existence and uniqueness of those optima, see [2] and [1]. Clearly, there should be no mutual forwarding in overall and individual optima.

3 Two-Node Systems of Linear Node Delays

3.1 The Model and Assumptions

The system considered here has two nodes (i. e. $m = 2$) and a communication means, of type (B), that connects both nodes. In order to avoid complexity due to suffixes, the following notation is used in this section.

Arrival rates, job forwarding rates:

$$\xi = \phi_1, \ \eta = \phi_2, \ x = x_{12}, \ y = x_{21}, \ \xi - x = x_{11}, \ \eta - y = x_{22},$$

mean response times: $C_1(x, y)/\xi = T_1(\boldsymbol{x}), \ \ C_2(x, y)/\eta = T_2(\boldsymbol{x})$.

Thus, x and y, respectively, are the variables controlled by decision makers 1 and 2. Define by $\mathbf{y}$ the vector (x, y). Note that

$$0 \le x \le \xi \quad \text{and} \quad 0 \le y \le \eta. \quad (4)$$

Consider a simple case where node delays are $D_1(u) = a + bu$ and $D_2(v) = c + dv$ $(a \ge 0, c \ge 0)$ when nodes 1 and 2 are, respectively, required to process the rates u and v of jobs. Thus, if decision makers 1 and 2, respectively, decide job forwarding rates x and y, the values $C_1(x, y) = \xi T_1$ and $C_2(x, y) = \eta T_2$ for the decision makers 1 and 2 are the following:

$$C_1(x, y) = (\xi - x)[a + b(\xi - x + y)] + x[c + d(\eta - y + x)] + xt, \quad (5)$$

$$C_2(x, y) = (\eta - y)[c + d(\eta - y + x)] + y[a + b(\xi - x + y)] + yt. \quad (6)$$

Therefore, by noting that the values of ξ and η are given, the optimal decisions by decision makers 1 and 2, respectively, are expressed as follows: Denote by $\tilde{\mathbf{y}} = (\tilde{x}, \tilde{y})$ such values of x and y that satisfy

$$C_1(\tilde{x}, \tilde{y}) = \min_{x} C_1(x, \tilde{y}) \ \text{ such that } \ 0 \le x \le \xi \ \text{ given } \ \tilde{y}, \quad (7)$$

$$C_2(\tilde{x}, \tilde{y}) = \min_{y} C_2(\tilde{x}, y) \ \text{ such that } \ 0 \le y \le \eta \ \text{ given } \ \tilde{x}. \quad (8)$$

If the system has a solution $\tilde{\mathbf{y}}$ of $\mathbf{y}$ that satisfies (7) and (8) at the same time, it is a *Nash equilibrium*. Denote, in particular,

$$C_{10} = \xi(a + b\xi) \quad \text{and} \quad C_{20} = \eta(c + d\eta). \quad (9)$$

3.2 The Properties of the Two-Node Linear Model

The following properties and their proofs are given in [12].

The Cases where Paradoxes do not Occur.

Proposition 1. *The system has a unique Nash equilibrium* $\tilde{\mathbf{y}}$ *as follows:*
(i) For $b+d>0$

$$(\tilde{x},\tilde{y}) = \begin{cases} (0,0) \quad \text{if } 2\check{x}-\check{y}<0 \;\&\; 2\check{y}-\check{x}<0 \quad \text{case (1-1)} \\ (0,\check{y}-\check{x}/2) \quad \text{if } \check{x}<0 \& \; 0\le 2\check{y}-\check{x}\le 2\eta \quad \text{case (1-2)} \\ (0,\eta) \quad \text{if } 2\check{x}-\check{y}+\eta<0 \& \; 2\eta<2\check{y}-\check{x} \quad \text{case (1-3)} \\ (\check{x}-\check{y}/2,0) \quad \text{if } \check{y}<0 \& \; 0\le 2\check{x}-\check{y}\le 2\xi \quad \text{case (2-1)} \\ (\check{x},\check{y}) \quad \text{if } 0\le\check{x}\le\xi \& \; 0\le\check{y}\le\eta \quad \text{case (2-2)} \\ (\check{x}-\check{y}/2+\eta/2,\eta) \quad \text{if } 0\le 2\check{x}-\check{y}+\eta\le 2\xi \;\&\; \eta<\check{y} \text{ case (2-3)} \\ (\xi,0) \quad \text{if } 2\check{y}-\check{x}+\xi<0 \& \; 2\xi<2\check{x}-\check{y} \quad \text{case (3-1)} \\ (\xi,\check{y}-\check{x}/2+\xi/2) \quad \text{if } 0\le 2\check{y}-\check{x}+\xi\le 2\eta \;\&\; \xi<\check{x} \text{ case (3-2)} \\ (\xi,\eta) \text{ if } 2\xi-\eta-(2\check{x}-\check{y})<0 \& 2\eta-\xi-(2\check{y}-\check{x})<0 \text{ case (3-3)} \end{cases} \tag{10}$$

where $\check{x}$ *and* $\check{y}$ *are defined:*

$$\check{x} = (b+d)^{-1}\left[\frac{1}{3}(a-c)+b\xi-t\right], \tag{11}$$

$$\check{y} = (b+d)^{-1}\left[\frac{1}{3}(c-a)+d\eta-t\right]. \tag{12}$$

(ii) For $b+d\le 0$

$$(\tilde{x},\tilde{y}) = (0,0),\ (0,\eta),\ (\xi,0) \text{ or } (\xi,\eta). \tag{13}$$

Now consider the degree to which the benefit of each decision maker is improved or degraded due to adding the communication means between the two nodes. Define $\tilde{C}_i \triangleq C_i(\tilde{\mathbf{y}}), i=1,2$. Let k_i denote the ratio $\tilde{C}_i/C_{i0}$ for $i=1,2$. If $k_1<1$ and $k_2<1$, both decision makers receive increased benefits. If $k_i\ge 1$ and $k_j<1$ for some i,j $(i\ne j)$, the situation is not paradoxical yet, and only from this it cannot be said that the situation is Pareto inefficient. If both $k_i>1$, $i=1,2$, hold, the situation is paradoxical like the Braess paradox and the prisoners' dilemma, and is clearly Pareto inefficient. Call this situation *paradoxical performance degradation.*

The following property is obtained.

Proposition 2. *In the case where* $b+d\le 0$*, no paradoxical performance degradation occurs.*

Proposition 3. *In the case where* $b+d>0$*, no paradoxical degradation occurs when the solution has either of* $\tilde{x}=0$ *or* $\tilde{y}=0$ *and when it is* (ξ,η)*.*

As to the cases where the solution is either $(\tilde{x}, \eta)$ with $0 < \tilde{x} < \xi$ [case (2-3)] or $(\xi, \tilde{y})$ with $0 < \tilde{y} < \eta$ [case (3-2)], it has been shown that the paradox should not occur, under the assumption that

$$b \geq 0, \quad d \geq 0, \quad 0 < \xi \leq 5\eta, \quad 0 < \eta \leq 5\xi\,. \tag{14}$$

For the cases other than the above assumption (14), nothing can be said definitely now.

Note the following: in case (2-2), it holds that

$$(b+d)\tilde{C}_1 = -\frac{(a-c)^2}{9} + (bc+ad)\xi + bd(\xi+\eta)\xi + \frac{2(c-a)t}{3} + b\xi t - t^2\,, \tag{15}$$

$$(b+d)\tilde{C}_2 = -\frac{(a-c)^2}{9} + (bc+ad)\eta + bd(\xi+\eta)\eta + \frac{2(c-a)t}{3} + d\eta t - t^2\,, \tag{16}$$

$$(b+d)(\tilde{C}_1 - C_{10}) = -\frac{(a-c)^2}{9} + b\xi[(c-a) + (d\eta - b\xi) + t] + \frac{2(c-a)t}{3} - t^2\,, \tag{17}$$

$$(b+d)(\tilde{C}_2 - C_{20}) = -\frac{(a-c)^2}{9} + d\eta[(a-c) + (b\xi - d\eta) + t] + \frac{2(c-a)t}{3} - t^2\,. \tag{18}$$

By noting definitions (11) and (12), eqs. (17) and (18) are expressed as follows:

$$\tilde{C}_1 - C_{10} = b\xi\breve{y} - (b+d)\breve{x}^2\,, \tag{19}$$

$$\tilde{C}_2 - C_{20} = d\eta\breve{x} - (b+d)\breve{y}^2\,. \tag{20}$$

Therefore, the paradoxical performance degradation occurs when $b\xi\breve{y} > (b+d)\breve{x}^2$ and $d\eta\breve{x} > (b+d)\breve{y}^2$ hold together.

The Case where the Paradoxical Performance Degradation is Worst.

Consider the paradoxical situation *worst* when the minimum value k_{min} among $\{k_1, k_2\}$ is largest (and it is greater than 1). It is natural to call k_{min} the *measure of the paradoxical performance degradation.* Denote by **a** the vector $(a, b, c, d, \xi, \eta, t)$ of the parameter values. Denote by $g_i = g_i(\mathbf{a})$ a function of **a** related to location i $(i = 1, 2)$. Assume $b + d > 0$ reflecting Proposition 2.

Lemma 1. *If $\sum_i g_i\tilde{C}_i$ has the largest value when (and, respectively, only when) the complete symmetry holds, k_{min} has the largest value in (and only in) complete symmetry if $\sum_i g_i C_{i0}$ remains unchanged between the completely symmetrical and asymmetrical cases.*

From (19) and (20) it follows that

$$\frac{(\tilde{C}_1 - C_{10})}{\breve{y}} + \frac{(\tilde{C}_2 - C_{20})}{\breve{x}} = 2t - \frac{(b+d)(\breve{x}+\breve{y})}{\breve{x}\breve{y}}(\breve{x}-\breve{y})^2 . \tag{21}$$

Call the situation where the values of all the parameters describing each node are identical *complete symmetry.*

Lemma 2. *Consider an arbitrary combination of the values of variables* $a(\geq 0)$, $c(\geq 0)$, b, d, $\xi(> 0)$, $\eta(> 0)$ *that satisfy the constraint* $C_{10}/\breve{y}+C_{20}/\breve{x} = R$. *(R is a constant.) Then* $\tilde{C}_1/\breve{y} + \tilde{C}_2/\breve{x}$ *has the largest value when complete symmetry holds, i. e.,* $a = c$, $b = d$ *and* $\xi = \eta$.

Proof. It is clear from (21) by noting that, in complete symmetry $\breve{x} = \breve{y}$. □

Proposition 4. *The measure of paradoxical performance degradation,* k_{min}, *of the completely symmetric system has the largest value among the systems that have the same value of* $C_{10}/\breve{y} + C_{20}/\breve{x}$ *with t fixed and that are in case (2-2).*

Proof. Follows directly from the two lemmas above. □

Theorem 1. *For any system considered here, there exists a completely symmetrical system for which the value,* k_{min}, *of the measure of paradoxical performance degradation is not less than that of the given system.*

The Worst Value of the Measure.

Lemma 3. *The worst value of the measure of paradoxical performance degradation for completely symmetrical systems is 9/8.*

Theorem 2. *The worst measure of paradoxical performance degradation for systems considered in this paper is 9/8.*

Proof. Follows directly from Theorem 1 and Lemma 3. □

Remarks on the Two-Node Linear Model.

The section has examined a model consisting of two nodes each with a job arrival process. In an analytically tractable model, a condition under which such degradation does not occur in the model was obtained. Furthermore, it was shown that the measure of such paradoxical performance degradation to all nodes is worst when the model is in complete symmetry.

4 Multiple-Node Systems of Nonlinear Node Delays

In the case where node delays are not linear functions, no analytic results have been obtained for asymmetric systems. There have been, however, some analytical results on symmetric systems [13]. In the following subsection 4.1, some properties based upon the results on symmetric systems, are given. In order to gain further insight into the problem, numerical examinations have been performed in the subsections after Subsection 4.1.

Throughout this section 4, consider systems consisting of m nodes each of which has the node delay $D(\beta_i) = 1/(\mu_i - \beta_i)$ if $\beta_i < \mu_i$ (and it is otherwise infinite).

4.1 Analytical Properties on a Multi-node Nonlinear Model

In this subsection, consider systems with communication means (A). Call such a system a *system (A)* here. The following property holds for the system.

Theorem 3. *For any system (A), there exist a completely symmetrical system for which the value, k_{min}, of the measure of paradoxical performance degradation is not less than that of the given system.*

Proof. It has been already shown that the ratio of paradoxical degradation in symmetric distributed systems (A) can increase without bound [13]. Therefore, for any asymmetric system (A), there exists a symmetric system with (A) for which the ratio is not better than the asymmetric system. □

4.2 Numerical Experiments

In this section, we present numerical results for the cases of $m = 2, 4$ as examples.

We use the following best reply algorithm to compute these Nash equilibria.

For some fixed parameters, t, ϕ_i, μ_i, $i = 1, 2, \ldots, m$,

— Initialize $\boldsymbol{x}^0 = (\boldsymbol{x}_1^0, \boldsymbol{x}_2^0, \cdots, \boldsymbol{x}_m^0) \in \boldsymbol{C}$.

— Define $\boldsymbol{x}^n$ $(n = (k-1)m + i,\ \ i = 1, 2, \ldots, m,\ \ k = 1, 2, \ldots)$ as $(\boldsymbol{x}_1^{n-1}, \cdots, \boldsymbol{x}_{i-1}^{n-1}, \boldsymbol{x}_i^n, \boldsymbol{x}_{i+1}^{n-1}, \cdots, \boldsymbol{x}_m^{n-1})$, where $\boldsymbol{x}_i^n = \arg\min_{\boldsymbol{x}_i} T_i(\boldsymbol{x}_1^{n-1}, \cdots, \boldsymbol{x}_{i-1}^{n-1}, \boldsymbol{x}_i, \boldsymbol{x}_{i+1}^{n-1}, \cdots, \boldsymbol{x}_m^{n-1})$.

This algorithm gives a class optimum for our model if it converges. As to the uniqueness, see [2] and [1]. For each set of data, μ_i and ϕ_i, $i = 1, 2, \ldots, m$, we can find some value t^∞ (depending upon the set of data) of the mean communication time such that the communication line is not used any more at equilibria if the mean communication time is larger than t^∞. For each set of data, μ_i and ϕ_i, $i = 1, 2, \ldots, m$, we increase the mean communication time from 0 to t^∞. For each t we compute the class optimum (Nash equilibrium).

We focus our attention on the degradation that may occur when increasing the communication capacity. To this aim, we say that a Braess-like paradox occurs if the following holds:

$$D_i(t_1, t_2) > 0 \text{ for all } i\,, \text{ for some } t_1, t_2 \text{ such that } 0 < t_1 < t_2\,, \tag{22}$$

where $D_i(t_1, t_2) = (\tilde{T}_i(t_1) - \tilde{T}_i(t_2))/\tilde{T}_i(t_2)$, and $\tilde{T}_i(t)$ denotes the mean response for class i jobs, computed at the (unique) Nash equilibrium, when the mean communication time is t.

For simplicity, we only consider the case where $t_2 = t^{\infty}$, i. e. , the system has no communication means, and we denote $D_i(t, t^{\infty})$ by $\Delta_i(t)$. Denote $\boldsymbol{\phi} = (\phi_1, \phi_2, \cdots, \phi_m)$ and $\boldsymbol{\mu} = (\mu_1, \mu_2, \cdots, \mu_m)$. Thus, we define the *worst ratio of performance degradation* in the paradox $\Gamma(\boldsymbol{\mu}, \boldsymbol{\phi})$ as follows:

$$\Gamma(\boldsymbol{\mu}, \boldsymbol{\phi}) = \max_t\{\min_i\{\Delta_i(t)\}\}\,. \tag{23}$$

Denote t_0 the mean communication time, such that the previous maximum is reached.

Notice that we can define a weaker paradox. For example we can think of some kind of *local paradox*, such that we can observe for some interval of t, $\tilde{T}_i(t)$, for some i, increases, but nevertheless the overall mean response time $\tilde{T}(t) = (1/\Phi)(\sum_i \phi_i \tilde{T}_i(t)$ decreases. This weaker paradox may occur even if (22) does not hold, and has been already discussed in [11].

Let us define three symmetries with respect to the system parameter setting.

Overall symmetry. If the following condition holds

$$\frac{\mu_i}{(\mu_i - \phi_i)^2} = \text{constant}, \text{ for all } i\,, \tag{24}$$

then there is no forwarding among nodes for any value of the communication line capacity $1/t$, for both of cases (A) and (B), when the system is at the overall optimum. This can be easily seen from the Kuhn-Tucker condition for the overall optimization problem (1). We say that we have an *overall symmetry* property among nodes in this case.

Individual symmetry. If the following condition holds

$$\frac{1}{\mu_i - \phi_i} = \text{constant}, \text{ for all } i\,, \tag{25}$$

it can be proved from the definition (2) that at the individual optimum there will be no forwarding among nodes for any value of the communication line capacity $1/t$, for both cases (A) and (B). We say that we have an *individual symmetry* property among nodes in this case.

Complete symmetry. If both conditions (24) and (25) hold or equivalently if μ_i and ϕ_i are constant, for all i, then we say that we have a *complete symmetry* among nodes. In the complete symmetry, no forwarding of jobs occurs both in the overall and individual optima.

In the following numerical results, we introduce some asymmetry. We start from a set of data such that the complete symmetry property holds.

— First we change μ_i and ϕ_i, $i = 1, 2, \dots, m$ in such a way that the property of overall symmetry, i. e. condition (24) is preserved.
— Starting again from the complete symmetric case, we change μ_i and ϕ_i, $i = 1, 2, \dots, m$ in such a way that the individual symmetry property, i. e. condition (25) is preserved.
— Starting from the complete symmetric case, we change μ_i and ϕ_i, $i = 1, 2, \dots, m$. In this case none of the properties of symmetry holds any more.

4.3 Numerical Results

The cases of $m = 2$ and $m = 4$ are presented in the following.

Two-Node Nonlinear Models. The system examined mainly here consists of two nodes and a communication means of type (A). The same system as this but with a communication means of type (B) showed similar properties. Note that the uniqueness of the solution $\boldsymbol{x}$ is assured for the communication means of types (A) and (B) [2].

Figure 2 shows how the measure, Γ, of paradoxical performance degradation depends on the combination of the values of μ_1 and μ_2, with ϕ_1, and ϕ_2 given by $\phi_i/(\mu_i - \phi_i) = 4$, in some case where no symmetry is maintained. (In fact, this case has still a kind of symmetry such that the expected number of jobs that stay in each node is identical when no communication means is available.) It can be observed, in the figure, that Γ is largest in complete symmetry where $\mu_1 = \mu_2$ and, thus, $\phi_1 = \phi_2$, and that it decreases as the values of μ_1 and μ_2 depart complete symmetry. More detailed studies [10] show the same trends as those given in the following on four-node models.

Four-Node Nonlinear Models. Here shown are typical numerical examples, mainly in the case (A) of the communication means, with changing parameter values from $\mu_i = 25$ and $\phi_i = 20$, for all i, in each of three directions: overall, individual, and no symmetry maintained. Note that the uniqueness of the solution $\boldsymbol{x}$ is assured for the communication means of type (A) [2]. In particular, we consider a family of systems for which the following holds for the cases with overall, individual, and no symmetry maintained:

$$\sum_i \frac{\phi_i}{\mu_i - \phi_i} = \text{constant} . \tag{26}$$

For case (B) of the communication means, we had similar observations, a part of which is mentioned below.

A part of the results are presented in the following tables. In these tables, the left and right tables, respectively, show the numerical data (the worst value Γ and the values of $\Delta_i(t_0)$, $i = 1, 2, 3, 4$) for wide and small ranges of

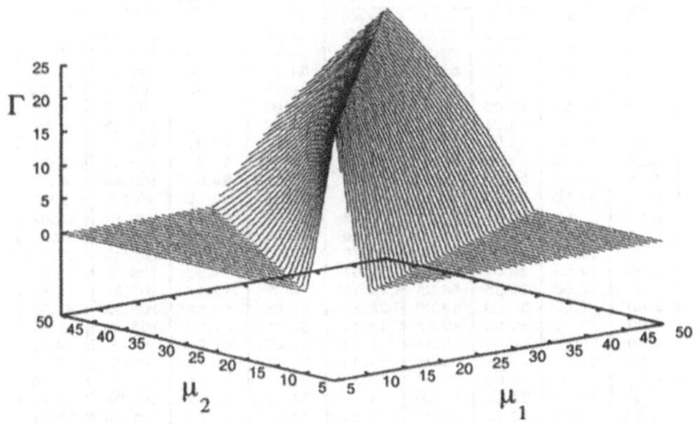

Fig. 2. Two-node nonlinear model (B).The vertical axis shows the values of $\Gamma(\%)$ given the values of μ_1 and μ_2 with ϕ_1 and ϕ_2 given by $\phi_i/(\mu_i - \phi_i) = 4, i = 1, 2$

parameter values of $\boldsymbol{\mu}$ in their large and small steps. The lowest row of each table shows the completely symmetrical case. The field that contains "–" shows the node where t_0 cannot obtained. Thus the rows that contain fields with "–" show the cases where no paradoxical cost degradation occurs. If more than one field that contains a positive value of Δ_i exists, the paradoxical cost degradation is observed only among the nodes i with the positive values of Δ_i for a certain value of t_1.

Overall symmetry maintained. Tables 1 and 2, respectively, show, for cases (A) and (B), the effect of changing μ_i and ϕ_i within the family satisfying relation (26) and with overall symmetry maintained.

Individual symmetry maintained. Table 3 shows for case (A) the effect of changing μ_i and ϕ_i within the family satisfying relation (26) and with individual symmetry maintained. The results for case (B) are quite similar to those for case (A) as shown above.

No symmetry maintained. Table 4 shows for case (A) the effect of changing μ_i and ϕ_i within the family satisfying relation (26) and with no symmetry maintained. The results for case (B) are quite similar to those for case (A) as shown above.

Overall Tendency. In these examples, it is observed that the worst ratio of performance degradation in the paradox is largest in completely symmetric case. The degradation decreases rapidly as the parameter setting departs the above-mentioned symmetric case without keeping any kind of symmetries. It decreases slowly with individual symmetry maintained where no forwarding of jobs occurs in the individual optimum. It decreases more slowly with overall symmetry maintained where no forwarding of jobs occurs in the overall opti-

Table 1. Overall symmetry maintained in systems with the communication means of type (A)

μ_1	μ_2	μ_3	μ_4	Δ_1(%)	Δ_2(%)	Δ_3(%)	Δ_4(%)	Γ(%)
6.25	6.25	36.00	81.00	11.70	11.70	91.41	97.33	11.70
6.25	6.25	42.25	72.25	9.40	9.40	94.55	104.80	9.40
6.25	9.00	36.00	72.25	10.32	23.10	87.53	103.47	10.32
6.25	9.00	42.25	64.00	7.78	24.55	92.70	107.46	7.78
6.25	12.25	30.25	72.25	10.99	36.12	79.36	102.50	10.99
6.25	12.25	36.00	64.00	8.15	38.03	87.08	105.46	8.15
6.25	16.00	30.25	64.00	8.64	50.62	80.06	104.53	8.64
6.25	16.00	36.00	56.25	5.29	51.93	87.27	106.55	5.29
6.25	20.25	30.25	56.25	5.96	63.64	81.34	105.37	5.96
6.25	20.25	36.00	49.00	4.27	63.38	89.04	103.46	4.27
6.25	25.00	36.00	42.25	2.83	72.90	90.56	98.58	2.83
9.00	9.00	30.25	72.25	24.03	24.03	79.61	102.17	24.03
9.00	9.00	36.00	64.00	22.96	22.96	85.56	107.99	22.96
9.00	12.25	30.25	64.00	21.79	36.07	77.29	106.43	21.79
9.00	12.25	36.00	56.25	22.07	37.83	87.60	106.55	22.07
9.00	16.00	30.25	56.25	19.85	50.37	79.43	107.91	19.85
9.00	16.00	36.00	49.00	20.80	50.86	89.66	103.82	20.80
9.00	20.25	30.25	49.00	18.45	61.80	81.70	105.40	18.45
9.00	20.25	36.00	42.25	19.12	62.02	91.61	99.49	19.12
9.00	25.00	30.25	42.25	17.83	71.31	82.13	99.42	17.83
9.00	25.00	36.00	36.00	18.30	71.50	91.67	91.67	18.30
12.25	12.25	25.00	64.00	34.71	34.71	66.99	109.94	34.71
12.25	12.25	30.25	56.25	35.81	35.81	79.52	107.82	35.81
12.25	16.00	36.00	42.25	34.66	49.21	92.01	99.81	34.66
12.25	20.25	25.00	49.00	32.18	59.73	71.96	106.32	32.18
12.25	20.25	30.25	42.25	33.28	60.44	83.34	100.44	33.28
12.25	25.00	30.25	36.00	34.37	73.10	83.78	93.02	34.37
16.00	16.00	25.00	49.00	46.75	46.75	72.54	106.85	46.75
16.00	16.00	30.25	42.25	47.63	47.63	83.97	101.04	47.63
16.00	20.25	25.00	42.25	48.18	61.37	73.45	101.14	48.18
16.00	25.00	25.00	36.00	49.08	74.03	74.03	93.83	49.08
20.25	20.25	25.00	36.00	62.38	62.38	74.38	94.21	62.38
20.25	25.00	25.00	30.25	62.87	74.77	74.77	85.35	62.87
25.00	**25.00**	**25.00**	**25.00**	75.00	75.00	75.00	75.00	**75.00**

μ_1	μ_2	μ_3	μ_4	Δ_1(%)	Δ_2(%)	Δ_3(%)	Δ_4(%)	Γ(%)
24.6	24.6	24.6	26.2144	74.08	74.08	74.08	77.66	74.08
24.6	24.6	24.7	26.1121	74.09	74.09	74.31	77.45	74.09
24.6	24.6	24.8	26.0100	74.09	74.09	74.54	77.23	74.09
24.6	24.6	25.0	25.8064	74.09	74.09	75.00	76.79	74.09
24.6	24.7	24.7	26.0100	74.09	74.31	74.31	77.23	74.09
24.6	24.7	24.8	25.9081	74.09	74.32	74.54	77.01	74.09
24.6	24.7	24.9	25.8064	74.09	74.32	74.77	76.79	74.09
24.6	24.7	25.0	25.7049	74.09	74.32	75.00	76.56	74.09
24.6	24.8	24.8	25.8064	74.09	74.54	74.54	76.79	74.09
24.6	24.8	24.9	25.7049	74.09	74.54	74.77	76.57	74.09
24.6	24.8	25.0	25.6036	74.09	74.54	75.00	76.34	74.09
24.6	24.9	24.9	25.6036	74.09	74.77	74.77	76.34	74.09
24.6	24.9	25.0	25.5025	74.09	74.77	75.00	76.12	74.09
24.6	25.0	25.0	25.4016	74.09	75.00	75.00	75.90	74.09
24.7	24.7	24.7	25.9081	74.32	74.32	74.32	77.01	74.32
24.7	24.7	24.8	25.8064	74.32	74.32	74.54	76.79	74.32
24.7	24.7	24.9	25.7049	74.32	74.32	74.77	76.57	74.32
24.7	24.7	25.0	25.6036	74.32	74.32	75.00	76.34	74.32
24.7	24.8	24.8	25.7049	74.32	74.54	74.54	76.57	74.32
24.7	24.8	24.9	25.6036	74.32	74.55	74.77	76.34	74.32
24.7	24.8	25.0	25.5025	74.32	74.55	75.00	76.12	74.32
24.7	24.9	24.9	25.5025	74.32	74.77	74.77	76.12	74.32
24.7	24.9	25.0	25.4016	74.32	74.77	75.00	75.90	74.32
24.7	25.0	25.0	25.3009	74.32	75.00	75.00	75.67	74.32
24.8	24.8	24.8	25.6036	74.55	74.55	74.55	76.34	74.55
24.8	24.8	24.9	25.5025	74.55	74.55	74.77	76.12	74.55
24.8	24.8	25.0	25.4016	74.55	74.55	75.00	75.90	74.55
24.8	24.9	24.9	25.4016	74.55	74.77	74.77	75.90	74.55
24.8	24.9	25.0	25.3009	74.55	74.77	75.00	75.67	74.55
24.8	25.0	25.0	25.2004	74.55	75.00	75.00	75.45	74.55
24.9	24.9	24.9	25.3009	74.77	74.77	74.77	75.67	74.77
24.9	24.9	25.0	25.2004	74.77	74.77	75.00	75.45	74.77
24.9	25.0	25.0	25.1001	74.77	75.00	75.00	75.23	74.77
25.0	**25.0**	**25.0**	**25.0000**	75.00	75.00	75.00	75.00	**75.00**

mum. These trends are the same as observed for two-node nonlinear models by Kameda et al. [10].

Table 2. Overall symmetry maintained in systems with the communication means of type (B)

μ_1	μ_2	μ_3	μ_4	$\Delta_1(\%)$	$\Delta_2(\%)$	$\Delta_3(\%)$	$\Delta_4(\%)$	$\Gamma(\%)$
6.25	6.25	36.00	81.00	-	-	30.72	39.13	-
6.25	6.25	42.25	72.25	-	-	33.75	42.70	-
12.25	12.25	16.00	81.00	4.31	4.31	11.27	45.93	4.31
12.25	12.25	20.25	72.25	3.27	3.27	16.56	50.62	3.27
12.25	12.25	25.00	64.00	2.55	2.55	23.55	58.76	2.55
12.25	12.25	30.25	56.25	3.53	3.53	32.08	55.40	3.53
12.25	12.25	36.00	49.00	4.13	4.13	39.47	51.09	4.13
12.25	12.25	42.25	42.25	4.34	4.34	45.80	45.80	4.34
12.25	16.00	16.00	72.25	3.41	10.35	10.35	50.82	3.41
12.25	16.00	20.25	64.00	2.54	9.46	15.80	55.13	2.54
12.25	16.00	25.00	56.25	2.65	11.80	26.60	57.29	2.65
12.25	16.00	30.25	49.00	3.51	12.54	34.89	53.22	3.51
12.25	16.00	36.00	42.25	3.95	12.91	42.06	48.16	3.95
12.25	20.25	20.25	56.25	2.09	20.00	20.00	59.39	2.09
12.25	20.25	25.00	49.00	3.53	21.17	29.14	55.24	3.53
12.25	20.25	30.25	42.25	4.42	21.89	37.18	50.19	4.42
12.25	20.25	36.00	36.00	4.71	22.14	44.18	44.18	4.71
12.25	25.00	25.00	42.25	4.73	30.07	30.07	50.42	4.73
12.25	25.00	30.25	36.00	5.34	30.55	37.88	44.65	5.34
12.25	30.25	30.25	30.25	5.65	38.11	38.11	38.11	5.65
16.00	16.00	16.00	64.00	9.89	9.89	9.89	63.08	9.89
16.00	16.00	20.25	56.25	11.64	11.64	20.30	59.81	11.64
16.00	16.00	25.00	49.00	12.93	12.93	29.46	55.66	12.93
16.00	16.00	30.25	42.25	13.72	13.72	37.53	50.59	13.72
16.00	16.00	36.00	36.00	13.99	13.99	44.56	44.56	13.99
16.00	20.25	20.25	49.00	13.22	21.75	21.75	55.90	13.22
16.00	20.25	25.00	42.25	14.30	22.75	30.67	51.08	14.30
16.00	20.25	30.25	36.00	14.85	23.26	38.49	45.28	14.85
16.00	25.00	25.00	36.00	15.15	31.41	31.41	45.54	15.15
16.00	25.00	30.25	30.25	15.43	31.66	38.99	38.99	15.43
20.25	20.25	20.25	42.25	23.05	23.05	23.05	51.53	23.05
20.25	20.25	25.00	36.00	23.84	23.84	31.75	45.96	23.84
20.25	20.25	30.25	30.25	24.11	24.11	39.37	39.37	24.11
20.25	25.00	25.00	30.25	24.38	32.27	32.27	39.63	24.38
25.00	**25.00**	**25.00**	**25.00**	32.57	32.57	32.57	32.57	**32.57**

μ_1	μ_2	μ_3	μ_4	$\Delta_1(\%)$	$\Delta_2(\%)$	$\Delta_3(\%)$	$\Delta_4(\%)$	$\Gamma(\%)$
24.5	24.8	25.0	25.7049	31.79	32.26	32.56	33.63	31.79
24.5	24.9	24.9	25.7049	31.79	32.41	32.41	33.63	31.79
24.5	24.9	25.0	25.6036	31.80	32.41	32.57	33.48	31.80
24.5	25.0	25.0	25.5025	31.80	32.57	32.57	33.33	31.80
24.6	24.6	24.7	26.1121	31.94	31.94	32.10	34.24	31.94
24.6	24.6	24.8	26.0100	31.95	31.95	32.25	34.09	31.95
24.6	24.7	24.7	26.0100	31.95	32.10	32.10	34.09	31.95
24.6	24.7	24.8	25.9081	31.95	32.10	32.26	33.94	31.95
24.6	24.7	25.0	25.7049	31.95	32.10	32.56	33.63	31.95
24.6	24.8	24.8	25.8064	31.95	32.26	32.26	33.79	31.95
24.6	24.8	24.9	25.7049	31.95	32.26	32.41	33.63	31.95
24.6	24.8	25.0	25.6036	31.95	32.26	32.57	33.48	31.95
24.6	24.9	24.9	25.6036	31.95	32.41	32.41	33.48	31.95
24.6	24.9	25.0	25.5025	31.95	32.41	32.57	33.33	31.95
24.6	25.0	25.0	25.4016	31.95	32.57	32.57	33.18	31.95
24.7	24.7	24.7	25.9081	32.10	32.10	32.10	33.94	32.10
24.7	24.7	24.8	25.8064	32.10	32.10	32.26	33.79	32.10
24.7	24.7	24.9	25.7049	32.10	32.10	32.41	33.64	32.10
24.7	24.7	25.0	25.6036	32.10	32.10	32.57	33.48	32.10
24.7	24.8	24.8	25.7049	32.10	32.26	32.26	33.64	32.10
24.7	24.8	24.9	25.6036	32.10	32.26	32.41	33.49	32.10
24.7	24.8	25.0	25.5025	32.11	32.26	32.57	33.33	32.11
24.7	24.9	24.9	25.5025	32.11	32.41	32.41	33.33	32.11
24.7	24.9	25.0	25.4016	32.11	32.41	32.57	33.18	32.11
24.7	25.0	25.0	25.3009	32.11	32.57	32.57	33.03	32.11
24.8	24.8	24.8	25.6036	32.26	32.26	32.26	33.49	32.26
24.8	24.8	24.9	25.5025	32.26	32.26	32.41	33.33	32.26
24.8	24.8	25.0	25.4016	32.26	32.26	32.57	33.18	32.26
24.8	24.9	24.9	25.4016	32.26	32.41	32.41	33.18	32.26
24.8	24.9	25.0	25.3009	32.26	32.41	32.57	33.03	32.26
24.8	25.0	25.0	25.2004	32.26	32.57	32.57	32.88	32.26
24.9	24.9	24.9	25.3009	32.41	32.41	32.41	33.03	32.41
24.9	24.9	25.0	25.2004	32.42	32.42	32.57	32.88	32.42
24.9	25.0	25.0	25.1001	32.42	32.57	32.57	32.72	32.42
25.0	**25.0**	**25.0**	**25.0000**	32.57	32.57	32.57	32.57	**32.57**

5 Concluding Remaks

In analytical and numerical examples, it is observed that the worst ratio of performance degradation in the paradox is largest in completely symmetrical

Table 3. Individual symmetry maintained in systems with the communication means of type (A)

μ_1	μ_2	μ_3	μ_4	Δ_1(%)	Δ_2(%)	Δ_3(%)	Δ_4(%)	Γ(%)
6.25	6.25	6.25	81.25	45.10	45.10	45.10	-	-
6.25	6.25	16.00	71.50	62.97	62.97	44.61	-	-
6.25	9.00	12.25	72.50	62.29	55.73	49.74	-	-
9.00	9.00	9.00	73.00	55.38	55.38	55.38	-	-
9.00	9.00	16.00	66.00	67.07	67.07	55.04	4.36	4.36
6.25	9.00	20.25	64.50	69.42	66.23	48.98	4.77	4.77
6.25	12.25	16.00	65.50	70.77	60.77	54.90	4.52	4.52
6.25	12.25	25.00	56.50	49.56	68.51	57.85	22.58	22.58
6.25	16.00	20.25	57.50	51.12	67.06	63.06	22.34	22.34
6.25	16.00	30.25	47.50	41.70	72.80	58.37	40.32	40.32
6.25	20.25	25.00	48.50	42.26	71.54	65.89	40.40	40.40
6.25	20.25	30.25	43.25	46.72	73.99	60.94	46.72	46.72
6.25	25.00	25.00	43.75	46.20	66.68	66.68	46.19	46.19
9.00	12.25	16.00	62.75	70.32	64.79	58.99	12.27	12.27
9.00	12.25	20.25	58.50	71.06	67.85	57.20	20.86	20.86
9.00	16.00	20.25	54.75	68.78	68.07	62.22	28.59	28.59
9.00	16.00	25.00	50.00	63.32	72.59	60.29	36.09	36.09
9.00	20.25	20.25	50.50	64.39	67.00	67.00	36.10	36.10
9.00	20.25	30.25	40.50	57.67	72.62	60.33	50.50	50.50
9.00	25.00	25.00	41.00	52.73	67.63	67.63	50.93	50.93
9.00	25.00	30.25	35.75	57.44	68.17	62.40	57.41	57.41
12.25	12.25	12.25	63.25	65.11	65.11	65.11	12.12	12.12
12.25	12.25	20.25	55.25	69.67	69.67	60.31	29.53	29.53
12.25	16.00	30.25	41.50	71.06	74.65	60.84	50.77	50.77
12.25	20.25	20.25	47.25	66.01	69.02	69.02	45.17	45.17
12.25	20.25	25.00	42.50	65.11	71.24	67.17	52.08	52.08
12.25	25.00	25.00	37.75	65.74	69.31	69.31	58.54	58.54
16.00	16.00	20.25	47.75	69.01	69.01	68.16	45.73	45.73
16.00	16.00	25.00	43.00	71.64	71.64	68.02	52.12	52.12
16.00	20.25	20.25	43.50	70.03	70.60	70.60	52.34	52.34
16.00	25.00	25.00	34.00	75.48	72.20	72.20	64.45	64.45
20.25	20.25	25.00	34.50	75.71	75.71	72.81	64.53	64.53
20.25	25.00	25.00	29.75	77.48	74.21	74.21	70.09	70.09
25.00	**25.00**	**25.00**	**25.00**	75.00	75.00	75.00	75.00	**75.00**

μ_1	μ_2	μ_3	μ_4	Δ_1(%)	Δ_2(%)	Δ_3(%)	Δ_4(%)	Γ(%)
24.6	24.6	24.7	26.1	75.31	75.31	75.22	74.01	74.01
24.6	24.6	24.8	26.0	75.31	75.31	75.14	74.11	74.11
24.6	24.6	24.9	25.9	75.32	75.32	75.06	74.20	74.20
24.6	24.6	25.0	25.8	75.32	75.32	74.98	74.29	74.29
24.6	24.7	24.7	26.0	75.31	75.23	75.23	74.11	74.11
24.6	24.7	24.8	25.9	75.32	75.23	75.15	74.20	74.20
24.6	24.7	24.9	25.8	75.32	75.24	75.07	74.29	74.29
24.6	24.7	25.0	25.7	75.33	75.24	74.99	74.38	74.38
24.6	24.8	24.8	25.8	75.32	75.15	75.15	74.29	74.29
24.6	24.8	24.9	25.7	75.33	75.16	75.07	74.38	74.38
24.6	24.8	25.0	25.6	75.33	75.16	74.99	74.47	74.47
24.6	24.9	24.9	25.6	75.33	75.08	75.08	74.47	74.47
24.6	24.9	25.0	25.5	75.33	75.08	74.99	74.56	74.56
24.6	25.0	25.0	25.4	75.34	74.99	74.99	74.65	74.65
24.7	24.7	24.7	25.9	75.23	75.23	75.23	74.20	74.20
24.7	24.7	24.8	25.8	75.24	75.24	75.15	74.29	74.29
24.7	24.7	24.9	25.7	75.24	75.24	75.07	74.38	74.38
24.7	24.7	25.0	25.6	75.25	75.25	74.99	74.47	74.47
24.7	24.8	24.8	25.7	75.24	75.16	75.16	74.38	74.38
24.7	24.8	24.9	25.6	75.25	75.16	75.08	74.47	74.47
24.7	24.8	25.0	25.5	75.25	75.16	74.99	74.56	74.56
24.7	24.9	24.9	25.5	75.25	75.08	75.08	74.56	74.56
24.7	24.9	25.0	25.4	75.25	75.08	75.00	74.65	74.65
24.7	25.0	25.0	25.3	75.25	75.00	75.00	74.74	74.74
24.8	24.8	24.8	25.6	75.16	75.16	75.16	74.47	74.47
24.8	24.8	24.9	25.5	75.16	75.16	75.08	74.56	74.56
24.8	24.8	25.0	25.4	75.17	75.17	75.00	74.65	74.65
24.8	24.9	24.9	25.4	75.17	75.08	75.08	74.65	74.65
24.8	24.9	25.0	25.3	75.17	75.08	75.00	74.74	74.74
24.8	25.0	25.0	25.2	75.17	75.00	75.00	74.83	74.83
24.9	24.9	24.9	25.3	75.08	75.08	75.08	74.74	74.74
24.9	24.9	25.0	25.2	75.09	75.09	75.00	74.83	74.83
24.9	25.0	25.0	25.1	75.09	75.00	75.00	74.91	74.91
25.0	**25.0**	**25.0**	**25.0**	75.00	75.00	75.00	75.00	**75.00**

cases. It looks quite strange that in these symmetrical cases, the conditions that Braess-like paradoxes occur are less tight than non-symmetrical situations and that *the paradox appears most strongly in complete symmetry*. In general, symmetric models are more analytically tractable than asymmetric ones. If the trends observed in this study hold generally, the analysis of

Table 4. No symmetry maintained in systems with the communication means of type (A)

μ_1	μ_2	μ_3	μ_4	$\Delta_1(\%)$	$\Delta_2(\%)$	$\Delta_3(\%)$	$\Delta_4(\%)$	$\Gamma(\%)$
6.25	6.25	36.00	81.00	-	-	0.00	16.68	-
6.25	6.25	42.25	72.25	-	-	-	18.82	-
6.25	9.00	36.00	72.25	-	-	0.00	18.15	-
6.25	9.00	42.25	64.00	-	-	-	20.58	-
6.25	12.25	30.25	72.25	-	-	-	16.57	-
6.25	12.25	36.00	64.00	-	-	0.00	18.71	-
6.25	16.00	30.25	64.00	-	-	-	15.89	-
6.25	16.00	36.00	56.25	-	-	0.00	17.92	-
6.25	20.25	30.25	56.25	-	-	-	13.68	-
6.25	20.25	36.00	49.00	-	-	0.00	15.30	-
6.25	25.00	36.00	42.25	-	0.00	2.98	11.56	-
9.00	9.00	30.25	72.25	-	-	45.25	128.11	-
9.00	9.00	36.00	64.00	-	-	66.65	132.99	-
9.00	12.25	30.25	64.00	-	-	51.13	138.04	-
9.00	12.25	36.00	56.25	-	-	75.64	134.54	-
9.00	16.00	30.25	56.25	-	2.41	60.23	141.82	-
9.00	16.00	36.00	49.00	-	7.67	93.35	139.55	-
9.00	20.25	30.25	49.00	-	28.11	70.07	136.44	-
9.00	20.25	36.00	42.25	-	32.91	102.79	127.17	-
9.00	25.00	30.25	42.25	-	58.01	82.47	132.69	-
9.00	25.00	36.00	36.00	-	59.22	109.37	109.37	-
12.25	12.25	25.00	64.00	-	-	37.13	153.13	-
12.25	12.25	30.25	56.25	-	-	66.84	157.39	-
12.25	16.00	36.00	42.25	-	11.41	105.88	130.48	-
12.25	20.25	25.00	49.00	-	33.10	57.64	157.70	-
12.25	20.25	30.25	42.25	-	36.48	87.12	138.42	-
12.25	25.00	30.25	36.00	-	65.39	91.48	117.72	-
16.00	16.00	25.00	49.00	10.65	10.65	58.84	159.29	10.65
16.00	16.00	30.25	42.25	13.63	13.63	88.50	140.01	13.63
16.00	20.25	25.00	42.25	15.63	39.89	65.31	143.38	15.63
16.00	25.00	25.00	36.00	18.76	69.38	69.38	122.52	18.76
20.25	20.25	25.00	36.00	44.71	44.71	70.74	124.17	44.71
20.25	25.00	25.00	30.25	47.21	73.57	73.57	100.55	47.21
25.00	**25.00**	**25.00**	**25.00**	75.00	75.00	75.00	75.00	**75.00**

μ_1	μ_2	μ_3	μ_4	$\Delta_1(\%)$	$\Delta_2(\%)$	$\Delta_3(\%)$	$\Delta_4(\%)$	$\Gamma(\%)$
24.8	25.2	25.2	25.2	73.39	75.53	75.53	75.53	73.39
24.8	25.1	25.2	25.2	73.52	75.13	75.67	75.67	73.52
24.8	25.0	25.2	25.2	73.65	74.73	75.80	75.80	73.65
24.8	25.1	25.1	25.2	73.65	75.27	75.27	75.80	73.65
24.8	24.9	25.2	25.2	73.79	74.32	75.94	75.94	73.79
24.8	25.0	25.1	25.2	73.79	74.86	75.40	75.94	73.79
24.8	25.1	25.1	25.1	73.79	75.40	75.40	75.40	73.79
24.8	24.8	25.2	25.2	73.92	73.92	76.07	76.07	73.92
24.8	24.9	25.1	25.2	73.92	74.46	75.54	76.07	73.92
24.8	25.0	25.0	25.2	73.92	75.00	75.00	76.07	73.92
24.8	25.0	25.1	25.1	73.92	75.00	75.54	75.54	73.92
24.8	24.8	25.1	25.2	74.05	74.05	75.67	76.21	74.05
24.8	24.9	25.0	25.2	74.05	74.59	75.13	76.21	74.05
24.8	24.9	25.1	25.1	74.05	74.59	75.67	75.67	74.05
24.8	25.0	25.0	25.1	74.05	75.13	75.13	75.67	74.05
24.8	24.8	25.0	25.2	74.19	74.19	75.27	76.34	74.19
24.8	24.8	25.1	25.1	74.19	74.19	75.81	75.81	74.19
24.8	24.9	24.9	25.2	74.19	74.73	74.73	76.34	74.19
24.8	24.9	25.0	25.1	74.19	74.73	75.27	75.81	74.19
24.8	25.0	25.0	25.0	74.19	75.27	75.27	75.27	74.19
24.8	24.8	24.9	25.2	74.32	74.32	74.86	76.48	74.32
24.8	24.8	25.0	25.1	74.32	74.32	75.40	75.94	74.32
24.8	24.9	24.9	25.1	74.32	74.86	74.86	75.94	74.32
24.8	24.9	25.0	25.0	74.32	74.86	75.40	75.40	74.32
24.8	24.8	24.8	25.2	74.46	74.46	74.46	76.62	74.46
24.8	24.8	24.9	25.1	74.46	74.46	75.00	76.08	74.46
24.8	24.8	25.0	25.0	74.46	74.46	75.54	75.54	74.46
24.8	24.9	24.9	25.0	74.46	75.00	75.00	75.54	74.46
24.8	24.8	24.8	25.1	74.59	74.59	74.59	76.21	74.59
24.8	24.8	24.9	25.0	74.59	74.59	75.13	75.68	74.59
24.8	24.9	24.9	24.9	74.59	75.14	75.14	75.14	74.59
24.8	24.8	24.8	25.0	74.73	74.73	74.73	75.81	74.73
24.8	24.8	24.9	24.9	74.73	74.73	75.27	75.27	74.73
25.0	**25.0**	**25.0**	**25.0**	75.00	75.00	75.00	75.00	**75.00**

symmetric Nash equilibria may be meaningful not only because of analytical tractability but also because they may present the best or worst cases.

References

1. Altman, E., Kameda, H. (2001): Equilibria for Multiclass Routing in Multi-Agent Networks. In: Proceedings of the 40th IEEE Conference on Decision and Control (December, 2001). Orlando, Florida, 604–609
2. Altman, E., Kameda, H., Hosokawa, Y. (2002): Nash Equilibria in Load Balancing in Distributed Computer Systems. International Game Theory Review (to appear)
3. Braess, D. (1968): Über ein Paradoxen aus der Verkehrsplanung. Unternehmensforschung **12**, 258–268
4. Cohen, J. E., Jeffries, C. (1997): Congestion Resulting from Increased Capacity in Single-Server Queueing Networks. IEEE/ACM Trans. Networking **5**, 1220–1225
5. Cohen, J. E., Kelly, F. P. (1990): A Paradox of Congestion in a Queuing Network. J. Appl. Prob. **27**, 730–734
6. Frank, M. (1981): The Braess Paradox. Mathematical Programming **20**, 283–302
7. Haurie, A., Marcotte, P. (1985): On the Relationship between Nash-Cournot and Wardrop Equilibria. Networks **15**, 295–308
8. Kameda, H. (2002): How Harmful the Paradox Can be in the Braess/Cohen–Kelly–Jeffries Networks. In: Proc. IEEE INFOCOM 2002. New York (to appear)
9. Kameda, H., Altman, E., Kozawa, T., Hosokawa, Y. (2000): Braess-Like Paradoxes in Distributed Computer Systems. IEEE Trans. Automatic Control **45**, 1687–1691
10. Kameda, H., Hosokawa, Y., Pourtallier, O. (2001): Effects of Symmetry on Braess-like Paradoxes in Distributed Computer Systems – A Numerical Study. In: Proceedings of the 40th IEEE Conference on Decision and Control (December, 2001). Orlando, Florida, 831–836
11. Kameda, H., Kozawa, T., Li, J. (1997): Anomalous Relations among Various Performance Objectives in Distributed Computer Systems. In: Proc. 1st World Congress on Systems Simulation. IEEE, 459–465
12. Kameda, H., Ohta, M., Hosokawa, Y. (2001): Effects of Symmetry on Paradoxical Cost Degradation in a Nash Non-Cooperative Network System. In: Proceedings of IFAC Symposium on Modeling and Control of Economic Systems (SME 2001). September, 2001, Klagenfurt, Austria
13. Kameda, H., Pourtallier, O. (2002): Paradoxes in Distributed Decisions on Optimal Load Balancing for Networks of Homogeneous Computers. J. ACM (to appear)
14. Korilis, Y. A., Lazar, A. A., Orda, A. (1995): Architecting Noncooperative Networks. IEEE J. Selected Areas in Communications **13**, 1241–1251
15. Korilis, Y. A., Lazar, A. A., Orda, A. (1999): Avoiding the Braess Paradox in Noncooperative Networks. J. Appl. Prob. **36**, 211–222
16. Murchland, J. D. (1970): Braess's Paradox of Traffic Flow. Transpn. Res. **4**, 391–394

Computation of Stackelberg Trajectories in a Class of Linear Differential Games on Plane

Sergei Osipov[1] and Anatolii Kleimenov[2]

[1] Ural State University, Chair of Theoretical Mechanics,
Lenin Ave. 53, Ekaterinburg, 620083, Russia
[2] Inst. of Mathematics and Mechanics of RAS,
Dynamical Systems Department,
S. Kovalevskaya Str. 16, Ekaterinburg, 620219, Russia
E-mail: `Kleimenov@imm.uran.ru`

Summary. The report suggests a numerical method for constructing Stackelberg trajectories in a linear two-person n-dimensional differential game with special terminal payoffs for players. The game is reduced to a game on the plane and the problem is transformed into a non-standard optimal control problem. For the approximation of the trajectories sets for this problem a set of computational geometry algorithms in plane is used, including convex hull construction, union and intersection of polygons and a Minkowski sum for polygons.

1 Introduction

There are various approaches for the solution of differential games, see, e. g., [1, 3, 4, 5]. Many of them suggest numeric methods for solution computation. Such algorithms proposed for antagonistic games are, for example, discussed in papers [2, 7], as well as in other studies of the same and other authors. Comparing to this, there are distinctly less studies concerning non-antagonistic games and they usually deal with linear quadratic games. The present paper describes an algorithm for the solution of more complicated hierarchical differential game, so it seems quiet new of its kind.

2 The Problem Statement

We consider a control system whose dynamics is described by the equation

$$\dot{x}(t) = Ax(t) + Bu(t) + Cv(t) , \quad x(t_0) = x_0 . \tag{1}$$

Here $t \in [t_0, \vartheta]$, where ϑ is given moment of time. Vector *realizationa* $\in I\!R^n$ describes phase state of the system, A, B and C are $n \times n$, $n \times k$ and $n \times l$ constant matrices,respectively. Vectors u and v satisfy the requirements

$$u(t) \in \mu(t)P\,, \qquad v(t) \in \nu(t)Q\,, \tag{2}$$

where the sets P and Q are polyhedrons in spaces $I\!R^k$ and $I\!R^l$, correspondingly; $\mu(t) = a + bt$ and $\nu(t) = c + dt$, where a, b, c and d are constants.

There are two players, controlling the system: player 1, using control u, and player 2, using control v. Assume that cost functionals of both the players depend only from two coordinates x_α and x_β of the phase vector x. Player 1 goal is the maximization of the cost functional $\sigma_1(x_\alpha(\vartheta), x_\beta(\vartheta))$. Player 2 maximizes the cost functional $\sigma_2(x_\alpha(\vartheta), x_\beta(\vartheta))$. We suppose, that the function $\sigma_1\colon I\!R^2 \to I\!R^1$ is continuous one and $\sigma_2\colon I\!R^2 \to I\!R^1$ is concave one.

We also assume that both the players know values $x_\alpha(t)$ and $x_\beta(t)$ at current moment of time $t \in [t_0, \vartheta)$. Then formalization of players' strategies in the game could be based on formalization and the results of positional antagonistic differential games theory [4, 5]. According to this formalization (see also [3]), strategies are described belonging to the class of *pure positional strategies* and are equated to pairs of functions. Strategy of player 1 U is equated to a pair $\{u(t, x, \varepsilon), \beta_1(\varepsilon)\}$ and strategy of player 2 V is equated to $\{v(t, x, \varepsilon), \beta_2(\varepsilon)\}$. Assume, that player 1 chooses his or her strategy U before the game and informs player 2 about this choice. Player 2 knowing the announced strategy U chooses a *rational strategy* V, providing maximization of cost functional σ_2. The game under these assumptions is called an *hierarchical* one. The problem is to find a such player 1 strategy, which paired with the rational player 2 strategy provides the maximum of the cost functional σ_1. This player 1 strategy is called *Stackelberg strategy* in the game. A trajectory, generated by Stackelberg strategy of player 1 and corresponding rational strategy of player 2, is called *Stackelberg trajectory*.

It was shown by Kleimenov [3], that the solution of considered problem could be reduced to solution of a *non-standard auxiliary problem of optimal control*. It is very difficult to solve the last one in general case.

3 Non-Standard Auxiliary Problem of Optimal Control

This problem is: a pair of measurable functions $u(t), v(t)$, $t_0 \le t \le \vartheta$ providing the maximum of cost functional $\sigma_1(x_\alpha(\vartheta), x_\beta(\vartheta))$ needs to be found with the following conditions

$$\sigma_2(x_\alpha(\vartheta), x_\beta(\vartheta)) = \gamma_2(\vartheta, x(\vartheta)) \ge \gamma_2(t, x(t)) \tag{3}$$

for $t \in [t_0, \vartheta)$. Here $x_\alpha(t), x_\beta(t)$ are two selected components of the trajectory $x(t)$ of system (1); the trajectory is generated by controls $u(\cdot)$ and $v(\cdot)$, and it starts from the position (t_0, x_0). The function $\gamma_2\colon [t_0, \vartheta] \times I\!R^n \to I\!R$ denotes *the value function* of the following *antagonistic differential game* Γ_2. The dynamics of the game Γ_2 is described by equation (1). In this game, player 2 maximizes his or her cost functional $\sigma_2(x_\alpha(\vartheta), x_\beta(\vartheta))$, and player 1 counteracts to this goal. It is known from [5] and from [4], that the value function of this game exists and it is continuous.

We would call trajectories of system (1), satisfying the conditions (3), *admissible trajectories*. The conditions (3) applied to an admissible trajectory per se are restrictions on the phase system states, that could happen during a motion. The distinctive feature of the restrictions, that they are formed by the state of the system at final time ϑ. An admissible trajectory, that provides the maximum $\max \sigma_1(x_\alpha(\vartheta), x_\beta(\vartheta))$, would be called an *optimal trajectory*. It was shown in [3] that these optimal trajectories are exactly Stackelberg trajectories in the considered game. Hence, the problem for the construction of Stackelberg trajectories is reduced to the optimal trajectories construction.

Next Sections present an algorithm for constructing approximate optimal trajectories.

4 General Considerations for the Algorithm

Consider an admissible trajectory $x(t)$. Let us call the pair $(x_\alpha(\vartheta), x_\beta(\vartheta))$ an *admissiblefinalstate*. Let symbol G denote the set consisting of all the admissible final states.

According to the definition, final state of an optimal trajectory $x(\vartheta) = x_{opt}$ provides the minimum $\min_{(x_\alpha, x_\beta) \in G} \sigma_1(x_\alpha(\vartheta), x_\beta(\vartheta))$. It would be convenient to construct the set G relying on the following considerations.

Let us consider the following *approach-evasion game* with final time ϑ [5]. The dynamics of the game is described by the equation (1) and the goal set of the player 2 is

$$M_c = \{(x_\alpha(\vartheta), x_\beta(\vartheta)) \colon \sigma_2(x_\alpha(\vartheta), x_\beta(\vartheta)) \geq c\} \ . \tag{4}$$

In another words, the player 2 tries to provide the inclusion:

$$(x_\alpha(\vartheta), x_\beta(\vartheta)) \in M_c \ . \tag{5}$$

The purpose of the player 1 is opposite. We need to find the greatest set $\widetilde{W}^c$ of all start states (t_*, x_*), so that for every of them the player 2 can guarantee the inclusion (5) at final time ϑ. According to the definition in [5], we would call $\widetilde{W}^c$ as *maximal stable bridge* for the set M_c.

Proposition 1. *A trajectory $x(\cdot)$ of the system (1) is admissible if and only if there exists a number c, such that $(t, x(t)) \notin int(\widetilde{W}^c)$ for all $t \in [t_0, \vartheta)$, and $(x_\alpha(\vartheta), x_\beta(\vartheta)) \in \partial M_c$, where ∂M_c denotes the boundary of the set M_c.*

Proposition 1 could be easily proven.

According to [5], the considered game could be reduced to a game on plane by substitution

$$z = X_{\alpha,\beta}(\vartheta, t)x \ , \tag{6}$$

where $X_{\alpha,\beta}(\vartheta, t)$ is matrix constructed from α and β rows of *fundamental Cauchy matrix* corresponding to matrix A. The dynamics of the reduced system would be described by the equation

$$\dot{z} = X_{\alpha,\beta}(\vartheta,t)Bu + X_{\alpha,\beta}(\vartheta,t)Cv\,, \quad z \in I\!R^2\,. \tag{7}$$

The restrictions on the controls u and v and the goals of the players are the same as in the initial game (1). The following correlations are true [5, 2]:

$$\widetilde{W}_t^c = \{x \in I\!R^n : X_{\alpha,\beta}(\vartheta,t)x \in W_{\vartheta-t}^c\}\,, \tag{8}$$

$$W_{\vartheta-t}^c = \{z \in I\!R^2 : (t,z) \in W^c\}\,, \tag{9}$$

$$\widetilde{W}_t^c = \{x \in I\!R^n : (t,x) \in \widetilde{W}^c\}\,, \tag{10}$$

$$t \leq \vartheta\,,$$

where W^c denotes maximal stable bridge in the game with dynamics (7).

According to the substitution, the conditions formulated in Proposition 1 for admissible trajectories $x(\cdot)$ of the system (1) could be transformed for trajectories $z(\cdot)$ of the system (7) in the following form

$$(t, z(t)) \notin int(W^c)\,, \quad t \in [t_0, \vartheta)\,, \tag{11}$$

$$z(\vartheta) \in \partial M_c\,. \tag{12}$$

In addition to this, $(x_\alpha(\vartheta), x_\beta(\vartheta)) = z(\vartheta)$, so the set G could be considered as the set of final states for the system (7) restricted by the conditions (11), (12).

5 Algorithm of Finding the Maximum of Cost Functional σ_1 over the Set G.

In general, the set G is non-convex and disconnected. We calculate the maximum of cost functional σ_1 over G using the approximate procedure described further. The procedure is based on comparison of values σ_1 over the grid built in set G. The output of the algorithm consists of the final state of optimal trajectory z_{opt} and the corresponding value $\sigma_1(z_{opt})$.

For the algorithm start we choose some value c_1. Because c must comply with the restriction $c \geq \underline{c}$, where $\underline{c}$ satisfies the condition that (t_0, z_0) belongs to the boundary of the bridge $W^{\underline{c}}$, so we choose $c_1 \approx \underline{c}$. The value of c_1 could be found as the result of successive approximations. At the beginning of the search, an arbitrary $c_0 > \underline{c}$ is chosen. For c_0, the boundary ∂M_{c_0} of the set M_{c_0} is approximated by convex polygon, represented by its vertices ordered counter-clockwise.

Further we choose the following subdivision of the interval $[0, \vartheta - t_0]$ of time (counted in reverse direction): $\tau_0 = 0$, $\tau_{i+1} = \tau_i + \Delta$, $0 \leq i \leq N-1$, $\tau_N = \vartheta - t_0$. Symbol $\Delta > 0$ denotes the step of the subdivision. (Below we will need direct time subdivision $\{t_i\}$, corresponding to previous one, where $t_i = \vartheta - \tau_i$.)

We construct polygons $W_{\tau_i}^{c_0}$, approximating the intersections of the maximal stable bridge W^{c_0} and the goal set M_{c_0} for the chosen subdivision, using the program described in [2].

After that we check, whether the $z_0 \in W^{c_0}_{\tau_N}$, where $z_0 = X(\vartheta - t_0)x_0$. An algorithm defining a location of a point relative to a non-convex polygon with n verges, based on the evenness verification for the number of intersections of a straight line containing the point and the verges, is used.

If $z_0 \notin W^{c_0}_{\tau_N}$ then we repeatedly decrease c_0 by some constant $\delta > 0$ and stop the process, if $z_0 \in W^{c_0-(k+1)\delta}_{\tau_N}$ for some integer $k > 0$. The existence of a such k is quiet obvious, so we assume $c_1 = c_0 - k\delta$. If $z_0 \in W^{c_0}_{\tau_N}$ then we repeatedly increase c_0 by δ and stop the process, if $z_0 \notin W^{c_0+k\delta}_{\tau_N}$. In this case we assume $c_1 = c_0 + k\delta$. Starting with the found c_1, the main algorithm looks as follows.

1. We built a polygon approximating an intersection of the set G and the set ∂M_c, where $c = c_1$. (The process of constructing the polygon will be described in details after the main algorithm.)
2. We choose points on this intersection section, equidistant from each other along ∂M_c.
3. We find a point z_1 providing the maximum of $\sigma_1(z)$ over points from step 2 by comparison of them.
4. We increase the value c by value $\xi > 0$ and repeat the search of the maximum analogously to steps 2 and 3 with the given $c = c_2 = c_1 + \xi$. Let us denote z_2 as the corresponding maximum point of $\sigma_1(z)$.
5. We put c_1 to the value c_2 and, if $\sigma_1(z_1) < \sigma_1(z_2)$, then we put z_1 to the value z_2, after that go to the step 4.

The algorithm stops, if $c \geq \overline{c}$, where whether $\overline{c} = \max_{z \in I\!R^2} \sigma_2(z)$, or it is equal to the value of c, providing an empty set as the result of intersection of the goal set M_c and the attainability set of the system (7).

Now let us describe how we could construct a polygon approximating the intersection of the set G and the set ∂M_c. We construct the set of positions (t, z) which are realized along the trajectories of (7), satisfying the condition (11) and the condition $z(\vartheta) \notin int(M_c)$. With that end in view we use the following iteration sequence of polygons. For the start of iterations we use $A_{t_0} = \{z_0\}$, then repeatedly

$$A_{t_{i+1}} = \{z \in I\!R^2 : \; z_{i+1} = z_i + \Delta X(\vartheta, t_i)(Bu + Cv)\,, \; z_{i+1} \notin int(W^c_{t_i}); \\ u \in \mu(t_i)P, \; v \in \nu(t_i)Q\,, \; z_i \in A_{t_i}\}\,.$$

Since in general case the sets A_{t_i} could be disconnected ones, we approximate every of them by collections of non-convex connected polygons B^i_j, $1 \leq j \leq L_i$, so that $A_{t_i} = \bigcup_{j=1}^{L_i} B^i_j$. For the polygon A_{t_0} we have $B^0_1 = \{z_0\}$, $L_0 = 1$. The collection of sets B^{i+1}_j, $1 \leq j \leq L_{i+1}$ is derived from the collection of sets B^i_j, $1 \leq j \leq L_{i+1}$ as the result of two operations applied to every B^i_j consequently.

1. A disconnected, in general, polygon

$$S_i = \{z \in IR^2 : z = b + d, \ b \in B_j^i, \ d \in V_i, \ 1 \le j \le L_{i+1}\}$$

is constructed, where

$$V_i = \{z \in IR^2 : z = \Delta X(\vartheta - t_i)(Bu + Cv) \ , \ u \in \mu(t_i)P, \ v \in \nu(t_i)Q\}$$

is convex polygon.

2. We find intersection of the polygon S_i and the set $IR^2 \setminus int(W^c_{\tau_i})$. Note, this the operation could produce more than one polygon from the collection B_j^{i+1}.

The first operation is Minkowski sum of non-convex connected polygon B_j^i and convex one V_i. The corresponding algorithm was proposed in [6]. The algorithm for the second operation is suggested in [7].

Further, having the set B_j^N, we can select a part of the boundary $\partial A_{t_N} = \{z \in A_{t_N} \colon z \in \partial M_c\}$. The set ∂A_{t_N} approximates the intersection of G and ∂M_c. Optimal trajectory $z(t)$ in the approach-evasion game for the system (7) is constructed with the help of backward movement procedure. It starts from the point z_{opt}, and is defined by condition $z(t_i) \in A_{t_i}$ at time moments t_i $(0 \le i \le N)$. So, the proposed algorithm allows to find an approximation of the optimal z-trajectory. An optimal x-trajectory is connected with the optimal z-trajectory via equation (6).

References

1. Basar, T., Olsder, G. J. (1982): Dynamic Noncooperative Game Theory. Acad. Press, N. Y.
2. Isakova, E. A., Logunova, G. V., Patsko, V. S. (1984): Stable Bridges Construction in Linear Differential Game with Fixed Final Time. In: Subbotin, A. I., Patsko, V. S. (Eds.): Algorithms and Programs for Linear Differential Games Solutions. Ural Sci. Center of Acad. Sci. of USSR, Sverdlovsk, 127–158 (in Russian)
3. Kleimenov, A. F. (1993): Positional Differential Nonantagonistic Games. Nauka, Urals Brunch, Ekaterinburg (in Russian)
4. Krasovskii, A. N., Krasovskii, N. N. (1995): Control under Lack of Information. Birkhäuser, Berlin
5. Krasovskii, N. N., Subbotin, A. I. (1988): Game-Theoretical Control Problems. Springer–Verlag, N. Y.
6. Osipov, S. I. (1989): A Minkowski Sum Algorithm for Non-Convex Connected and Convex Polygons. In: Problems of Theoretical and Applied Mathematics. Ural Sci. Center of Acad. Sci. of USSR, Sverdlovsk, 11–12 (in Russian)
7. Vahrushev, V. A., Tarasiev, A. M., Ushakov, V. N. (1987): An Algorithm of Union and Intersection of Sets in Plane. In: Subbotin, A. I., Ushakov, V. N. (Eds.): Control with Guaranteed Result. Ural Sci. Center of Acad. Sci. of USSR, Sverdlovsk, 28–36 (in Russian)

A Game Theoretic Approach to Value Information in Data Mining

Yücel Saygin[1], Arnold Reisman[2], and Yun Tong Wang[3]

[1] Faculty of Engineering and Natural Sciences, Sabanci University, Orhanli 81474 Tuzla, Istanbul, Turkey
E-mail: ysaygin@sabanciuniv.edu

[2] Graduate School of Management, Sabanci University, Orhanli 81474 Tuzla, Istanbul, Turkey
and Reisman and Associates, 18428 Parkland Drive, Shaker Heights, OH., 44122, USA
E-mail: reismana@cs.com

[3] Faculty of Arts and Social Sciences, Sabanci University, Orhanli 81474 Tuzla, Istanbul, Turkey
E-mail: wang@sabanciuniv.edu

Summary. This paper applies a game-theoretic framework to suggest a fair value for information extracted via data mining and shared between two retail market competitor firms. Neither firm has a dominant position in that market. Two players, each owning a privileged information set (a collection of data) may wish to share or pool that information for mutual benefit. We assume that each player is equipped with a mining technique which extracts information from the data. We first model information sharing as a cooperative game. Then we use results from the cost sharing literature to construct information sharing methods when data can be quantified either as discrete or as continuous variables. In the latter case, we supply a method to obtain decision rules for price shared information.

1 Introduction

Advances in data collection and storage technology have made it possible to store large collections of data regarding customers and business transactions. The collected data contain useful information. Data mining research deals with the extraction of useful and valuable information, that cannot be otherwise (via standard querying tools) uncovered from large collections of data. The tools thus created allow companies to uncover interesting patterns deeply buried within the data. Such patterns facilitate the making of strategic decisions. In the context of data mining itself, the interesting problems can be categorized as:

- *classification and clustering*, where we try to partition the data items into disjoint groups or classes [1];
- *associations*, where some correlations among data items are sought [17];
- *sequences*, where we try to find sequences among data items [3].

Associations, in particular, in the context of retail sales, provide information about those items that are often bought together. Finding associations among the items attracted much attention among the data mining community due to the vast accumulation of transactional data recently gathered at the point of sale in various businesses. These association rules are very relevant to, to name a few, pricing, promotion package, and store organization, decisions. Many companies perform data mining on their local data. IT giants like Oracle and IBM have developed and marketed data mining products for extracting associations. The rationale for seeking associations is obvious. Motivations range from product promotions, customer demand analysis, market competition, and more. Thus, this paper focuses on the associations resulting from data mining.

Our basic hypothesis begins with the assumption that information sharing is beneficial. The motivation for cooperation may be due to the fact that the individual data bases may not yield reliable information, or more reliable information can be extracted by data mining when parties pool their data. Or, parties simply provide the information (part of their data mining results) for a price and/or other considerations.

A simple example may help illustrate the point. Consider the following. An observation (formally called an association rule, or simply a rule) *buys*(*IBM Server*) ⇒ *buys*(*HP laser printer*) may appear frequently in Company A's database (meaning that it is a significant rule). But the same rule may have a very low occurrence rate in Company B's database, which is just setting out to sell printers. Nevertheless, if Company A can share this information with Company B, it might be beneficial to both. For example, Company B may have a comparative advantage in selling printers and, Company A, on the other hand, has a comparative advantage in selling computers. Consider another case. A company may decide to discontinue selling a certain set of items. Therefore the rules regarding those items are no longer important. But the information may still be valuable to another company. Thus, companies, even competitors, are at times willing to share data mining results (information sharing).

One way of sharing the results of data mining is to share the data and do the data mining at the local site in each company. However, because companies' databases are often confidential, the parties involved may not want to share their entire database with others. In such cases, each party may selectively contribute its data mining results (a set of rules) rather than the very database. This allows a company to decide whether to provide more or less information by taking into account any effect such sharing might have on its market share or profitability. Therefore, data mining takes on a strategic role in companies' decision making.

We approach the information sharing problem from an economic point of view. We use the axiomatic approach to valuing information. We regard data mining as a strategic input or decision variable for a company in a competitive economic background. Particularly, we consider information to be a valuable commodity. We first model the information sharing as a cooperative game. This immediately allows us to use many solution concepts from game theory. We focus first on the case when the information can be quantified as integers. We use recent results in the axiomatic cost-sharing literature [9] to provide a model for valuing the information. In the last part, we assume that information can be quantified in terms of continuous variables. We then use the well-known Aumann–Shapley [18] pricing formula to suggest prices for information. We conclude the paper with a brief discussion.

2 Association Rule Mining

Since we will treat information as a bundle of commodities, it helps to explain how these commodities are produced. Namely, we want to see the supply side of these goods before they are traded. We use data mining as our technology in the extraction (production) of information. In the following, we briefly explain some of the essential concepts in data mining.

We discussed classification, clustering and finding associations as possible data mining techniques. As indicated, among these, finding associations is very popular especially in the retail sector. Toward that end, many powerful systems have been developed to store and analyze customer transactions such as IBM's Intelligent Miner [19]. We therefore focus on associations, an output of data mining, that is going to be shared. Before embarking into a detailed discussion of fair data-mining result sharing, following are some definitions related to stored transactions data; for obtaining associations, and the properties of the associations (represented as association rules).

The problem of finding association rules among items is formally defined by Agrawal et al. in [2] as follows. Let $I = i_1, i_2, \ldots, i_m$ be a set of literals, called items. Let D be a family of subsets of I called transactions, i. e., if $T \in D$, then $T \subseteq I$. Call D a database (of transactions). A unique identifier, which we call it TID, is associated with each transaction. We say that a transaction T supports X, a set of items in I, if $X \subseteq T$. We assume that the items in a transaction or an itemset, can be and are sorted in certain lexicographic order.

Example 1. A sample database of transactions is shown in Table 1. We next illustrate the database in bit map notation, where each column is a transaction and a value of 1 in a row indicates that the corresponding item is purchased in this transaction, 0 indicates otherwise. There are 3 different items, and 5 transactions consisting of these items in the database. {"IBM Server", "MS Visual Studio"} is an itemset, and transactions T_1, T_2 and T_5 support that itemset.

An itemset X has support s if $s\%$ of the transactions support X. Support of X is denoted as $Supp(X)$. All possible itemsets and their supports obtained from the database in Table 1 are listed in Table 2. Itemset {"IBM Server", "MS Visual Studio"} is supported by 3 transactions out of 5, and therefore has 60% support.

TID	IBM Server	MS Visual Studio	Digital Camera
T1	1	1	1
T2	1	1	0
T3	1	0	1
T4	1	0	0
T5	1	1	0

Table 1. Sample database D

Itemset	Support
"IBM Server"	100%
"MS Visual Studio"	60%
"Digital Camera"	40%
"IBM Server", "MS Visual Studio"	60%
"IBM Server", "Digital Camera"	40%
"MS Visual Studio", "Digital Camera"	20%
"IBM Server", "MS Visual Studio", "Digital Camera"	20%

Table 2. All possible itemsets obtained from D

An association rule is an implication of the form $X \Rightarrow Y$, where $X \subset I$, $Y \subset I$ and $X \cap Y = \emptyset$. We say that the rule $X \Rightarrow Y$ holds in the database D with *confidence* c if $(|X \cup Y|)/(|X|) \geq c\%$ (where $|X|$ is the number of occurrences of the set of items X in the set of transactions D, and X occurs in a transaction T, if and only if $X \subseteq T$). We also say that the rule $X \Rightarrow Y$ has *support* s if $(|X \cup Y|)/(N) \geq s\%$, where N is the number of transactions in D. Note that while the support is a measure of the frequency of a rule, the confidence is a measure of the strength of the relation between sets of items. A high support for a rule indicates that this rule appears in many transactions in the database, therefore it is significant in terms of its frequency in the database. Confidence on the other hand mimics the relationship between the left and right part of the rule. If the right part of a rule appears in many transactions that contain the left part, then the rule will have a high confidence. The confidence can be thought as the probability of the appearance of the right part of the rule in a transaction given that the left hand side appears in the transaction. However, the confidence values can not be used as probabilities.

Example 2. All possible rules obtained from the large itemsets in Table 2 are shown in Table 3. Rule "MS Visual Studio" ⇒ "IBM Server" has support 60%, since it appears in 3 out of 5 transactions in D and it has confidence 100% since all the transactions that contain "IBM Server" also contain "MS Visual Studio".

Rules	Confidence	Support
"IBM Server" ⇒ "MS Visual Studio"	60%	60%
"MS Visual Studio" ⇒ "IBM Server"	100%	60%
"IBM Server" ⇒ "Digital Camera"	100%	40%
"Digital Camera" ⇒ "IBM Server"	100%	40%
"MS Visual Studio" ⇒ "Digital Camera"	33%	20%
"Digital Camera" ⇒ "MS Visual Studio"	50%	20%
"IBM Server" ⇒ "MS Visual Studio","Digital Camera"	20%	20%
"MS Visual Studio" ⇒ "IBM Server", "Digital Camera"	33%	20%
"Digital Camera" ⇒ "IBM Server", "MS Visual Studio"	50%	20%
"IBM Server","MS Visual Studio" ⇒ "Digital Camera"	33%	20%
"IBM Server","Digital Camera" ⇒ "MS Visual Studio"	50%	20%
"MS Visual Studio","Digital Camera" ⇒ "IBM Server"	100%	20%

Table 3. The rules derived from the large itemsets of Table 2

We have a set of items and a database consisting of transactions. Each transaction contains the items bought by the customer. The problem is to find non-obvious rules such as "if somebody buys diapers then he/she buys a bottle of beer as well", using the database of customer transactions. Association rule mining algorithms scan the database of transactions and calculate the support and confidence of the candidate rules to determine if they are significant or not. In Example 2, we have shown all possible itemsets that can be obtained from the database. However, in practice we are more interested in significant rules which have high confidence and support values. For that purpose, threshold values are used by the algorithms to prune the insignificant rules. A rule is significant if its support and confidence is higher than the user specified minimum support and minimum confidence threshold. In this way, algorithms do not retrieve all the association rules that could possibly be generated from a database, instead only a very small subset of rules each satisfies the minimum support and minimum confidence requirements set by the users. In summary, an association rule-mining algorithm works as follows: it finds all the sets of items that appear frequently enough to be considered relevant, then it derives from them the association rules that are strong enough to be considered interesting.

3 The Economic Value of a Rule

3.1 What is the Worth of a Rule to a Company?

Support and confidence of the association rules mimic the importance of these rules as implied by their statistical significance. However they do not contain any semantic information such as the strategic importance for the company of the items contained in the rule. A rule's significance from the data mining expert's perspective is its support and confidence. However, the economic value of a rule (e. g., from the salesman's viewpoint) depends on other parameters as well. A rule may be more important if the items included in the rule have a high sales volume, and or profit margin. Also, a rule may be more significant if it appears frequently in more recent transactions. Therefore, the value of a rule is a function of the rule support, confidence, the items on the right-hand side of the rule, and the time frame in which the rule holds. More formally, for a rule r of the form $X \Rightarrow Y$, where X and Y are sets of items, $worth\,(r) = f(s, c, X, p(Y), t)$, where s is the support, and c is the confidence of r, and Y is the set of items in the right hand side of the rule, $p(Y)$ is the cumulative profit resulting from the sales of items in Y, and t is the time frame the rule holds. We are interested in the items on the right-hand side of the rule. This is so because the rule is an implication with confidence allowing consideration of the items on the right-hand side called the *target items*. Target items, are those that affect the company's market share. The experts in the company will determine the function of a rule's value for the company by giving appropriate weights to the worth function's parameters. For example, the item "IBM Server" can be a target item since it appears in the right hand side of the rule "MS Visual Studio" ⇒ "IBM Server" shown in Table 3.

Indeed, the information revealed from a given rule is very important to a firm, in our case a retailer. Any information about customers' demands is useful for a retailer in making pricing decisions. For example, it is a common practice in retailing to bundle two or more products and selling them as a package. This is called *bundling* [10]. To see how bundling can increase retailer's profit, consider the example in the last paragraph. Assume that customer A wants to buy an IBM Server for which his willingness to pay is $ 1200 but he may consider buying MS Visual Studio if the price is $ 300 or lower. Customer B, on the other hand, also wants to buy an IBM Server with a relatively lower willingness to pay, say, $ 1000, but also needs MS Visual Studio with relatively higher willingness to pay, say, $ 400. If the retailer sells them separately, the maximum price that can be charged for IBM Server is $ 1000, and for MS Visual Studio is $ 300. The total revenue would be $ 2600. If the retailer sells them as a package priced, for example, at $ 1350, which is below each customer's total willingness to pay for both items ($ 1500 for A, $ 1400 for B), both A and B will buy the package. Thus, the retailer takes in $ 2700 total. Clearly, the retailer earns more through the bundling of products.

Generally, bundling makes sense when customers have heterogeneous demands and cannot be classified or otherwise discriminated.In practice, mixed bundling is often used, namely selling two or more items both as a package and individually. It can be shown that when demands are negatively related and/or when marginal costs (of the products) are significant, mixed bundling is a profitable pricing strategy (see [10] for detail).

Nevertheless, the economic values of a rule can be approximately determined by $worth(r) = f(s, c, X, Y)$ (where t is fixed). Determining the value of a set of rules is more complicated since there may be overlaps in the rules and different rules may imply the same set of items. we leave the determining of the value of a set of rules as future work. The value of a set R of rules can be approximately calculated as follows:

$$V(D) = \sum_{r \in D} worth(r) = \sum_{r=\{X \Rightarrow Y\} \in D} f_r(s, c, X, Y) \,. \qquad (1)$$

3.2 The Degradation of Economic Value for a Rule

It is important to point out that the value of a rule may change when the parameters determining its value change over time. In other words, all rules have a temporal dimension. Indeed, the support and confidence of the rules in the database will change over time as new transactions are added to the database. When the rule set R migrates to R', some of the rules in R may no longer appear in R' since they have lost their significance in terms of their support and confidence. However, new rules may appear in R' that do not exist in R. For example, some rules may have a shelf value and if the items are no longer sold by the company then they can just be removed from R. More specifically, let D_{t_1} be the set of transactions collected up to time t_1, R_{t_1} be the set of rules obtained from D_{t_1} with minimum support threshold s, and minimum confidence threshold c. At time t_2, where $t_2 > t_1$, a set, δD, of new transactions will be added to the database and thus $D_{t_2} = D_{t_1} \cup \delta D$. The set of rules that are going to be obtained form D_{t_2}, denoted by R_{t_2} with the same confidence and support thresholds may be different than R_{t_1}. Some rules may lose their statistical significance, i. e., their support and confidence may go below the respective minimum thresholds for both confidence and support. These rules are in the set $R_{t_1} - R_{t_2}$. Some rules that were previously insignificant will now be significant, and they are in the set $R_{t_2} - R_{t_1}$. Therefore, the economic value of the extracted information may be time limited .

4 Rule Sharing Among Companies

Given, that rules have economic values and that these values are measurable, we now consider how parties each having a set of rules can share their information or learn from each other so as to improve their respective profit-maximizing performance. Before modeling the benefits of information sharing and its *fair* attribution to the contributors, we first look at how parties can learn from each other through information sharing. That is, each party makes sure that it is beneficial to pool or share information (here rules) in the first place. Since this is important, we approach it step by step.

For simplicity, suppose that there are two companies that are willing to share some of their respective rules. Let's denote these companies by A and B. The database of transactions of A is denoted by D_A, and the transactions of B is denoted by D_B. Denote D_{AB} the hypothetical database that is constructed by combining D_A and D_B. We say D_{AB} is hypothetical because companies do not actually share their data but the data mining results, namely the association rules. Both companies first agree on the support and confidence thresholds defining the rule's minimum significance requirements. Assume these fractions are, for example, $s\%$, and $c\%$. After determining these threshold values, both companies A and B apply the data mining algorithm to find the association rules that meet both the minimum support and the minimum confidence requirements. Then, a new common set of rules, with no risk of confusion, say D_{AB}, will be generated. And this D_{AB} will be shared between them. (Remember that they have the same data mining algorithm and given that agreed confidence and support requirements they will produce the same D_{AB}.) Also note that our general sharing model in Section 5.1 does allow each party having different local interpretation based on its own data mining with different confidence and/or support. Aside from this issue, in the following subsections, we will discuss in detail the benefits and the process of rule sharing with a few examples that further motivate the topic of this paper.

4.1 The Benefits of Rule Sharing

In fact, the benefit of rule sharing is two fold. First, companies may obtain rules which are significant but cannot be extracted with predefined thresholds, from their individual databases Second, using the items it offers, a company may learn about the items that the other is selling. The following examples clearly explain how these two cases may occur.

Assume that the databases of two companies are as shown in Table 4 and Table 5 respectively. Assume also that the minimum support and confidence thresholds are set to be 50% and 85% respectively.[4] With these threshold values, in database D_A, only the itemset {"IBM Server", "MS Visual Studio"} is frequent (or significant) since its support is $\frac{4}{7} = 57\%$ which is above 50%. The possible candidate rules that can be generated from this itemset are "IBM Server" ⇒ "MS Visual Studio" and "MS Visual Studio" ⇒ "IBM Server". The first rule has confidence $\frac{4}{5} = 80\%$, and the second rule has confidence $\frac{4}{6} = 66\%$. Both of the rules fail to satisfy the minimum confidence

[4] The minimum support and confidence thresholds are usually smaller in real databases. For illustrative purposes, we used a small database and higher support and confidence thresholds. The same results could be shown on a larger database with smaller confidence and support thresholds.

TID	IBM Server	MS Visual Studio	Digital Camera
T1	1	1	1
T2	1	1	0
T3	1	0	1
T4	0	1	0
T5	1	1	0
T6	0	1	1
T7	1	1	0

Table 4. Sample database D_A

TID	IBM Server	MS Visual Studio	Digital Camera
T1	1	1	1
T2	1	1	0
T3	1	1	1
T4	0	1	0
T5	0	0	1
T6	0	1	0
T7	0	1	0

Table 5. Sample database D_B

threshold which is 85%. Therefore the mining algorithm will return no rules. In the database of the second firm (D_B), no itemset satisfies the minimum support threshold. Although the rule "IBM Server" $\Rightarrow$ "MS Visual Studio" meets the minimum confidence requirement, the itemset that is supposed to generate this rule (i. e., {"IBM Server", "MS Visual Studio"}) does not meet the minimum support requirement, therefore it can not be extracted by the data mining algorithm. However in the combined database which is shown in Table 6, both the frequent itemset {"IBM Server", "MS Visual Studio"} and the rule , "IBM Server" $\Rightarrow$ "MS Visual Studio" can be obtained with the same threshold values, since their support and confidence are $\frac{8}{14} = 57\%$ and $\frac{7}{8} = 88\%$ respectively, and they are above the thresholds. Both companies may then reason about this rule which tells them that people who purchase server type computers will probably do a lot of programming, and therefore they will also purchase the MS Visual studio programming environment.

From the above example we can see that the combined database reveals a significant rule which could not have been obtained from the individual databases given the agreed upon minimum support and confidence thresholds. We have shown the combined version of the database as well. The rules can be obtained without actually combining the databases using just the frequent

TID	IBM Server	MS Visual Studio	Digital Camera
T1	1	1	1
T2	1	1	0
T3	1	0	1
T4	0	1	0
T5	1	1	0
T6	0	1	1
T7	1	1	0
T1'	1	1	1
T2'	1	1	0
T3'	1	1	1
T4'	0	1	0
T5'	0	0	1
T6'	0	1	0
T7'	0	1	0

Table 6. Combination of databases D_A and D_B

itemsets in the original databases. We will discuss how this could be achieved in Section 4.2.

Now, let's consider the databases in Table 7 and Table 8. Assume that the minimum support threshold is 25% and minimum confidence threshold is 100%. The rules that can be obtained from Table 7 with the given threshold values are, "IBM server", "MS Visual Studio" $\Rightarrow$ "Art of C++" and "HP Desktop" $\Rightarrow$ "MS Office Guide". The first rule mimics the fact that customers who purchased IBM server and MS visual studio have also placed the book "Art of C++ programming" into their shopping chart. The second rule on the other hand, tells us that customers who bought a desk top which is usually used for less CPU intensive jobs like word processing, will probably buy a guide for MS Office as well. On the other hand, the rule "Color printer" $\Rightarrow$ "Digital Camera" can be obtained from the database shown in Table 8. From the combination of the rules, the first company may realize that it does not need to sell MS Office Guide, since only the users that buy less powerful computers need such a book. People who buy server machines are probably more knowledgeable about computers. So it may replace the item MS Office by the book "Art of C++". The database $D_{B'}$ has a rule that tells us that people who buy color printers are also likely to buy digital cameras. Since the company owning $D_{A'}$ is selling a lot of color printers it may consider selling digital Cameras as well.

TID	HP Desk Top	IBM Server	MS Visual Studio	Art of C++	MS Office Guide	Color Printer
T1	0	1	1	1	0	0
T2	0	1	1	1	0	0
T3	0	1	1	1	0	1
T4	0	1	1	1	0	1
T5	0	1	1	1	0	1
T6	0	1	0	0	0	0
T7	0	0	1	0	0	1
T8	0	0	0	1	0	1
T9	0	1	0	0	0	1
T10	0	0	1	1	1	0
T11	1	0	0	0	0	1
T12	1	0	0	0	1	1
T13	1	0	0	0	1	1
T14	0	0	1	0	1	1
T15	1	0	0	1	0	1
T16	1	0	0	0	1	1
T17	1	0	0	0	1	1
T18	1	0	0	0	1	1
T19	1	0	1	0	1	1
T20	0	0	0	0	1	0

Table 7. Sample database $D_{A'}$

The above example illustrates the case where a company may not be in the business of selling a given item but, based on the information revealed by the rules, it may consider entering that market to enhance its position.

4.2 Rule Sharing Process

To formally define the process of rule sharing, some additional notation follows. Let D_A, D_B be two databases. Denote D_{AB} the union of D_A and D_B. The support of a rule r in database D_A is denoted by $sup_A(r)$, and the support r in D_B and D_{AB} are denoted similarly as $sup_B(r)$ and $sup_{AB}(r)$. The set of rules that are shared by these companies, are denoted by R_A and R_B respectively. These rules are obtained by giving minimum support and confidence parameters, but we will assume that these parameters are time invariant. We also assume that both A and B are compatible in terms of the items they sell and their customer profiles, i. e., they are not in completely different domains but they are complementary in a way. The sets of items that A and B offer

TID	color printer	IBM Server	MS Visual Studio	DVD	MS Office Guide	Digital Camera
T1	0	1	1	1	0	1
T2	0	1	1	1	0	0
T3	1	1	1	0	0	1
T4	0	0	0	0	0	0
T5	1	1	1	1	0	1
T6	0	1	0	0	0	0
T7	0	0	1	0	0	0
T8	1	0	0	1	0	1
T9	0	1	0	0	0	0
T10	0	0	1	1	1	0
T11	1	0	0	0	0	1
T12	1	0	0	0	1	1
T13	1	1	0	0	1	1
T14	0	0	1	1	1	0
T15	1	1	1	0	0	1
T16	1	0	0	0	1	1
T17	0	1	1	1	1	1
T18	1	1	0	0	1	1
T19	1	0	1	0	1	1
T20	0	0	0	0	1	0

Table 8. Sample database $D_{B'}$

to their customers, denoted by I_A and I_B respectively. Note that I_A may not be exactly same as I_B.

Denote R_{AB} the combined version of the rules. Calculating the supports as well as the confidences of rules in the combined set is not trivial. As there may be many overlapping rules. The supports of a rule $r \in R_{AB}$ in D_{AB} is the combination of the supports of r in D_A and D_B and it is calculated by the formula $sup_{AB}(r) = (sup_A(r) \times |D_A| + sup_B(r) \times |D_B|)/(|D_{AB}|)$. Note that the support of some of the rules may go below the support threshold if they are not frequent in both databases. Some of the rules that appear in R_A or R_B may no longer appear in R_{AB}. Calculation of the confidence of the rules in the combined database is more tricky. In order to do that, we define the following: The itemset that contain a rule is called its generating itemset. For example the generating itemset of $B \Rightarrow AC$ is ABC and in order to find the global set of rules we need to have all the frequent itemsets from both companies. This set will cover the generating itemsets of all the rules and will be used for confidence calculation. The frequent itemsets are the sets of items that has the occurrence frequency higher than the support threshold. Let the frequent itemsets from A be F_A, and from B be F_B. For a rule r: $X \Rightarrow Y$, where $r \in R_{AB}$, the confidence of $r = (sup_{AB}(XY))/(sup_{AB}(X))$ where XY is the generating itemset. Therefore the companies should share the generating itemsets and from the generating itemsets they should obtain the rules with confidence higher than the minimum confidence threshold. In the new database, some rules may loose their significance and some new rules may appear.

To avoid the computational complication and ambiguity in generating a new set of rules from any two given sets, we regard the sets R_A and R_B as two information sets, each containing a certain amount of information as it is. If we combine them, namely $R_A \cup R_B$, we have a new information set as it is. We do not make further modification (data mining) which will definitely change the information it contains. However, we do allow the possibility for the two data contributing parties to have differing interpretations of the collective data ($R_A \cup R_B$). This may lead each to additional data mining of their own within the collective data pool. In fact, our rationality assumption for each player will imply that each player will make best use of the data available and thus take into account how the data are used.

5 Information Sharing Framework

Until now we have shown that sharing data mining results is beneficial to both parties. We also have shown how to combine data mining results coming from two firms and constructed the combined set of rules. The immediate question is, how will the companies share the benefits of their contributions. In this section we provide an axiomatic framework for fair sharing of the benefits emanating from the collective data mining results. The framework consists of three models. The first one is a general model which is independent of how rules are represented. The second and the third models are discrete and continuous models.

5.1 The General Model

Let A and B be two players. Let $\mathcal{D}_i = \{r_1, r_2, \ldots, r_{n_i}\}$, $i \in \{A, B\}$ be the data set of player i, where n_i is a positive integer. Let $S(R_A, R_B) \in R_+$, $R_A \subseteq \mathcal{D}_A$, $R_B \subseteq \mathcal{D}_B$ be the surplus generated from data profile (R_A, R_B). Apparently, S can be considered as a game on the player set $\mathcal{D}_A \cup \mathcal{D}_B$. Denote $\mathcal{S}$ all such games. A data sharing problem is a pair $((R_A, R_B); S)$. A solution for a problem $((R_A, R_B); S)$ is a vector $x = (x_A, x_B) \in R^2_+$ such that

$$x_A((R_A, R_B); S) + x_B((R_A, R_B); S) = S(R_A, R_B) .$$

A data sharing method is a mapping x which assigns to each problem $((R_A, R_B); S)$ a solution $x((R_A, R_B); S)$.

5.2 The Discrete Model

In this model, we assume that data can be measured in integer amounts. This occurs when we do not have information about the individual values of the rules, and the worth of each rule is assumed to be one unit.

Denote N the set of nonnegative integers. A data profile is a vector $(n_1, n_2) \in N^2$. Total surplus or benefit (e. g., in terms of market shares) of the collective data $n = (n_1, n_2)$ is a function $S(n_1, n_2) \in R_+$. Denote $\mathcal{S}$ the set of all possible surplus functions. A problem of data sharing is a pair $(n; S)$. A solution for a problem $(n; S)$ is a vector $x = (x_1, x_2)$ such that

$$x_1 + x_2 = S(n_1, n_2) .$$

A surplus sharing method is a mapping x which assigns to each problem $(n; S)$ a solution $x(n; S)$.

We impose the following four properties on the sharing method. These properties are well-known in the cost sharing literature and the cooperative game theory [8, 15].

1. *Additivity* (ADD). $x(n; S^1 + S^2) = x(n; S^1) + x(n; S^2)$ for any $n \in \mathbf{N}^2$ and any $S^1, S^2 \in \mathcal{S}$.

In words, if the surplus generated from the data inputs consists of two parts (e. g., from the two divisions of a player), then the final shares will be the same whether the method is applied to each surplus component separately or to the total surplus once. In the accounting practice, Additivity is often required.

2. *Dummy* (DUM). For any i $(i = 1, 2)$, if $\partial_i S(n) = 0$, $\forall n \in \mathbf{N}^2$ $(n_i > 0)$, then

$$x_i(n; S) = 0 ,$$

where $\partial_i S(n) = S(n) - S(n_i - 1, n_j)$, $n_i > 0$.

Dummy property requires that if a player's data input contributes no increase in the surplus at all possible levels, then this player should get zero surplus. In other words, if the player's information is worthless, this player should share no surplus.

3. *Symmetry*. x is called *symmetric* if $x_i(n; S) = x_j(n; S)$ whenever $n_i = n_j$ and S is a symmetric function of i and j's inputs, i. e., $S(n_1, n_2) = S(n_2, n_1)$, $\forall n_1, n_2$.

This symmetry property says that equals should be treated equally.

4. *Monotonicity*. For any i $(i = 1, 2)$, any S, any n and n', if $n'_i > n_i$, then

$$x_i(n'; S) \geq x_i(n; S) .$$

Monotonicity says that if a player increases his data inputs his share of surplus will never decrease. This is an *incentive compatibility* property. Without this property, there will be no incentive for a player to contribute and share the data information cooperatively.

5. *Coherence* [16]. Suppose that the data inputs have been increased in size, which we call a refinement. For example, more detailed rules or information have been extracted from the data. If such a refinement brings no additional surplus, we may regard the value of the information or data in terms of total surplus is the same as before. A surplus sharing method may use this richer information or data inputs to revise the previous surplus shares. If a method is *coherent*, then there should exist at least one such a refinement (of the data inputs and the unchanged surplus) for which the surplus shares remain unchanged. To define this property formally, let $(n; S)$ be a given problem and $\lambda \in \mathbf{N}_+^2$ (where N_+ is the set of positive integers). Say a surplus function S^λ λ-refines S if $S^\lambda(\lambda \otimes t) = S(t)$ for every $t \in \mathbf{N}_+^2$, where $\lambda \otimes t = (\lambda_1 t_1, \lambda_2 t_2)$. Note that the set of surplus functions that λ-refine S is nonempty. Now formally, *Coherence.* For every problem $(n; S)$ and every $\lambda \in \mathbf{N}^2$, there exists a surplus function S^λ which λ refines S such that $x(n; S) = x(\lambda \otimes n; S^\lambda)$.

Let $\Gamma(S; \lambda)$ be the set of surplus functions which λ refine S. Then, $\Gamma(S; \lambda)$ is a nonempty convex set, so that its relative interior is nonempty. The strict Coherence is defined similarly as Coherence.
Strict Coherence. For every problem $(n; S)$ and every $\lambda \in \mathbf{N}^2$, there exists a surplus function S^λ in the relative interior of $\Gamma(S; \lambda)$ which λ refines S such that $x(n; S) = x(\lambda \otimes n; S^\lambda)$.

Proposition 1 (Sprumont [16]). *There exists a unique surplus sharing method satisfying the properties of Additivity, Dummy, Symmetry, Monotonicity, and (Strict) Coherence. It is the Shapley–Shubik method.*

The Shapley–Shubik method [8, 15] for the two-player case is defined below. For any given problem $(n; S)$, the surplus shares are

$$x_1(n; S) = \frac{1}{2} S(n_1, 0) + \frac{1}{2}(S(n_1, n_2) - S(0, n_2)) ,$$

and

$$x_2(n; S) = \frac{1}{2} S(0, n_2) + \frac{1}{2}(S(n_1, n_2) - S(n_1, 0)) .$$

Consider the two companies A and B with the databases shown in Table 7 and Table 8 respectively. Company A has two rules and company B has one rule in this example. Assume that Company A with its own database D_A has the market value or share of 6%. Company B has the market share of 8% with its own database D_B. Now suppose that each contribute one rule to the pool of rules they are going to share which can increase their joint market share up to 20%. According to the Shapley–Shubik method, company A should get a (increased) market share of

$$x_A((1,1);S) = \frac{1}{2}S(1,0) + \frac{1}{2}(S(1,0) - S(0,1)) = \frac{1}{2}6\% + \frac{1}{2}(20\% - 8\%) = 9\,,$$

and company B should get a (increased) market share of

$$x_B((1,1);S) = \frac{1}{2}S(0,1) + \frac{1}{2}(S(1,1) - S(1,0)) = \frac{1}{2}8\% + \frac{1}{2}(20\% - 6\%) = 11\,.$$

Thus we say that the data inputs from company B are (relatively) more valuable than those from company A. The reason being that even though both contribute the same number of rules, the stand-alone value of Company B's database (8%) is higher than the stand-alone value of Company A's database. Therefore, the relative values or contributions, obtained by our fair-sharing axioms or conditions, are different.

5.3 The Continuous Model

In the discrete model, we assumed that the rule counts represent the only available information having a discrete value. However the value of the rules shared could be measured as a function of the properties of the rules as explained in Section 3. The value of a rule is a continuous number, and so are the values of a set of rules. Therefore we also modeled the case where the shared information is continuous in value.

Now assume that the data inputs are two continuous variables $(c_1, c_2) \in R^2_+$. Assume that the surplus function $S\colon R^2_+ \to R_+$ is continuously differentiable. The surplus sharing problem and the sharing method are similarly defined as in the discrete model.

In the continuous model, the following two properties on the sharing method are relevant.

Scale Invariance. For any $c \in R^2_+$, and any $\lambda \in R^2_{++}$,

$$x(c; S) = x(\lambda \otimes c; S^\lambda)\,,$$

where $\lambda \otimes c = (\lambda_1 c_1, \lambda_2 c_2)$ and

$$S^\lambda(c) = S(\lambda_1^{-1} c_1, \lambda_2^{-1} c_2)\,, \quad \forall c \in R^2_+\,.$$

In words, changing the measuring units of the data will not affect the surplus shares.

Proportionality. If S is homogeneous, i.e., there is a mapping $s\colon R_+ \to R_+$ such that

$$S(c) = s(c_1 + c_2)\,, \quad \forall c \in R^2_+\,,$$

then

$$x_i(c; S) = \frac{c_i}{c_1 + c_2} s(c_1 + c_2)\,, \quad i = 1, 2\,. \tag{2}$$

Alternately stated, if the data are homogeneous and additively enter into the surplus function, then the corresponding share of the surplus should be proportional to the data's contribution.

Proposition 2 (Samet and Tauman [14]). *There exists a unique method satisfying the properties of Additivity, Dummy, Scale Invariance, and Proportionality. It is the Aumann–Shapley method.*

The Aumann–Shapley method [14] is defined below.

$$x_i^{AS}(c;S) = c_i \int_0^1 \partial_i S(tc)\,dt\,, \quad i = 1,2\,. \tag{3}$$

We call

$$p_1(c;S) = \int_0^1 \partial_1 S(tc)\,dt\,, \tag{4}$$

and

$$p_2(c;S) = \int_0^1 \partial_2 S(tc)\,dt\,, \tag{5}$$

the Aumann–Shapley prices of data c_1 and c_2 respectively. Indeed,

$$p_1(c;S)c_1 + p_2(c;S)c_2 = S(c)\,.$$

Consequently, $(p_1(c;S), p_2(c;S))$ defined by (4) and (5) are the unit prices for the data (c_1, c_2).

Below we give a numerical example to clarify the concepts. Let A and B be two companies with their databases shown in Tables 7, and Table 8 respectively. Assume that each of them contributes one rule. Specifically, the contributed rules are respectively "IBM server", "MS Visual Studio" ⇒ "Art of C++" from company A, and "Color printer" ⇒ "Digital Camera" from company B. As we demonstrated in Section 3, the value of a rule to each company is a function of the support, confidence and the average profit obtained from the items of the rule. In general, different value functions can be assigned by domain experts. However in order to illustrate our point, with no loss of generality here we assume the value of a rule is a linear function of its parameters (e. g., support, confidence, items). To be concrete, assume that for the rule "IBM server", "MS Visual Studio" ⇒ "Art of C++", the support is 25%, and confidence is 100% and the average profit of selling the items in this rules is 200 \$. Then the value of this rule can be calculated by multiplying these values, i. e., $\frac{100}{100} \times \frac{25}{100} \times 200 = 40$. Similarly, for the rule "Color printer" ⇒ "Digital Camera", the support is 50 %, and the confidence is 100 %, and assume that the average profit obtained from these items is 100 \$, then the value of this rule is calculated as 50. Further, assume for illustrative purposes that the joint surplus function from shared inputs between A and B is 1

$$S(c_1, c_2) = c_1 + c_2 + c_1\sqrt{c_2}\,,$$

According to their inputs $c_1 = 40$, $c_2 = 50$, now we calculate their relative values using the formula recommended by the above Aumann–Shapley pricing. Their Aumann–Shapley prices are

$$p_A((40,50);S) = \int_0^1 (1+\sqrt{50t})\,dt = 1 + \frac{2}{3}\cdot\sqrt{50} \approx 6\,, \tag{6}$$

and

$$p_B((40,50);S) = \int_0^1 \left(1 + 40t\cdot\frac{1}{2}\frac{1}{\sqrt{50t}}\right) dt = 1 + \frac{2}{3}\frac{20}{\sqrt{50}} \approx 3\,. \tag{7}$$

Where the budget balance condition holds as shown in (11)

$$p_A \cdot 40 + p_B \cdot 50 = S(40,50)\,. \tag{8}$$

Note that the calculated price of the rule from A is approximately twice that of the rule from B. We say that the rule from A is more (information) valuable than the rule from B, even though Company B values its rule at 50.

6 Conclusions and Future Work

Data mining and information sharing have increasingly gained importance in corporate decision making. This study attempts to gain insights on how, in a competitive and cooperative economic environment, firms might view their respective private information in terms of being a valuable input to strategic decision-making. We provide analytical formulae to value the information in the context of information sharing. Results showing that otherwise competing firms can each benefit from sharing market sensitive information are indeed counterintuitive. Moreover, they recognize the possibility of asymmetry in contributions to the market share synergy by the individual player's information and suggest the relative worth of each. Of course, information has different value in different contexts. Our point is that game theory can be useful in providing guidance for this particular case of information sharing. Clearly, this approach can be expanded to multi player participation and or tested in real environments with real data.

Producers of innovative technologic products wishing to create a market in a developing economy often partner with indigenous firms having marketing expertise within their own country. The approach suggested in this paper may well be used to address corporate considerations of such cross-cultural initiatives. As a concrete example, Turkey represents fertile ground for such possibilities. Over several decades much technology transfer has thusly taken place in the automotive, household appliance, telecommunication devices, computer hardware and software, and power tool, sectors. At the time of this writing, tech transfer in the do-it-yourself sector is in its early stages. On the sales side it currently involves at least two very sophisticated hypermarket firms.

Lastly, several strategies can be used in pursuing research [11, 12]. Bridging of disciplines is worthwhile as it often tends to enlarge the knowledge domain of each of the constituent disciplines. This paper uses Game Theory to essentially help Data Mining justify its use by competing corporations for mutual benefit. Further research should be pursued to show if and how Data Mining can help Game Theory overcome its greatest weakness – a paucity, as is shown in [13], of real world applications.

References

1. Agrawal, R., Ghosh, S., Imielinski, T., Iyer, B., Swami, A. (1992): An Interval Classifier for Database Mining Applications. In: Proc. of the 18th Int'l Conference on Very Large Databases. Vancouver
2. Agrawal, R., Srikant, R. (1994): Fast Algorithms for Mining Association Rules. In: Proc. of the 20th Int'l Conference on Very Large Databases. Santiago, Chile
3. Agrawal, R., Srikant, R. (1995): Mining Sequential Patterns. In: Proc. of the Int'l Conference on Data Engineering. Taipei, Taiwan
4. Billera, L., Heath, D. (1982): Allocation of Shared Costs: A Set of Axioms Yielding a Unique Procedure. Math. Oper. Res. **7** 32–39
5. Friedman, E., Moulin, H. (1999): Three Additive Methods to Share Joint Costs or Surplus. J. Econ. Theory **87**, 275–312
6. Kleinberg, J., Papadimitriou, C. H., Raghavan, P. (2002): On the Value of Private Information. Working paper, `http://www.cs.berkeley.edu/~christos/`
7. Mirman, L., Tauman, Y. (1982): Demand Compatible Equitable Cost Sharing Prices. Math. Oper. Res. **7**, 40–56
8. Moulin, H. (1995): On Additive Methods to Share Joint Costs. Japanese Econ. Rev. **46**, 303–332
9. Moulin, H. (1999): Axiomatic Cost and Surplus Sharing. In: Arrow, Sen, Suzumura (Eds.): Handbook of Social Choice and Welfare.
10. Pindyck, R. S., Rubinfeld, D. L. (2001): Microeconomics. Prentice Hall, New Jersey
11. Reisman, A. (1988): On Alternative Strategies for Doing Research in the Management and Social Sciences. IEEE Transactions on Engineering Management **35(4)**, 215–221
12. Reisman, A. (1992): Management Science Knowledge: It's Creation, Generalization and Consolidation. Quorum Books Publishing Company, Westport CT
13. Reisman, A., Kumar, A., Motwani, J. (2001): A Meta Review of Game Theory Publications in the Flagship US-Based OR/MS Journals. Management Decision **39(2)**, 147–155
14. Samet, D., Tauman, Y. (1982): The Determination of Marginal Cost Prices under a Set of Axioms. Econometrica **50**, 895–909
15. Shapley, L. S. (1953): A Value for n-Person Games. In: Kuhn, H. W., Tucker, A. W. (Eds.): Contributions to the Theory of Games. Annals of Mathematics Studies. Vol. II, No. 28, 307–317
16. Sprumont, Y. (2000): Coherent Cost-Sharing Rules. Games and Economic Behavior **33**, 126–144
17. Srikant, R., Agrawal, R. (1995): Mining Generalized Association Rules. In: Proc. of the 21st Int'l Conference on Very Large Databases. Zurich, Switzerland

18. Tauman, Y. (1988): The Aumann–Shapley Prices: A Survey. In: Roth, A. (Ed.): The Shapley Value. Cambridge Univ. Press, Cambridge
19. IBM Almaden Quest Project (2002): http://www.almaden.ibm.com/cs/quest

GPSR Compliance

The European Union's (EU) General Product Safety Regulation (GPSR) is a set of rules that requires consumer products to be safe and our obligations to ensure this.

If you have any concerns about our products, you can contact us on ProductSafety@springernature.com

In case Publisher is established outside the EU, the EU authorized representative is:

Springer Nature Customer Service Center GmbH
Europaplatz 3
69115 Heidelberg, Germany

Batch number: 10704030

Printed by Printforce, the Netherlands